ALGEBRA

This book is in the

ADDISON-WESLEY SERIES IN MATHEMATICS

LYNN H. LOOMIS

Consulting Editor

ALGEBRA

SERGE LANG
Columbia University, New York, New York

ADDISON-WESLEY PUBLISHING COMPANY

READING, MASSACHUSETTS · PALO ALTO · LONDON · DON MILLS, ONTARIO

Printed in the United States of America

Library of Congress Catalog Card No. 65–23677

Second Printing — March 1967

Foreword

Je préfère la nommer ainsi [algèbre abstraite] plutôt qu'algèbre moderne, parcequ'elle vivra sans doute longtemps et finira donc par devenir l'algèbre ancienne.

F. SEVERI
Liège, 1949

The present book is meant as a basic text for a one year course in algebra, at the graduate level.

Unfortunately, the amount of algebra which one should ideally absorb during this first year in order to have a proper background (irrespective of the subject in which one eventually specializes), exceeds the amount which can be covered physically by a lecturer during a one year course. Hence more material must be included than can actually be handled in class.

Many shortcuts can be taken in the presentation of the topics, which admits many variations. For instance, one can proceed into field theory and Galois theory immediately after giving the basic definitions for groups, rings, fields, polynomials in one variable, and vector spaces. Since the Galois theory gives very quickly an impression of depth, this is very satisfactory in many respects.

One can also treat first the linear algebra, after covering the basic definitions, and postpone the field theory. The chapters have been so written as to allow maximal flexibility in this respect, and I have frequently committed the crime of lèse-Bourbaki by repeating short arguments or definitions to make certain sections or chapters logically independent of each other.

I have followed Artin in the treatment of Galois theory, except for minor modifications. The reader can profitably consult Artin's short book on the subject to see the differences. Furthermore, the reader would also profit from seeing an exposition based on the Jacobson-Bourbaki theorem, which is useful in the inseparable case. However, the standard case is sufficiently important in most applications to warrant the classical treatment which I have chosen here.

Since Artin taught me algebra, my indebtedness to him is all-pervasive. It is perhaps least in the section on linear algebra and representations, where the influence of Bourbaki is more decisive (in content, not expository style). However, in choice of subject matter, I am more selective than Bourbaki, with the resulting advantages and disadvantages of being less encyclopaedic.

Granting the material which under no circumstances can be omitted from a basic course, there exist several options for leading the course in various directions. It is impossible to treat all of them with the same degree of thoroughness. The precise point at which one is willing to stop in any given direction will depend on time, place, and mood. The chapters on real fields and absolute values, for instance, can be omitted safely, or can be read by students independently of the class. The chapter on group representations also. The Witt theorem on quadratic forms can also be omitted. However, any book with the aims of the present one must include a choice of these topics, pushing ahead in deeper waters, while stopping short of full involvement, and keeping the number of pages within reasonable bounds. There can be no universal agreement on these matters, not even between the author and himself. Thus the concrete decisions as to what to include and what not to include are finally taken on grounds of general coherence and aesthetic balance. For instance, I have deliberately avoided getting too involved in commutative algebra. I could not write a basic course in algebra as an exclusive training ground for future algebraic geometers. However, anyone teaching the course will want to impress his own personality on the material, and may push certain topics with more vigor than I have, at the expense of others. Nothing in the present book is meant to inhibit this.

The order of the book is still remarkably like that given by Artin-Noether-Van der Waerden some thirty years ago. I agree wholeheartedly with Van der Waerden's inclusion of the representation theory of finite groups in a basic text. In view of progress made by Brauer during the past thirty years, it has been possible to give a much more complete treatment than Van der Waerden could at that time.

There is some reason to include more on linear groups and their representations than I could do and still have a reasonably sized book. This can be done especially with students who have a proper background in linear algebra from their undergraduate days. Fortunately, several texts dealing with Lie algebras and Lie groups are now becoming available, so that I did not feel too guilty in omitting these topics. (Cf. in particular Serre's notes, *Lie Algebras and Lie Groups.*)

As prerequisites, I assume only that the reader is acquainted with the basic language of mathematics (i.e. essentially sets and mappings), and the integers and rational numbers. A more specific description of what is assumed will be summarized below. On a few occasions, we use determinants before treating these formally in the text. Most readers will already be acquainted with determinants, and we feel it is better for the organization of the whole book to allow ourselves such minor deviations from a total ordering of the logic involved.

New York, 1965 SERGE LANG

Prerequisites

We assume that the reader is familiar with sets, and the symbols $\cap$, $\cup$, $\supset$, $\subset$, $\in$. If A, B are sets, we use the symbol $A \subset B$ to mean that A is contained in B but may be equal to B. Similarly for $A \supset B$.

If $f : A \to B$ is a mapping of one set into another, we write

$$x \mapsto f(x)$$

to denote the effect of f on an element x of A. We distinguish between the arrows $\to$ and $\mapsto$.

Let $f : A \to B$ be a mapping (also called a map). We say that f is *injective* if $x \neq y$ implies $f(x) \neq f(y)$. We say f is *surjective* if given $b \in B$ there exists $a \in A$ such that $f(a) = b$. We say that f is *bijective* if it is both surjective and injective.

A subset A of a set B is said to be *proper* if $A \neq B$.

Let $f : A \to B$ be a map, and A' a subset of A. The restriction of f to A' is a map of A' into B denoted by $f|A'$.

If $f : A \to B$ and $g : B \to C$ are maps, then we have a composite map $g \circ f$ such that $(g \circ f)(x) = g(f(x))$ for all $x \in A$.

Let $f : A \to B$ be a map, and B' a subset of B. By $f^{-1}(B')$ we mean the subset of A consisting of all $x \in A$ such that $f(x) \in B'$. We call it the *inverse image* of B'. We call $f(A)$ the *image* of f.

A *diagram*

$$\begin{array}{ccc} A & \xrightarrow{f} & B \\ & {}_h\searrow \quad \swarrow_g & \\ & C & \end{array}$$

is said to be *commutative* if $g \circ f = h$. Similarly, a *diagram*

$$\begin{array}{ccc} A & \xrightarrow{f} & B \\ {\scriptstyle \varphi}\downarrow & & \downarrow{\scriptstyle g} \\ C & \xrightarrow[\psi]{} & D \end{array}$$

is said to be *commutative* if $g \circ f = \psi \circ \varphi$. We deal sometimes with more complicated diagrams, consisting of arrows between various objects. Such diagrams are called commutative if, whenever it is possible to go from one object to another by means of two sequences of arrows, say

$$A_1 \xrightarrow{f_1} A_2 \xrightarrow{f_2} \cdots \xrightarrow{f_n} A_n$$

and

$$A_1 \xrightarrow{g_1} B_2 \xrightarrow{g_2} \cdots \xrightarrow{g_m} B_m = A_n,$$

then

$$f_n \circ f_{n-1} \circ \cdots \circ f_1 = g_m \circ g_{m-1} \circ \cdots \circ g_1,$$

in other words, the composite maps are equal. Most of our diagrams are composed of triangles or squares as above, and to verify that a diagram consisting of triangles or squares is commutative, it suffices to verify that each triangle and square in it is commutative.

We assume that the reader is acquainted with the integers and rational numbers, denoted respectively by **Z** and **Q**. For many of our examples, we also assume that the reader knows the real and complex numbers, denoted by **R** and **C**.

Let A and I be two sets. By a family of elements of A, indexed by I, one means a map $f : I \to A$. Thus for each $i \in I$ we are given an element $f(i) \in A$. Although a family does not differ from a map, we think of it as determining a collection of objects from A, and write it often as

$$\{f(i)\}_{i \in I}$$

or

$$\{a_i\}_{i \in I},$$

writing a_i instead of $f(i)$. We call I the indexing set.

We assume that the reader knows what an equivalence relation is. Let A be a set with an equivalence relation, let E be an equivalence class of elements of A. We sometimes try to define a map of the equivalence classes into some set B. To define such a map f on the class E, we sometimes first give its value on an element $x \in E$ (called a representative of E), and then show that it is independent of the choice of representative $x \in E$. In that case we say that f is *well defined*.

We have products of sets, say finite products $A \times B$, or $A_1 \times \cdots \times A_n$, and products of families of sets.

We shall use Zorn's lemma, which we now describe.

A set A is said to be (partially) ordered if we are given a relation $x \leqq y$ between certain pairs of elements satisfying the following conditions:

For all $x, y, z \in A$:

We have $x \leqq x$.

If $x \leqq y$ and $y \leqq z$ then $x \leqq z$.

If $x \leqq y$ and $y \leqq x$ then $y = x$.

A subset T of A is said to be totally ordered if given $x, y \in T$ we have $x \leqq y$ or $y \leqq x$.

If S is a subset of A, an upper bound b for S in A is an element $b \in A$ such that $x \leqq b$ for all $x \in S$.

The ordered set A is said to be *inductively ordered* if every totally ordered subset of A has an upper bound in A.

A maximal element in A is an element $a \in A$ such that whenever $x \in A$ and $a \leqq x$ then $a = x$. (Thus maximal means "relatively maximal" not "absolutely maximal".)

Zorn's lemma asserts: *If A is an ordered set, and is inductively ordered, and is non-empty, then there exists a maximal element in A.*

We shall also use cardinality statements like the following:

Let A be an infinite set. The set of all finite subsets of A has the same cardinality as A. If D is denumerable, then $A \times D$ has the same cardinality as A. We sometimes abbreviate cardinality by card. We have

$$\text{card}(A) \leqq \text{card}(B) \text{ and } \text{card}(B) \leqq \text{card}(A) \quad \text{imply} \quad \text{card}(A) = \text{card}(B).$$

Bibliography

[1] E. Artin, *Galois Theory*, Notre Dame Mathematical Lectures, No. 2, 1946.

[2] ——, *Geometric Algebra*, Interscience, New York, 1957.

[3] N. Bourbaki, *Algèbre commutative*, Hermann, Paris, 1962.

[4] ——, *Formes sesquilinéaires et formes quadratiques*, Hermann, Paris, 1959.

[5] ——, *Algèbre*, Hermann, Paris.

[6] R. Godement, *Cours d'algèbre*, Hermann, Paris, 1963.

[7] N. Jacobson, *Lectures in abstract algebra*, Van Nostrand, Princeton, N. J., Vol. 1 (1951), Vol. 2 (1953), Vol. 3 (1964).

[8] S. Lang, *Algebraic numbers*, Addison-Wesley, Reading, Mass., 1964.

[9] ——, *Diophantine geometry*, Interscience, New York, 1960.

[10] Van der Waerden, *Moderne Algebra*, Springer Verlag, Berlin, 1931.

[11] H. Weber, *Lehrbuch der Algebra*, 1898 (reprinted by Chelsea, 1963).

[12] O. Zariski and P. Samuel, *Commutative algebra*, Van Nostrand, Princeton, N. J., Vol. 1 (1958), Vol. 2 (1960).

The above is a very brief list of texts and treatises in algebra. Bourbaki is always the most complete, and is excellent as a reference. Jacobson treats the Galois theory from the point of view of the Jacobson-Bourbaki theorem, which is useful among other things when dealing with purely inseparable extensions. The reader should browse through all the above books to be aware of points of view different from those given in the present book.

Contents

Part One
Groups, Rings, and Modules

Chapter I

Groups

Chapter II

Rings

Chapter III

Modules

CHAPTER IV

Homology

CHAPTER V

Polynomials

CHAPTER VI

Noetherian Rings and Modules

Part Two
Field Theory

CHAPTER VII

Algebraic Extensions

Chapter VIII

Galois Theory

Chapter IX

Extensions of Rings

Chapter X

Transcendental Extensions

Chapter XI

Real Fields

Chapter XII

Absolute Values

Part Three
Linear Algebra and Representations

Chapter XIII

Matrices and Linear Maps

Chapter XIV

Structure of Bilinear Forms

Chapter XV

Representation of One Endomorphism

Chapter XVI

Multilinear Products

Chapter XVII

Semisimplicity

Chapter XVIII

Representations of Finite Groups

PART ONE

GROUPS, RINGS and MODULES

This part introduces the basic notions of algebra, and the main difficulty for the beginner is to absorb a reasonable vocabulary in a short time. None of the concepts is difficult, but there is an accumulation of new concepts which may sometimes seem heavy.

To understand the next parts of the book, the reader needs to know essentially only the basic definitions of this first part. Of course, a theorem may be used later for some specific and isolated applications, but on the whole, we have avoided making long logical chains of interdependence.

CHAPTER I

Groups

§1. Monoids

Let S be a set. A mapping

$$S \times S \to S$$

is sometimes called a *law of composition* (of S into itself). If x, y are elements of S, the image of the pair (x, y) under this mapping is also called their *product* under the law of composition, and will be denoted by xy. (Sometimes, we also write $x \cdot y$, and in many cases it is also convenient to use an additive notation, and thus to write $x + y$. In that case, we call this element the *sum* of x and y. It is customary to use the notation $x + y$ only when the relation $x + y = y + x$ holds.)

Let S be a set with a law of composition. If x, y, z are elements of S, then we may form their product in two ways: $(xy)z$ and $x(yz)$. If $(xy)z = x(yz)$ for all x, y, z in S then we say that the law of composition is *associative*.

An element e of S such that $ex = x = xe$ for all $x \in S$ is called a *unit element*. (When the law of composition is written additively, the unit element is denoted by 0, and is called a *zero element*.) A unit element is unique, for if e' is another unit element, we have

$$e = ee' = e'$$

by assumption. In most cases, the unit element is written simply 1 (instead of e). For most of this chapter, however, we shall write e so as to avoid confusion in proving the most basic properties.

A *monoid* is a set G, with a law of composition which is associative, and having a unit element (so that in particular, G is not empty).

Let G be a monoid, and $x_1, \ldots, x_n$ elements of G (where n is an integer > 1). We define their product inductively:

$$\prod_{\nu=1}^{n} x_\nu = x_1 \cdots x_n = (x_1 \cdots x_{n-1})x_n.$$

We then have the following rule:

$$\prod_{\mu=1}^{m} x_\mu \cdot \prod_{\nu=1}^{n} x_{m+\nu} = \prod_{\nu=1}^{m+n} x_\nu,$$

which essentially asserts that we can insert parentheses in any manner in our product without changing its value. The proof is easy by induction, and we shall leave it as an exercise.

One also writes

$$\prod_{m+1}^{m+n} x_\nu \qquad \text{instead of} \qquad \prod_{\nu=1}^{n} x_{m+\nu}$$

and we define

$$\prod_{\nu=1}^{0} x_\nu = e.$$

As a matter of convention, we agree also that the empty product is equal to the unit element.

It would be possible to define more general laws of composition, i.e. maps $S_1 \times S_2 \to S_3$ using arbitrary sets. One can then express associativity and commutativity in any setting for which they make sense. For instance, for commutativity we need a law of composition

$$f : S \times S \to T$$

where the two sets of departure are the same. *Commutativity* then means $f(x, y) = f(y, x)$, or $xy = yx$ if we omit the mapping f from the notation. For associativity, we leave it to the reader to formulate the most general combination of sets under which it will work. We shall meet special cases later, for instance arising from maps

$$S \times S \to S \qquad \text{and} \qquad S \times T \to T.$$

Then a product $(xy)z$ makes sense with $x \in S$, $y \in S$, and $z \in T$. The product $x(yz)$ also makes sense for such elements x, y, z and thus it makes sense to say that our law of composition is associative, namely to say that for all x, y, z as above we have $(xy)z = x(yz)$.

If the law of composition of G is commutative, we also say that G is *commutative* (or *abelian*).

Let G be a commutative monoid, and $x_1, \ldots, x_n$ elements of G. Let ψ be a bijection of the set of integers $(1, \ldots, n)$ onto itself. Then

$$\prod_{\nu=1}^{n} x_{\psi(\nu)} = \prod_{\nu=1}^{n} x_\nu.$$

We prove this by induction, it being obvious for $n = 1$. We assume it for $n - 1$. Let k be an integer such that $\psi(k) = n$. Then

$$\prod_1^n x_{\psi(\nu)} = \prod_1^{k-1} x_{\psi(\nu)} \cdot x_{\psi(k)} \cdot \prod_1^{n-k} x_{\psi(k+\nu)}$$

$$= \prod_1^{k-1} x_{\psi(\nu)} \cdot \prod_1^{n-k} x_{\psi(k+\nu)} \cdot x_{\psi(k)}.$$

Define a map φ of $(1, \ldots, n-1)$ into itself by the rule

$$\varphi(\nu) = \psi(\nu) \qquad \text{if} \quad \nu < k,$$
$$\varphi(\nu) = \psi(\nu + 1) \qquad \text{if} \quad \nu \geqq k.$$

Then

$$\prod_1^n x_{\psi(\nu)} = \prod_1^{k-1} x_{\varphi(\nu)} \prod_1^{n-k} x_{\varphi(k-1+\nu)} \cdot x_n$$

$$= \prod_1^{n-1} x_{\varphi(\nu)} \cdot x_n,$$

which, by induction, is equal to $x_1 \cdots x_n$, as desired.

Let G be a commutative monoid, let I be a set, and let $f : I \to G$ be a mapping such that $f(i) = e$ for almost all $i \in I$. (Here and thereafter, *almost all* will mean *all but a finite number*.) Let I_0 be the subset of I consisting of those i such that $f(i) \neq e$. By

$$\prod_{i \in I} f(i)$$

we shall mean the product

$$\prod_{i \in I_0} f(i)$$

taken in any order (the value does not depend on the order, according to the preceding remark). It is understood that the empty product is equal to e.

When G is written additively, then instead of a product sign, we write the sum sign $\sum$.

There are a number of formal rules for dealing with products which it would be tedious to list completely. We give one example. Let I, J be two sets, and $f : I \times J \to G$ a mapping into a commutative monoid which takes the value e for almost pairs (i, j). Then

$$\prod_{i \in I} \left[\prod_{j \in J} f(i, j) \right] = \prod_{j \in J} \left[\prod_{i \in I} f(i, j) \right].$$

We leave the proof as an exercise.

As a matter of notation, we sometimes write $\prod f(i)$, omitting the signs $i \in I$, if the reference to the indexing set is clear.

Let x be an element of a monoid G. For every integer $n \geqq 0$ we define x^n to be

$$\prod_{1}^{n} x,$$

so that in particular we have $x^0 = e$, $x^1 = x$, $x^2 = xx$, ... We obviously have $x^{(n+m)} = x^n x^m$ and $(x^n)^m = x^{nm}$. Furthermore, from our preceding rules of associativity and commutativity, if x, y are elements of G such that $xy = yx$, then $(xy)^n = x^n y^n$. We leave the formal proof as an exercise.

If S, S' are two subsets of a monoid G, then we define SS' to be the subset consisting of all elements xy, with $x \in S$ and $y \in S'$. Inductively, we can define the product of a finite number of subsets, and we have associativity. For instance, if S, S', S'' are subsets of G, then $(SS')S'' = S(S'S'')$. Observe that $GG = G$ (because G has a unit element). If $x \in G$, then we define xS to be $\{x\}S$, where $\{x\}$ is the set consisting of the single element x. Thus xS consists of all elements xy, with $y \in S$.

By a *submonoid* of G, we shall mean a subset H of G containing the unit element e, and such that, if $x, y \in H$ then $xy \in H$ (we say that H is *closed* under the law of composition). It is then clear that H is itself a monoid, under the law of composition induced by that of G.

If x is an element of a monoid G, then the subset of powers x^n $(n = 0, 1, \ldots)$ is a submonoid of G.

Example of a monoid. We assume that the reader is familiar with the terminology of elementary topology. Let M be the set of homeomorphism classes of compact (connected) surfaces. We shall define an addition in M. Let S, S' be compact surfaces. Let D be a small disc in S, and D' a small disc in S'. Let C, C' be the circles which form the boundaries of D and D' respectively. Let D_0, D_0' be the interiors of D and D' respectively, and glue $S - D_0$ to $S' - D_0'$ by identifying C with C'. It can be shown that the resulting surface is independent, up to homeomorphism, of the various choices made in the preceding construction. If σ, σ' denote the homeomorphism classes of S and S' respectively, we define $\sigma + \sigma'$ to be the class of the surface obtained by the preceding gluing process. It can be shown that this addition defines a monoid structure on M, whose unit element is the class of the ordinary 2-sphere. Furthermore, if τ denotes the class of the torus, and π denotes the class of the projective plane, then every element σ of M has a unique expression of the form

$$\sigma = n\tau + m\pi$$

where n is an integer $\geqq 0$ and $m = 0, 1$, or 2. We have $3\pi = \tau + \pi$.

(The reasons for inserting the preceding example are twofold: First to relieve the essential dullness of the section. Second to show the reader that monoids exist in nature. Needless to say, the example will not be used in any way throughout the rest of the book.)

§2. *Groups*

A *group* G is a monoid, such that for every element $x \in G$ there exists an element $y \in G$ such that $xy = yx = e$. Such an element y is called an *inverse* for x. Such an inverse is unique, because if y' is also an inverse for x, then

$$y' = y'e = y'(xy) = (y'x)y = ey = y.$$

We denote this inverse by x^{-1} (or by $-x$ when the law of composition is written additively).

For any positive integer n, we let $x^{-n} = (x^{-1})^n$. Then the usual rules for exponentiation hold for all integers, not only for integers $\geqq 0$ (as we pointed out for monoids in §1). The trivial proofs are left to the reader.

In the definitions of unit elements and inverses, we could also define left units and left inverses (in the obvious way). One can easily prove that these are also units and inverses respectively. Namely:

Let G be a set with an associative law of composition, let e be a left unit for that law, and assume that every element has a left inverse. Then e is a unit, and each left inverse is also an inverse. In particular, G is a group.

To prove this, let $a \in G$ and let $b \in G$ be such that $ba = e$. Then

$$bab = eb = b.$$

Multiplying on the left by a left inverse for b yields

$$ab = e,$$

or in other words, b is also a right inverse for a. One sees also that a is a left inverse for b. Furthermore,

$$ae = aba = ea = a,$$

whence e is a right unit.

Example. Let G be a group and S a non-empty set. The set of maps $M(S, G)$ is itself a group; namely for two maps f, g of S into G we define fg to be the map such that

$$(fg)(x) = f(x)g(x),$$

and we define f^{-1} to be the map such that $f^{-1}(x) = f(x)^{-1}$. It is then trivial to verify that $M(S, G)$ is a group. If G is commutative, so is

$M(S, G)$, and when the law of composition in G is written additively, so is the law of composition in $M(S, G)$, so that we would write $f + g$ instead of fg, and $-f$ instead of f^{-1}.

Example. Let S be a non-empty set. Let G be the set of bijective mappings of S onto itself. Then G is a group, the law of composition being ordinary composition of mappings. The unit element of G is the identity map of S, and the other group properties are trivially verified. The elements of G are called *permutations* of S.

Example. The set of rational numbers forms a group under addition. The set of non-zero rational numbers forms a group under multiplication. Similar statements hold for the real and complex numbers.

Let G be a group. A *subgroup* H of G is a subset of G containing the unit element, and such that H is closed under the law of composition and inverse (i.e. it is a submonoid, such that if $x \in H$ then $x^{-1} \in H$). A subgroup is called *trivial* if it consists of the unit element alone. The intersection of an arbitrary non-empty family of subgroups is a subgroup (trivial verification).

Let G, G' be monoids. A *monoid-homomorphism* (or simply homomorphism) of G into G' is a mapping $f : G \to G'$ such that $f(xy) = f(x)f(y)$ for all $x, y \in G$, and mapping the unit element of G into that of G'. If G, G' are groups, a *group-homomorphism* of G into G' is simply a monoid-homomorphism.

We sometimes say: "Let $f : G \to G'$ be a group-homomorphism" to mean: "Let G, G' be groups, and let f be a homomorphism from G into G'".

Let $f : G \to G'$ be a group-homomorphism. Then

$$f(x^{-1}) = f(x)^{-1}$$

because if e, e' are the unit elements of G, G' respectively, then

$$e' = f(e) = f(xx^{-1}) = f(x)f(x^{-1}).$$

Furthermore, if G, G' are groups and $f : G \to G'$ is a map such that $f(xy) = f(x)f(y)$ for all x, y in G, then $f(e) = e'$ because $f(ee) = f(e)$ and also $= f(e)f(e)$. Multiplying by the inverse of $f(e)$ shows that $f(e) = e'$.

Let G, G' be monoids. A homomorphism $f : G \to G'$ is called an *isomorphism* if there exists a homomorphism $g : G' \to G$ such that $f \circ g$ and $g \circ f$ are the identity mappings (in G' and G respectively). It is trivially verified that f is an isomorphism if and only if f is bijective. The existence of an isomorphism between two groups G and G' is sometimes denoted by $G \approx G'$. If $G = G'$, we say that the isomorphism is an *automorphism*. A homomorphism of G into itself is also called an *endomorphism*.

Example. Let G be a monoid and x an element of G. Let $\mathbf{N}$ denote the (additive) monoid of integers $\geqq 0$. Then the map $f : \mathbf{N} \to G$ such that $f(n) = x^n$ is a homomorphism. If G is a group, we can extend f to a homomorphism of $\mathbf{Z}$ into G (x^n is defined for all $n \in \mathbf{Z}$, as pointed out previously). The trivial proofs are left to the reader.

Let n be a fixed integer and let G be a *commutative* group. Then one verifies easily that the map

$$x \mapsto x^n$$

from G into itself is a homomorphism. So is the map $x \mapsto x^{-1}$. The map $x \mapsto x^n$ is called the n-th *power map*.

Let G be a group and S a subset of G. We shall say that S *generates* G, or that S is a set of *generators* for G, if every element of G can be expressed as a product of elements of S or inverses of elements of S, i.e. as a product $x_1 \cdots x_n$ where each x_i or x_i^{-1} is in S. It is clear that the set of all such products is a subgroup of G (the empty product is the unit element), and is the smallest subgroup of G containing S. Thus S generates G if and only if the smallest subgroup of G containing S is G itself.

Let G be a group, S a set of generators for G, and G' another group. Let $f : S \to G'$ be a map. If there exists a homomorphism $\bar{f}$ of G into G' whose restriction to S is f, then there is only one, i.e. f has at most one extension to a homomorphism of G into G'. This is obvious, but will be used many times in the sequel.

Let $f : G \to G'$ and $g : G' \to G''$ be two group-homomorphisms. Then the composite map $g \circ f$ is a group-homomorphism. If f, g are isomorphisms, then so is $g \circ f$. Furthermore $f^{-1} : G' \to G$ is also an isomorphism. In particular, the set of all automorphisms of G is itself a group, denoted by $\mathrm{Aut}(G)$.

Let $f : G \to G'$ be a group-homomorphism. Let e, e' be the respective unit elements of G, G'. We define the *kernel* of f to be the subset of G consisting of all x such that $f(x) = e'$. From the definitions, it follows at once that the kernel H of f is a subgroup of G. (Let us prove for instance that H is closed under the inverse mapping. Let $x \in H$. Then

$$f(x^{-1})f(x) = f(e) = e'.$$

Since $f(x) = e'$, we have $f(x^{-1}) = e'$, whence $x^{-1} \in H$. We leave the other verifications to the reader.)

Let $f : G \to G'$ be a group-homomorphism again. Let H' be the *image* of f. Then H' is a subgroup of G', because it contains e', and if $f(x)$, $f(y) \in H'$, then $f(xy) = f(x)f(y)$ lies also in H'. Furthermore, $f(x^{-1}) = f(x)^{-1}$ is in H', and hence H' is a subgroup of G'.

The kernel and image of f are sometimes denoted by $\mathrm{Ker}\, f$ and $\mathrm{Im}\, f$.

A homomorphism $f : G \to G'$ which establishes an isomorphism between G and its image in G' will also be called an *embedding*.

A homomorphism whose kernel is trivial is injective.

To prove this, suppose that the kernel of f is trivial, and let $f(x) = f(y)$ for some $x, y \in G$. Multiplying by $f(y^{-1})$ we obtain

$$f(xy^{-1}) = f(x)f(y^{-1}) = e'.$$

Hence xy^{-1} lies in the kernel, hence $xy^{-1} = e$, and $x = y$. If in particular f is also surjective, then f is an isomorphism. Thus a surjective homomorphism whose kernel is trivial must be an isomorphism. We note that an injective homomorphism is an embedding.

Let G be a group and H a subgroup. A *left coset* of H in G is a subset of G of type aH, for some element a of G. An element of aH is called a *coset representative* of aH. The map $x \mapsto ax$ induces a bijection of H onto aH. Hence any two left cosets have the same cardinality.

Observe that if a, b are elements of G and aH, bH are cosets having one element in common, then they are equal. Indeed, let $ax = by$ with x, $y \in H$. Then $a = byx^{-1}$. But $yx^{-1} \in H$. Hence $aH = b(yx^{-1})H = bH$, because for any $z \in H$ we have $zH = H$.

We conclude that G is the disjoint union of the left cosets of H. A similar remark applies to *right cosets* (i.e. subsets of G of type Ha). The number of left cosets of H in G is denoted by $(G : H)$, and is called the (left) *index* of H in G. The index of the trivial subgroup is called the *order* of G and is written $(G : 1)$. From the above conclusion, we get:

PROPOSITION 1. *Let G be a group and H a subgroup. Then*

$$(G : H)(H : 1) = (G : 1),$$

in the sense that if two of these indices are finite, so is the third and equality holds as stated. If $(G : 1)$ is finite, the order of H divides the order of G.

More generally, let H, K be subgroups of G and let $H \supset K$. Let $\{x_i\}$ be a set of (left) coset representatives of K in H and let $\{y_j\}$ be a set of coset representatives of H in G. Then we contend that $\{y_j x_i\}$ is a set of coset representatives of K in G.

To prove this, note that

$$H = \bigcup_i x_i K \qquad \text{(disjoint)},$$

$$G = \bigcup_j y_j H \qquad \text{(disjoint)}.$$

Hence

$$G = \bigcup_{i,j} y_j x_i K.$$

We must show that this union is disjoint, i.e. that the $y_j x_i$ represent distinct cosets. Suppose

$$y_j x_i K = y_{j'} x_{i'} K$$

for a pair of indices (j, i) and (j', i'). Multiplying by H on the right, and noting that $x_i, x_{i'}$ are in H, we get

$$y_j H = y_{j'} H,$$

whence $y_j = y_{j'}$. From this it follows that $x_i K = x_{i'} K$ and therefore that $x_i = x_{i'}$, as was to be shown.

The formula of Proposition 1 may therefore be generalized by writing

$$(G : K) = (G : H)(H : K),$$

with the understanding that if two of the three indices appearing in this formula are finite, then so is the third and the formula holds.

§3. *Cyclic groups*

The integers $\mathbf{Z}$ form an additive group. We shall determine its subgroups. Let H be a subgroup of $\mathbf{Z}$. If H is not trivial, let a be the smallest positive integer in H. We contend that H consists of all elements na, with $n \in \mathbf{Z}$. To prove this, let $y \in H$. There exist integers n, r with $0 \leqq r < a$ such that

$$y = na + r.$$

Since H is a subgroup and $r = y - na$, we have $r \in H$, whence $r = 0$, and our assertion follows.

Let G be a group. We shall say that G is *cyclic* if there exists an element a of G such that every element x of G can be written in the form a^n for some $n \in \mathbf{Z}$ (in other words, if the map $f : \mathbf{Z} \to G$ such that $f(n) = a^n$ is surjective). Such an element a of G is then called a *generator* of G.

Let G be a group and $a \in G$. The subset of all elements a^n $(n \in \mathbf{Z})$ is obviously a subgroup of G, which is cyclic. If m is an integer such that $a^m = e$ and $m > 0$ then we shall call m an *exponent* of a. We shall say that $m > 0$ is an *exponent* of G if $x^m = e$ for all $x \in G$.

Let G be a group and $a \in G$. Let $f : \mathbf{Z} \to G$ be the homomorphism such that $f(n) = a^n$ and let H be the kernel of f. Two cases arises:

(i) The kernel is trivial. Then f is an isomorphism of $\mathbf{Z}$ onto the cyclic subgroup of G generated by a, and this subgroup is infinite cyclic. If a generates G, then G is cyclic. We also say that a has *infinite period.*

(ii) The kernel is not trivial. Let d be the smallest positive integer in the kernel. Then d is called the *period* of a. If m is an integer such that $a^m = e$ then $m = ds$ for some integer s. We observe that the elements $e, a, \ldots, a^{d-1}$ are distinct. Indeed, if $a^r = a^s$ with $0 \leqq r, s \leqq d - 1$, and say $r \leqq s$, then $a^{s-r} = e$. Since $0 \leqq s - r < d$ we must have $s - r = 0$. The cyclic subgroup generated by a has order d. Hence:

PROPOSITION 2. *Let G be a finite group of order $n > 1$. Let a be an element of G, $a \neq e$. Then the period of a divides n. If the order of G is a prime number p, then G is cyclic and the period of any generator is equal to p.*

Furthermore:

PROPOSITION 3. *Let G be a cyclic group. Then every subgroup of G is cyclic. If f is a homomorphism of G, then the image of f is cyclic.*

Proof. If G is infinite cyclic, it is isomorphic to $\mathbf{Z}$, and we determined above all subgroups of $\mathbf{Z}$, finding that they are all cyclic. If G is finite cyclic, and a is a generator, and H is a subgroup, we let m be the smallest positive integer such that a^m lies in H. It is then easily verified that a^m is a generator of H. Finally, if $f : G \to G'$ is a homomorphism, and a is a generator of G, then $f(a)$ is obviously a generator of $f(G)$, which is therefore cyclic.

We shall leave the proofs of the following statements concerning cyclic groups as exercises.

(i) *An infinite cyclic group has exactly two generators* (if a is a generator, then a^{-1} is the only other generator).

(ii) Let G be a finite cyclic group of order n, and let x be a generator. *The set of generators of G consists of those powers x^ν of x such that ν is relatively prime to n.*

(iii) Let G be a cyclic group, and let a, b be two generators. *Then there exists an automorphism of G mapping a onto b. Conversely, any automorphism of G maps a on some generator of G.*

§4. Normal subgroups

We have already observed that the kernel of a group-homomorphism is a subgroup. We now wish to characterize such subgroups.

Let $f : G \to G'$ be a group-homomorphism, and let H be its kernel. If x is an element of G, then $xH = Hx$, as one verifies at once from the definitions. We can also rewrite this relation as $xHx^{-1} = H$.

Conversely, let G be a group, and let H be a subgroup. Assume that for all elements x of G we have $xH \subset Hx$ (or equivalently, $xHx^{-1} \subset H$). If we write x^{-1} instead of x, we get $H \subset xHx^{-1}$, whence $xHx^{-1} = H$. Thus our condition is equivalent to the condition $xHx^{-1} = H$ for all $x \in G$. A subgroup H satisfying this condition will be called *normal*. We shall now see that a normal subgroup is the kernel of a homomorphism.

Let G' be the set of cosets of H. (By assumption, a left coset is equal to a right coset, so we need not distinguish between them.) If xH and yH

are cosets, then their product $(xH)(yH)$ is also a coset, because

$$xHyH = xyHH = xyH.$$

By means of this product, we have therefore defined a law of composition on G' which is associative. It is clear that the coset H itself is a unit element for this law of composition, and that $x^{-1}H$ is an inverse for the coset xH. Hence G' is a group.

Let $f : G \to G'$ be the mapping such that $f(x)$ is the coset xH. Then f is clearly a homomorphism, and (the subgroup) H is contained in its kernel. If $f(x) = H$, then $xH = H$. Since H contains the unit element, it follows that $x \in H$. Thus H is equal to the kernel, and we have obtained our desired homomorphism.

The group of cosets of a normal subgroup H is denoted by G/H (which we read G modulo H, or $G \bmod H$). The map f of G onto G/H constructed above is called the *canonical map*, and G/H is called the *factor group* of G by H.

Remarks

(1) Let $\{H_i\}_{i \in I}$ be a family of normal subgroups of G. Then the subgroup

$$H = \bigcap_{i \in I} H_i$$

is a normal subgroup. Indeed, if $y \in H$, and $x \in G$, then xyx^{-1} lies in each H_i, whence in H.

(2) Let S be a subset of G and let $N = N_S$ be the set of all elements $x \in G$ such that $xSx^{-1} = S$. Then N is obviously a subgroup of G, called the *normalizer* of S. If S consists of one element a, then N is also called the *centralizer* of a. More generally, let Z_S be the set of all elements $x \in G$ such that $xyx^{-1} = y$ for all $y \in S$. Then Z_S is called the *centralizer* of S. The centralizer of G itself is called the *center* of G. It is the subgroup of G consisting of all elements of G commuting with all other elements, and is obviously a normal subgroup of G.

Let H be a subgroup of G. Then H is obviously a normal subgroup of its normalizer N_H. We leave the following statements as exercises:

If K is any subgroup of G containing H and such that H is normal in K, then $K \subset N_H$.

If K is a subgroup of N_H, then KH is a group and H is normal in KH.

The normalizer of H is the largest subgroup of G in which H is normal.

Let G be a group and H a normal subgroup. Let $x, y \in G$. We shall write

$$x \equiv y \pmod{H}$$

if x and y lie in the same coset of H, or equivalently if xy^{-1} (or $y^{-1}x$) lie in H. We read this relation "x and y are congruent modulo H".

When G is an additive group, then

$$x \equiv 0 \pmod{H}$$

means that x lies in H, and

$$x \equiv y \pmod{H}$$

means that $x - y$ (or $y - x$) lies in H. This notation of congruence is used mostly for additive groups.

Let

$$G' \xrightarrow{f} G \xrightarrow{g} G''$$

be a sequence of homomorphisms. We shall say that this sequence is *exact* if $\operatorname{Im} f = \operatorname{Ker} g$. For example, if H is a subgroup of G then the sequence

$$H \xrightarrow{j} G \xrightarrow{\varphi} G/H$$

is exact (where $j =$ inclusion and $\varphi =$ canonical map). A sequence of homomorphisms having more than one term, like

$$G_1 \xrightarrow{f_1} G_2 \xrightarrow{f_2} G_3 \to \cdots \xrightarrow{f_{n-1}} G_n,$$

is called *exact* if it is exact at each joint, i.e. if

$$\operatorname{Im} f_i = \operatorname{Ker} f_{i+1}$$

for each $i = 1, \dots, n - 2$. For example to say that

$$0 \to G' \xrightarrow{f} G \xrightarrow{g} G'' \to 0$$

is exact means that f is injective, that $\operatorname{Im} f = \operatorname{Ker} g$, and that g is surjective. If $H = \operatorname{Ker} g$ then this sequence is essentially the same as the exact sequence

$$0 \to H \to G \to G/H \to 0.$$

Next we describe some homomorphisms, all of which are called canonical.

(i) Let G, G' be groups and $f : G \to G'$ a homomorphism whose kernel is H. Let $\varphi : G \to G/H$ be the canonical map. Then there exists a unique homomorphism $f_* : G/H \to G'$ such that $f = f_* \circ \varphi$, and f_* is injective.

To define f_*, let xH be a coset of H. Since $f(xy) = f(x)$ for all $y \in H$, we define $f_*(xH)$ to be $f(x)$. This value is independent of the choice of coset representative x, and it is then trivially verified that f_* is a homomorphism, is injective, and is the unique homomorphism satisfying our requirements. We shall say that f_* is *induced* by f.

Our homomorphism f_* induces an isomorphism

$$\lambda : G/H \to \operatorname{Im} f$$

of G/H onto the image of f, and thus f can be factored into the following succession of homomorphisms:

$$G \xrightarrow{\varphi} G/H \xrightarrow{\lambda} \operatorname{Im} f \xrightarrow{j} G'.$$

Here, j is the inclusion of $\operatorname{Im} f$ in G'.

(ii) Let G be a group and H a subgroup. Let N be the intersection of all normal subgroups containing H. Then N is normal, and hence is the smallest normal subgroup of G containing H. Let $f : G \to G'$ be a homomorphism whose kernel contains H. Then the kernel of f contains N, and there exists a unique homomorphism $f_* : G/N \to G'$, said to be induced by f, making the following diagram commutative:

$$\begin{array}{ccc} G & \xrightarrow{f} & G' \\ & {}_{\varphi}\searrow \quad \nearrow_{f_*} & \\ & G/N & \end{array}$$

As before, φ is the canonical map.

We can define f_* as in (i) by the rule

$$f_*(xN) = f(x).$$

This is well defined, and is trivially verified to satisfy all our requirements.

(iii) Let G be a group and $H \supset K$ two normal subgroups of G. Then K is normal in H, and we can define a map of G/K into G/H by associating with each coset xK the coset xH. It is immediately verified that this map is a homomorphism, and that its kernel consists of all cosets xK such that $x \in H$. Thus we have a canonical *isomorphism*

$$(G/K)/(H/K) \approx G/H.$$

One could also describe this isomorphism using (i) and (ii). We leave it to the reader to show that we have a commutative diagram

$$\begin{array}{ccccccccc} 0 & \to & H & \to & G & \to & G/H & \to & 0 \\ & & \downarrow \text{can} & & \downarrow \text{can} & & \downarrow \text{id} & & \\ 0 & \to & H/K & \to & G/K & \to & G/H & \to & 0 \end{array}$$

where the rows are exact.

(iv) Let G be a group and let H, K be two subgroups. Assume that H is contained in the normalizer of K. Then $H \cap K$ is obviously a normal subgroup of H, and equally obviously $HK = KH$ is a subgroup of G. There is a surjective homomorphism

$$H \to HK/K$$

associating with each $x \in H$ the coset xK of K in the group HK. The reader will verify at once that the kernel of this homomorphism is exactly $H \cap K$. Thus we have a canonical isomorphism

$$H/(H \cap K) \approx HK/K.$$

(v) Let $f : G \to G'$ be a group homomorphism, let H' be a normal subgroup of G', and let $H = f^{-1}(H')$.

$$\begin{array}{ccc} G & \longrightarrow & G' \\ \uparrow & & \uparrow \\ f^{-1}(H') & \to & H' \end{array}$$

Then $f^{-1}(H')$ is normal in G. [Proof: If $x \in G$, then $f(xHx^{-1}) = f(x)f(H)f(x)^{-1}$ is contained in H', so $xHx^{-1} \subset H$.] We then obtain a homomorphism

$$G \to G' \to G'/H'$$

composing f with the canonical map of G' onto G'/H', and the kernel of this composite is H. Hence we get an injective homomorphism

$$\bar{f} : G/H \to G'/H'$$

again called canonical, giving rise to the commutative diagram

$$\begin{array}{ccccccccc} 0 & \to & H & \to & G & \to & G/H & \to & 0 \\ & & \downarrow & & \downarrow f & & \downarrow \bar{f} & & \\ 0 & \to & H' & \to & G' & \to & G'/H' & \to & 0. \end{array}$$

If f is surjective, then $\bar{f}$ is an isomorphism.

We shall now describe some applications of our homomorphism statements.

Let G be a group. A sequence of subgroups

$$G = G_0 \supset G_1 \supset G_2 \supset \cdots \supset G_m$$

is called a *tower* of subgroups. The tower is said to be *normal* if each G_{i+1} is normal in G_i ($i = 0, \ldots, m - 1$). It is said to be *abelian* (resp. *cyclic*) if it is normal and if each factor group G_i/G_{i+1} is abelian (resp. cyclic).

Let $f : G \to G'$ be a homomorphism, and let

$$G' = G'_0 \supset G'_1 \supset \cdots \supset G'_m$$

be a normal tower in G'. Let $G_i = f^{-1}(G'_i)$. Then the G_i ($i = 0, \ldots, m$) form a normal tower. If the G'_i form an abelian tower (resp. cyclic tower) then the G_i form an abelian tower (resp. cyclic tower), because we have an injective homomorphism

$$G_i/G_{i+1} \to G'_i/G'_{i+1}$$

for each i, and because a subgroup of an abelian group (resp. a cyclic group) is abelian (resp. cyclic).

A *refinement* of a tower

$$G = G_0 \supset G_1 \supset \cdots \supset G_m$$

is a tower which can be obtained by inserting a finite number of subgroups in the given tower. A group is said to be *solvable* if it has an abelian tower, whose last element is the trivial subgroup (i.e. $G_m = \{e\}$ in the above notation).

PROPOSITION 4. *Let G be a finite group. An abelian tower of G admits a cyclic refinement. Let G be a finite solvable group. Then G admits a cyclic tower, whose last element is $\{e\}$.*

Proof. The second assertion is an immediate consequence of the first, and it clearly suffices to prove that if G is finite, abelian, then G admits a cyclic tower. We use induction on the order of G. Let x be an element of G. We may assume that $x \neq e$. Let X be the cyclic group generated by x. Let $G' = G/X$. By induction, we can find a cyclic tower in G', and its inverse image is a cyclic tower in G whose last element is X. If we refine this tower by inserting $\{e\}$ at the end, we obtain the desired cyclic tower.

§5. *Operations of a group on a set*

Let S be a set and G a monoid. By an *operation* of G on S (on the left) we shall mean a mapping $G \times S \to S$, such that, denoting by xs the image of the pair (x, s) under our mapping ($x \in G$ and $s \in S$), we have for all $x, y \in G$ and $s \in S$:

$$(xy)s = x(ys) \qquad \text{and} \qquad es = s.$$

We then say that G operates on S (on the left), and also that S is a *G-set*.

Consider a G-set S. Let $x \in G$. Then x induces a mapping $T_x : S \to S$ of S into itself, given by

$$T_x(s) = xs$$

for all $s \in S$. Furthermore, we have by definition

$$T_{xy} = T_x T_y$$

for all $x, y \in G$.

If G is a group, then T_x has an inverse, namely $T_{x^{-1}}$, and hence each T_x is a permutation of S. In this way, the map $x \mapsto T_x$ is obviously a homomorphism of G into the group of permutations of S, and we say that G is *represented* as a group of permutations.

For the rest of this section, we assume that G is a group. The two most important examples of representations of G as a group of permutations are the following:

(i) *Conjugation.* For each x in G we let $\sigma_x : G \to G$ be the map such that $\sigma_x(y) = xyx^{-1}$. The mapping

$$(x, y) \mapsto xyx^{-1}$$

defines an operation of G on itself, called *conjugation.* (The required conditions for an operation are trivially verified.) In fact, each σ_x is an automorphism of G, i.e. for all $y, z \in G$ we have

$$\sigma_x(yz) = \sigma_x(y)\sigma_x(z),$$

and σ_x has an inverse, namely $\sigma_{x^{-1}}$. In the present case, *we see therefore that the map*

$$x \mapsto \sigma_x$$

is a homomorphism of G into its group of automorphisms. The kernel of this homomorphism is a normal subgroup of G, which consists of all $x \in G$ such that $xyx^{-1} = y$ for all $y \in G$, i.e. all $x \in G$ which commute with every element of G. This kernel is called the *center* of G.

To avoid confusion about the operation on the left, we don't write xy for $\sigma_x(y)$. Sometimes, one writes

$$\sigma_{x^{-1}}(y) = x^{-1}yx = y^x,$$

i.e. one uses an exponential notation, so that we have the rules

$$y^{(xz)} = (y^x)^z \qquad \text{and} \qquad y^e = y$$

for all $x, y, z \in G$.

We note that G also operates by conjugation on the set of subsets of G. Indeed, let S be the set of subsets of G, and let $A \in S$ be a subset of G. Then xAx^{-1} is also a subset of G which may be denoted by $\sigma_x(A)$, and one verifies trivially that the map

$$(x, A) \mapsto xAx^{-1}$$

of $G \times S \to S$ is an operation of G on S. We note in addition that if A is a subgroup of G then xAx^{-1} is also a subgroup, so that G operates on the set of subgroups by conjugation.

If A, B are two subsets of G, we say that they are *conjugate* if there exists $x \in G$ such that $B = xAx^{-1}$.

(ii) *Translation.* For each $x \in G$ we define the translation $T_x : G \to G$ by $T_x(y) = xy$. Then the map

$$(x, y) \mapsto xy = T_x(y)$$

defines an operation of G on itself. *Warning:* T_x is not a group-homomorphism! Only a permutation of G.

Similarly, G operates by translation on the set of subsets, for if A is a subset of G, then $xA = T_x(A)$ is also a subset. If H is a subgroup of G, then $T_x(H) = xH$ is of course not a subgroup but a coset of H, and hence we see that G operates by translation on the set of cosets of H. We denote the set of left cosets of H by G/H. Thus even though H need not be normal, G/H is a G-set. It has become customary to denote the set of *right* cosets by $H\backslash G$.

The above two representations of G as a group of permutation will be used frequently in the sequel. In particular, the representation by conjugation will be used throughout the next section, in the proof of the Sylow theorems.

Let S, S' be two G-sets, and $f : S \to S'$ a map. We say that f is a morphism of G-sets, or a G-map, if

$$f(xs) = xf(s)$$

for all $x \in G$ and $s \in S$. (We shall soon define categories, and see that G-sets form a category.)

We now return to the general situation, and consider a group operating on a set S. Let $s \in S$. The set of elements $x \in G$ such that $xs = s$ is obviously a subgroup of G, called the *isotropy* group of s in G, and denoted by G_s.

When G operates on itself by conjugation, then the isotropy group of an element is none other than the normalizer of this element. Similarly, when G operates on the set of subgroups by conjugation, the isotropy group of a subgroup is again its normalizer.

Let G operate on a set S. Let s, s' be elements of S, and y an element of G such that $ys = s'$. Then

$$G_{s'} = yG_sy^{-1}$$

Indeed, one sees at once that yG_sy^{-1} leaves s' fixed, and that $y^{-1}G_{s'}y$

leaves s fixed, whence the indicated equality follows. In other words, the isotropy groups of s, s' are conjugate.

Let G operate on a set S. Let $s \in S$. The subset of S consisting of all elements xs (with $x \in G$) is denoted by Gs, and is called the *orbit* of s under G. If x and y are in the same coset of the subgroup $H = G_{s'}$, then $xs = ys$, and conversely (obvious). In this manner, we get a mapping

$$f : G/H \to S$$

given by $f(xH) = xs$, and it is clear that this map is a morphism of G-sets. In fact, one sees at once that it induces a bijection of G/H onto the orbit Gs. *Consequently, if G is a group operating on a set S, and $s \in S$, then the order of the orbit Gs is equal to the index $(G : G_s)$.*

In particular, when G operates by conjugation on the set of subgroups, and H is a subgroup, then *the number of conjugate subgroups to H is equal to the index of the normalizer of H.*

Example. Let G be a group and H a subgroup of index 2. Then H is normal in G. Proof: Note that H is contained in its normalizer N_H; so the index of N_H in G is 1 or 2. If it is 1, then we are done. Suppose it is 2. Let G operate by conjugation on the set of subgroups. The orbit of H has 2 elements, and G operates on this orbit. In this way we get a homomorphism of G into the group of permutations of 2 elements. Since there is one conjugate of H unequal to H, then the kernel of our homomorphism is normal, of index 2, hence equal to H, which is normal, a contradiction which concludes the proof.

Let G operate on a set S. Then two orbits of G are either disjoint or are equal. Indeed, if Gs_1 and Gs_2 are two orbits with an element s in common, then $s = xs_1$ for some $x \in G$, and hence $Gs = Gxs_1 = Gs_1$. Similarly, $Gs = Gs_2$. Hence S is the disjoint union of the distinct orbits, and we can write

$$S = \bigcup_{i \in I} Gs_i \qquad \text{(disjoint)},$$

where I is some indexing set, and the s_i are elements of distinct orbits. If S is finite, this gives a decomposition of the order of S as a sum of orders of orbits, which we call the *orbit decomposition formula*, namely

$$\operatorname{card}(S) = \sum_{i \in I} (G : G_{s_i}).$$

Let x, y be elements of a group (or monoid) G. They are said to commute if $xy = yx$. If G is a group, the set of all elements $x \in G$ which commute with all elements of G is a subgroup of G which we called the *center* of G. Let G act on itself by conjugation. Then x is in the center if and only if the orbit of x is x itself, and thus has one element. In general, the order of the orbit of x is equal to the index of the normalizer of x. Thus when G is a finite group, the above formula reads

$$(G : 1) = \sum_{x \in C} (G : G_x)$$

where C is a set of representatives for the distinct conjugacy classes, and the sum is taken over all $x \in C$. This formula is also called the *class formula*.

§6. Sylow subgroups

Let p be a prime number. By a *p-group*, we mean a finite group whose order is a power of p (i.e. p^n for some integer $n \geqq 0$). Let G be a finite group and H a subgroup. We call H a *p-subgroup* of G if H is a p-group. We call H a *p-Sylow* subgroup if the order of H is p^n and if p^n is the highest power of p dividing the order of G. We shall prove below that such subgroups always exist. For this we need a lemma.

LEMMA. *Let G be a finite abelian group of order m, let p be a prime number dividing m. Then G has a subgroup of order p.*

Proof. We first prove by induction that if G has exponent n then the order of G divides some power of n. Let $b \in G$, $b \neq 1$, and let H be the cyclic subgroup generated by b. Then the order of H divides n since $b^n = 1$, and n is an exponent for G/H. Hence the order of G/H divides a power of n by induction, and consequently so does the order of G because

$$(G : 1) = (G : H)(H : 1).$$

Let G have order divisible by p. By what we have just seen, there exists an element x in G whose period is divisible by p. Let this period be ps for some integer s. Then $x^s \neq 1$ and obviously x^s has period p, and generates a subgroup of order p, as was to be shown.

THEOREM 1. *Let G be a finite group and p a prime number dividing the order of G. Then there exists a p-Sylow subgroup of G.*

Proof. By induction on the order of G. If the order of G is prime, our assertion is obvious. We now assume given a finite group G, and assume the theorem proved for all groups of order smaller than that of G. If there exists a proper subgroup H of G whose index is prime to p, then a p-Sylow subgroup of H will also be one of G, and our assertion follows by induction. We may therefore assume that every proper subgroup has an index divisible by p. We now let G act on itself by conjugation. From the class formula we obtain

$$(G : 1) = (Z : 1) + \sum (G : G_x).$$

Here, Z is the center of G, and the term $(Z : 1)$ corresponds to the orbits having one element, namely the elements of Z. The sum on the right is taken over the other orbits, and each index $(G : G_x)$ is then > 1, hence divisible by p. Since p divides the order of G, it follows that p divides the order of Z, hence in particular that G has a non-trivial center.

Let a be an element of order p in Z, and let H be the cyclic group generated by a. Since H is contained in Z, it is normal. Let $f : G \to G/H$ be the canonical map. Let p^n be the highest power of p dividing $(G : 1)$. Then p^{n-1} divides the order of G/H. Let K' be a p-Sylow subgroup of G/H (by induction) and let $K = f^{-1}(K')$. Then $K \supset H$ and f maps K onto K'. Hence we have an isomorphism $K/H \approx K'$. Hence K has order $p^{n-1}p = p^n$, as desired.

THEOREM 2. *Let G be a finite group.*

(i) *If H is a p-subgroup of G, then H is contained in some p-Sylow subgroup.*

(ii) *All p-Sylow subgroups are conjugate.*

(iii) *The number of p-Sylow subgroups of G is $\equiv 1 \bmod p$.*

Proof. All proofs are applications of the technique of the class formula. We let S be the set of p-Sylow subgroups of G. Then G operates on S by conjugation. Let P be one of the p-Sylow subgroups. The isotropy group G_P of P contains P, and hence the orbit of P has order prime to p. Let H be a p-subgroup of order > 1. Then H operates by conjugation on this orbit, which we shall denote by S_0, and S_0 itself breaks up into a disjoint union of orbits under H. Since the order of H is a power of p, the index of any proper subgroup is divisible by p, and hence one of the H-orbits in S_0 must consist of only one element, namely a certain Sylow subgroup P'. Then H is contained in the normalizer of P', and hence HP' is a subgroup of G. Furthermore P' is normal in HP'. Since

$$HP'/P' \approx H/(H \cap P'),$$

the order of HP'/P' is a power of p, and therefore so is the order of HP'. Since P' is a maximal p-subgroup of G, we must have $HP' = P'$, and hence $H \subset P'$, which proves (i).

In particular, let H be any p-Sylow subgroup of G. We have shown that H is contained in some conjugate of P, and is therefore equal to that conjugate (because the orders are the same). This proves (ii). Finally, we take $H = P$ itself. Then one orbit under H has exactly one element (P itself), and all other orbits have more than one element, and in fact the cardinalities of these orbits are divisible by p, being indices of proper subgroups of P. This proves (iii).

THEOREM 3. *Let G be a finite p-group. Then G is solvable. If its order is > 1, then G has a non-trivial center.*

Proof. The first assertion follows from the second, since if G has center Z, and we have an abelian tower for G/Z by induction, we can lift this abelian tower to G to show that G is solvable. To prove the second assertion, we use the class equation

$$(G : 1) = \text{card}(Z) + \sum (G : G_x),$$

the sum being taken over certain x for which $(G : G_x) \neq 1$. Then p divides $(G : 1)$ and also divides every term in the sum, so that p divides the order of the center, as was to be shown.

COROLLARY. *Let G be a p-group which is not of order* 1. *Then there exists a sequence of subgroups*

$$\{e\} = G_0 \subset G_1 \subset G_2 \subset \cdots \subset G_n = G$$

such that G_i is normal in G and G_{i+1}/G_i is cyclic of order p.

Proof. Since G has a non-trivial center, there exists an element $a \neq e$ in the center of G, and such that a has order p. Let H be the cyclic group generated by a. By induction, if $G \neq H$, we can find a sequence of subgroups as stated above in the factor group G/H. Taking the inverse image of this tower in G gives us the desired sequence in G.

§7. *Categories and functors*

Before proceeding further, it will now be convenient to introduce some new terminology. We have met already several kinds of objects: sets, monoids, groups. We shall meet many more, and for each such kind of objects we define special kinds of maps between them (e.g. homomorphisms). Some formal behavior will be common to all of these, namely the existence of identity maps of an object onto itself, and the associativity of maps when such maps occur in succession. We introduce the notion of category to give a general abstract setting for all of these.

A *category* $\mathfrak{A}$ consists of a collection of objects $\text{Ob}(\mathfrak{A})$; and for two objects $A, B \in \text{Ob}(\mathfrak{A})$ a set $\text{Mor}(A, B)$ called the set of *morphisms* of A into B; and for three objects $A, B, C \in \text{Ob}(\mathfrak{A})$ a law of composition (i.e. a map)

$$\text{Mor}(B, C) \times \text{Mor}(A, B) \to \text{Mor}(A, C)$$

satisfying the following axioms:

CAT 1. Two sets $\text{Mor}(A, B)$ and $\text{Mor}(A', B')$ are disjoint unless $A = A'$ and $B = B'$, in which case they are equal.

CAT 2. For each object A of $\mathfrak{A}$ there is a morphism $id_A \in \text{Mor}(A, A)$ which acts as left and right identity for the elements of $\text{Mor}(A, B)$ and $\text{Mor}(B, A)$ respectively, for all objects $B \in \text{Ob}(\mathfrak{A})$.

CAT 3. The law of composition is associative (when defined), i.e. given $f \in \text{Mor}(A, B)$, $g \in \text{Mor}(B, C)$ and $h \in \text{Mor}(C, D)$ then

$$(h \circ g) \circ f = h \circ (g \circ f),$$

for all objects A, B, C, D of $\mathcal{A}$.

Here we write the composition of an element g in $\text{Mor}(B, C)$ and an element f in $\text{Mor}(A, B)$ as $g \circ f$, to suggest composition of mappings. In practice, in this book we shall see that most of our morphisms are actually mappings, or closely related to mappings.

The collection of all morphisms in a category $\mathcal{A}$ will be denoted by $\text{Ar}(\mathcal{A})$ ("arrows of $\mathcal{A}$"). We shall sometimes use the symbols "$f \in \text{Ar}(\mathcal{A})$" to mean that f is a morphism of $\mathcal{A}$, i.e. an element of some set $\text{Mor}(A, B)$ for some $A, B \in \text{Ob}(\mathcal{A})$.

By abuse of language, we sometimes refer to the collection of objects as the category itself, if it is clear what the morphisms are meant to be.

An element $f \in \text{Mor}(A, B)$ is also written $f : A \to B$ or

$$A \xrightarrow{f} B.$$

A morphism f is called an *isomorphism* if there exists a morphism $g : B \to A$ such that $g \circ f$ and $f \circ g$ are the identities in $\text{Mor}(A, A)$ and $\text{Mor}(B, B)$ respectively. If $A = B$, then we also say that the isomorphism is an *automorphism.*

A morphism of an object A into itself is called an *endomorphism.* The set of endomorphisms of A is denoted by $\text{End}(A)$. It follows at once from our axioms that $\text{End}(A)$ is a monoid.

Let A be an object of a category $\mathcal{A}$. We denote by $\text{Aut}(A)$ the set of automorphisms of A. This set is in fact a group, because all of our definitions are so adjusted so as to see immediately that the group axioms are satisfied (associativity, unit element, and existence of inverse). Thus we now begin to see some feed-back between abstract categories and more concrete ones.

Examples. Let $\mathcal{S}$ be the category whose objects are sets, and whose morphisms are maps between sets. We say simply that $\mathcal{S}$ is the category of sets. The three axioms CAT 1, 2, 3 are trivially satisfied.

Let *Grp* be the category of groups, i.e. the category whose objects are groups, and whose morphisms are group-homomorphisms. Here again the three axioms are trivially satisfied. Similarly, we have a category of monoids, denoted by *Mon*.

It is also clear that if G is a group, then the G-sets form a category (with obvious morphisms).

More generally, we can now define the notion of an operation of a group G on an object in any category. Indeed, let $\mathcal{A}$ be a category and $A \in \mathrm{Ob}(\mathcal{A})$. By an *operation* of G on A we shall mean a homomorphism of G into the group $\mathrm{Aut}(A)$. In practice, an object A is a set with elements, and an automorphism in $\mathrm{Aut}(A)$ operates on A as a set, i.e. induces a permutation of A. Thus, if we have a homomorphism

$$\sigma : G \to \mathrm{Aut}(A),$$

then for each $x \in G$ we have an automorphism σ_x of A which is a permutation of A.

Consider the special case where $\mathcal{A}$ is the category of abelian groups, which we may denote by Ab. Let A be an abelian group and G a group. Given an operation of G on the abelian group A, i.e. a homomorphism

$$\sigma : G \to \mathrm{Aut}(A),$$

let us denote by $x \cdot a$ the element $\sigma_x(a)$. Then we see that for all $x \in G$, $a, b \in A$, we have:

$$\begin{aligned} x \cdot (y \cdot a) &= (xy) \cdot a, & x \cdot (a + b) &= x \cdot a + x \cdot b, \\ e \cdot a &= a, & x \cdot 0 &= 0. \end{aligned}$$

We observe that when a group G operates on itself by conjugation, then not only does G operate on itself as a set but also operates on itself as an object in the category of groups, i.e. the permutations induced by the operation are actually group-automorphisms.

Similarly, we shall introduce later other categories (rings, modules, fields) and we have given a general definition of what it means for a group to operate on an object in any one of these categories.

Let $\mathcal{A}$ be a category. We may take as objects of a new category $\mathcal{C}$ the morphisms of $\mathcal{A}$. If $f : A \to B$ and $f' : A' \to B'$ are two morphisms in $\mathcal{A}$ (and thus objects of $\mathcal{C}$), then we define a morphism $f \to f'$ (in $\mathcal{C}$) to be a pair of morphisms (φ, ψ) in $\mathcal{A}$ making the following diagram commutative:

$$\begin{array}{ccc} A & \xrightarrow{f} & B \\ \varphi \downarrow & & \downarrow \psi \\ A' & \xrightarrow[f']{} & B' \end{array}$$

In that way, it is clear that $\mathcal{C}$ is a category. Strictly speaking, as with maps of sets, we should index (φ, ψ) by f and f' (otherwise CAT 1 is not necessarily satisfied), but such indexing is omitted in practice.

There are many variations on this example. For instance, we could restrict our attention to morphisms in $\mathcal{A}$ which have a fixed object of departure, or those which have a fixed object of arrival.

Thus let A be an object of $\mathcal{A}$, and let $\mathcal{A}_A$ be the category whose objects are morphisms

$$f : X \to A$$

in $\mathcal{A}$, having A as object of arrival. A morphism in $\mathcal{A}_A$ from $f : X \to A$ to $g : Y \to A$ is simply a morphism

$$h : X \to Y$$

in $\mathcal{A}$ such that the diagram is commutative:

$$\begin{array}{ccc} X & \xrightarrow{h} & Y \\ & {\scriptstyle f}\searrow \quad \swarrow{\scriptstyle g} & \\ & A & \end{array}$$

Let $\mathcal{A}$, $\mathcal{B}$ be categories. A *covariant functor* F of $\mathcal{A}$ into $\mathcal{B}$ is a rule which to each object A in $\mathcal{A}$ associates an object $F(A)$ in $\mathcal{B}$, and to each morphism $f : A \to B$ associates a morphism $F(f) : F(A) \to F(B)$ such that:

FUN 1. For all A in $\mathcal{A}$ we have $F(id_A) = id_{F(A)}$.

FUN 2. If $f : A \to B$ and $g : B \to C$ are two morphisms of $\mathcal{A}$ then $F(g \circ f) = F(g) \circ F(f)$.

Example. If to each group G we associate its set (stripped of the group structure) we obtain a functor from the category of groups into the category of sets, provided that we associate with each group-homomorphism itself, viewed only as a set-theoretic map. Such a functor is called a *stripping* functor.

We observe that a functor transforms isomorphisms into isomorphisms, because $f \circ g = id$ implies $F(f) \circ F(g) = id$ also.

We can define the notion of a *contravariant functor* from $\mathcal{A}$ into $\mathcal{B}$ by using the same condition FUN 1, and reversing the arrows in condition FUN 2, i.e. to each morphism $f : A \to B$ the contravariant functor associates a morphism

$$F(f) : F(B) \to F(A)$$

(going in the opposite direction), such that, if

$$f : A \to B \quad \text{and} \quad g : B \to C$$

are morphisms in $\mathcal{A}$, then

$$F(g \circ f) = F(f) \circ F(g).$$

Sometimes a functor is denoted by writing f_* instead of $F(f)$ in the case of a covariant functor, and by writing f^* in the case of a contravariant functor.

Example. Let $\mathcal{A}$ be a category and A a fixed object in $\mathcal{A}$. Then we obtain a covariant functor

$$M_A : \mathcal{A} \to \mathcal{S}$$

by letting $M_A(X) = \mathrm{Mor}(A, X)$ for any object X of $\mathcal{A}$. If $\varphi : X \to X'$ is a morphism, we let

$$M_A(\varphi) : \mathrm{Mor}(A, X) \to \mathrm{Mor}(A, X')$$

be the map given by the rule

$$g \to \varphi \circ g$$

for any $g \in \mathrm{Mor}(A, X)$,

$$A \xrightarrow{g} X \xrightarrow{\varphi} X'.$$

The axioms FUN 1 and FUN 2 are trivially verified.

Similarly, for each object B of $\mathcal{A}$, we have a contravariant functor

$$M^B : \mathcal{A} \to \mathcal{S}$$

such that $M^B(Y) = \mathrm{Mor}(Y, B)$. If $\psi : Y' \to Y$ is a morphism, then

$$M^B(\psi) : \mathrm{Mor}(Y, B) \to \mathrm{Mor}(Y', B)$$

is the map given by the rule

$$f \to f \circ \psi$$

for any $f \in \mathrm{Mor}(Y, B)$,

$$Y' \xrightarrow{\psi} Y \xrightarrow{f} B.$$

The preceding two functors are called the *representation functors.*

Consider the important special case when we deal with the category of groups. If S is a set and G a group, then the set of maps $M(S, G)$ is itself a group, as we noticed previously. If G, G' are two groups, the set of morphisms $\mathrm{Mor}(G, G')$ in the category of groups is simply the set of homomorphisms of G into G', and will be denoted by $\mathrm{Hom}(G, G')$. We note that $\mathrm{Hom}(G, G')$ is not necessarily a group if G' is not commutative.

We observe in addition the important fact that the representation functors give rise to *homomorphisms.* For instance, consider the covariant representation functor.

Let S be a set, X, X' groups, and $\varphi : X \to X'$ a group-homomorphism. We have an induced map

$$M_S(\varphi) : M(S, X) \to M(S, X')$$

given by $g \mapsto \varphi \circ g$. If g, $h \in M(S, X)$, then for $x \in X$,

$$\varphi \circ (gh)(x) = \varphi\big((gh)(x)\big) = \varphi\big(g(x)h(x)\big) = \varphi\big(g(x)\big)\varphi\big(h(x)\big).$$

Hence $M_S(\varphi)$ is a homomorphism. A similar statement holds for the contravariant representation functor.

The fact that $\operatorname{Hom}(G, X)$ is a group when G, X are commutative is of special significance. We shall study more closely the commutative case when we deal with the dual group, and later when we discuss the duality of vector spaces. These provide good further examples for the present discussion, and the reader may read them at once if he wishes.

As Grothendieck pointed out, one can use the representation functor to transport the notions of certain structures on sets to arbitrary categories. For instance, let $\mathcal{A}$ be a category and G an object of $\mathcal{A}$. We say that G is a *group object* in $\mathcal{A}$ if for each object X of $\mathcal{A}$ we are given a group structure on the set $\operatorname{Mor}(X, G)$ in such a way that the association

$$X \mapsto \operatorname{Mor}(X, G)$$

is functorial (i.e. is a functor from $\mathcal{A}$ into the category of groups). One sometimes denotes the set $\operatorname{Mor}(X, G)$ by $G(X)$, and thinks of it as the set of points of G in X. To justify this terminology, the reader is referred to Chapter X, §3.

As another example, we have the notion of product defined in the category of sets. We shall extend this notion to arbitrary categories in such a way that it is compatible with the representation functor.

Products and coproducts

Let $\mathcal{A}$ be a category and let A, B be objects of $\mathcal{A}$. By a *product* of A, B in $\mathcal{A}$ one means a triple (P, f, g) consisting of an object P in $\mathcal{A}$ and two morphisms

$$\begin{array}{ccc} & P & \\ {}^{f}\swarrow & & \searrow^{g} \\ A & & B \end{array}$$

satisfying the following condition: Given two morphisms

$$\varphi : C \to A \qquad \text{and} \qquad \psi : C \to B$$

in $\mathcal{A}$, there exists a unique morphism $h : C \to P$ which makes the following diagram commutative:

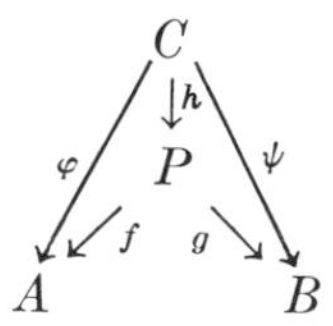

in other words, $\varphi = f \circ h$ and $\psi = g \circ h$.

More generally, given a family of objects $\{A_i\}_{i \in I}$ in $\mathcal{A}$, a *product* for this family consists of $(P, \{f_i\}_{i \in I})$, where P is an object in $\mathcal{A}$ and $\{f_i\}_{i \in I}$ is a family of morphisms

$$f_i : P \to A_i,$$

satisfying the following condition: Given a family of morphisms

$$g_i : C \to A_i,$$

there exists a unique morphism $h : C \to P$ such that $f_i \circ h = g_i$ for all i.

Example. Let $\mathcal{A}$ be the category of sets, and let $\{A_i\}_{i \in I}$ be a family of sets. Let $P = \prod_{i \in I} A_i$ be their cartesian product, and let $f_i : P \to A_i$ be the projection on the i-th factor. Then $(P, \{f_i\})$ clearly satisfies the requirements of a product in the category of sets.

As a matter of notation, we shall usually write $A \times B$ for the product of two objects in a category, and $\prod_{i \in I} A_i$ for the product of an arbitrary family in a category, following the same notation as in the category of sets. In the next section, we shall investigate the product in the category of groups.

We shall also meet the dual notion: Let $\{A_i\}_{i \in I}$ be a family of objects in a category $\mathcal{A}$. By their *coproduct* one means a pair $(S, \{f_i\}_{i \in I})$ consisting of an object S and a family of morphisms

$$\{f_i : A_i \to S\},$$

satisfying the following property. Given a family of morphisms $\{g_i : A_i \to C\}$, there exists a unique morphism $h : S \to C$ such that $h \circ f_i = g_i$ for all i.

In the product and coproduct, the morphism h will be said to be the morphism *induced* by the family $\{g_i\}$.

Examples. Let $\mathcal{S}$ be the category of sets. *Then coproducts exist.* For instance, let S, S' be sets. Let T be a set having the same cardinality as S' and disjoint from S. Let $f_1 : S \to S$ be the identity, and $f_2 : S' \to T$ be a bijection. Let U be the union of S and T. Then (U, f_1, f_2) is a coproduct for S, S', viewing f_1, f_2 as maps into U.

Let $\mathcal{S}_0$ be the category of pointed sets. Its objects consist of pairs (S, x) where S is a set and x is an element of S. A morphism of (S, x) into (S', x') in this category is a map $g : S \to S'$ such that $g(x) = x'$. *Then the coproduct of (S, x) and (S', x') exists in this category*, and can be constructed as follows. Let T be the union of x and a set whose cardinality is the same as that of the complement of x' in S', and such that $T \cap S = \{x\}$. Let $U = S \cup T$, and let

$$f_1 : (S, x) \to (U, x)$$

be the map which induces the identity on S. Let

$$f_2 : (S', x') \to (U, x)$$

be a map sending x' on x and inducing a bijection of $S' - \{x'\}$ on $T - \{x\}$. Then the triple

$$((U, x), f_1, f_2)$$

is a coproduct for (S, x) and (S', x') in the category of pointed sets.

Similar constructions can be made for the coproduct of arbitrary families of sets or pointed sets. The category of pointed sets is especially important in homotopy theory.

As a final example, let $\mathcal{A}$ be a category, and A an object in $\mathcal{A}$. We let $\mathcal{A}_A$ be the category whose objects are morphisms $f : X \to A$ in $\mathcal{A}$ (X ranging over the objects of $\mathcal{A}$). If $f : X \to A$ and $g : Y \to A$ are two such objects, we define a *morphism* $h : f \to g$ in $\mathcal{A}_A$ to be a morphism $h : X \to Y$ in $\mathcal{A}$ such that the following diagram is commutative:

$$\begin{array}{ccc} X & \xrightarrow{h} & Y \\ & {\scriptstyle f}\searrow \quad \swarrow{\scriptstyle g} & \\ & A & \end{array}$$

Let $\mathcal{C}$ be a category. An object P of $\mathcal{C}$ is called *universally attracting* if there exists a unique morphism of each object of $\mathcal{C}$ into P, and is called *universally repelling* if for every object of $\mathcal{C}$ there exists a unique morphism of P into this object.

When the context makes our meaning clear, we shall call objects P as above *universal*. Since a universal object P admits the identity morphism into itself, it is then clear that if P, P' are two universal objects in $\mathcal{C}$, then there exists a unique isomorphism between them.

Let us see how this applies to say, the coproduct. Let $\mathcal{A}$ be a category, and let $\{A_i\}$ be a family of objects in $\mathcal{A}$. We now define $\mathcal{C}$. We let the objects of $\mathcal{C}$ be the families of morphisms $\{f_i : A_i \to B\}_{i \in I}$ and given two such families,

$$\{f_i : A_i \to B\} \qquad \text{and} \qquad \{f'_i : A_i \to B'\},$$

we define a morphism from the first into the second to be a morphism $\varphi : B \to B'$ in $\mathcal{A}$ such that $\varphi \circ f_i = f'_i$ for all i. Then a coproduct of $\{A_i\}$ is simply a universal object in $\mathcal{C}$.

The coproduct of $\{A_i\}$ will be denoted by

$$\coprod_{i \in I} A_i.$$

The coproduct of two objects A, B will also be denoted by

$$A \amalg B.$$

By the general uniqueness statement, we see that it is uniquely determined, up to a unique isomorphism. Similarly for the direct product.

§8. *Free groups*

Let I be a set, and for each $i \in I$, let G_i be a group. Let $G = \prod G_i$ be the set-theoretic product of the sets G_i. Then G is the set of all families $(x_i)_{i \in I}$ with $x_i \in G_i$. We can define a group structure on G by componentwise multiplication, namely, if $(x_i)_{i \in I}$ and $(y_i)_{i \in I}$ are two elements of G, we define their product to be $(x_i y_i)_{i \in I}$. We define the inverse of $(x_i)_{i \in I}$ to be $(x_i^{-1})_{i \in I}$. It is then obvious that G is a group, and that the projection mappings

$$f_i : G \to G_i$$

are homomorphisms. Since G is a set-theoretic product for the G_i, we obtain:

PROPOSITION 5. *The group $\prod G_i$, together with the projection homomorphisms, constitute a product of the family $\{G_i\}_{i \in I}$ in the category of groups.*

Indeed, if $\{g_i : G' \to G_i\}_{i \in I}$ is a family of homomorphisms, there is a unique homomorphism $g : G' \to \prod G_i$ which makes the required diagram commutative. It is the homomorphism such that $g(x')_i = g_i(x')$ for $x' \in G'$ and each $i \in I$.

We observe that each G_j admits an injective homomorphism into the product, on the j-th component, namely the map $\lambda_j : G_j \to \prod_i G_i$ such that for x in G_j, the i-th component of $\lambda_j(x)$ is the unit element of G_i if $i \neq j$, and is equal to x itself if $i = j$. This embedding will be called the *canonical* one.

There is a useful criterion for a group to be a direct product of subgroups:

PROPOSITION 6. *Let G be a group and let H, K be two subgroups such that $H \cap K = e$, $HK = G$, and such that $xy = yx$ for all $x \in H$ and $y \in K$. Then the map*

$$H \times K \to G$$

such that $(x, y) \mapsto xy$ is an isomorphism.

Proof. It is obviously a homomorphism, which is surjective since $HK = G$. If (x, y) is in its kernel, then $x = y^{-1}$, whence x lies in both H and K, and $x = e$, so that $y = e$ also, and our map is an isomorphism.

We observe that Proposition 6 generalizes by induction to a finite number of subgroups $H_1, \ldots, H_n$ which commute with each other, such that $H_1 \cdots H_n = G$, and such that

$$H_{i+1} \cap (H_1 \cdots H_i) = e.$$

In that case, G is isomorphic to the direct product

$$H_1 \times \cdots \times H_n.$$

Let G be a group and S a subset of G. We recall that G is *generated* by S if every element of G can be written as a finite product of elements of S and their inverses (the empty product being always taken as the unit element of G). Elements of S are then called *generators*. If there exists a finite set of generators for G we call G *finitely generated*. If S is a set and $\varphi : S \to G$ is a map, we say that φ *generates* G if its image generates G.

Let S be a set, and $f : S \to F$ a map into a group. Let $g : S \to G$ be another map. If $f(S)$ (or as we also say, f) generates F, then it is obvious that there exists at most one homomorphism ψ of F into G which makes the following diagram commutative:

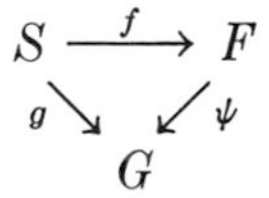

$$\begin{array}{ccc} S & \xrightarrow{f} & F \\ & {\scriptstyle g}\searrow \quad \swarrow{\scriptstyle \psi} & \\ & G & \end{array}$$

We now consider the category $\mathfrak{C}$ whose objects are the maps of S into groups. If $f : S \to G$ and $f' : S \to G'$ are two objects in this category, we define a morphism from f to f' to be a homomorphism $\varphi : G \to G'$ such that $\varphi \circ f = f'$, i.e. the diagram is commutative:

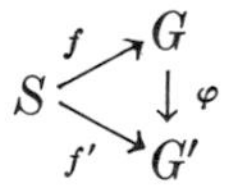

$$\begin{array}{ccc} & \overset{f}{\nearrow} & G \\ S & & \downarrow{\scriptstyle \varphi} \\ & \underset{f'}{\searrow} & G' \end{array}$$

By a *free group* determined by S, we shall mean a universal element in this category.

PROPOSITION 7. *Let S be a set. Then there exists a free group (F, f) determined by S. Furthermore, f is injective, and F is generated by the image of f.*

Proof. (I owe this proof to J. Tits.) For simplicity, we shall first carry out the proof when S is finite. Let T be an infinite denumerable set. Let

Γ be the set of all group structures on T, and for each $\gamma \in \Gamma$ let T_γ be the corresponding group. Let M_γ be the set of all maps of S into T_γ. We let $T_{\gamma,\varphi}$ be the set-theoretic product of T_γ and the set with one element $\{\varphi\}$, using therefore φ as an index, so that $T_{\gamma,\varphi}$ is the "same" group as T_γ, indexed by φ. We let

$$F_0 = \prod_{\gamma \in \Gamma} \prod_{\varphi \in M_\gamma} T_{\gamma,\varphi}$$

be the Cartesian product of the groups $T_{\gamma,\varphi}$. We define a map

$$f_0 : S \to F_0$$

by sending S on the factor $T_{\gamma,\varphi}$ by means of φ itself. We contend that given a map $g : S \to G$ of S into a group G, there exists a homomorphism $g_* : F_0 \to G$ making the usual diagram commutative:

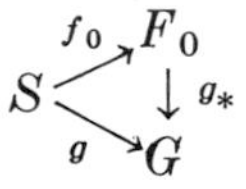

$$\begin{array}{ccc} & \overset{f_0}{\nearrow} & F_0 \\ S & & \downarrow g_* \\ & \underset{g}{\searrow} & G \end{array}$$

i.e. $g_* \circ f_0 = g$. To prove this, we first note that we may assume that g generates G, simply by restricting our attention to the subgroup of G generated by the image of g. Then the cardinality of G is $\leqq$ the cardinality of T. Let $\overline{G}$ be the product of G and the group of integers $\mathbf{Z}$, so that $\text{card}(\overline{G}) = \text{card}(T)$. Then there exists an isomorphism

$$\lambda : \overline{G} \to T_\gamma$$

for some $\gamma \in \Gamma$, and G has a natural injection in $\overline{G} = G \times \mathbf{Z}$ as a factor. Let us call this injection h, so that $h(G) = G \times \{0\}$. We then have the following sequence of homomorphisms and maps:

$$S \xrightarrow{g} G \xrightarrow{h} \overline{G} = G \times \mathbf{Z} \xrightarrow{\lambda} T_\gamma.$$

We let $\psi = \lambda \circ h \circ g$ be the composite map. Then $\psi \in M_\gamma$, and we can view ψ as a map of S into $T_{\gamma,\psi}$. We define $\psi_* = \text{pr}_G \circ \lambda^{-1} \circ \text{pr}_{\gamma,\psi}$, where $\text{pr}_{\gamma,\psi}$ is the projection of F_0 on its factor $T_{\gamma,\psi}$. It follows at once from the definitions that the following diagram is commutative:

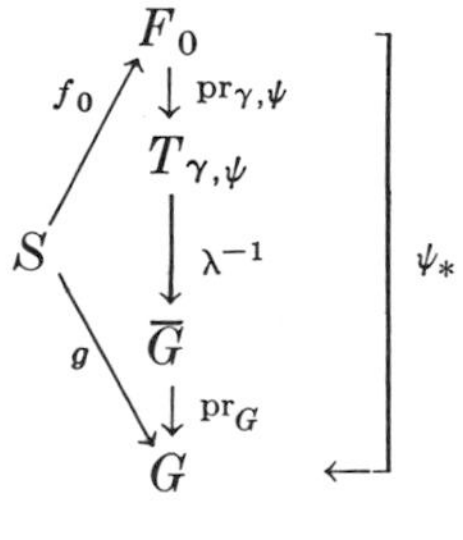

$$\begin{array}{ccc} & \overset{f_0}{\nearrow} & F_0 \\ & & \downarrow \text{pr}_{\gamma,\psi} \\ & & T_{\gamma,\psi} \\ S & & \downarrow \lambda^{-1} \\ & & \overline{G} \\ & \underset{g}{\searrow} & \downarrow \text{pr}_G \\ & & G \end{array} \qquad \psi_* : F_0 \to G$$

We let F be the subgroup of F_0 generated by the image of f_0, and we let f simply be equal to f_0, viewed as a map of S into F. We let g_* be the restriction of ψ_* to F. In this way, we see at once that the map g_* is the unique one making our diagram commutative, and thus that (F, f) is the required free group. Furthermore, it is clear that f is injective.

Suppose that S is not finite. It is then easy to adjust the cardinalities to make the proof valid, namely we let $T = S$, and we let $\overline{G}$ be the product of G with enough copies of $\mathbf{Z}$ to make again $\operatorname{card}(\overline{G}) = \operatorname{card}(T)$. The rest of the proof runs as before.

For each set S we select one free group determined by S, and denote it by $(F(S), f_S)$ or briefly by $F(S)$. It is generated by the image of f_S. One may view S as contained in $F(S)$, and the elements of S are called *free* generators of $F(S)$. If $g : S \to G$ is a map, we denote by $g_* : F(S) \to G$ the homomorphism realizing the universality of our free group $F(S)$.

If $\lambda : S \to S'$ is a map of one set into another, we let $F(\lambda) : F(S) \to F(S')$ be the map $(f_{S'} \circ \lambda)_*$.

$$\begin{array}{ccc} S & \xrightarrow{f_S} & F(S) \\ \lambda \downarrow & \searrow & \downarrow \lambda_* F(\lambda) \\ S' & \xrightarrow[f_{S'}]{} & F(S') \end{array}$$

Then we may regard F as a functor from the category of sets to the category of groups (the functorial properties are trivially verified, and will be left to the reader).

If λ is surjective, then $F(\lambda)$ is also surjective. We again leave the proof to the reader.

If two sets S, S' have the same cardinality, then they are isomorphic in the category of sets (an isomorphism being in this case a bijection!), and hence $F(S)$ is isomorphic to $F(S')$. If S has n elements, we call $F(S)$ the *free group on n generators.*

Let G be a group, and let S be the same set as G (i.e. G viewed as a set, without group structure). We have the identity map $g : S \to G$, and hence a surjective homomorphism

$$g_* : F(S) \to G$$

which will be called *canonical.* Thus every group is a factor group of a free group.

One can also construct groups by what is called *generators and relations.* Let S be a set, and $F(S)$ the free group. We assume that $f : S \to F(S)$ is an inclusion. Let R be a set of elements of $F(S)$. Each element of R can be written as a finite product

$$\prod_{\nu=1}^{n} x_\nu$$

where each x_ν is an element of S or an inverse of an element of S. Let N be the smallest normal subgroup of $F(S)$ containing R, i.e. the intersection of all normal subgroups of $F(S)$ containing R. Then $F(S)/N$ will be called the group *determined by the generators S and the relations R.*

Example. One shows easily that the group determined by one generator a, and the relation $\{a^2\}$, has order 2. Exercises at the end of the chapter will suggest less trivial examples.

The canonical homomorphism $\varphi : F(S) \to F(S)/N$ satisfies (clearly) the universal mapping property for homomorphisms ψ of $F(S)$ into groups G such that $\psi(x) = 1$ for all $x \in R$. In view of this, one sometimes calls the group $F(S)/N$ the group determined by the generators S, and the relations $x = 1$ (for all $x \in R$). For instance, the group in the preceding example would be called the group determined by the generator a, and the relation $a^2 = 1$.

PROPOSITION 8. *Coproducts exist in the category of groups.*

Proof. Let $\{G_i\}_{i \in I}$ be a family of groups. We let $\mathfrak{C}$ be the category whose objects are families of group-homomorphisms

$$\{g_i : G_i \to G\}_{i \in I}$$

and whose morphisms are the obvious ones. We must find a universal element in this category. For each index i, we let S_i be the same set as G_i if G_i is infinite, and we let S_i be denumerable if G_i is finite. We let S be a set having the same cardinality as the set-theoretic disjoint union of the sets S_i (i.e. their coproduct in the category of sets). We let Γ be the set of group structures on S, and for each $\gamma \in \Gamma$, we let Φ_γ be the set of all families of homomorphisms

$$\varphi = \{\varphi_i : G_i \to S_\gamma\}.$$

Each pair (S_γ, φ), where $\varphi \in \Phi_\gamma$, is then a group, using φ merely as an index. We let

$$F_0 = \prod_{\gamma \in \Gamma} \prod_{\varphi \in \Phi_\gamma} (S_\gamma, \varphi),$$

and for each i, we define a homomorphism $f_i : G_i \to F_0$ by prescribing the component of f_i on each factor (S_γ, φ) to be the same as that of φ_i.

Let now $g = \{g_i : G_i \to G\}$ be a family of homomorphisms. Replacing G if necessary by the subgroup generated by the images of the g_i, we see that $\text{card}(G) \leqq \text{card}(S)$, because each element of G is a *finite* product of elements in these images. Embedding G as a factor in a product with sufficiently many copies of $\mathbf{Z}$, we may then assume that $\text{card}(G) = \text{card}(S)$. There exists a homomorphism $g_* : F_0 \to G$ such that

$$f_i \circ g_* = g_i$$

for all i. Indeed, we may assume without loss of generality that $G = S_\gamma$ for some γ and that $g = \psi$ for some $\psi \in \Phi_\gamma$. We let g_* be the projection of F_0 on the factor (S_γ, ψ).

Let F be the subgroup of F_0 generated by the union of the images of the maps f_i for all i. The restriction of g_* to F is the unique homomorphism satisfying $f_i \circ g_* = g_i$ for all i, and we have thus constructed our universal object.

I am indebted to Eilenberg for the neat arrangement of the proof of the next proposition.

Proposition 9. *Let A, B be two groups whose set-theoretic intersection is $\{1\}$. There exists a group $A \circ B$ containing A, B as subgroups, such that $A \cap B = \{1\}$, and having the following property. Every element $\neq 1$ of G has a unique expression as a product*

$$a_1 \cdots a_n \qquad (n \geqq 1, a_i \neq 1 \text{ all } i)$$

with $a_i \in A$ or $a_i \in B$, and such that if $a_i \in A$ then $a_{i+1} \in B$ and if $a_i \in B$ then $a_{i+1} \in A$.

Proof. Let $A \circ B$ be the set of sequences

$$a = (a_1, \ldots, a_n) \qquad (n \geqq 0)$$

such that either $n = 0$, and the sequence is empty, or $n \geqq 1$, and then elements in the sequence belong to A or B, are $\neq 1$, and two consecutive elements of the sequence do not belong both to A or both to B. If $b = (b_1, \ldots, b_m)$, we define the product ab to be the sequence

$$(a_1, \ldots, a_n, b_1, \ldots, b_m) \quad \text{if } a_n \in A, b_1 \in B \text{ or } a_n \in B, b_1 \in A,$$

$$(a_1, \ldots, a_n b_1, \ldots, b_m) \quad \text{if } a_n, b_1 \in A \text{ or } a_n, b_1 \in B, \text{ and } a_n b_1 \neq 1,$$

$$(a_1, \ldots, a_{n-1})(b_2, \ldots, b_m) \text{ by induction,} \quad \text{if } a_n, b_1 \in A \text{ or } a_n, b_1 \in B \text{ and } a_n b_1 = 1.$$

The case when $n = 0$ or $m = 0$ is included in the first case, and the empty sequence is the unit element of $A \circ B$. Clearly,

$$(a_1, \ldots, a_n)(a_n^{-1}, \ldots, a_1^{-1}) = \text{unit element},$$

so only associativity need be proved. Let $c = (c_1, \ldots, c_r)$.

First consider the case $m = 0$, i.e. b is empty. Then clearly $(ab)c = a(bc)$ and similarly if $n = 0$ or $r = 0$. Next consider the case $m = 1$. Let $b = (x)$ with $x \in A$, $x \neq 1$. We then verify in each possible case that $(ab)c = a(bc)$. These cases are as follows:

$$\begin{array}{ll}
(a_1, \ldots, a_n, x, c_1, \ldots, c_r) & \text{if } a_n \in B \text{ and } c_1 \in B, \\
(a_1, \ldots, a_n x, c_1, \ldots, c_r) & \text{if } a_n \in A,\ a_n x \neq 1,\ c_1 \in B, \\
(a_1, \ldots, a_n, x c_1, \ldots, c_r) & \text{if } a_n \in B,\ c_1 \in A,\ x c_1 \neq 1, \\
(a_1, \ldots, a_{n-1})(c_1, \ldots, c_r) & \text{if } a_n = x^{-1} \text{ and } c_1 \in B, \\
(a_1, \ldots, a_n)(c_2, \ldots, c_r) & \text{if } a_n \in B \text{ and } c_1 = x^{-1}, \\
(a_1, \ldots, a_{n-1}, a_n x c_1, c_2, \ldots, c_r) & \text{if } a_n, c_1 \in A,\ a_n x c_1 \neq 1, \\
(a_1, \ldots, a_{n-1})(c_2, \ldots, c_r) & \text{if } a_n, c_1 \in A \text{ and } a_n x c_1 = 1.
\end{array}$$

If $m > 1$, then we proceed by induction. Write $b = b'b''$ with b' and b'' shorter. Then

$$\begin{aligned}
(ab)c &= (a(b'b''))c = ((ab')b'')c = (ab')(b''c), \\
a(bc) &= a((b'b'')c) = a(b'(b''c)) = (ab')(b''c)
\end{aligned}$$

as was to be shown.

We have obvious injections of A and B into $A \circ B$, and identifying A, B with their images in $A \circ B$ we obtain a proof of our proposition.

We can prove the similar result for several factors. In particular, we get the following corollary for the free group.

COROLLARY 1. *Let $F(S)$ be the free group on a set S, and let $x_1, \ldots, x_n$ be distinct elements of S. Let $\nu_1, \ldots, \nu_r$ be integers $\neq 0$ and let $i_1, \ldots, i_r$ be integers,*

$$1 \leqq i_1, \ldots, i_r \leqq n$$

such that $i_j \neq i_{j+1}$ for $j = 1, \ldots, r - 1$. Then

$$x_{i_1}^{\nu_1} \cdots x_{i_r}^{\nu_r} \neq 1.$$

Proof. Let $G_1, \ldots, G_n$ be the cyclic groups generated by $x_1, \ldots, x_n$. Let $G = G_1 \circ \cdots \circ G_n$. Let

$$F(S) \to G$$

be the homomorphism sending each x_i on x_i, and all other elements of S on the unit element of G. Our assertion follows at once.

COROLLARY 2. *Let S be a set with n elements $x_1, \dots, x_n$, $n \geqq 1$. Let $G_1, \dots, G_n$ be the infinite cyclic groups generated by these elements. Then the map*

$$F(S) \to G_1 \circ \cdots \circ G_n$$

sending each x_i on itself is an isomorphism.

Proof. It is obviously surjective and injective.

COROLLARY 3. *Let $G_1, \dots, G_n$ be groups. The homomorphism*

$$G_1 \amalg \cdots \amalg G_n \to G_1 \circ \cdots \circ G_n$$

of their coproduct into $G_1 \circ \cdots \circ G_n$ induced by the natural inclusion $G_i \to G_1 \circ \cdots \circ G_n$ is an isomorphism.

Proof. Again, it is obviously injective and surjective.

§9. Direct sums and free abelian groups

Abelian groups form a category which could be denoted by Ab. We note that if $\{A_i\}_{i \in I}$ is a family of abelian groups, then the product in the category of groups is also a product in the category of abelian groups, i.e. if we form the set-theoretic product

$$\prod_{i \in I} A_i$$

and give it a group structure by composing its elements componentwise, then it is an abelian group having the desired universal property.

The coproduct in the category of abelian groups is usually called the *direct sum*.

PROPOSITION 10. *Direct sums exist in the category of abelian groups.*

Proof. Let $\{A_i\}_{i \in I}$ be a family of abelian groups. We consider the subset A of the direct product $\prod A_i$ consisting of all families $(x_i)_{i \in I}$ with $x_i \in A_i$ such that $x_i = 0$ for all but a finite number of indices i. Then it is clear that A is a subgroup of the product. For each index $j \in I$, we map

$$\lambda_j : A_j \to A$$

by letting $\lambda_j(x)$ be the element whose j-th component is x, and having all other components equal to 0. Then λ_j is an injective homomorphism. We contend that A, together with this family of maps $\{\lambda_i\}_{i \in I}$, is a direct sum for the family $\{A_i\}$. Let $\{f_i : A_i \to B\}$ be a family of homomorphisms into an abelian group B. We define a map

$$f : A \to B$$

by the rule

$$f((x_i)_{i \in I}) = \sum_{i \in I} f_i(x_i).$$

The sum on the right is actually finite since all but a finite number of terms are 0. It is immediately verified that our map f is a homomorphism. Furthermore, we clearly have $f \circ \lambda_j(x) = f_j(x)$ for each j and each $x \in A_j$. Thus f has the desired commutativity property. It is also clear that the map f is uniquely determined, as was to be shown.

If I is a finite set, then we note that the direct sum and direct product coincide.

Let A be an abelian group and B, C subgroups. If $B + C = A$ and $B \cap C = 0$, then the map

$$B \times C \to A$$

given by $(x, y) \mapsto x + y$ is an isomorphism (as we already noted in the non-commutative case). Instead of writing $A = B \times C$ we shall write

$$A = B \oplus C$$

and say that A is the *direct sum* of B and C. We use a similar notation for the direct sum of a finite number of subgroups $B_1, \ldots, B_n$ such that $B_1 + \cdots + B_n = A$ and

$$B_{i+1} \cap (B_1 + \cdots + B_i) = 0.$$

In that case we write

$$A = B_1 \oplus \cdots \oplus B_n.$$

Next let S be a set. Let $\mathfrak{C}$ be the category whose objects are maps $f : S \to A$ of S into abelian groups, and whose morphisms are the obvious ones: If $f : S \to A$ and $f' : S \to A'$ are two maps into abelian groups, then a morphism of f into f' is a (group) homomorphism $g : A \to A'$ such that the usual diagram is commutative, namely $g \circ f = f'$. A universal element in this category $\mathfrak{C}$ is called a *free abelian group* generated by S. *We shall see that it always exists.*

In fact, let $\mathbf{Z}\langle S \rangle$ be the set of all maps $\varphi : S \to \mathbf{Z}$ such that $\varphi(x) = 0$ for almost all $x \in S$. Then $\mathbf{Z}\langle S \rangle$ is an abelian group (addition being the usual addition of maps). If k is an integer and x is an element of S, we denote by $k \cdot x$ the map φ such that $\varphi(x) = k$ and $\varphi(y) = 0$ if $y \neq x$. Then it is obvious that every element φ of $\mathbf{Z}\langle S \rangle$ can be written in the form

$$\varphi = k_1 \cdot x_1 + \cdots + k_n \cdot x_n$$

for some integers k_i and elements $x_i \in S$ $(i = 1, \ldots, n)$, all the x_i being distinct. Furthermore, φ *admits a unique such expression*, because if we have

$$\varphi = \sum_{x \in S} k_x \cdot x = \sum_{x \in S} k'_x \cdot x$$

then

$$0 = \sum_{x \in S} (k_x - k'_x) \cdot x,$$

whence $k'_x = k_x$ for all $x \in S$.

We map S into $\mathbf{Z}\langle S \rangle$ by the map $f_S = f$ such that $f(x) = 1 \cdot x$. It is then clear that f is injective, and that $f(S)$ generates $\mathbf{Z}\langle S \rangle$. If $g : S \to B$ is a mapping of S into some abelian group B, then we can define a map

$$g_* : \mathbf{Z}\langle S \rangle \to B$$

such that

$$g_*\left(\sum_{x \in S} k_x \cdot x\right) = \sum_{x \in S} k_x g(x).$$

This map is a homomorphism (trivial) and makes the required diagram commutative, i.e. $g_* \circ f = g$ (also trivial). It is the only homomorphism which has this property, for any such homomorphism g_* must be such that $g_*(1 \cdot x) = g(x)$. Thus we have constructed our universal object.

It is customary to identify S in $\mathbf{Z}\langle S \rangle$, and we sometimes omit the dot when we write $k_x x$ or a sum $\sum k_x x$.

If $\lambda : S \to S'$ is a mapping of sets, there is a unique homomorphism $\bar{\lambda}$ making the following diagram commutative:

$$\begin{array}{ccc} S & \xrightarrow{f_S} & \mathbf{Z}\langle S \rangle \\ \lambda \downarrow & & \downarrow \bar{\lambda} \\ S' & \xrightarrow[f_{S'}]{} & \mathbf{Z}\langle S' \rangle \end{array}$$

In fact, $\bar{\lambda}$ is none other than $(f_{S'} \circ \lambda)_*$, with the notation of the preceding paragraph. The proof of this statement is left as a trivial exercise.

If we write $F_{ab}(S) = \mathbf{Z}\langle S \rangle$ and $\bar{\lambda} = F_{ab}(\lambda)$, then we see that F_{ab} is a functor from the category of sets to the category of abelian groups.

As an exercise, show that every abelian group A is a factor group of a free abelian group F. If A is finitely generated, show that one can select F to be finitely generated also.

If the set S above consists of n elements, then we say that the free abelian group $F_{ab}(S)$ is the free abelian group on n generators. If S is the set of n letters $x_1, \ldots, x_n$, we say that $F_{ab}(S)$ is the free abelian group with free generators $x_1, \ldots, x_n$.

An abelian group is said to be *free* if it is isomorphic to a free abelian group $F_{ab}(S)$ for some set S. Let A be an abelian group, and let S be a subset of A such that, given any $z \in A$, there exist a unique integer n_x for each $x \in S$, such that almost all $n_x = 0$, and

$$z = \sum_{x \in S} n_x x.$$

Then it is clear that A is isomorphic to the free abelian group $F_{ab}(S)$, and we call S a set of *free generators* for A, or also a *basis* for A. One defines a family $\{x_i\}_{i \in I}$ to be a basis in a similar way.

As a matter of notation, if A is an abelian group and T a subset of elements of A, we denote by $\langle T \rangle$ the subgroup generated by the element of T, i.e. the smallest subgroup of A containing T.

Example. The Grothendieck group. Let M be a commutative monoid, written additively. There exists a commutative group $K(M)$ and a monoid-homomorphism

$$\gamma : M \to K(M)$$

having the universal property with respect to homomorphisms of M into commutative groups.

Proof. Let $F_{ab}(M)$ be the free abelian group generated by M. We denote the generator of $F_{ab}(M)$ corresponding to an element $x \in M$ by $[x]$. Let B be the subgroup generated by all elements of type

$$[x + y] - [x] - [y]$$

where $x, y \in M$. We let $K(M) = F_{ab}(M)/B$, and let

$$\gamma : M \to K(M)$$

be the map obtained by composing the injection of M into $F_{ab}(M)$ given by $x \mapsto [x]$, and the canonical map

$$F_{ab}(M) \to F_{ab}(M)/B.$$

It is then clear that γ is a homomorphism, and satisfies the desired universal property.

We shall say that the *cancellation law* holds in M if, whenever $x, y, z \in M$, and $x + z = y + z$, we have $x = y$.

We then have an important criterion when the universal map γ above is injective:

If the cancellation law holds in M, then the canonical map γ of M into its Grothendieck group is injective.

Proof. This is essentially the same proof as when one constructs the negative integers from the natural numbers. We consider pairs (x, y) with $x, y \in M$ and say that (x, y) is equivalent to (x', y') if $y + x' = x + y'$. We define addition of pairs componentwise. Then the equivalence classes of pairs form a group, whose 0 element is the class of $(0, 0)$ [or the class of (x, x) for any $x \in M$]. The negative of an element (x, y) is (y, x). We have a homomorphism

$$x \mapsto \text{class of } (0, x)$$

which is injective, as one sees immediately by applying the cancellation law. Thus we have constructed a homomorphism of M into a group, which is injective. It follows that the universal homomorphism must also be injective.

We shall see later several examples of the universal group $K(M)$, which will be called the *Grothendieck group* of M.

Given an abelian group A and a subgroup B, it is sometimes desirable to find a subgroup C such that $A = B \oplus C$. The next lemma gives us a condition under which this is true.

LEMMA. *Let $A \xrightarrow{f} A'$ be a surjective homomorphism of abelian groups, and assume that A' is free. Let B be the kernel of f. Then there exists a subgroup C of A such that the restriction of f to C induces an isomorphism of C with A', and such that $A = B \oplus C$.*

Proof. Let $\{x'_i\}_{i \in I}$ be a basis of A', and for each $i \in I$, let x_i be an element of A such that $f(x_i) = x'_i$. Let C be the subgroup of A generated by all elements x_i, $i \in I$. If we have a relation

$$\sum_{i \in I} n_i x_i = 0$$

with integers n_i, almost all of which are equal to 0, then applying f yields

$$0 = \sum_{i \in I} n_i f(x_i) = \sum_{i \in I} n_i x'_i,$$

whence all $n_i = 0$. Hence our family $\{x_i\}_{i \in I}$ is a basis of C. Similarly, one sees that if $z \in C$ and $f(z) = 0$ then $z = 0$. Hence $B \cap C = 0$. Let $x \in A$. Since $f(x) \in A'$ there exist integers n_i, $i \in I$, such that

$$f(x) = \sum_{i \in I} n_i x'_i.$$

Applying f to $x - \sum_{i \in I} n_i x_i$, we find that this element lies in the kernel of f, say

$$x - \sum_{i \in I} n_i x_i = b \in B.$$

From this we see that $x \in B + C$, and hence finally that $A = B \oplus C$ is a direct sum, as contended.

THEOREM 4. *Let A be a free abelian group, and let B be a subgroup. Then B is also a free abelian group, and the cardinality of a basis of B is $\leqq$ the cardinality of a basis for A. Any two bases of B have the same cardinality, which is called the rank of B.*

Proof. We shall give the proof only when A is finitely generated, say by a basis $\{x_1, \ldots, x_n\}$ $(n \geqq 1)$, and give the proof by induction on n. We have an expression of A as direct sum:

$$A = \mathbf{Z}x_1 \oplus \cdots \oplus \mathbf{Z}x_n.$$

Let $f : A \to \mathbf{Z}x_1$ be the projection, i.e. the homomorphism such that

$$f(m_1x_1 + \cdots + m_nx_n) = m_1x_1$$

whenever $m_i \in \mathbf{Z}$. Let B_1 be the kernel of $f \mid B$. Then B_1 is contained in the free subgroup $\langle x_2, \ldots, x_n \rangle$. By induction, B_1 is free and has a basis with $\leqq n - 1$ elements. By the lemma, there exists a subgroup C_1 isomorphic to a subgroup of $\mathbf{Z}x_1$ (namely the image of $f \mid B$) such that

$$B = B_1 \oplus C_1.$$

Since $f(B)$ is either 0 or infinite cyclic, i.e. free on one generator, this proves that B is free.

(When A is not finitely generated, one can use a similar transfinite argument, which we leave to the reader.)

We also observe that our proof shows that there exists at least one basis of B whose cardinality is $\leqq n$. We shall therefore be finished when we prove the last statement, that any two bases of B have the same cardinality. Let S be one basis, with a finite number of elements m. Let T be another basis, and suppose that T has at least r elements. It will suffice to prove that $r \leqq m$ (one can then use symmetry). Let p be a prime number. Then B/pB is a direct sum of cyclic groups of order p, with m terms in the sum. Hence its order is p^m. Using the basis T instead of S, we conclude that B/pB contains an r-fold product of cyclic groups of order p, whence $p^r \leqq p^m$, and $r \leqq m$, as was to be shown. (Note that we did not assume *a priori* that T was finite.)

§10. *Finitely generated abelian groups*

The groups referred to in the title of this section occur so frequently that it is worth while to state a theorem which describes their structure completely. Throughout this section we write our abelian groups additively.

Let A be an abelian group. An element $a \in A$ is said to be a *torsion* element if it has finite period. The subset of all torsion elements of A is a subgroup of A called the *torsion subgroup* of A. (If a has period m and b has period n then, writing the group law additively, we see that $a \pm b$ has a period dividing mn.)

A finitely generated torsion abelian group is obviously finite. We shall begin by studying finite abelian groups. If A is an abelian group and p a prime number, we denote by $A(p)$ the subgroup of all elements $x \in A$ whose period is a power of p. Then $A(p)$ is a torsion group, and is a p-group if it is finite.

THEOREM 5. *Let A be a finite abelian group. Then A is the direct product of its subgroups $A(p)$ for all primes p such that $A(p) \neq 0$.*

Proof. We first consider the case of an abelian group A of exponent n such that n can be written as a product, $n = mm'$, where m, m' are integers > 1, and relatively prime. There exist integers r, s such that

$$rm + sm' = 1.$$

Then

$$A = rmA + sm'A \subset mA + m'A \subset A$$

whence equality holds everywhere. If $a \in mA \cap m'A$ then $m'a = 0$ and $ma = 0$, whence $a = rma + sm'a = 0$. Consequently A is the direct product of mA and $m'A$.

Let A_m denote the subgroup of A consisting of all x such that $mx = 0$. Then $m'A \subset A_m$ because $mm'A = 0$. Conversely, if $x \in A_m$, then $x = rmx + sm'x = m'sx$, so $x \in m'A$. Hence we obtain $m'A = A_m$, and similarly $mA = A_{m'}$, so that finally

$$A = A_m \times A_{m'}.$$

Inductively, we conclude that A is the direct product of its subgroups $A(p)$ as stated in the theorem.

Our next task is to describe the structure of finite abelian p-groups. Let $r_1, \ldots, r_s$ be integers $\geqq 1$. A finite p-group A is said to be of *type* $(p^{r_1}, \ldots, p^{r_s})$ if A is isomorphic to the product of cyclic groups of orders p^{r_i} $(i = 1, \ldots, s)$.

THEOREM 6. *Every finite abelian p-group is isomorphic to a product of cyclic p-groups. If it is of type $(p^{r_1}, \ldots, p^{r_s})$ with*

$$r_1 \geqq r_2 \geqq \cdots \geqq r_s \geqq 1,$$

then the sequence of integers $(r_1, \ldots, r_s)$ is uniquely determined.

Proof. Let A be a finite abelian p-group. We shall need the following remark: Let b be an element of A, $b \neq 0$. Let k be an integer $\geqq 0$ such that $p^k b \neq 0$, and let p^m be the period of $p^k b$. Then b has period p^{k+m}. Proof: We certainly have $p^{k+m}b = 0$, and if $p^n b = 0$ then first $n \geqq k$, and second $n \geqq k + m$, otherwise the period of $p^k b$ would be smaller than p^m.

We shall now prove the existence of the desired product by induction. Let $a_1 \in A$ be an element of maximal period. We may assume without loss of generality that A is not cyclic. Let A_1 be the cyclic subgroup generated by a_1, say of period p^{r_1}. We need a lemma.

LEMMA. *Let $\bar{b}$ be an element of A/A_1, of period p^r. Then there exists a representative a of $\bar{b}$ in A which also has period p^r.*

Proof. Let b be any representative of $\bar{b}$ in A. Then $p^r b$ lies in A_1, say $p^r b = na_1$ with some integer $n \geqq 0$. We note that the period of $\bar{b}$ is $\leqq$ the period of b. Write $n = p^k \mu$ where μ is prime to p. Then μa_1 is also a generator of A_1, and hence has period p^{r_1}. We may assume $k \leqq r_1$. Then $p^k \mu a_1$ has period $p^{r_1 - k}$. By our previous remark, the element b has period

$$p^{r+r_1-k}$$

whence by hypothesis, $r + r_1 - k \leqq r_1$ and $r \leqq k$. This proves that there exists an element $c \in A_1$ such that $p^r b = p^r c$. Let $a = b - c$. Then a is a representative for $\bar{b}$ in A and $p^r a = 0$. Since period $(a) \leqq p^r$ we conclude that a has period equal to p^r.

We return to the main proof. By induction, the factor group A/A_1 has a product expression

$$A/A_1 = \bar{A}_2 \times \cdots \times \bar{A}_s$$

into cyclic subgroups of orders $p^{r_2}, \ldots, p^{r_s}$ respectively, and we may assume $r_2 \geqq \cdots \geqq r_s$. Let $\bar{a}_i$ be a generator for $\bar{A}_i$ $(i = 2, \ldots, s)$ and let a_i be a representative in A of the same period as $\bar{a}_i$. Let A_i be the cyclic subgroup generated by a_i. We contend that A is the direct product of $A_1, \ldots, A_s$.

Given $x \in A$, let $\bar{x}$ denote its residue class in A/A_1. There exist integers $m_i \geqq 0$ $(i = 2, \ldots, s)$ such that

$$\bar{x} = m_2\bar{a}_2 + \cdots + m_s\bar{a}_s.$$

Hence $x - m_2a_2 - \cdots - m_sa_s$ lies in A_1, and there exists an integer $m_1 \geqq 0$ such that

$$x = m_1a_1 + m_2a_2 + \cdots + m_sa_s.$$

Hence $A_1 + \cdots + A_s = A$.

Conversely, suppose that $m_1, \dots, m_s$ are integers $\geqq 0$ such that

$$0 = m_1 a_1 + \cdots + m_s a_s.$$

Since a_i has period p^{r_i} $(i = 1, \dots, s)$, we may suppose that $m_i < p^{r_i}$. Putting a bar on this equation yields

$$0 = m_2 \bar{a}_2 + \cdots + m_s \bar{a}_s.$$

Since A/A_1 is a direct product of $\bar{A}_2, \dots, \bar{A}_s$ we conclude that each $m_i = 0$ for $i = 2, \dots, s$. But then $m_1 = 0$ also, and hence all $m_i = 0$ $(i = 1, \dots, s)$. From this it follows at once that

$$(A_1 + \cdots + A_i) \cap A_{i+1} = 0$$

for each $i \geqq 1$, and hence that A is the direct product of $A_1, \dots, A_s$, as desired.

We prove uniqueness, by induction. Suppose that A is written in two ways as a product of cyclic groups, say of type

$$(p^{r_1}, \dots, p^{r_s}) \qquad \text{and} \qquad (p^{m_1}, \dots, p^{m_k})$$

with $r_1 \geqq \cdots \geqq r_s \geqq 1$ and $m_1 \geqq \cdots \geqq m_k \geqq 1$. Then pA is also a p-group, of order strictly less than the order of A, and is of type

$$(p^{r_1 - 1}, \dots, p^{r_s - 1}) \qquad \text{and} \qquad (p^{m_1 - 1}, \dots, p^{m_k - 1}),$$

it being understood that if some exponent r_i or m_j is equal to 1, then the factor corresponding to

$$p^{r_i - 1} \qquad \text{or} \qquad p^{m_j - 1}$$

in pA is simply the trivial group 0. By induction, the subsequence of

$$(r_1 - 1, \dots, r_s - 1)$$

consisting of those integers $\geqq 1$ is uniquely determined, and is the same as the corresponding subsequence of

$$(m_1 - 1, \dots, m_k - 1).$$

In other words, we have $r_i - 1 = m_i - 1$ for all those integers i such that $r_i - 1$ or $m_i - 1 \geqq 1$. Hence $r_i = m_i$ for all these integers i, and the two sequences

$$(p^{r_1}, \dots, p^{r_s}) \qquad \text{and} \qquad (p^{m_1}, \dots, p^{m_k})$$

can differ only in their last components which can be equal to p. These correspond to factors of type $(p, \dots, p)$ occurring say ν times in the first sequence and μ times in the second sequence. Thus for some integer n, A is of type

$$(p^{r_1}, \ldots, p^{r_n}, \underbrace{p, \ldots, p}_{\nu \text{ times}}) \quad \text{and} \quad (p^{r_1}, \ldots, p^{r_n}, \underbrace{p, \ldots, p}_{\mu \text{ times}}).$$

Thus the order of A is equal to

$$p^{r_1+\cdots+r_n}p^{\nu} = p^{r_1+\cdots+r_n}p^{\mu},$$

whence $\nu = \mu$, and our theorem is proved.

A group G is said to be *torsion free*, or without torsion, if whenever an element x of G has finite period, then x is the unit element.

THEOREM 7. *Let A be a finitely generated torsion free abelian group. Then A is free.*

Proof. Assume $A \neq 0$. Let S be a finite set of generators, and let $x_1, \ldots, x_n$ be a maximal subset of S having the property that whenever $\nu_1, \ldots, \nu_n$ are integers such that

$$\nu_1 x_1 + \cdots + \nu_n x_n = 0,$$

then $\nu_j = 0$ for all j. (Note that $n \geqq 1$ since $A \neq 0$.) Let B be the subgroup generated by $x_1, \ldots, x_n$. Then B is free. Given $y \in A$ there exist integers $m_1, \ldots, m_n, m$ not all zero such that

$$my + m_1 x_1 + \cdots + m_n x_n = 0,$$

by the assumption of maximality on $x_1, \ldots, x_n$. Furthermore, $m \neq 0$, otherwise all $m_j = 0$. Hence my lies in B. This is true for every one of a finite set of generators y of A, whence there exists an integer $m \neq 0$ such that $mA \subset B$. The map

$$x \mapsto mx$$

of A into itself is a homomorphism, having trivial kernel since A is torsion free. Hence it is an isomorphism of A onto a subgroup of B. By Theorem 4 of the preceding section, we conclude that mA is free, whence A is free.

THEOREM 8. *Let A be a finitely generated abelian group, and let A_t be the subgroup consisting of all elements of A having finite period. Then A_t is finite, and A/A_t is free. There exists a free subgroup B of A such that A is the direct sum of A_t and B.*

Proof. We recall that a finitely generated torsion abelian group is obviously finite. Let A be finitely generated by n elements, and let F be the free abelian group on n generators. By the universal property, there exists a surjective homomorphism

$$F \xrightarrow{\varphi} A$$

of F onto A. The subgroup $\varphi^{-1}(A_t)$ of F is finitely generated by Theorem 4. Hence A_t itself is finitely generated, hence finite.

Next, we prove that A/A_t has no torsion. Let $\bar{x}$ be an element of A/A_t such that $m\bar{x} = 0$ for some integer $m \neq 0$. Then for any representative x of $\bar{x}$ in A, we have $mx \in A_t$, whence $qmx = 0$ for some integer $q \neq 0$. Then $x \in A_t$, so $\bar{x} = 0$, and A/A_t is torsion free. By Theorem 7, A/A_t is free. We now use the lemma of Theorem 4 to conclude the proof.

The rank of A/A_t is also called the *rank of* A.

§11. The dual group

Let A be an abelian group of exponent $m \geqq 1$. This means that for each element $x \in A$ we have $mx = 0$. Let Z_m be a cyclic group of order m. We denote by A^*, or $\mathrm{Hom}(A, Z_m)$ the group of homomorphisms of A into Z_m, and call it the *dual* of A.

Let $f : A \to B$ be a homomorphism of abelian groups, and assume both have exponent m. Then f induces a homomorphism

$$f^* : B^* \to A^*.$$

Namely, for each $\psi \in B^*$ we define $f^*(\psi) = \psi \circ f$. It is trivially verified that f^* is a homomorphism. One may view $\mathrm{Hom}(A, Z_m)$ as a contravariant functor on the category of abelian groups of exponent m, since the properties

$$id^* = id \qquad \text{and} \qquad (f \circ g)^* = g^* \circ f^*$$

are trivially verified.

THEOREM 9. *If A is a finite abelian group, expressed as a product $A = B \times C$, then A^* is isomorphic to $B^* \times C^*$ (under the mapping described below). A finite abelian group is isomorphic to its own dual.*

Proof. Consider the two projections

$$\begin{array}{ccc} & B \times C & \\ {}^{f}\swarrow & & \searrow^{g} \\ B & & C \end{array}$$

of $B \times C$ on its two components. We get homomorphisms

$$\begin{array}{ccc} & (B \times C)^* & \\ {}^{f^*}\nearrow & & \nwarrow^{g^*} \\ B^* & & C^* \end{array}$$

and we contend that these homomorphisms induce an isomorphism of $B^* \times C^*$ onto $(B \times C)^*$.

In fact, let ψ_1, ψ_2 be in $\mathrm{Hom}(B, Z_m)$ and $\mathrm{Hom}(C, Z_m)$ respectively. Then $(\psi_1, \psi_2) \in B^* \times C^*$, and we have a corresponding element of $(B \times C)^*$ by defining

$$(\psi_1, \psi_2)(x, y) = \psi_1(x) + \psi_2(y),$$

for $(x, y) \in B \times C$. In this way we get a homomorphism

$$B^* \times C^* \to (B \times C)^*.$$

Conversely, let $\psi \in (B \times C)^*$. Then

$$\psi(x, y) = \psi(x, 0) + \psi(0, y).$$

The function ψ_1 on B such that $\psi_1(x) = \psi(x, 0)$ is in B^*, and similarly the function ψ_2 on C such that $\psi_2(y) = \psi(0, y)$ is in C^*. Thus we get a homomorphism

$$(B \times C)^* \to B^* \times C^*,$$

which is obviously inverse to the one we defined previously. Hence we obtain an isomorphism, thereby proving the first assertion in our theorem.

We can write any finite abelian group as a product of cyclic groups. Thus to prove the second assertion, it will suffice to deal with a cyclic group.

Let A be cyclic, generated by one element x of period n. Then $n|m$, and Z_m has precisely one subgroup of order n, Z_n, which is cyclic (Exercise 20). If $\psi : A \to Z_m$ is a homomorphism, and x is a generator for A, then the period of x is an exponent for $\psi(x)$, so that $\psi(x)$, and hence $\psi(A)$, is contained in Z_n. Let y be a generator for Z_n. We have an isomorphism

$$\psi_1 : A \to Z_n$$

such that $\psi_1(x) = y$. For each integer k with $0 \leqq k < n$ we have the homomorphism $k\psi_1$ such that

$$(k\psi_1)(x) = k \cdot \psi_1(x) = \psi_1(kx).$$

In this way we get a cyclic subgroup of A^* consisting of the n elements $k\psi_1$ $(0 \leqq k < n)$. Conversely, any element ψ of A^* is uniquely determined by its effect on the generator x, and must map x on one of the n elements kx $(0 \leqq k < n)$ of Z_n. Hence ψ is equal to one of the maps $k\psi_1$. These maps constitute the full group A^*, which is therefore cyclic of order n, generated by ψ_1. This proves our theorem.

In considering the dual group, we take various cyclic groups Z_m. There are many applications where such groups occur, for instance the group of m-th roots of unity in the complex numbers, the subgroup of order m of $\mathbf{Q}/\mathbf{Z}$, etc.

Let A, A' be two abelian groups. A *bilinear* map of $A \times A'$ into an abelian group C is a map

$$A \times A' \to C$$

denoted by

$$(x, x') \mapsto \langle x, x' \rangle$$

having the following property. For each $x \in A$ the function $x' \mapsto \langle x, x' \rangle$ is a homomorphism, and similarly for each $x' \in A'$ the function $x \mapsto \langle x, x' \rangle$ is a homomorphism.

As a special case of a bilinear map, we have the one given by

$$A \times \mathrm{Hom}(A, C) \to C$$

which to each pair (x, f) with $x \in A$ and $f \in \mathrm{Hom}(A, C)$ associates the element $f(x)$ in C.

A bilinear map is also called a *pairing*.

An element $x \in A$ is said to be *orthogonal* (or *perpendicular*) to a subset S' of A' if $\langle x, x' \rangle = 0$ for all $x' \in S'$. It is clear that the set of $x \in A$ orthogonal to S' is a subgroup of A. We make similar definitions for elements of A', orthogonal to subsets of A.

The *kernel* of our bilinear map on the left is the subgroup of A which is orthogonal to all of A'. We define its kernel on the right similarly.

Given a bilinear map $A \times A' \to C$, let B, B' be the respective kernels of our bilinear map on the left and right. An element x' of A' gives rise to an element of $\mathrm{Hom}(A, C)$ given by $x \mapsto \langle x, x' \rangle$, which we shall denote by $\psi_{x'}$. Since $\psi_{x'}$ vanishes on B we see that $\psi_{x'}$ is in fact a homomorphism of A/B into C. Furthermore, $\psi_{x'} = \psi_{y'}$ if x', y' are elements of A' such that

$$x' \equiv y' \pmod{B'}.$$

Hence ψ is in fact a homomorphism

$$0 \to A'/B' \to \mathrm{Hom}(A/B, C),$$

which is injective since we defined B' to be the group orthogonal to A. Similarly, we get an injective homomorphism

$$0 \to A/B \to \mathrm{Hom}(A'/B', C).$$

Assume that C is cyclic of order m. Then for any $x' \in A'$ we have $m\psi_{x'} = \psi_{mx'} = 0$, whence A'/B' has exponent m. Similarly, A/B has exponent m.

THEOREM 10. *Let $A \times A' \to C$ be a bilinear map of two abelian groups into a cyclic group C of order m. Let B, B' be its respective kernels on the right and left. Assume that A'/B' is finite. Then A/B is finite, and A'/B' is isomorphic to the dual group of A/B (under our map ψ).*

Proof. The injection of A/B into $\mathrm{Hom}(A'/B', C)$ shows that A/B is finite. Furthermore, we get the inequalities

$$\mathrm{ord}\, A/B \leqq \mathrm{ord}(A'/B')^* = \mathrm{ord}\, A'/B'$$

and

$$\mathrm{ord}\, A'/B' \leqq \mathrm{ord}(A/B)^* = \mathrm{ord}\, A/B.$$

From this it follows that our map ψ is bijective, hence an isomorphism.

COROLLARY. *Let A be a finite abelian group, B a subgroup, A^* the dual group, and $B^{\perp}$ the set of $\varphi \in A^*$ such that $\varphi(B) = 0$. Then we have a natural isomorphism of $A^*/B^{\perp}$ with B^*.*

Proof. This is a special case of Theorem 10.

EXERCISES

1. Show that every group of order $\leqq 5$ is abelian.

2. Show that there are two non-isomorphic groups of order 4, namely the cyclic one, and the product of two cyclic groups of order 2.

3. Let p be the smallest prime number dividing the order of a finite group G. Let H be a subgroup of index p. Show that H is normal in G.

4. Show that there are exactly two non-isomorphic non-abelian groups of order 8. (One of them is given by generators σ, τ with the relations

$$\sigma^4 = 1, \qquad \tau^2 = 1, \qquad \tau\sigma\tau = \sigma^3.$$

The other is the quaternion group.)

5. Let G be a group, and A a normal abelian subgroup. Show that G/A operates on A by conjugation, and in this manner get a homomorphism of G/A into $\mathrm{Aut}(A)$.

6. Show that every group of order 15 is cyclic.

7. Determine all groups of order $\leqq 10$ up to isomorphism.

8. A group G is called a *torsion group* if for every $x \in G$ there exists an integer $n \geqq 1$ such that $x^n = 1$. Show that infinite direct products exist in the category of torsion abelian groups.

9. (a) Let σ be a permutation of a finite set I having n elements. Define the *sign* of σ, written $\epsilon(\sigma)$, to be $(-1)^m$ where

$$m = n - \text{number of orbits of } \sigma.$$

If $I_1, \ldots, I_r$ are the orbits of σ, then m is also equal to the sum

$$m = \sum_{\nu=1}^{r} [\text{card}(I_\nu) - 1].$$

A permutation τ of I is called a *transposition* if there exist two elements $i \neq j$ in I such that $\tau(i) = j$, $\tau(j) = i$, and $\tau(x) = x$ for $x \in I$, $x \neq i, j$. If τ is a transposition, show that $\epsilon(\sigma\tau) = -\epsilon(\sigma)$ by considering the two cases when i, j lie in the same orbit of σ, or lie in different orbits. In the first case, $\sigma\tau$ has one more orbit and in the second case one less orbit than σ. In particular, the sign of a transposition is -1. (b) Prove by induction that the transpositions generate the group of permutations of I (called the *symmetric group*, also denoted by S_n). If $\sigma = \tau_1 \cdots \tau_m$ where τ_i are transpositions, then $\epsilon(\sigma) = (-1)^m$. If σ, σ' are two permutations, show that $\epsilon(\sigma\sigma') = \epsilon(\sigma)\epsilon(\sigma')$. (c) Let I be the set of integers $(1, \ldots, n)$. Show that for any permutation σ,

$$\prod_{1 \leqq i < j \leqq n} [\sigma(j) - \sigma(i)] = \epsilon(\sigma) \prod_{1 \leqq i < j \leqq n} (j - i).$$

10. If $n \geqq 5$ then the symmetric group S_n is not solvable. [*Hint:* Prove that if H, N are two subgroups of S_n such that N is normal in H and H/N is abelian, and if H contains every 3-cycle, then N also contains every 3-cycle, as follows. Let i, j, k, r, s be 5 distinct integers between 1 and n. Let $\sigma = (ijk)$ and $\tau = (krs)$. Then $(kjs) = \sigma^{-1}\tau^{-1}\sigma\tau \in N$.]

11. Prove: A subgroup and factor group of a solvable group is solvable.

12. Let G be a group and H a subgroup of finite index. Show that there exists a normal subgroup N of G contained in H and also of finite index. [*Hint:* If $(G : H) = n$, find a homomorphism of G into S_n whose kernel is contained in H.]

13. Let $f : A \to A'$ be a homomorphism of abelian groups. Let B be a subgroup of A. Denote by A^f and A_f the image and kernel of f in A respectively, and similarly for B^f and B_f. Show that $(A : B) = (A^f : B^f)(A_f : B_f)$, in the sense that if two of these three indices are finite, so is the third, and the stated equality holds.

14. Let G be a finite cyclic group of order n, generated by an element σ. Assume that G operates on an abelian group A, and let $f, g : A \to A$ be the endomorphisms of A given by

$$f(x) = \sigma x - x \quad \text{and} \quad g(x) = x + \sigma x + \cdots + \sigma^{n-1}x.$$

Define the *Herbrand quotient* by the expression $q(A) = (A_f : A^g)/(A_g : A^f)$, provided both indices are finite. Assume now that B is a subgroup of A such that $GB \subset B$. (a) Define in a natural way an operation of G on A/B. (b) Prove that

$$q(A) = q(B)q(A/B)$$

in the sense that if two of these quotients are finite, so is the third, and the stated equality holds. (c) If A is finite, show that $q(A) = 1$.

(This exercise is a special case of the general theory of Euler characteristics discussed in Chapter IV. After reading this chapter, the present exercise becomes trivial. Why?)

15. Let I be a set of indices. Suppose given a relation of partial ordering in I, namely for some pairs (i, j) we have a relation $i \leqq j$ satisfying the conditions: For all i, j, k in I, we have $i \leqq i$; if $i \leqq j$ and $j \leqq k$ then $i \leqq k$; if $i \leqq j$ and $j \leqq i$ then $i = j$. We say that I is *directed* if given $i, j \in I$, there exists k such that $i \leqq k$ and $j \leqq k$. Assume that I is directed. Let $\mathfrak{A}$ be a category, and $\{A_i\}$ a family of objects in $\mathfrak{A}$. For each pair i, j such that $i \leqq j$ assume given a morphism

$$f_j^i : A_i \to A_j$$

such that, whenever $i \leqq j \leqq k$, we have $f_k^j \circ f_j^i = f_k^i$ and $f_i^i = \text{id}$. A *direct limit* for the family $\{f_j^i\}$ is a universal object in the following category $\mathfrak{C}$. $\text{Ob}(\mathfrak{C})$ consists of pairs $(A, (f^i))$ where $A \in \text{Ob}(\mathfrak{A})$ and (f^i) is a family of morphisms $f^i : A_i \to A$, $i \in I$, such that for all $i \leqq j$ the following diagram is commutative:

$$\begin{array}{ccccccc} \cdots \to & A_i & \xrightarrow{f_j^i} & A_j & \to \cdots \\ & & {\scriptstyle f^i}\searrow \quad \swarrow{\scriptstyle f^j} & & \\ & & A & & \end{array}$$

(Universal of course means universally repelling.)

Show that direct limits exist in the category of abelian groups. [*Hint:* Factor out the direct sum by the relations imposed by the f_j^i.]

16. Reversing the arrows in the preceding exercise, define the notion of *inverse limit or projective limit*. Prove that inverse limits exist in the category of groups. [*Hint:* Obtain the inverse limit as a subgroup of the product, consisting of all vectors (x_i) satisfying the compatibility relations imposed by the f_j^i.]

17. Let H, G, G' be groups, and let

$$f : H \to G, \qquad g : H \to G'$$

be two homomorphisms. Define the notion of coproduct of these two homomorphisms over H, and show that it exists.

18. Let A be a torsion abelian group. Show that A is the direct sum of its subgroups $A(p)$ for all primes p.

19. Viewing $\mathbf{Z}$, $\mathbf{Q}$ as additive groups, show that $\mathbf{Q}/\mathbf{Z}$ is a torsion group, which has one and only one subgroup of order n for each integer $n \geqq 1$, and that this subgroup is cyclic.

20. Show that if A is a cyclic group of order n, and if d is a positive integer, $d \mid n$, then A contains exactly one subgroup of order d, and that this subgroup is cyclic.

21. Show that a finite abelian group which is not cyclic contains a subgroup of type (p, p) for some prime p.

22. Let G be a cyclic group of order n and H a cyclic group of order m. If m, n are relatively prime, show that $G \times H$ is cyclic (of order mn).

CHAPTER II

Rings

§1. Rings and homomorphisms

A *ring* A is a set, together with two laws of composition called multiplication and addition respectively, and written as a product and as a sum respectively, satisfying the following conditions:

RI 1. With respect to addition, A is a commutative group.
RI 2. The multiplication is associative, and has a unit element.
RI 3. For all $x, y, z \in A$ we have

$$(x + y)z = xz + yz \qquad \text{and} \qquad z(x + y) = zx + zy.$$

(This is called *distributivity*.)

As usual, we denote the unit element for addition by 0, and the unit element for multiplication by 1. We do not assume that $1 \neq 0$. We observe that $0x = 0$ for all $x \in A$. Proof: We have $0x + x = (0 + 1)x = 1x = x$. Hence $0x = 0$. In particular, if $1 = 0$, then A consists of 0 alone.

For any $x, y \in A$ we have $(-x)y = -(xy)$. Proof: We have

$$xy + (-x)y = (x + (-x))y = 0y = 0,$$

so $(-x)y$ is the additive inverse of xy.

Other standard laws relating addition and multiplication are easily proved, for instance $(-x)(-y) = xy$. We leave these as exercises.

Let A be a ring, and let U be the set of elements of A which have both a right and left inverse. Then U is a multiplicative group. Indeed, if a has a right inverse b, so that $ab = 1$, and a left inverse c, so that $ca = 1$, then $cab = b$, whence $c = b$, and we see that c (or b) is a two-sided inverse, and that c itself has a two-sided inverse, namely a. Therefore U satisfies all the axioms of a multiplicative group, and is called the group of *units* of A. It is sometimes denoted by A^*, and is also called the group of *invertible* elements of A. A ring A such that $1 \neq 0$, and such that every non-zero element is invertible is called a *division ring*.

A ring A is said to be *commutative* if $xy = yx$ for all $x, y \in A$. A commutative division ring is called a *field*. We observe that by definition, a field contains at least two elements, namely 0 and 1.

A subset B of a ring A is called a *subring* if it is an additive subgroup, if it contains the multiplicative unit, and if $x, y \in B$ implies $xy \in B$. If that is the case, then B itself is a ring, the laws of operation in B being the same as the laws of operation in A.

For example, the *center* of a ring A is the subset of A consisting of all elements $a \in A$ such that $ax = xa$ for all $x \in A$. One sees immediately that the center of A is a subring.

Just as we proved general associativity from the associativity for three factors, one can prove general distributivity. If $x, y_1, \ldots, y_n$ are elements of a ring A, then by induction one sees that

$$x(y_1 + \cdots + y_n) = xy_1 + \cdots + xy_n.$$

If x_i $(i = 1, \ldots, n)$ and y_j $(j = 1, \ldots, m)$ are elements of A, then it is also easily proved that

$$\left(\sum_{i=1}^{n} x_i\right)\left(\sum_{j=1}^{m} y_j\right) = \sum_{i=1}^{n} \sum_{j=1}^{m} x_i x_j.$$

Furthermore, distributivity holds for subtraction, e.g.

$$x(y_1 - y_2) = xy_1 - xy_2.$$

We leave all the proofs to the reader.

Examples. Let S be a set and A a ring. Let $\mathfrak{M}(S, A)$ be the set of mappings of S into A. Then $\mathfrak{M}(S, A)$ is a ring if for $f, g \in \mathfrak{M}(S, A)$ we define

$$(fg)(x) = f(x)g(x) \qquad \text{and} \qquad (f + g)(x) = f(x) + g(x)$$

for all $x \in A$. The multiplicative unit is the constant map whose value is the multiplicative unit of A. The additive unit is the constant map whose value is the additive unit of A, namely 0. The verification that $\mathfrak{M}(S, A)$ is a ring under the above laws of composition is trivial and left to the reader.

Let M be an additive abelian group, and let A be the set $\mathrm{End}(M)$ of group-homomorphisms of A into itself. We define addition in A to be the addition of mappings, and we define multiplication to be *composition* of mappings. Then it is trivially verified that A is a ring. Its unit element is of course the identity mapping. In general, A is not commutative.

A *left* ideal $\mathfrak{a}$ in a ring A is a subset of A which is a subgroup of the additive group of A, and such that $A\mathfrak{a} \subset \mathfrak{a}$ (and hence $A\mathfrak{a} = \mathfrak{a}$ since A contains 1). To define a right ideal, we require $\mathfrak{a}A = \mathfrak{a}$, and a *two-sided ideal* is a subset which is both a left and a right ideal. A two-sided ideal is called simply an *ideal* in this section.

If A is a ring and $a \in A$, then Aa is a left ideal, called *principal*. We say that a is a generator of $\mathfrak{a}$ (over A). Similarly, AaA is a principal two-sided ideal.

A ring A is said to be *commutative* if $xy = yx$ for all $x, y \in A$. In that case, every left or right ideal is two-sided.

A *commutative* ring such that every ideal is principal and such that $1 \neq 0$ is called a *principal* ring.

Example. The integers $\mathbf{Z}$ form a ring, which is commutative. Let $\mathfrak{a}$ be an ideal $\neq \mathbf{Z}$ and $\neq 0$. If $n \in \mathfrak{a}$, then $-n \in \mathfrak{a}$. Let d be the smallest integer > 0 lying in $\mathfrak{a}$. If $n \in \mathfrak{a}$ then there exist integers q, r with $0 \leqq r < d$ such that

$$n = dq + r.$$

Since $\mathfrak{a}$ is an ideal, it follows that r lies in $\mathfrak{a}$, hence $r = 0$. Hence $\mathfrak{a}$ consists of all multiples qd of d, with $q \in \mathbf{Z}$, and $\mathbf{Z}$ *is a principal ring.*

Let A be a ring. Then (0) and A itself are ideals.

Let $\mathfrak{a}, \mathfrak{b}$ be ideals of A. We define $\mathfrak{ab}$ to be the set of all sums

$$x_1y_1 + \cdots + x_ny_n$$

with $x_i \in \mathfrak{a}$ and $y_i \in \mathfrak{b}$. Then one verifies immediately that $\mathfrak{ab}$ is an ideal, and that the set of ideals forms a multiplicative monoid, the unit element being the ring itself. This unit element is called the unit ideal, and is often written (1). If $\mathfrak{a}, \mathfrak{b}$ are left ideals, we define their product $\mathfrak{ab}$ as above. It is also a left ideal, and if $\mathfrak{a}, \mathfrak{b}, \mathfrak{c}$ are left ideals, then we again have associativity $(\mathfrak{ab})\mathfrak{c} = \mathfrak{a}(\mathfrak{bc})$.

If $\mathfrak{a}, \mathfrak{b}$ are left ideals of A, then $\mathfrak{a} + \mathfrak{b}$ (the sum being taken as additive subgroup of A) is obviously a left ideal. Similarly for right and two-sided ideals. Thus ideals also form a monoid under addition. We also have distributivity: If $\mathfrak{a}_1, \ldots, \mathfrak{a}_n, \mathfrak{b}$ are ideals of A, then clearly

$$\mathfrak{b}(\mathfrak{a}_1 + \cdots + \mathfrak{a}_n) = \mathfrak{b}\mathfrak{a}_1 + \cdots + \mathfrak{b}\mathfrak{a}_n,$$

and similarly on the other side. (However, the set of ideals does not form a ring!)

If $\{\mathfrak{a}_i\}_{i \in I}$ is a family of ideals, then their intersection

$$\bigcap_{i \in I} \mathfrak{a}_i$$

is also an ideal. Similarly for left ideals. We note that if $\mathfrak{a}, \mathfrak{b}$ are two ideals of a ring A, then $\mathfrak{ab} \subset \mathfrak{a} \cap \mathfrak{b}$.

If $a_1, \ldots, a_n$ are elements of a ring A, then we denote by $(a_1, \ldots, a_n)$ the intersection of all ideals of A containing these elements, or the intersection of all left ideals of A containing these elements. It will always be specified in the context which is meant (i.e. two-sided or left ideal). We call $a_1, \ldots, a_n$ the *generators* of the ideal. A principal ideal (or a principal left ideal) is therefore generated by one element. We see at once that the left ideal generated by $a_1, \ldots, a_n$ consists of all elements which can be written in the form

$$x_1a_1 + \cdots + x_na_n$$

with $x_i \in A$.

By a *ring-homomorphism* one means a mapping $f : A \to B$ where A, B are rings, and such that f is a monoid-homomorphism for the multiplicative structures on A and B, and also a monoid-homomorphism for the additive structures. In other words, f must satisfy:

$$f(a + a') = f(a) + f(a'), \qquad f(aa') = f(a)f(a')$$
$$f(1) = 1, \qquad f(0) = 0$$

for all $a, a' \in A$. Its *kernel* is defined to be the kernel of f viewed as additive homomorphism.

The kernel of a ring-homomorphism $f : A \to B$ is an ideal of A, as one verifies at once.

Conversely, let $\mathfrak{a}$ be an ideal of the ring A. We can construct the *factor ring* $A/\mathfrak{a}$ as follows. Viewing A and $\mathfrak{a}$ as additive groups, let $A/\mathfrak{a}$ be the factor group. We define a multiplicative law of composition on $A/\mathfrak{a}$: If $x + \mathfrak{a}$ and $y + \mathfrak{a}$ are two cosets of $\mathfrak{a}$, we define $(x + \mathfrak{a})(y + \mathfrak{a})$ to be the coset $(xy + \mathfrak{a})$. This coset is well defined, for if x_1, y_1 are in the same coset as x, y respectively, then one verifies at once that x_1y_1 is in the same coset as xy. Our multiplicative law of composition is then obviously associative, has a unit element, namely the coset $1 + \mathfrak{a}$, and the distributive law is satisfied since it is satisfied for coset representatives. We have therefore defined a ring structure on $A/\mathfrak{a}$, and the canonical map

$$f : A \to A/\mathfrak{a}$$

is then clearly a ring-homomorphism.

If $g : A \to A'$ is a ring-homomorphism whose kernel contains $\mathfrak{a}$, then there exists a unique ring-homomorphism $g_* : A/\mathfrak{a} \to A'$ making the following diagram commutative:

$$\begin{array}{ccc} A & \xrightarrow{g} & A' \\ & {\scriptstyle f}\searrow \quad \nearrow{\scriptstyle g_*} & \\ & A/\mathfrak{a} & \end{array}$$

Indeed, viewing f, g as group-homomorphisms (for the additive structures), there is a unique group-homomorphism g_* making our diagram commutative. We contend that g_* is in fact a ring-homomorphism. We could leave the trivial proof to the reader, but we carry it out in full. If $x \in A$, then $g(x) = g_*f(x)$. Hence for $x, y \in A$,

$$\begin{aligned} g_*(f(x)f(y)) &= g_*(f(xy)) = g(xy) = g(x)g(y) \\ &= g_*f(x)g_*f(y). \end{aligned}$$

Given $\xi, \eta \in A/\mathfrak{a}$, there exist $x, y \in A$ such that $\xi = f(x)$ and $\eta = f(y)$. Since $f(1) = 1$, we get $g_*f(1) = g(1) = 1$, and hence the two conditions that g_* be a multiplicative monoid-homomorphism are satisfied, as was to be shown.

The statement we have just proved is equivalent to saying that the canonical map $f : A \to A/\mathfrak{a}$ is universal in the category of homomorphisms whose kernel contains $\mathfrak{a}$.

Let A be a ring and B a subring. Let S be a subset of A commuting with B. We denote by $B[S]$ the intersection of all subrings of A containing B and S. It is obviously a ring, consisting of all elements of type

$$\sum b_{i_1 \cdots i_n} s_1^{i_1} \cdots s_n^{i_n},$$

the sum ranging over a finite number of n-tuples $(i_1, \ldots, i_n)$ of integers $\geqq 0$, and $b_{i_1 \cdots i_n} \in B$, $s_1, \ldots, s_n \in S$. If $A = B[S]$, we say that S is a set of generators (or more precisely, *ring generators*) for A over B, or that A is generated by S over B. If S is finite, we say that A is *finitely generated as a ring over* B.

As with groups, we observe that a homomorphism is uniquely determined by its effect on generators. In other words, let $f : B \to B'$ be a ring-homomorphism, and let $A = B[S]$ as above. Then there exists at most one extension of f to a ring-homomorphism of A having prescribed values on S.

Let A be a ring, $\mathfrak{a}$ an ideal, and S a subset of A. We write

$$S \equiv 0 \pmod{\mathfrak{a}}$$

if $S \subset \mathfrak{a}$. If $x, y \in A$, we write

$$x \equiv y \pmod{\mathfrak{a}}$$

if $x - y \in \mathfrak{a}$. If $\mathfrak{a}$ is principal, equal to (a), then we also write

$$x \equiv y \pmod{a}.$$

If $f : A \to A/\mathfrak{a}$ is the canonical homomorphism, then $x \equiv y \pmod{\mathfrak{a}}$ means that $f(x) = f(y)$. The congruence notation is sometimes convenient when we want to avoid writing explicitly the canonical map f.

The factor ring $A/\mathfrak{a}$ is also called a *residue class ring*. Cosets of $\mathfrak{a}$ in A are called *residue classes* modulo $\mathfrak{a}$, and if $x \in A$, then the coset $x + \mathfrak{a}$ is called the *residue class of x modulo $\mathfrak{a}$*.

Any ring-homomorphism $f : A \to B$ which is bijective is an isomorphism. Indeed, there exists a set-theoretic inverse $g : B \to A$, and it is trivial to verify that g is a ring-homomorphism.

Instead of saying "ring-homomorphism" we sometimes say simply "homomorphism" if the reference to rings is clear. We note that rings form a category (the morphisms being the homomorphisms).

Let $f : A \to B$ be a ring-homomorphism. Then the image $f(A)$ of f is a subring of B. Proof obvious.

It is clear that an injective ring-homomorphism $f : A \to B$ establishes a ring-isomorphism between A and its image. Such a homomorphism will be called an *embedding* (of rings).

Let $f : A \to A'$ be a ring-homomorphism, and let $\mathfrak{a}'$ be an ideal of A'. Then $f^{-1}(\mathfrak{a}')$ is an ideal $\mathfrak{a}$ in A, and we have an induced injective homomorphism

$$A/\mathfrak{a} \to A'/\mathfrak{a}'.$$

The trivial proof is left to the reader.

PROPOSITION 1. *Products exist in the category of rings.*

In fact, let $\{A_i\}_{i \in I}$ be a family of rings, and let $A = \prod A_i$ be their product as additive abelian groups. We define a multiplication in A in the obvious way: If $(x_i)_{i \in I}$ and $(y_i)_{i \in I}$ are two elements of A, we define their product to be $(x_i y_i)_{i \in I}$, i.e. we define multiplication componentwise, just as we did for addition. The multiplicative unit is simply the element of the product whose i-th component is the unit element of A_i. It is then clear that we obtain a ring structure on A, and that the projection on the i-th factor is a ring-homomorphism. Furthermore, A together with these projections clearly satisfies the required universal property.

Note that the usual inclusion of A_i on the i-th factor is *not* a ring-homomorphism because it does not map the unit element e_i of A_i on the unit element of A. Indeed, it maps e_i on the element of A having e_i as i-th component, and 0 $(= 0_i)$ as all other components.

Let A be a ring. Elements x, y of A are said to be *zero divisors* if $x \neq 0$, $y \neq 0$, and $xy = 0$. Most of the rings without zero divisors which we consider will be commutative. In view of this, we define a ring A to be *entire* if $1 \neq 0$, if A is commutative, and if there are no zero divisors in the ring.

Examples. The ring of integers $\mathbf{Z}$ is without zero divisors, i.e. is entire. If S is a set with more than 2 elements, and A is a ring with $1 \neq 0$, then the ring of mappings $M(S, A)$ has zero divisors. (Proof?)

Let m be a positive integer $\neq 1$. The ring $\mathbf{Z}/m\mathbf{Z}$ has zero divisors if and only if m is not a prime number. (Proof left as an exercise.)

The next criterion is used very frequently.

Let A be an entire ring, and let a, b be non-zero elements of A. Then a, b generate the same ideal if and only if there exists a unit u of A such that $b = au$.

Proof. If such a unit exists we have $Ab = Aua = Aa$. Conversely, assume $Aa = Ab$. Then we can write $a = bc$ and $b = ad$ with some elements $c, d \in A$. Hence $a = adc$, whence $a(1 - dc) = 0$, and therefore $dc = 1$. Hence c is a unit.

§2. *Commutative rings*

Throughout this section, the word "ring" will mean "commutative ring".

Let A be a ring. A *prime* ideal in A is an ideal $\mathfrak{p} \neq A$ such that $A/\mathfrak{p}$ is entire. Equivalently, we could say that it is an ideal $\mathfrak{p} \neq A$ such that, whenever $x, y \in A$ and $xy \in \mathfrak{p}$, then $x \in \mathfrak{p}$ or $y \in \mathfrak{p}$.

Let $\mathfrak{m}$ be an ideal. We say that $\mathfrak{m}$ is a *maximal* ideal if $\mathfrak{m} \neq A$ and if there is no ideal $\mathfrak{a} \neq A$ containing $\mathfrak{m}$ and $\neq \mathfrak{m}$.

Every maximal ideal is prime. Proof: Let $\mathfrak{m}$ be maximal and let $x, y \in A$ be such that $xy \in \mathfrak{m}$. Suppose $x \notin \mathfrak{m}$. Then $\mathfrak{m} + Ax$ is an ideal properly containing $\mathfrak{m}$, hence equal to A. Hence we can write

$$1 = u + ax$$

with $u \in \mathfrak{m}$ and $a \in A$. Multiplying by y we find

$$y = yu + axy,$$

whence $y \in \mathfrak{m}$ and $\mathfrak{m}$ is therefore prime.

Let A be a ring, and let $\mathfrak{a}$ be an ideal $\neq A$. Then $\mathfrak{a}$ is contained in some maximal ideal $\mathfrak{m}$. Proof: The set of ideals containing $\mathfrak{a}$ and $\neq A$ is inductively ordered by ascending inclusion. Indeed, if $\{\mathfrak{b}_i\}$ is a totally ordered set of such ideals, then $1 \notin \mathfrak{b}_i$ for any i, and hence 1 does not lie in the ideal $\mathfrak{b} = \bigcup \mathfrak{b}_i$, which dominates all $\mathfrak{b}_i$. If $\mathfrak{m}$ is a maximal element in our set, then $\mathfrak{m} \neq A$ and $\mathfrak{m}$ is a maximal ideal, as desired.

Let A be a ring. Then $\{0\}$ is a prime ideal if and only if A is entire. (Proof obvious.)

We defined a *field* K to be a ring such that $1 \neq 0$, and such that the multiplicative monoid of non-zero elements of K is a group (i.e. such that whenever $x \in K$ and $x \neq 0$ then there exists an inverse for x). We note that the only ideals of a field K are K and the zero ideal.

If A is a ring and $\mathfrak{m}$ is a maximal ideal, then $A/\mathfrak{m}$ is a field. Proof: If $x \in A$, we denote by $\bar{x}$ its residue class mod $\mathfrak{m}$. Since $\mathfrak{m} \neq A$ we note that $A/\mathfrak{m}$ has a unit element $\neq 0$. Any non-zero element of $A/\mathfrak{m}$ can be written as $\bar{x}$ for some $x \in A$, $x \notin \mathfrak{m}$. To find its inverse, note that $m + Ax$ is an ideal of $A \neq \mathfrak{m}$ and hence equal to A. Hence we can write

$$1 = u + yx$$

with $u \in \mathfrak{m}$ and $y \in A$. This means that $\bar{y}\bar{x} = 1$ (i.e. $= \bar{1}$) and hence that $\bar{x}$ has an inverse, as desired.

Conversely, we leave it as an exercise to the reader to prove that *if A is a ring and $\mathfrak{m}$ is an ideal such that $A/\mathfrak{m}$ is a field, then $\mathfrak{m}$ is maximal.*

Let $f : A \to A'$ be a homomorphism (of commutative rings, according to the convention in force). Let $\mathfrak{p}'$ be a prime ideal of A', and let $\mathfrak{p} = f^{-1}(\mathfrak{p}')$. Then $\mathfrak{p}$ is prime.

To prove this, let $x, y \in A$, and $xy \in \mathfrak{p}$. Suppose $x \notin \mathfrak{p}$. Then $f(x) \notin \mathfrak{p}'$. But $f(x)f(y) = f(xy) \in \mathfrak{p}'$. Hence $f(y) \in \mathfrak{p}'$, as desired.

As an exercise, prove that if f is surjective, and if $\mathfrak{m}'$ is maximal in A', then $f^{-1}(\mathfrak{m}')$ is maximal in A.

Example. Let $\mathbf{Z}$ be the ring of integers. Since an ideal is also an additive subgroup of $\mathbf{Z}$, every ideal is principal, of the form $n\mathbf{Z}$ for some integer $n > 0$ (uniquely determined by the ideal). Let $\mathfrak{p}$ be a prime ideal, $\mathfrak{p} = n\mathbf{Z}$. Then n must be a prime number, as follows essentially directly from the definition of a prime ideal. Conversely, if p is a prime number, then $p\mathbf{Z}$ is a prime ideal (trivial exercise). Furthermore, $p\mathbf{Z}$ is a maximal ideal. Indeed, suppose $p\mathbf{Z}$ contained in some ideal $n\mathbf{Z}$. Then $p = nm$ for some integer m, whence $n = p$ or $n = 1$, thereby proving $p\mathbf{Z}$ maximal.

If n is an integer, the factor ring $\mathbf{Z}/n\mathbf{Z}$ is called the ring of *integers modulo n*. If n is a prime number p, then the ring of integers modulo p is in fact a field, denoted by $\mathbf{F}_p$. In particular, the multiplicative group of $\mathbf{F}_p$ is called the group of non-zero integers modulo p. From the elementary properties of groups, we get a standard fact of elementary number theory: If x is an integer $\not\equiv 0 \pmod{p}$, then $x^{p-1} \equiv 1 \pmod{p}$. (For simplicity, it is customary to write mod p instead of mod $p\mathbf{Z}$, and similarly to write mod n instead of mod $n\mathbf{Z}$ for any integer n.) Similarly, given an integer $n > 1$, the units in the ring $\mathbf{Z}/n\mathbf{Z}$ consist of those residue classes mod $n\mathbf{Z}$ which are represented by integers $m \neq 0$ and prime to n. The order of the group of units in $\mathbf{Z}/n\mathbf{Z}$ is called by definition $\varphi(n)$ (where φ is known as the *Euler phi-function*). Consequently, if x is an integer prime to n, then $x^{\varphi(n)} \equiv 1 \pmod{n}$.

CHINESE REMAINDER THEOREM. *Let A be a ring and let $\mathfrak{a}_1, \ldots, \mathfrak{a}_n$ be ideals such that $\mathfrak{a}_i + \mathfrak{a}_j = A$ for all $i \neq j$. Given elements $x_1, \ldots, x_n \in A$, there exists $x \in A$ such that $x \equiv x_i \pmod{\mathfrak{a}_i}$ for all i.*

Proof. By induction. If $n = 2$, we have an expression

$$1 = a_1 + a_2$$

for some elements $a_i \in \mathfrak{a}_i$, and we let $x = x_2 a_1 + x_1 a_2$.

Assume the theorem proved for $n - 1$ ideals. For each $i \geqq 2$ we can find elements $a_i \in \mathfrak{a}_1$ and $b_i \in \mathfrak{a}_i$ such that

$$a_i + b_i = 1, \qquad i \geqq 2.$$

The product $\prod_{i=2}^{n} (a_i + b_i)$ is equal to 1, and lies in $\mathfrak{a}_1 + \prod_{i=2}^{n} \mathfrak{a}_i$, i.e. in $\mathfrak{a}_1 + \mathfrak{a}_2 \cdots \mathfrak{a}_n$. Hence

$$\mathfrak{a}_1 + \prod_{i=2}^{n} \mathfrak{a}_i = A.$$

By the theorem for $n = 2$, we can find an element $y_1 \in A$ such that

$$y_1 \equiv 1 \pmod{\mathfrak{a}_1},$$

$$y_1 \equiv 0 \left(\operatorname{mod} \prod_{i=2}^{n} \mathfrak{a}_i\right).$$

We find similarly elements $y_2, \ldots, y_n$ such that

$$y_j \equiv 1 \pmod{\mathfrak{a}_j} \quad \text{and} \quad y_j \equiv 0 \pmod{\mathfrak{a}_i} \quad \text{for } i \neq j.$$

Then $x = x_1 y_1 + \cdots + x_n y_n$ satisfies our requirements.

In the same vein as above, we observe that if $\mathfrak{a}_1, \ldots, \mathfrak{a}_n$ are ideals of a ring A such that

$$\mathfrak{a}_1 + \cdots + \mathfrak{a}_n = A,$$

and if $\nu_1, \ldots, \nu_n$ are positive integers, then

$$\mathfrak{a}_1^{\nu_1} + \cdots + \mathfrak{a}_n^{\nu_n} = A.$$

The proof is trivial, and is left as an exercise.

COROLLARY. *Let A be a ring and $\mathfrak{a}_1, \ldots, \mathfrak{a}_n$ ideals of A. Assume that $\mathfrak{a}_i + \mathfrak{a}_j = A$ for $i \neq j$. Let*

$$f : A \to \prod_{i=j}^{n} A/\mathfrak{a}_i = (A/\mathfrak{a}_1) \times \cdots \times (A/\mathfrak{a}_n)$$

be the map of A into the product induced by the canonical map of A onto $A/\mathfrak{a}_i$ for each factor. Then the kernel of f is $\bigcap_{i=1}^{n} \mathfrak{a}_i$, and f is surjective, thus giving an isomorphism

$$A/\bigcap \mathfrak{a}_i \xrightarrow{\approx} \prod A/\mathfrak{a}_i.$$

Proof. That the kernel of f is what we said it is, is obvious. The surjectivity follows from the theorem.

The theorem and its corollary are frequently applied to the ring of integers $\mathbf{Z}$, and to distinct prime ideals $(p_1), \ldots, (p_n)$. These satisfy the hypothesis of the theorem since they are maximal. Similarly, one could take integers $m_1, \ldots, m_n$ which are relatively prime in pairs, and apply the theorem to the principal ideals $(m_1) = m_1\mathbf{Z}, \ldots, (m_n) = m_n\mathbf{Z}$. This is the ultraclassical case of the Chinese remainder theorem.

In particular, let m be an integer > 1, and let

$$m = \prod_i p_i^{r_i}$$

be a factorization of m into primes, with exponents $r_i \geqq 1$. Then we have a ring-isomorphism:

$$\mathbf{Z}/m\mathbf{Z} \approx \prod_i \mathbf{Z}/p_i^{r_i}\mathbf{Z}.$$

If A is a ring, we denote as usual by A^* the multiplicative group of invertible elements of A. We leave the following assertions as exercises:

The preceding ring-isomorphism of $\mathbf{Z}/m\mathbf{Z}$ onto the product induces a group-isomorphism

$$(\mathbf{Z}/m\mathbf{Z})^* \approx \prod_i (\mathbf{Z}/p_i^{r_i}\mathbf{Z})^*.$$

In view of our isomorphism, we have

$$\varphi(m) = \prod_i \varphi(p_i^{r_i}).$$

If p is a prime number and r an integer $\geqq 1$, then

$$\varphi(p^r) = (p-1)p^{r-1}.$$

One proves this last formula by induction. If $r = 1$, then $\mathbf{Z}/p\mathbf{Z}$ is a field, and the multiplicative group of that field has order $p - 1$. Let r be $\geqq 1$, and consider the canonical ring-homomorphism

$$\mathbf{Z}/p^{r+1}\mathbf{Z} \to \mathbf{Z}/p^r\mathbf{Z},$$

arising from the inclusion of ideals $(p^{r+1}) \subset (p^r)$. We get an induced group-homomorphism

$$\lambda : (\mathbf{Z}/p^{r+1}\mathbf{Z})^* \to (\mathbf{Z}/p^r\mathbf{Z})^*,$$

which is surjective because any integer a which represents an element of $\mathbf{Z}/p^r\mathbf{Z}$ and is prime to p will represent an element of $(\mathbf{Z}/p^{r+1}\mathbf{Z})^*$. Let a be an integer representing an element of $(\mathbf{Z}/p^{r+1}\mathbf{Z})^*$, such that $\lambda(a) = 1$. Then

$$a \equiv 1 \pmod{p^r\mathbf{Z}},$$

and hence we can write

$$a \equiv 1 + xp^r \pmod{p^{r+1}\mathbf{Z}}$$

for some $x \in \mathbf{Z}$. Letting $x = 0, 1, \ldots, p - 1$ gives rise to p distinct elements of $(\mathbf{Z}/p^{r+1}\mathbf{Z})^*$, all of which are in the kernel of λ. Furthermore, the element x above can be selected to be one of these p integers because every integer is congruent to one of these p integers modulo (p). Hence the kernel of λ has order p, and our formula is proved.

Note that the kernel of λ is isomorphic to $\mathbf{Z}/p\mathbf{Z}$. (Proof?)

Let A be a ring, and denote its unit element by e for the moment. The map

$$\lambda : \mathbf{Z} \to A$$

such that $\lambda(n) = ne$ is a ring-homomorphism (obvious), and its kernel is an ideal (n), generated by an integer $n \geqq 0$. We have a canonical injective homomorphism $\mathbf{Z}/n\mathbf{Z} \to A$, which is a (ring) isomorphism between $\mathbf{Z}/n\mathbf{Z}$ and a subring of A. If A is entire, then $n\mathbf{Z}$ is a prime ideal, and hence $n = 0$ or $n = p$ for some prime number p. In the first case, A contains as a subring a ring which is isomorphic to $\mathbf{Z}$, and which is often identified with $\mathbf{Z}$. In that case, we say that A has *characteristic* 0. If on the other hand $n = p$, then we say that A has *characteristic* p, and A contains (an isomorphic image of) $\mathbf{F}_p$ as a subring.

If K is a field, then K has characteristic 0 or $p > 0$. In the first case, K contains as a subfield an isomorphic image of the rational numbers, and in the second case, it contains an isomorphic image of $\mathbf{F}_p$. In either case, this subfield will be called the *prime field* (contained in K). Since this prime field is the smallest subfield of K containing 1 and has no automorphism except the identity, it is customary to identify it with $\mathbf{Q}$ or $\mathbf{F}_p$ as the case may be. By the *prime ring* (in K) we shall mean either the integers $\mathbf{Z}$ if K has characteristic 0, or $\mathbf{F}_p$ if K has characteristic p.

§3. *Localization*

We continue to assume that "ring" means "commutative ring".

Let A be a ring. By a *multiplicative subset* of A we shall mean a submonoid of A (viewed as a multiplicative monoid according to RI 2). In other words, it is a subset S containing 1, and such that, if $x, y \in S$, then $xy \in S$.

We shall now construct the *quotient ring of A by S*, also known as the *ring of fractions of A by S*.

We consider pairs (a, s) with $a \in A$ and $s \in S$. We define a relation

$$(a, s) \sim (a', s')$$

between such pairs, by the condition that there exists an element $s_1 \in S$ such that

$$s_1(s'a - sa') = 0.$$

It is then trivially verified that this is an equivalence relation, and the equivalence class containing a pair (a, s) is denoted by a/s. The set of equivalence classes is denoted by $S^{-1}A$.

Note that if $0 \in S$, then $S^{-1}A$ has precisely one element, namely $0/1$.

We define a multiplication in $S^{-1}A$ by the rule

$$(a/s)(a'/s') = aa'/ss'.$$

It is trivially verified that this is well defined. This multiplication has a unit element, namely $1/1$, and is clearly associative.

We define an addition in $S^{-1}A$ by the rule

$$\frac{a}{s} + \frac{a'}{s'} = \frac{s'a + sa'}{ss'}.$$

It is trivially verified that this is well defined. As an example, we give the proof in detail. Let $a_1/s_1 = a/s$, and let $a_1'/s_1' = a'/s'$. We must show that

$$(s_1'a_1 + s_1a_1')/s_1s_1' = (s'a + sa')/ss'.$$

There exist $s_2, s_3 \in S$ such that

$$s_2(sa_1 - s_1a) = 0,$$

$$s_3(s'a_1' - s_1'a') = 0.$$

We multiply the first equation by $s_3s's_1'$ and the second by s_2ss_1. We then add, and obtain

$$s_2s_3[s's_1'(sa_1 - s_1a) + ss_1(s'a_1' - s_1'a')] = 0.$$

By definition, this amounts to what we want to show, namely that there exists an element of S (e.g. s_2s_3) which when multiplied with

$$ss'(s_1'a_1 + s_1a_1') - s_1s_1'(s'a + sa')$$

yields 0.

We observe that given $a \in A$ and $s, s' \in S$ we have

$$a/s = s'a/s's.$$

Thus this aspect of the elementary properties of fractions still remains true in our present general context.

Finally, it is also trivially verified that our two laws of composition on $S^{-1}A$ define a ring structure.

We let

$$\varphi_S : A \to S^{-1}A$$

be the map such that $\varphi(a) = a/1$. Then one sees at once that φ_S is a ring-homomorphism. Furthermore, every element of $\varphi_S(S)$ is invertible in $S^{-1}A$ (the inverse of $s/1$ is $1/s$).

Let $\mathfrak{C}$ be the category whose objects are ring-homomorphisms

$$f : A \to B$$

such that for every $s \in S$, the element $f(s)$ is invertible in B. If $f : A \to B$ and $f' : A \to B'$ are two objects of $\mathfrak{C}$, a morphism g of f into f' is a homomorphism

$$g : B \to B'$$

making the diagram commutative:

$$\begin{array}{ccc} A & \xrightarrow{f} & B \\ & {\scriptstyle f'}\searrow \quad \swarrow{\scriptstyle g} & \\ & B' & \end{array}$$

We contend that φ_S is a universal object in this category $\mathfrak{C}$.

Proof. Suppose that $a/s = a'/s'$, or in other words that the pairs (a, s) and (a', s') are equivalent. There exists $s_1 \in S$ such that

$$s_1(s'a - sa') = 0.$$

Let $f : A \to B$ be an object of $\mathfrak{C}$. Then

$$f(s_1)[f(s')f(a) - f(s)f(a')] = 0.$$

Multiplying by $f(s_1)^{-1}$, and then by $f(s')^{-1}$ and $f(s)^{-1}$, we obtain

$$f(a)f(s)^{-1} = f(a')f(s')^{-1}.$$

Consequently, we can define a map

$$h : S^{-1}A \to B$$

such that $h(a/s) = f(a)f(s)^{-1}$, for all $a/s \in S^{-1}A$. It is trivially verified that h is a homomorphism, and makes the usual diagram commutative. It is also trivially verified that such a map h is unique, and hence that φ_S is the required universal object.

Let A be an entire ring, and let S be a multiplicative subset which does not contain 0. *Then*

$$\varphi_S : A \to S^{-1}A$$

is injective.

Indeed, by definition, if $a/1 = 0$ then there exists $s \in S$ such that $sa = 0$, and hence $a = 0$.

The most important cases of a multiplicative set S are the following:

(i) Let A be a ring, and let S be the set of invertible elements of A (i.e. the set of units). Then S is obviously multiplicative, and is denoted frequently by A^*. If A is a field, then A^* is the multiplicative group of non-zero elements of A. In that case, $S^{-1}A$ is simply A itself.

(ii) Let A be an entire ring, and let S be the set of non-zero elements of A. Then S is a multiplicative set, and $S^{-1}A$ is then a field, called the *quotient field* or the *field of fractions* of A. It is then customary to identify A as a subset of $S^{-1}A$, and we can write

$$a/s = s^{-1}a$$

for $a \in A$ and $s \in S$.

(iii) A ring A is called a *local ring* if it has a unique maximal ideal. If A is a local ring and $\mathfrak{m}$ is its maximal ideal, and $x \in A$, $x \notin \mathfrak{m}$, then x is a unit (otherwise x generates a proper ideal, contained in $\mathfrak{m}$, which is impossible). Let A be a ring and $\mathfrak{p}$ a prime ideal. Let S be the complement of $\mathfrak{p}$ in A. Then S is a multiplicative subset of A, and $S^{-1}A$ is denoted by $A_{\mathfrak{p}}$. It is a local ring (cf. Exercise 3) and is called *the local ring of A at $\mathfrak{p}$.*

Let A be a ring and S a multiplicative subset. Denote by $J(A)$ the set of ideals of A. Then we can define a map

$$\psi_S : J(A) \to J(S^{-1}A),$$

namely we let $\psi_S(\mathfrak{a}) = S^{-1}\mathfrak{a}$ be the subset of $S^{-1}A$ consisting of all fractions a/s with $a \in \mathfrak{a}$ and $s \in S$. The reader will easily verify that $S^{-1}\mathfrak{a}$ is an $S^{-1}A$-ideal, and that ψ_S is a homomorphism for both the additive and multiplicative monoid structures on the set of ideals $J(A)$. Furthermore, ψ_S also preserves intersections and inclusions; in other words, for ideals $\mathfrak{a}$, $\mathfrak{b}$ of A we have:

$$S^{-1}(\mathfrak{a} + \mathfrak{b}) = S^{-1}\mathfrak{a} + S^{-1}\mathfrak{b}, \qquad S^{-1}(\mathfrak{a}\mathfrak{b}) = (S^{-1}\mathfrak{a})(S^{-1}\mathfrak{b})$$
$$S^{-1}(\mathfrak{a} \cap \mathfrak{b}) = S^{-1}\mathfrak{a} \cap S^{-1}\mathfrak{b}.$$

As an example, we prove this last relation. Let $x \in \mathfrak{a} \cap \mathfrak{b}$. Then x/s is in $S^{-1}\mathfrak{a}$ and also in $S^{-1}\mathfrak{b}$, so the inclusion is trivial. Conversely, suppose we have an element of $S^{-1}A$ which can be written as $a/s = b/s'$ with $a \in \mathfrak{a}$, $b \in \mathfrak{b}$, and $s, s' \in S$. Then there exists $s_1 \in S$ such that

$$s_1 s' a = s_1 s b,$$

and this element lies in both $\mathfrak{a}$ and $\mathfrak{b}$. Hence

$$a/s = s_1 s' a / s_1 s' s$$

lies in $S^{-1}(\mathfrak{a} \cap \mathfrak{b})$, as was to be shown.

§4. *Principal rings*

Throughout this section, "ring" means "commutative ring".

Let A be an entire ring. An element $a \neq 0$ is called *irreducible* if it is not a unit, and if whenever one can write $a = bc$ with $b \in A$ and $c \in A$ then b or c is a unit.

Let $a \neq 0$ be an element of A and assume that the principal ideal (a) is prime. Then a is irreducible. Indeed, if we write $a = bc$, then b or c lies in (a), say b. Then we can write $b = ad$ with some $d \in A$, and hence $a = acd$. Since A is entire, it follows that $cd = 1$, in other words, that c is a unit.

The converse of the preceding assertion is not always true. We shall discuss under which conditions it is true. An element $a \in A$, $a \neq 0$, is said to have a *unique factorization into irreducible elements* if there exists a unit u and there exist irreducible elements p_i $(i = 1, \ldots, r)$ in A such that

$$a = u \prod_{i=1}^{r} p_i,$$

and if given two factorizations into irreducible elements,

$$a = u \prod_{i=1}^{r} p_i = u' \prod_{j=1}^{s} q_j,$$

we have $r = s$, and after a permutation of the indices i, we have $p_i = u_i q_i$ for some unit u_i in A, $i = 1, \ldots, r$.

We note that if p is irreducible and u is a unit, then up is also irreducible, so we must allow multiplication by units in a factorization. In the ring of integers $\mathbf{Z}$, the ordering allows us to select a representative irreducible element (a prime number) out of two possible ones differing by a unit, namely $\pm p$, by selecting the positive one. This is, of course, impossible in more general rings.

Taking $r = 0$ above, we adopt the convention that a unit of A has a factorization into irreducible elements.

A ring is called *factorial* (or a unique factorization ring) if it is entire and if every element $\neq 0$ has a unique factorization into irreducible elements. We shall prove below that a principal entire ring is factorial.

Let A be an entire ring and $a, b \in A$, $ab \neq 0$. We say that *a divides b* and write $a \mid b$ if there exists $c \in A$ such that $ac = b$. We say that $d \in A$, $d \neq 0$, is a *greatest common divisor* of a and b if $d \mid a$, $d \mid b$, and if any element e of A, $e \neq 0$, which divides both a and b also divides d.

PROPOSITION 2. *Let A be a principal entire ring and $a, b \in A$, $a, b \neq 0$. Let $(a, b) = (c)$. Then c is a greatest common divisor of a and b.*

Proof. Since b lies in the ideal (c), we can write $b = xc$ for some $x \in A$, so that $c \mid b$. Similarly, $c \mid a$. Let d divide both a and b, and write $a = dy$, $b = dz$ with $y, z \in A$. Since c lies in (a, b) we can write

$$c = wa + tb$$

with some $w, t \in A$. Then $c = w\,dy + t\,dz = d(wy + tz)$, whence $d \mid c$, and our proposition is proved.

THEOREM 1. *Let A be a principal entire ring. Then A is factorial.*

Proof. We first prove that every non-zero element of A has a factorization into irreducible elements. Let S be the set of principal ideals $\neq 0$ whose generators do not have a factorization into irreducible elements, and suppose S is not empty. Let (a_1) be in S. Consider an ascending chain

$$(a_1) \subsetneqq (a_2) \subsetneqq \cdots \subsetneqq (a_n) \subsetneqq \cdots$$

of ideals in S. We contend that such a chain cannot be infinite. Indeed, the union of such a chain is an ideal of A, which is principal, say equal to (a). The generator a must already lie in some element of the chain, say (a_n), and then we see that

$$(a_n) \subset (a) \subset (a_n),$$

whence the chain stops at (a_n). Hence any ideal of A containing (a_n) and $\neq (a_n)$ has a generator admitting a factorization.

We note that a_n cannot be irreducible (otherwise it has a factorization), and hence we can write $a_n = bc$ with neither b nor c equal to a unit. But then $(b) \neq (a_n)$ and $(c) \neq (a_n)$ and hence both b, c admit factorizations into irreducible elements. The product of these factorizations is a factorization for a_n, contradicting the assumption that S is not empty.

To prove uniqueness, we first remark that if p is an irreducible element of A and $a, b \in A$, $p \mid ab$, then $p \mid a$ or $p \mid b$. Proof: If $p \nmid a$, then the g.c.d. of p, a is 1 and hence we can write

$$1 = xp + ya$$

with some $x, y \in A$. Then $b = bxp + yab$, and since $p \mid ab$ we conclude that $p \mid b$.

Suppose that a has two factorizations

$$a = p_1 \cdots p_r = q_1 \cdots q_s$$

into irreducible elements. Since p_1 divides the product farthest to the right, p_1 divides one of the factors, which we may assume to be q_1 after renumbering these factors. Then there exists a unit u_1 such that $q_1 = u_1 p_1$. We can now cancel p_1 from both factorizations and get

$$p_2 \cdots p_r = u_1 q_2 \cdots q_s.$$

The argument is completed by induction.

We could call two elements $a, b \in A$ equivalent if there exists a unit u such that $a = bu$. Let us select one irreducible element p out of each equivalence class belonging to such an irreducible element, and let us denote by P the set of such representatives. Let $a \in A$, $a \neq 0$. Then there exists a unit u and integers $\nu(p) \geqq 0$, equal to 0 for almost all $p \in P$ such that

$$a = u \prod_{p \in P} p^{\nu(p)}.$$

Furthermore, the unit u and the integers $\nu(p)$ are uniquely determined by a. We call $\nu(p)$ the *order* of a at p, also written $\text{ord}_p\, a$.

If A is a factorial ring, then an irreducible element p generates a prime ideal (p). Thus in a factorial ring, an irreducible element will also be called a *prime element*, or simply a *prime*.

We observe that one can define the notion of *least common multiple* of a finite number of non-zero elements of A in the usual manner: If

$$a_1, \ldots, a_n \in A$$

are such elements, we define a l.c.m. for these elements to be any $c \in A$ such that for all primes p of A we have

$$\text{ord}_p\, c = \max_i \text{ord}_p\, a_i.$$

This element c is well defined up to a unit.

If $a, b \in A$ are non-zero elements, we say that a, b are *relatively prime* if $(a, b) = (1)$. This means that the g.c.d. of a and b is a unit.

Example. The ring of integers $\mathbf{Z}$ is factorial. Its group of units consists of 1 and -1. It is natural to take as representative prime element the positive prime element (what is called a prime number) p from the two possible choices p and $-p$. Similarly, we shall show later that the ring of polynomials in one variable over a field is factorial, and one selects representatives for the prime elements to be the irreducible polynomials with leading coefficient 1.

EXERCISES

All rings are assumed commutative.

1. Let A be a ring with $1 \neq 0$. Let S be a multiplicative subset not containing 0. Let $\mathfrak{p}$ be a maximal element in the set of ideals of A whose intersection with S is empty. Show that $\mathfrak{p}$ is prime.

2. Let $f : A \to A'$ be a surjective homomorphism of rings, and assume that A is local. Show that A' is local.

3. Let A be a ring and $\mathfrak{p}$ a prime ideal. Show that $A_{\mathfrak{p}}$ has a unique maximal ideal, consisting of all elements a/s with $a \in \mathfrak{p}$ and $s \notin \mathfrak{p}$.

4. Let A be a principal ring and S a multiplicative subset. Show that $S^{-1}A$ is principal.

5. Let A be a factorial ring and S a multiplicative subset. Show that $S^{-1}A$ is factorial, and that the prime elements of $S^{-1}A$ are those primes p of A such that $(p) \cap S$ is empty.

6. Let A be a principal ring and $a_1, \ldots, a_n$ non-zero elements of A. Let $(a_1, \ldots, a_n) = (d)$. Show that d is a greatest common divisor for the a_i $(i = 1, \ldots, n)$.

7. Let p be a prime number, and let A be the ring $\mathbf{Z}/p^r\mathbf{Z}$ (r = integer $\geqq 1$). Let G be the group of units in A, i.e. the group of integers prime to p, modulo p^r. Show that G is cyclic, except in case when

$$p = 2, \qquad r \geqq 3,$$

in which case it is of type $(2, 2^{r-2})$.

[*Hint:* In the general case, show that G is the product of a cyclic group generated by $1 + p$, and a cyclic group of order $p - 1$. In the exceptional case, show that G is the product of the group $\{\pm 1\}$ with the cyclic group generated by the residue class of 5 mod 2^r.]

8. Let i be the complex number $\sqrt{-1}$. Show that the ring $\mathbf{Z}[i]$ is principal, and hence factorial. What are the units?

9. Let A be the ring of entire functions on the complex plane. Show that every finitely generated ideal of A is principal. What are the principal prime ideals of A? What are the units of A? Show that A itself is not factorial.

CHAPTER III

Modules

§1. *Basic definitions*

Let A be a ring. A *left module* over A, or a left A-module M is an abelian group, usually written additively, together with an operation of A on M (viewing A as a multiplicative monoid by RI 2), such that, for all $a, b \in A$ and $x, y \in M$ we have

$$(a + b)x = ax + bx \qquad \text{and} \qquad a(x + y) = ax + ay.$$

We leave it as an exercise to prove that $a(-x) = -(ax)$ and that $0x = 0$. By definition of an operation, we have $1x = x$.

In a similar way, one defines a *right* A-module. We shall deal only with left A-modules, unless otherwise specified, and hence call these simply A-modules, or even modules if the reference is clear.

Examples

We note that A is a module over itself.

Any commutative group is a $\mathbf{Z}$-module.

An additive group consisting of 0 alone is a module over any ring.

Any left ideal of A is a module over A.

Let S be a non-empty set and M an A-module. Then the set of maps $\mathfrak{M}(S, M)$ is an A-module. We have already noted previously that it is a commutative group, and for $f \in \mathfrak{M}(S, M)$, $a \in A$ we define af to be the map such that $(af)(s) = af(s)$. The axioms for a module are then trivially verified.

For the rest of this section, we deal with a fixed ring A, and hence may omit the prefix A-.

Let M be an A-module. By a *submodule* N of M we mean an additive subgroup such that $AN \subset N$. Then N is a module (with the operation induced by that of A on M).

Let $\mathfrak{a}$ be a left ideal, and M a module. We define $\mathfrak{a}M$ to be the set of all elements

$$a_1x_1 + \cdots + a_nx_n$$

with $a_i \in \mathfrak{a}$ and $x_i \in M$. It is obviously a submodule of M. If $\mathfrak{a}$, $\mathfrak{b}$ are left ideals, then we have associativity, namely

$$\mathfrak{a}(\mathfrak{b}M) = (\mathfrak{a}\mathfrak{b})M.$$

We also have some obvious distributivities, like $(\mathfrak{a} + \mathfrak{b})M = \mathfrak{a}M + \mathfrak{b}M$. If N, N' are submodules of M, then $\mathfrak{a}(N + N') = \mathfrak{a}N + \mathfrak{a}N'$.

Let M be an A-module, and N a submodule. We shall define a module structure on the factor group M/N (for the additive group structure). Let $x + N$ be a coset of N in M, and let $a \in A$. We define $a(x + N)$ to be the coset $ax + N$. It is trivial to verify that this is well defined (i.e. if y is in the same coset as x, then ay is in the same coset as ax), and that this is an operation of A on M/N satisfying the required condition, making M/N into a module, called the *factor module* of M by N.

By a *module-homomorphism* one means a map

$$f : M \to M'$$

of one module into another (over the same ring A), which is an additive group-homomorphism, and such that

$$f(ax) = af(x)$$

for all $a \in A$ and $x \in M$. It is then clear that the collection of A-modules is a category, whose morphisms are the module-homomorphisms usually also called homomorphisms for simplicity, if no confusion is possible. If we wish to refer to the ring A, we also say that f is an *A-homomorphism*, or also that it is an *A-linear map*.

If M is a module, then the identity map is a homomorphism. For any module M', the map $\zeta : M \to M'$ such that $\zeta(x) = 0$ for all $x \in M$ is a homomorphism, called *zero*.

Let M be a module and N a submodule. We have the canonical additive group-homomorphism

$$f : M \to M/N$$

and one verifies trivially that it is a module-homomorphism.

Equally trivially, one verifies that it is universal in the category of homomorphisms of M whose kernel contains N.

If $f : M \to M'$ is a module-homomorphism, then its kernel and image are submodules of M and M' respectively (trivial verification). Canonical homomorphisms discussed in Chapter I, §4 apply to modules *mutatis mutandis*. For the convenience of the reader, we summarize these homomorphisms:

Let N, N' be two submodules of a module M. Then $N + N'$ is also a submodule, and we have an isomorphism

$$N/(N \cap N') \approx (N + N')/N'.$$

If $M \supset M' \supset M''$ are modules, then

$$(M/M'')/(M'/M'') \approx M/M'.$$

If $f : M \to M'$ is a module-homomorphism, and N' is a submodule of M', then $f^{-1}(N')$ is a submodule of M and we have a canonical injective homomorphism

$$\bar{f} : M/f^{-1}(N') \to M'/N'.$$

If f is surjective, then $\bar{f}$ is a module-isomorphism.

The proofs are obtained by verifying that all homomorphisms which appeared when dealing with abelian groups are now A-homomorphisms of modules. We leave the verification to the reader.

As with groups, we observe that a module-homomorphism which is bijective is a module-isomorphism. Here again, the proof is the same as for groups, adding only the observation that the inverse map, which we know is a group-isomorphism, actually is a module-isomorphism. Again, we leave the verification to the reader.

As with abelian groups, we define a sequence of module-homomorphisms

$$M' \xrightarrow{f} M \xrightarrow{g} M''$$

to be *exact* if $\operatorname{Im} f = \operatorname{Ker} g$. We have an exact sequence associated with a submodule N of a module M, namely

$$0 \to N \to M \to M/N \to 0,$$

the map of N into M being the inclusion, and the subsequent map being the canonical map. The notion of exactness is due to Eilenberg-Steenrod.

§2. *The group of homomorphisms*

Let A be a ring, and let X, X' be A-modules. We denote by $\operatorname{Hom}_A(X', X)$ the set of A-homomorphisms of X' into X. Then $\operatorname{Hom}_A(X', X)$ is an abelian group, the law of addition being that of addition for mappings into an abelian group.

If A is commutative then we can make $\operatorname{Hom}_A(X', X)$ into an A-module, by defining af for $a \in A$ and $f \in \operatorname{Hom}_A(X', X)$ to be the map such that

$$(af)(x) = af(x).$$

The verification that the axioms for an A-module are satisfied is trivial. However, if A is not commutative, then we view $\operatorname{Hom}_A(X', X)$ simply as an abelian group.

We also view Hom_A as a functor. It is actually a functor of two variables, contravariant in the first and covariant in the second. Indeed, let

Y be an A-module, and let

$$X' \xrightarrow{f} X$$

be an A-homomorphism. Then we get an induced homomorphism

$$\operatorname{Hom}_A(f, Y) : \operatorname{Hom}_A(X, Y) \to \operatorname{Hom}_A(X', Y)$$

(reversing the arrow!) given by

$$g \mapsto g \circ f.$$

This is illustrated by the following sequence of maps:

$$X' \xrightarrow{f} X \xrightarrow{g} Y.$$

The fact that $\operatorname{Hom}_A(f, Y)$ is a homomorphism is simply a rephrasing of the property $(g_1 + g_2) \circ f = g_1 \circ f + g_2 \circ f$, which is trivially verified. If $f = id$, then composition with f acts as an identity mapping on g, i.e. $g \circ id = g$.

If we have a sequence of A-homomorphisms

$$X' \to X \to X'',$$

then we get an induced sequence

$$\operatorname{Hom}_A(X', Y) \leftarrow \operatorname{Hom}_A(X, Y) \leftarrow \operatorname{Hom}_A(X'', Y).$$

Given an exact sequence

$$X' \xrightarrow{\lambda} X \longrightarrow X'' \longrightarrow 0,$$

then the induced sequence

$$\operatorname{Hom}_A(X', Y) \leftarrow \operatorname{Hom}_A(X, Y) \leftarrow \operatorname{Hom}_A(X'', Y) \leftarrow 0$$

is exact.

This is an important fact, whose proof is trivial. For instance, if $g : X'' \to Y$ is an A-homomorphism, its image in $\operatorname{Hom}_A(X, Y)$ is obtained by composing g with the surjective map of X on X''. If this composition is 0, it follows that $g = 0$ because $X \to X''$ is surjective. As another example, consider a homomorphism $g : X \to Y$ such that the composition

$$X' \xrightarrow{\lambda} X \xrightarrow{g} Y$$

is 0. Then g vanishes on the image of λ. Hence we can factor g through the factor module,

$$\begin{array}{ccc} & X/\operatorname{Im}\lambda & \\ \nearrow & & \searrow \\ X & \xrightarrow[g]{} & Y \end{array}$$

Since $X \to X''$ is surjective, we have an isomorphism

$$X/\mathrm{Im}\,\lambda \leftrightarrow X''.$$

Hence we can factor g through X'', thereby showing that the kernel of

$$\mathrm{Hom}_A(X', Y) \leftarrow \mathrm{Hom}_A(X, Y)$$

is contained in the image of

$$\mathrm{Hom}_A(X, Y) \leftarrow \mathrm{Hom}_A(X'', Y).$$

The other conditions needed to verify exactness are left to the reader.

We have a similar situation with respect to the second variable, but then the functor is covariant. Thus if X is fixed, and we have a sequence of A-homomorphisms

$$Y' \to Y \to Y'',$$

then we get an induced sequence

$$\mathrm{Hom}_A(X, Y') \to \mathrm{Hom}_A(X, Y) \to \mathrm{Hom}_A(X, Y'').$$

Given an exact sequence

$$0 \to Y' \to Y \to Y'',$$

the induced sequence

$$0 \to \mathrm{Hom}_A(X, Y') \to \mathrm{Hom}_A(X, Y) \to \mathrm{Hom}_A(X, Y'')$$

is exact.

The verification will be left to the reader. It follows at once from the definitions.

We note that to say that

$$0 \to Y' \to Y$$

is exact means that Y' is embedded in Y, i.e. is isomorphic to a submodule of Y. A homomorphism into Y' can be viewed as a homomorphism into Y if we have $Y' \subset Y$. This corresponds to the injection

$$0 \to \mathrm{Hom}_A(X, Y') \to \mathrm{Hom}_A(X, Y).$$

Let M be an A-module. From the relations

$$(g_1 + g_2) \circ f = g_1 \circ f + g_2 \circ f$$

and its analogue on the right, namely

$$g \circ (f_1 + f_2) = g \circ f_1 + g \circ f_2,$$

and the fact that there is an identity for composition, namely id_M, we conclude that $\mathrm{Hom}_A(M, M)$ is a ring, the multiplication being defined as composition of mappings. If n is an integer $\geqq 1$, we can write f^n to mean the iteration of f with itself n times, and define f^0 to be id. According to the general definition of endomorphisms in a category, we also write $\mathrm{End}_A(M)$ instead of $\mathrm{Hom}_A(M, M)$.

Since an A-module M is an abelian group, we see that $\mathrm{Hom}_{\mathbf{Z}}(M, M)$ (= set of group-homomorphisms of M into itself) is a ring, and that we could have defined an operation of A on M to be a ring-homomorphism $A \to \mathrm{Hom}_{\mathbf{Z}}(M, M)$.

§3. Direct products and sums of modules

Let A be a ring. As with abelian groups, a coproduct in the category of A-modules is called a direct sum.

Proposition 1. *Direct products and direct sums exist in the category of A-modules.*

Proof. We shall leave the proof of the product as an exercise. We discuss the sum as an example, following the construction given for the direct sum of abelian groups. Let $\{M_i\}_{i \in I}$ be a family of A-modules, and let

$$M = \coprod_{i \in I} M_i$$

be their direct sum as abelian groups. We define on M a structure of A-module: If $(x_i)_{i \in I}$ is an element of M, i.e. a family of elements $x_i \in M_i$ such that $x_i = 0$ for almost all i, and if $a \in A$, then we define

$$a(x_i)_{i \in I} = (ax_i)_{i \in I},$$

that is we define multiplication by a componentwise. It is trivially verified that this is an operation of A on M which makes M into an A-module. If one refers back to the proof given for the existence of direct sums in the category of abelian groups, one sees immediately that this proof now extends in the same way to show that M is a direct sum of the family $\{M_i\}_{i \in I}$ as A-modules. (For instance, the map

$$\lambda_j : M_j \to M$$

such that $\lambda_j(x)$ has j-th component equal to x and i-th component equal to 0 for $i \neq j$ is now seen to be an A-homomorphism.) Given a family of A-homomorphisms $\{f_i : M_i \to N\}$, the map f defined as in the proof for abelian groups is also an A-isomorphism and has the required properties.

When I is a finite set, there is a useful criterion for a module to be a direct product.

PROPOSITION 2. *Let M be an A-module and n an integer $\geqq 1$. For each $i = 1, \dots, n$ let $\varphi_i : M \to M$ be an A-homomorphism such that*

$$\sum_{i=1}^{n} \varphi_i = id \qquad \text{and} \qquad \varphi_i \circ \varphi_j = 0 \qquad \text{if} \quad i \neq j.$$

Then $\varphi_i^2 = \varphi_i$ for all i. Let $M_i = \varphi_i(M)$, and let $\varphi : M \to \prod M_i$ be such that

$$\varphi(x) = (\varphi_1(x), \dots, \varphi_n(x)).$$

Then φ is an A-isomorphism of M onto the direct product $\prod M_i$.

Proof. For each j, we have

$$\varphi_j = \varphi_j \circ id = \varphi_j \circ \sum_{i=1}^{n} \varphi_i = \varphi_j \circ \varphi_j = \varphi_j^2,$$

thereby proving the first assertion. It is clear that φ is an A-homomorphism. Let x be in its kernel. Since

$$x = id(x) = \sum_{i=1}^{n} \varphi_i(x)$$

we conclude that $x = 0$, so φ is injective. Given elements $y_i \in M_i$ for each $i = 1, \dots, n$, let $x = y_1 + \cdots + y_n$. We obviously have $\varphi_j(y_i) = 0$ if $i \neq j$. Hence

$$\varphi_j(x) = y_j$$

for each $j = 1, \dots, n$. This proves that φ is surjective, and concludes the proof of our proposition.

We observe that when I is a finite set, the direct sum and the direct product are equal.

Just as with abelian groups, we use the symbol $\oplus$ to denote direct sum.

Let M be a module over a ring A and let S be a subset of M. By a *linear combination* of elements of S (with coefficients in A) one means a sum

$$\sum_{x \in S} a_x x$$

where $\{a_x\}$ is a set of elements of A, almost all of which are equal to 0. These elements a_x are called the *coefficients* of the linear combination. Let N be the set of all linear combinations of elements of S. Then N is a submodule of M, for if

$$\sum_{x \in S} a_x x \qquad \text{and} \qquad \sum_{x \in S} b_x x$$

are two linear combinations, then their sum is equal to

$$\sum_{x \in S} (a_x + b_x)x,$$

and if $c \in A$, then

$$c\left(\sum_{x \in S} a_x x\right) = \sum_{x \in S} c a_x x,$$

and these elements are again linear combinations of elements of S. We shall call N the submodule *generated* by S, and we call S a set of *generators* for N. We sometimes write $N = A\langle S \rangle$. If S consists of one element x, the module generated by x is also written Ax, or simply (x), and sometimes we say that (x) is a *principal module.*

A module M is said to be *finitely generated*, or of *finite type*, if it has a finite number of generators.

A *subset* S of a module M is said to be *linearly independent* (over A) if whenever we have a linear combination

$$\sum_{x \in S} a_x x$$

which is equal to 0, then $a_x = 0$ for all $x \in S$. If S is linearly independent and if two linear combinations

$$\sum a_x x \qquad \text{and} \qquad \sum b_x x$$

are equal, then $a_x = b_x$ for all $x \in S$. Indeed, subtracting one from the other yields $\sum (a_x - b_x)x = 0$, whence $a_x - b_x = 0$ for all x. If S is linearly independent we shall also say that its elements are linearly independent. Similarly, a *family* $\{x_i\}_{i \in I}$ of elements of M is said to be linearly independent if whenever we have a linear combination

$$\sum_{i \in I} a_i x_i = 0,$$

then $a_i = 0$ for all i. A subset S (resp. a family $\{x_i\}$) is called *linearly dependent* if it is not linearly independent, i.e. if there exists a relation

$$\sum_{x \in S} a_x x = 0 \qquad \text{resp.} \qquad \sum_{i \in I} a_i x_i = 0$$

with not all a_x (resp. a_i) $= 0$. *Warning.* Let x be a single element of M which is linearly independent. Then the family $\{x_i\}_{i=1,\ldots,n}$ such that $x_i = x$ for all i is linearly dependent if $n > 1$, but the set consisting of x itself is linearly independent.

Let M be an A-module, and let $\{M_i\}_{i \in I}$ be a family of submodules. Since we have inclusion-homomorphisms

$$\lambda_i : M_i \to M$$

we have an induced homomorphism

$$\lambda_* : \coprod M_i \to M$$

which is such that for any family of elements $(x_i)_{i \in I}$, all but a finite number of which are 0, we have

$$\lambda_*((x_i)) = \sum_{i \in I} x_i.$$

If λ_* is an isomorphism, then we say that the family $\{M_i\}_{i \in I}$ is a *direct sum decomposition* of M. This is obviously equivalent to saying that every element of M has a unique expression as a sum

$$\sum x_i$$

with $x_i \in M_i$, and almost all $x_i = 0$. By abuse of notation, we also write

$$M = \coprod M_i$$

in this case.

If the family $\{M_i\}$ is such that every element of M has *some* expression as a sum $\sum x_i$ (not necessarily unique), then we write $M = \sum M_i$. In any case, if $\{M_i\}$ is an arbitrary family of submodules, the image of the homomorphism λ_* above is a submodule of M, which will be denoted by $\sum M_i$.

If M is a module and N, N' are two submodules such that $N + N' = M$ and $N \cap N' = 0$, then we have a module-isomorphism

$$M \approx N \oplus N',$$

just as with abelian groups, and similarly with a finite number of submodules.

We note, of course, that our discussion of abelian groups is a special case of our discussion of modules, simply by viewing abelian groups as modules over $\mathbf{Z}$. However, it seems usually desirable (albeit inefficient) to develop first some statements for abelian groups, and then point out that they are valid (obviously) for modules in general.

Let M, M', N be modules. Then we have an isomorphism of abelian groups

$$\boxed{\operatorname{Hom}_A(M \oplus M', N) \xleftrightarrow{\approx} \operatorname{Hom}_A(M, N) \times \operatorname{Hom}_A(M', N)}\ ,$$

and similarly

$$\boxed{\operatorname{Hom}_A(N, M \times M') \xleftrightarrow{\approx} \operatorname{Hom}_A(N, M) \oplus \operatorname{Hom}_A(N, M')} \; .$$

The first one is obtained as follows. If $f : M \oplus M' \to N$ is a homomorphism, then f induces a homomorphism $f_1 : M \to N$ and a homomorphism $f_2 : M' \to N$ by composing f with the injections of M and M' into their direct sum respectively:

$$M \to M \oplus \{0\} \subset M \oplus M' \xrightarrow{f} N,$$
$$M' \to \{0\} \oplus M' \subset M \oplus M' \xrightarrow{f} N.$$

We leave it to the reader to verify that the association

$$f \mapsto (f_1, f_2)$$

gives an isomorphism as in the first box. The isomorphism in the second box is obtained in a similar way. Given homomorphisms $f_1 : N \to M$ and $f_2 : N \to M'$ we have a homomorphism $f : N \to M \times M'$ defined by

$$f(x) = (f_1(x), f_2(x)).$$

It is trivial to verify that the association

$$(f_1, f_2) \mapsto f$$

gives an isomorphism as in the second box.

Of course, the direct sum and direct product of two modules are isomorphic, but we distinguished them in the notation for the sake of functoriality.

PROPOSITION 3. *Let $0 \to M' \xrightarrow{f} M \xrightarrow{g} M'' \to 0$ be an exact sequence of modules. The following conditions are equivalent:*

(1) There exists a homomorphism $\varphi : M'' \to M$ such that $g \circ \varphi = id$.

(2) There exists a homomorphism $\psi : M \to M'$ such that $\psi \circ f = id$.

If these conditions are satisfied, then we have isomorphisms:

$$M = \operatorname{Im} f \oplus \operatorname{Ker} \psi, \qquad M = \operatorname{Ker} g \oplus \operatorname{Im} \varphi,$$
$$M \approx M' \oplus M''.$$

Proof. Let us write the homomorphisms on the right:

$$M \underset{\varphi}{\overset{g}{\rightleftarrows}} M'' \to 0.$$

Let $x \in M$. Then

$$x - \varphi(g(x))$$

is in the kernel of g, and hence $M = \operatorname{Ker} g + \operatorname{Im} \varphi$.

This sum is direct, for if

$$x = y + z$$

with $y \in \operatorname{Ker} g$ and $z \in \operatorname{Im} \varphi$, $z = \varphi(w)$ with $w \in M''$, and applying g yields $g(x) = w$. Thus w is uniquely determined by x, and therefore z is uniquely determined by x. Hence so is y, thereby proving the sum is direct.

The arguments concerning the other side of the sequence are similar and will be left as exercises, as well as the equivalence between our conditions. When these conditions are satisfied, the exact sequence of Proposition 3 is said to *split*.

§4. *Free modules*

Let M be a module over a ring A and let S be a subset of M. We shall say that S is a *basis* of M if S is not empty, if S generates M, and if S is linearly independent. If S is a basis of M, then in particular $M \neq \{0\}$ if $A \neq \{0\}$ and every element of M has a unique expression as a linear combination of elements of S. Similarly, let $\{x_i\}_{i \in I}$ be a non-empty family of elements of M. We say that it is a *basis* of M if it is linearly independent and generates M.

If A is a ring, then as a module over itself, A admits a basis, consisting of the unit element 1.

Let I be a non-empty set, and for each $i \in I$, let $A_i = A$, viewed as an A-module. Let

$$F = \coprod_{i \in I} A_i.$$

Then F admits a basis, which consists of the elements e_i of F whose i-th component is the unit element of A_i, and having all other components equal to 0.

By a *free* module we shall mean a module which admits a basis, or the zero module.

THEOREM 1. *Let A be a ring and M a module over A. Let I be a non-empty set, and let $\{x_i\}_{i \in I}$ be a basis of M. Let N be an A-module, and let $\{y_i\}_{i \in I}$ be a family of elements of N. Then there exists a unique homomorphism $f : M \to N$ such that $f(x_i) = y_i$ for all i.*

Proof. Let x be an element of M. There exists a unique family $\{a_i\}_{i \in I}$ of elements of A such that

$$x = \sum_{i \in I} a_i x_i.$$

We define

$$f(x) = \sum a_i y_i.$$

It is then clear that f is a homomorphism satisfying our requirements, and that it is the unique such, because we must have

$$f(x) = \sum a_i f(x_i).$$

COROLLARY 1. *Let the notation be as in the theorem, and assume that* $\{y_i\}_{i \in I}$ *is a basis of* N. *Then the homomorphism* f *is an isomorphism, i.e. a module-isomorphism.*

Proof. By symmetry, there exists a unique homomorphism

$$g : N \to M$$

such that $g(y_i) = x_i$ for all i, and $f \circ g$ and $g \circ f$ are the respective identity mappings.

COROLLARY 2. *Two modules having bases whose cardinalities are equal are isomorphic.*

Proof. Clear.

We shall leave the proofs of the following statements as exercises.

Let M be a free module over A, with basis $\{x_i\}_{i \in I}$, so that

$$M = \coprod_{i \in I} Ax_i.$$

Let $\mathfrak{a}$ be a left ideal of A. Then $\mathfrak{a}M$ is a submodule of M. Each $\mathfrak{a}x_i$ is a submodule of Ax_i. *We have an isomorphism* (of A-modules)

$$\boxed{M/\mathfrak{a}M \approx \coprod_{i \in I} Ax_i/\mathfrak{a}x_i}\ .$$

Furthermore, each $Ax_i/\mathfrak{a}x_i$ is isomorphic to $A/\mathfrak{a}$, as A-module.

Suppose in addition that A *is commutative. Then* $A/\mathfrak{a}$ *is a ring. Furthermore* $M/\mathfrak{a}M$ *is a free module over* $A/\mathfrak{a}$, *and each* $Ax_i/\mathfrak{a}x_i$ *is free over* $A/\mathfrak{a}$. *If* $\bar{x}_i$ *is the image of* x_i *under the canonical homomorphism*

$$Ax_i \to Ax_i/\mathfrak{a}x_i,$$

then the single element $\bar{x}_i$ *is a basis of* $Ax_i/\mathfrak{a}x_i$ *over* $A/\mathfrak{a}$.

§5. *Vector spaces*

A module over a field is called a *vector space.*

THEOREM 2. *Let* V *be a vector space over a field* K, *and assume that* $V \neq \{0\}$. *Let* Γ *be a set of generators of* V *over* K *and let* S *be a subset of* Γ *which is linearly independent. Then there exists a basis* $\mathfrak{B}$ *of* V *such that* $S \subset \mathfrak{B} \subset \Gamma$.

Proof. Let $\mathfrak{T}$ be the set whose elements are subsets T of Γ which contain S and are linearly independent. Then $\mathfrak{T}$ is not empty (it contains S), and we contend that $\mathfrak{T}$ is inductively ordered. Indeed, if $\{T_i\}$ is a totally ordered subset of $\mathfrak{T}$ (by ascending inclusion), then $\bigcup T_i$ is again linearly independent and contains S. By Zorn's lemma, let $\mathcal{B}$ be a maximal element of $\mathfrak{T}$. Then $\mathcal{B}$ is linearly independent. Let W be the subspace of V generated by $\mathcal{B}$. If $W \neq V$, there exists some element $x \in \Gamma$ such that $x \notin W$. Then $\mathcal{B} \cup \{x\}$ is linearly independent, for given a linear combination

$$\sum_{y \in \mathcal{B}} a_y y + bx = 0, \qquad a_y, b \in K,$$

we must have $b = 0$, otherwise we get

$$x = -\sum_{y \in \mathcal{B}} b^{-1} a_y y \in W.$$

By construction, we now see that $a_y = 0$ for all $y \in \mathcal{B}$, thereby proving that $\mathcal{B} \cup \{x\}$ is linearly independent, and contradicting the maximality of $\mathcal{B}$. It follows that $W = V$, and furthermore that $\mathcal{B}$ is not empty since $V \neq \{0\}$. This proves our theorem.

If V is a vector space $\neq \{0\}$, then in particular, we see that every set of linearly independent elements of V can be extended to a basis, and that a basis may be selected from a given set of generators.

THEOREM 3. *Let V be a vector space over a field K. Then two bases of V over K have the same cardinality.*

Proof. Let us first assume that there exists a basis of V with a finite number of elements, say $\{v_1, \ldots, v_m\}$, $m \geqq 1$. We shall prove that any other basis must also have m elements. For this it will suffice to prove: If $w_1, \ldots, w_n$ are elements of V which are linearly independent over K, then $n \leqq m$ (for we can then use symmetry). We proceed by induction. There exist elements $c_1, \ldots, c_m$ of K such that

$$w_1 = c_1 v_1 + \cdots + c_m v_m, \tag{1}$$

and some c_i, say c_1, is not equal to 0. Then v_1 lies in the space generated by $w_1, v_2, \ldots, v_m$ over K, and this space must therefore be equal to V itself. Furthermore, $w_1, v_2, \ldots, v_m$ are linearly independent, for suppose $b_1, \ldots, b_m$ are elements of K such that

$$b_1 w_1 + b_2 v_2 + \cdots + b_m v_m = 0.$$

If $b_1 \neq 0$, divide by b_1 and express w_1 as a linear combination of $v_2, \ldots, v_m$. Subtracting from (1) would yield a relation of linear dependence among the v_i, which is impossible. Hence $b_1 = 0$, and again we must have all $b_i = 0$ because the v_i are linearly independent.

Suppose inductively that after a suitable renumbering of the v_i, we have found $w_1, \ldots, w_r$ $(r < n)$ such that

$$\{w_1, \ldots, w_r, v_{r+1}, \ldots, v_m\}$$

is a basis of V. We express w_{r+1} as a linear combination

$$w_{r+1} = c_1 w_1 + \cdots + c_r w_r + c_{r+1} v_{r+1} + \cdots + c_m v_m \tag{2}$$

with $c_i \in K$. The coefficients of the v_i in this relation cannot all be 0, otherwise there would be a linear dependence among the w_j. Say $c_{r+1} \neq 0$. Using an argument similar to that used above, we can replace v_{r+1} by w_{r+1} and still have a basis of V. This means that we can repeat the procedure until $r = n$, and therefore that $n \leqq m$, thereby proving our theorem.

We shall leave the general case of an infinite basis as an exercise to the reader. [*Hint:* Use the fact that a finite number of elements in one basis is contained in the space generated by a finite number of elements in another basis.]

If a vector space V admits one basis with a finite number of elements, say m, then we shall say that V is *finite dimensional* and that m is its *dimension*. In view of Theorem 3, we see that m is the number of elements in *any* basis of V. If $V = \{0\}$, then we define its dimension to be 0, and say that V is 0-dimensional. We abbreviate "dimension" by "dim" or "$\dim_K$" if the reference to K is needed for clarity.

When dealing with vector spaces over a field, we use the words subspace and factor space instead of submodule and factor module.

THEOREM 4. *Let V be a vector space over a field K, and let W be a subspace. Then*

$$\dim_K V = \dim_K W + \dim_K V/W.$$

If $f : V \to U$ is a homomorphism of vector spaces over K, then

$$\dim V = \dim \operatorname{Ker} f + \dim \operatorname{Im} f.$$

Proof. The first statement is a special case of the second, taking for f the canonical map. Let $\{u_i\}_{i \in I}$ be a basis of $\operatorname{Im} f$, and let $\{w_j\}_{j \in J}$ be a basis of $\operatorname{Ker} f$. Let $\{v_i\}_{i \in I}$ be a family of elements of V such that $f(v_i) = u_i$ for each $i \in I$. We contend that

$$\{v_i, w_j\}_{i \in I,\ j \in J}$$

is a basis for V. This will obviously prove our assertion.

Let x be an element of V. Then there exist elements $\{a_i\}_{i\in I}$ of K almost all of which are 0 such that

$$f(x) = \sum_{i\in I} a_i u_i.$$

Hence $f(x - \sum a_i v_i) = f(x) - \sum a_i f(v_i) = 0$. Thus

$$x - \sum a_i v_i$$

is in the kernel of f, and there exist elements $\{b_j\}_{j\in J}$ of K almost all of which are 0 such that

$$x - \sum a_i v_i = \sum b_j w_j.$$

From this we see that $x = \sum a_i v_i + \sum b_j w_j$, and that $\{v_i, w_j\}$ generates V. It remains to be shown that the family $\{v_i, w_j\}$ is linearly independent. Suppose that there exist elements c_i, d_j such that

$$0 = \sum c_i v_i + \sum d_j w_j.$$

Applying f yields

$$0 = \sum c_i f(v_i) = \sum c_i u_i,$$

whence all $c_i = 0$. From this we conclude at once that all $d_j = 0$, and hence that our family $\{v_i, w_j\}$ is a basis for V over K, as was to be shown.

Corollary. *Let V be a vector space and W a subspace. Then*

$$\dim W \leqq \dim V.$$

If V is finite dimensional and $\dim W = \dim V$ *then* $W = V$.

Proof. Clear.

§6. *The dual space*

Let V be a vector space over a field K. We view K as a 1-dimensional space over itself. By the *dual space* V^* of V we shall mean the space $\mathrm{Hom}_K(V, K)$. Its elements will be called *functionals*. Thus a functional on V is a K-linear map $f : V \to K$. If $x \in V$ and $f \in V^*$, we sometimes denote $f(x)$ by $\langle x, f\rangle$. Keeping x fixed, we see that the symbol $\langle x, f\rangle$ as a function of $f \in V^*$ is K-linear in its second argument, and hence that x induces a linear map on V^*, which is 0 if and only if $x = 0$. Hence we get an injection $V \to V^{**}$ which is not always a surjection.

Let $\{x_i\}_{i\in I}$ be a basis of V. For each $i \in I$ let f_i be the unique functional such that $f_i(x_j) = \delta_{ij}$ (in other words, 1 if $i = j$ and 0 if $i \neq j$). Such a linear map exists by general properties of bases (Theorem 1 of §4).

THEOREM 5. *Let V be a vector space over the field K, of finite dimension n. Then* $\dim V^* = n$. *If $\{x_1, \ldots, x_n\}$ is a basis for V, and f_i is the functional such that $f_i(x_j) = \delta_{ij}$, then $\{f_1, \ldots, f_n\}$ is a basis for V^*.*

Proof. Let $f \in V^*$, and let $a_i = f(x_i)$ $(i = 1, \ldots, n)$. We have

$$f(a_1x_1 + \cdots + a_nx_n) = a_1f(x_1) + \cdots + a_nf(x_n).$$

Hence $f = a_1f_1 + \cdots + a_nf_n$, and we see that the f_i generate V^*. Furthermore, they are linearly independent, for if

$$a_1f_1 + \cdots + a_nf_n = 0$$

with $a_i \in K$, then evaluating the left-hand side on x_i yields

$$a_if_i(x_i) = 0,$$

whence $a_i = 0$ for all i. This proves our theorem.

COROLLARY. *When V is finite dimensional, then the map $V \to V^{**}$ which to each $x \in V$ associates the functional $f \mapsto \langle x, f \rangle$ on V^* is an isomorphism of V onto V^{**}.*

Proof. The map is an injective homomorphism. Its image is therefore a subspace of V^{**} of dimension n, and hence must be all of V^{**}.

Given a basis $\{x_i\}$ $(i = 1, \ldots, n)$ as in the theorem, we call the basis $\{f_i\}$ the *dual basis*. In terms of these bases, we can express an element A of V with coordinates $(a_1, \ldots, a_n)$, and an element B of V^* with coordinates $(b_1, \ldots, b_n)$, such that

$$A = a_1x_1 + \cdots + a_nx_n, \qquad B = b_1f_1 + \cdots + b_nf_n.$$

Then in terms of these coordinates, we see that

$$\langle A, B \rangle = a_1b_1 + \cdots + a_nb_n = A \cdot B$$

is the usual dot product of n-tuples.

Let V be a vector space over the field K, and let

$$0 \to W \xrightarrow{\lambda} V \xrightarrow{\varphi} U \to 0$$

be an exact sequence of K-linear maps. We contend that the induced sequence

$$0 \leftarrow \operatorname{Hom}_K(W, K) \leftarrow \operatorname{Hom}_K(V, K) \leftarrow \operatorname{Hom}_K(U, K) \leftarrow 0$$

i.e.

$$0 \leftarrow W^* \leftarrow V^* \leftarrow U^* \leftarrow 0$$

is also exact.

The exactness at all joints other than the one farthest to the left is a general fact having nothing to do with vector spaces, and holding for arbitrary modules. The essential point here is that we have the surjectivity of V^* on W^*. To see this, let g be a functional on W. There exists a subspace T of V such that

$$V = \lambda(W) + T$$

is a direct sum. We may view W in fact as a subspace of V since λ is an embedding. Any element of V then has a unique expression as a sum $w + t$ with $w \in W$ and $t \in T$. We define a functional f on V by putting $f(w + t) = g(w)$ for all $w \in W$ and $t \in T$. Then the restriction of f to $W(= \lambda(W))$ is g. This means that our induced map on the left is surjective.

Let V, V' be two vector spaces, and suppose given a mapping

$$V \times V' \to K$$

denoted by

$$(x, x') \mapsto \langle x, x' \rangle$$

for $x \in V$ and $x' \in V'$. We call the mapping *bilinear* if for each $x \in V$ the function $x' \mapsto \langle x, x' \rangle$ is linear, and similarly for each $x' \in V'$ the function $x \mapsto \langle x, x' \rangle$ is linear. An element $x \in V$ is said to be *orthogonal* (or *perpendicular*) to a subset S' of V' if $\langle x, x' \rangle = 0$ for all $x' \in S'$. We make a similar definition in the opposite direction. It is clear that the set of $x \in V$ orthogonal to S' is a subspace of V.

We define the *kernel* of the bilinear map on the left to be the subspace of V which is orthogonal to V', and similarly for the kernel on the right.

Given a bilinear map as above,

$$V \times V' \to K,$$

let W' be its kernel on the right and let W be its kernel on the left. Let x' be an element of V'. Then x' gives rise to a functional on V, by the rule $x \mapsto \langle x, x' \rangle$, and this functional obviously depends only on the coset of x' modulo W'; in other words, if $x_1' \equiv x_2' \pmod{W'}$, then the functionals $x \mapsto \langle x, x_1' \rangle$ and $x \mapsto \langle x, x_2' \rangle$ are equal. Hence we get a homomorphism

$$V' \to V^*$$

whose kernel is precisely W' by definition, whence an injective homomorphism

$$0 \to V'/W' \to V^*.$$

Since all the functionals arising from elements of V' vanish on W, we can view them as functionals on V/W, i.e. as elements of $(V/W)^*$. So we actually get an injective homomorphism

$$0 \to V'/W' \to (V/W)^*.$$

One could give a name to the homomorphism

$$g : V' \to V^*$$

such that

$$\langle x, x' \rangle = \langle x, g(x') \rangle$$

for all $x \in V$ and $x' \in V'$. However, it will usually be possible to describe it by an arrow and call it the induced map, or the natural map. Giving a name to it would tend to make the terminology heavier than necessary.

THEOREM 6. *Let $V \times V' \to K$ be a bilinear map, let W, W' be its kernels on the left and right respectively, and assume that V'/W' is finite dimensional. Then the induced homomorphism $V'/W' \to (V/W)^*$ is an isomorphism.*

Proof. By symmetry, we have an induced homomorphism

$$V/W \to (V'/W')^*$$

which is injective. Since

$$\dim(V'/W')^* = \dim V'/W'$$

it follows that V/W is finite dimensional. From the above injective homomorphism and the other, namely

$$0 \to V'/W' \to (V/W)^*,$$

we get the inequalities

$$\dim V/W \leqq \dim V'/W'$$

and

$$\dim V'/W' \leqq \dim V/W,$$

whence an equality of dimensions. Hence our homomorphisms are surjective and inverse to each other, thereby proving the theorem.

EXERCISES

1. Show that a module over a ring A is always a homomorphic image of a free module.

2. Generalize the dimension statement of Theorem 3 to free modules over a commutative ring. [*Hint:* Recall how an analogous statement was proved for free abelian groups, and use a maximal ideal instead of a prime number.]

3. Prove in detail that the conditions given in Proposition 3 for a sequence to split are equivalent. Show that a sequence $0 \to M' \xrightarrow{f} M \xrightarrow{g} M'' \to 0$ splits if and only if there exists a submodule N of M such that M is equal to the direct sum $\operatorname{Im} f \oplus N$, and that if this is the case, then N is isomorphic to M''. Complete all the details of the proof of Proposition 3.

4. Let A be a commutative ring and let M be an A-module. Let S be a multiplicative subset of A. Define $S^{-1}M$ in a manner analogous to the one we used to define $S^{-1}A$, and show that $S^{-1}M$ is an $S^{-1}A$-module.

5. Let A, S, be as in Exercise 4. If $0 \to M' \to M \to M'' \to 0$ is an exact sequence, show that $0 \to S^{-1\prime}M' \to S^{-1}M \to S^{-1}M'' \to 0$ is exact.

6. Let V be a vector space over a field K, and let U, W be subspaces. Show that

$$\dim U + \dim W = \dim(U + W) + \dim(U \cap W).$$

7. Let E and E_i $(i = 1, \ldots, m)$ be modules over a ring. Let $\varphi_i : E_i \to E$ and $\psi_i : E \to E_i$ be homomorphisms having the following properties:

$$\psi_i \circ \varphi_i = id, \qquad \psi_i \circ \varphi_j = 0 \quad \text{if} \quad i \neq j,$$

$$\sum_{i=1}^{m} \varphi_i \circ \psi_i = id.$$

Show that the map $x \mapsto (\psi_1 x, \ldots, \psi_m x)$ is an isomorphism of E onto the direct product of the E_i $(i = 1, \ldots, m)$, and that the map

$$(x_1, \ldots, x_m) \mapsto \varphi_1 x_1 + \cdots + \varphi_m x_m$$

is an isomorphism of this direct product onto E.

Conversely, if E is equal to a direct product (or sum) of submodules E_i $(i = 1, \ldots, m)$, if we let ψ_i be the inclusion of E_i in E, and φ_i the projection of E on E_i, then these maps satisfy the above-mentioned properties.

8. *Projective modules.* Let A be a ring. A module P is called *projective* if given any homomorphism $f : P \to M''$ and a surjective homomorphism $g : M \to M''$, there exists a homomorphism $h : P \to A$ such that the following diagram is commutative:

$$\begin{array}{ccccc} & {\scriptstyle h}\swarrow & P & & \\ & & \downarrow{\scriptstyle f} & & \\ M & \xrightarrow[g]{} & M'' & \longrightarrow & 0 \end{array}$$

Prove:

(a) A direct sum of modules is projective if and only if each summand is projective.

(b) A module P is projective if and only if there exists a module M such that $P \oplus M$ is free.

(c) Every module M is part of an exact sequence $0 \to N \to F \to M \to 0$ where F is projective (cf. Exercise 1).

(d) A module P is projective if and only if every exact sequence

$$0 \to M' \to M \to P \to 0$$

splits.

9. *Injective modules.* Let A be a ring. A module Q is called *injective* if given any module N and a submodule N', and a homomorphism $N' \to Q$, there exists an extension of this homomorphism to N, i.e. there exists a homomorphism $N \to Q$ making the following diagram commutative:

$$\begin{array}{ccccc} 0 \to & N' & \to & N \\ & \downarrow & \swarrow & \\ & Q & & \end{array}$$

Prove:

(a) A direct product of modules is injective if and only if each factor is injective.

(b) As a module over the integers $\mathbf{Z}$, the abelian group $\mathbf{Q}/\mathbf{Z}$ is injective. (Use Zorn's lemma.) If $\mathbf{R}$ is the group of real numbers, so is $\mathbf{R}/\mathbf{Z}$.

(c) Let Q be a module, over A. Assume that for every left ideal J of A, every homomorphism $\varphi : J \to Q$ can be extended to a homomorphism of A into Q. Then Q is injective. [*Hint:* Given $N' \subset N$ and $f : N' \to Q$, let $x_0 \in N$, $x_0 \notin N'$. Let J be the left ideal of $a \in A$ such that $ax_0 \in N'$. Let $\varphi(a) = f(ax_0)$ be extended to A, and extend f by the formula $f(x' + bx_0) = f(x') + \varphi(b)$ for $x' \in N'$ and $b \in A$. Use Zorn's lemma.]

(d) Let $A_0 = \text{Hom}_{\mathbf{Z}}(A, \mathbf{R}/\mathbf{Z})$, and make A_0 into an A-module: If $a \in A$ and $f \in A_0$, define $(af)(x) = f(xa)$. Show that A_0 is injective, using (c).

(e) Every module is a submodule of an injective module. [*Hint:* Let M be an A-module; let $x \in M$, $x \neq 0$. Show that there exists a homomorphism $f_x : M \to A_0$ such that $f_x(x) \neq 0$. First let J be the ideal of A annihilating x, and let $\varphi : A \to \mathbf{R}/\mathbf{Z}$ be a homomorphism vanishing on J such that $\varphi(1) \neq 0$. Construct f_x such that $f_x(x) = \varphi$. Then take the product of all f_x.]

(f) A module Q is injective if and only if every exact sequence

$$0 \to Q \to N \to M \to 0$$

splits.

10. Let A be an additive subgroup of Euclidean space $\mathbf{R}^n$, and assume that in every bounded region of space, there is only a finite number of elements of A. Show that A is a free abelian group on $\leqq n$ generators. [Hint: Induction on the maximal number of linearly independent elements of A over $\mathbf{R}$. Let $v_1, \ldots, v_m$ be a maximal set of such elements, and let A_0 be the subgroup of A contained in the $\mathbf{R}$-space generated by $v_1, \ldots, v_{m-1}$. By induction, one may assume that any element of A_0 is a linear integral combination of $v_1, \ldots, v_{m-1}$. Let S be the subset of elements $v \in A$ of the form $v = a_1v_1 + \cdots + a_mv_m$ with real coefficients a_i satisfying

$$\begin{aligned} &0 \leqq a_i < 1 \quad \text{if} \quad i = 1, \ldots, m-1 \\ &0 \leqq a_m \leqq 1. \end{aligned}$$

If v'_m is an element of S with the smallest $a_m \neq 0$, show that $\{v_1, \ldots, v_{m-1}, v'_m\}$ is a basis of A over $\mathbf{Z}$.

CHAPTER IV

Homology

§1. Complexes

Let A be a ring. By an *open complex* of A-modules, one means a sequence of modules and homomorphisms $\{(E_i, d_i)\}$,

$$\to E_{i-1} \xrightarrow{d_{i-1}} E_i \xrightarrow{d_i} E_{i+1} \to$$

where i ranges over all integers, and d_i maps E_i into E_{i+1}, and such that

$$d_i \circ d_{i-1} = 0$$

for all i.

One frequently considers a finite sequence of homomorphisms, say

$$E_1 \to \cdots \to E_r$$

such that the composite of two successive ones is 0, and one can make this sequence into a complex by inserting 0 at each end:

$$\to 0 \to 0 \to E_1 \to \cdots \to E_r \to 0 \to 0 \to$$

A *closed complex* of A-modules is a sequence of modules and homomorphisms $\{(E_i, d_i)\}$ where i ranges over the set of integers mod n for some $n \geqq 2$ and otherwise satisfying the same properties as above. Thus a closed complex looks like this:

$$E_1 \to E_2 \to \cdots \to E_n$$

(with an arrow from E_n back to E_1)

We call n the *length* of the closed complex.

Without fear of confusion, one can omit the index i on d_i and write just d. We also write (E, d) for the complex $\{(E_i, d_i)\}$, or even more briefly, we write simply E.

Let (E, d) and (E', d') be complexes (both open or both closed). Let r be an integer. A *morphism* (of complexes)

$$f : (E', d') \to (E, d)$$

of *degree* r is a sequence

$$f_i : E'_i \to E_{i+r}$$

of homomorphisms such that for all i the following diagram is commutative:

$$\begin{array}{ccc} E'_{i-1} & \xrightarrow{f_{i-1}} & E_{i-1+r} \\ d' \downarrow & & \downarrow d \\ E'_i & \xrightarrow[f_i]{} & E_{i+r} \end{array}$$

Just as we write d instead of d_i, we shall also write f instead of f_i. If the complexes are closed, we define a morphism from one into the other only if they have the same length.

It is clear that complexes form a category.

It will be useful to have another notion to deal with objects indexed by a monoid. Let G be a monoid, which we assume commutative and additive to fit the applications we have in mind here. Let $\{M_i\}_{i \in G}$ be a family of modules indexed by G. The direct sum

$$M = \coprod_{i \in G} M_i$$

will be called the *G-graded module associated with the family* $\{M_i\}_{i \in G}$. Let $\{M_i\}_{i \in G}$ and $\{M'_i\}_{i \in G}$ be families indexed by G, and let M, M' be their associated G-graded modules. Let $r \in G$. By a *G-graded morphism* $f : M' \to M$ of degree r we shall mean a homomorphism such that f maps M'_i into M_{i+r} for each $i \in G$ (identifying M_i with the corresponding submodule of the direct sum on the i-th component). Thus f is nothing else than a family of homomorphisms $f_i : M'_i \to M_{i+r}$.

If (E, d) is a complex we may view E as a G-graded module (taking the direct sum of the components of the complex), and we may view d as a G-graded morphism of degree 1, letting G be $\mathbf{Z}$ or $\mathbf{Z}/n\mathbf{Z}$.

Conversely, if G is $\mathbf{Z}$ or $\mathbf{Z}/n\mathbf{Z}$, one may view a G-graded module as a complex, by defining d to be the zero map.

For simplicity, we shall often omit the prefix "G-graded" in front of the word "morphism", when dealing with G-graded morphisms.

§2. Homology sequence

Let (E, d) be a complex. We let

$$Z_i(E) = \operatorname{Ker} d_i$$

and call $Z_i(E)$ the module of *i-cycles*. We let

$$B_i(E) = \operatorname{Im} d_{i-1}$$

and call $B_i(E)$ the module of *i-boundaries*. We frequently write Z_i and B_i instead of $Z_i(E)$ and $B_i(E)$ respectively. We let

$$H_i(E) = Z_i/B_i = \text{Ker } d_i/\text{Im } d_{i-1},$$

and call $H_i(E)$ the i-th *homology group* of the complex. The graded module associated with the family $\{H_i\}$ will be denoted by $H(E)$, and will be called the *homology* of E. One sometimes writes $H_*(E)$ instead of $H(E)$.

If $f : E' \to E$ is a morphism of complexes, say of degree 0, then we get an induced canonical homomorphism of degree 0,

$$f_* : H(E') \to H(E),$$

on the homology. This is immediately clear from the commutative diagrams describing the morphism of complexes. Indeed, the reader will verify at once that $f_i(Z_i') \subset Z_i$ and $f_i(B_i') \subset B_i$, whence the induced homomorphism $Z_i'/B_i' \to Z_i/B_i$. (The reader should carry out the details once, and only once, in his life.) Thus H is a functor from the category of complexes into the category of graded modules. We could write $H(f)$ instead of f_*, and also $H_i(f)$ or f_{i*} for the induced map on H_i'.

We shall consider short exact sequences of complexes with morphisms of degree 0:

$$0 \to E' \xrightarrow{f} E \xrightarrow{g} E'' \to 0,$$

which written out in full look like this (writing d instead of d' or d''):

$$\begin{array}{ccccccccc}
 & & \downarrow & & \downarrow & & \downarrow & & \\
0 & \to & E'_{i-1} & \longrightarrow & E_{i-1} & \longrightarrow & E''_{i-1} & \to & 0 \\
 & & \downarrow & & \downarrow & & \downarrow & & \\
0 & \to & E'_i & \xrightarrow{f} & E_i & \xrightarrow{g} & E''_i & \to & 0 \\
 & & d\downarrow & & d\downarrow & & d\downarrow & & \\
0 & \to & E'_{i+1} & \xrightarrow{f} & E_{i+1} & \xrightarrow{g} & E''_{i+1} & \to & 0 \\
 & & \downarrow & & \downarrow & & \downarrow & & \\
0 & \to & E'_{i+2} & \longrightarrow & E_{i+2} & \longrightarrow & E''_{i+2} & \to & 0 \\
 & & \downarrow & & \downarrow & & \downarrow & &
\end{array}$$

One can define a morphism

$$\delta : H(E'') \to H(E')$$

of degree 1, *in other words, a family of homomorphisms*

$$\delta_i : H_i'' \to H_{i+1}'$$

as follows.

Let z'' be in Z_i''. Since g is surjective, there exists an element $z \in E_i$ such that $gz = z''$. We now move vertically downwards by d, and take

dz. Using commutativity $gd = dg$ we find that $gdz = 0$, whence dz is in the kernel of g in E_{i+1}. By exactness, there exists an element $z' \in E'_{i+1}$ such that $fz' = dz$. In brief, we can write

$$z' = f^{-1}\, dg^{-1}z''.$$

We leave it as a routine exercise to the reader to verify that z' is an element of Z'_{i+1}, in other words, is a cycle, and that its class modulo B'_{i+1} is independent of the choice of element z such that $gz = z''$. Furthermore, the map

$$z \mapsto f^{-1}\, dg^{-1}z \quad \text{modulo } B'_{i+1}$$

induces a homomorphism

$$\delta_i : Z''_i/B''_i \to Z'_{i+1}/B'_{i+1}$$

which is the i-th component of our desired morphism δ.

THEOREM 1. *Let*

$$0 \to E' \xrightarrow{f} E \xrightarrow{g} E'' \to 0$$

be an exact sequence of complexes with f, g *of degree* 0. *Then the sequence*

$$\begin{array}{ccc} H(E') & \xrightarrow{f_*} & H(E) \\ {}^{\delta}\nwarrow & & \swarrow{}^{g_*} \\ & H(E'') & \end{array}$$

is exact.

Proof. The proof is essentially routine and consists in chasing around diagrams. It should be carried out in full detail by the reader who wishes to acquire a feeling for this type of triviality. As an example, we shall prove that

$$\operatorname{Ker} \delta \subset \operatorname{Im} g_*.$$

We use the notation preceding our theorem, describing the definition of δ. If z'' represents a class whose image under δ is 0, then this means that z' is a boundary, in other words, there exists an element $u' \in E'_i$ such that $z' = du'$. Then using the notation defining δ, we have

$$dz = fz' = fdu' = dfu'$$

by commutativity. Hence

$$d(z - fu') = 0,$$

and $z - fu'$ is a cycle in E_i. But $g(z - fu') = gz = z''$. This means that the class of z'' is in the image of g_*, as desired.

If one writes out in full the homology sequence in the theorem, then it looks like this:

$$\xrightarrow{\delta} H'_i \to H_i \to H''_i \xrightarrow{\delta} H'_{i+1} \to H_{i+1} \to H''_{i+1} \xrightarrow{\delta}$$

It is clear that our map δ is functorial (in an obvious sense), and hence that our whole structure (H, δ) is a functor from the category of short exact sequences of complexes into the category of complexes.

§3. *Euler characteristic*

We continue to consider A-modules. Let Γ be an abelian group, written additively. Let φ be a rule which to certain modules associates an element of Γ, subject to the following condition:

If $0 \to M' \to M \to M'' \to 0$ *is exact, then* $\varphi(M)$ *is defined if and only if* $\varphi(M')$ *and* $\varphi(M'')$ *are defined, and in that case, we have*

$$\varphi(M) = \varphi(M') + \varphi(M'').$$

Furthermore $\varphi(0)$ *is defined and equal to* 0.

Such a rule φ will be called an *Euler-Poincaré mapping* on the category of A-modules. If M' is isomorphic to M, then from the exact sequence

$$0 \to M' \to M \to 0 \to 0$$

we conclude that $\varphi(M')$ is defined if $\varphi(M)$ is defined, and that $\varphi(M') = \varphi(M)$. Thus if $\varphi(M)$ is defined for a module M, φ is defined on every submodule and factor module of M. In particular, if we have an exact sequence of modules

$$M' \to M \to M''$$

and if $\varphi(M')$ and $\varphi(M'')$ are defined, then so is $\varphi(M)$, as one sees at once by considering the kernel and image of our two maps, and using the definition.

Examples. We could let $A = \mathbf{Z}$, and let φ be defined for all finite abelian groups, and be equal to the order of the group. The value of φ is in the multiplicative group of positive rational numbers.

As another example, we consider the category of vector spaces over a field k. We let φ be defined for finite dimensional spaces, and be equal to the dimension. The values of φ are then in the additive group of integers.

Consider the general case again, and let E be a complex, such that almost all H_i are equal to 0. Assume that E is an open complex. Let φ be an Euler-Poincaré mapping on the category of modules (i.e. A-modules).

We define the *Euler-Poincaré characteristic* $\chi_\varphi(E)$ (or more briefly the Euler characteristic) with respect to φ, to be

$$\chi_\varphi(E) = \sum (-1)^i \varphi(H_i)$$

provided $\varphi(H_i)$ is defined for all H_i, in which case we say that χ_φ is *defined* for the complex E.

If E is a closed complex, we select a definite order $(E_1, \ldots, E_n)$ for the integers mod n and define the Euler characteristic by the formula

$$\chi_\varphi(E) = \sum_{i=1}^{n} (-1)^i \varphi(H_i)$$

provided again all $\varphi(H_i)$ are defined. If n is even, then the ordering we take for the integers mod n is irrelevant.

For an example, the reader may refer to Exercise 14 of Chapter I.

One may view H as a complex, defining d to be the zero map. In that case, we see that $\chi_\varphi(H)$ is the alternating sum given above. More generally:

THEOREM 2. *Let F be a complex, which is of even length if it is closed. Assume that $\varphi(F_i)$ is defined for all i, $\varphi(F_i) = 0$ for almost all i, and $H_i(F) = 0$ for almost all i. Then $\chi_\varphi(F)$ is defined, and*

$$\chi_\varphi(F) = \sum_i (-1)^i \varphi(F_i).$$

Proof. Let Z_i and B_i be the groups of i-cycles and i-boundaries in F_i respectively. We have an exact sequence

$$0 \to Z_i \to F_i \to B_{i+1} \to 0.$$

Hence $\chi_\varphi(H)$ is defined, and

$$\varphi(F_i) = \varphi(Z_i) + \varphi(B_{i+1}).$$

Taking the alternating sum, our conclusion follows at once.

A complex whose homology is trivial is called *acyclic*.

COROLLARY. *Let F be an acyclic complex, such that $\varphi(F_i)$ is defined for all i, and equal to 0 for almost all i. If F is closed, we assume that F has even length. Then*

$$\chi_\varphi(F) = 0.$$

In many applications, an open complex F is such that $F_i = 0$ for almost all i, and one can then treat this complex as a closed complex by defining an additional map going from a zero on the far right to a zero on the far left. Thus in this case, the study of such an open complex is reduced to the study of a closed complex.

THEOREM 3. *Let*

$$0 \to E' \to E \to E'' \to 0$$

be an exact sequence of complexes, with morphisms of degree 0. *If the complexes are closed, assume that their length is even. Let* φ *be an Euler-Poincaré mapping on the category of modules. If* χ_φ *is defined for two of the above three complexes, then it is defined for the third, and we have*

$$\chi_\varphi(E) = \chi_\varphi(E') + \chi_\varphi(E'').$$

Proof. We have an exact homology sequence

$$\to H''_{i-1} \to H'_i \to H_i \to H''_i \to H'_{i+1} \to$$

This homology sequence is nothing but a complex whose homology is trivial. Furthermore, each homology group belonging say to E is between homology groups of E' and E''. Hence if χ_φ is defined for E' and E'' it is defined for E. Similarly for the other two possibilities. If our complexes are closed of even length n, then this homology sequence has even length $3n$. We can therefore apply the corollary of Theorem 2 to get what we want.

For certain applications, it is convenient to construct a universal Euler mapping. Let $\mathfrak{A}$ be the set of isomorphism classes of certain modules. If E is a module, let $[E]$ denote its isomorphism class. We require that $\mathfrak{A}$ satisfy the Euler-Poincaré condition, i.e. if we have an exact sequence

$$0 \to E' \to E \to E'' \to 0,$$

then $[E]$ is in $\mathfrak{A}$ if and only if $[E']$ and $[E'']$ are in $\mathfrak{A}$. Furthermore, the zero module is in $\mathfrak{A}$. *We contend that there is a map*

$$\gamma : \mathfrak{A} \to K(\mathfrak{A})$$

of $\mathfrak{A}$ *into an abelian group* $K(\mathfrak{A})$ *having the universal property with respect to Euler-Poincaré maps defined on* $\mathfrak{A}$.

To construct this, let $F_{ab}(\mathfrak{A})$ be the free abelian group generated by the set of such $[E]$. Let B be the subgroup generated by all elements of type

$$[E] - [E'] - [E'']$$

where

$$0 \to E' \to E \to E'' \to 0$$

is an exact sequence whose members are in $\mathfrak{A}$. We let $K(\mathfrak{A})$ be the factor group $F_{ab}(\mathfrak{A})/B$, and let $\gamma : \mathfrak{A} \to K(\mathfrak{A})$ be the natural map. It is clear that γ has the universal property.

We observe the similarity of construction with the Grothendieck group of a monoid. In fact, the present group is known as the *Euler-Grothendieck group of* $\mathfrak{A}$.

Important generalization. It is clear from what precedes that most of what we have done is purely arrow-theoretic. In fact, we need merely the notion of kernel and cokernel (factor module) to define homology. We used elements only to define δ.

One can axiomatize the special notion of a category in which all the preceding arguments are valid. Consider first a category $\mathfrak{A}$ such that $\mathrm{Mor}(E, F)$ is an abelian group for each pair of objects E, F of $\mathfrak{A}$, satisfying the following two conditions:

AB 1. The law of composition of morphisms is bilinear, and there exists a zero object 0, i.e. such that $\mathrm{Mor}(0, E)$ and $\mathrm{Mor}(E, 0)$ have precisely one element for each object E.

AB 2. Finite products and finite coproducts exist in the category.

Then we say that $\mathfrak{A}$ is an *additive category*.

Given a morphism $E \xrightarrow{f} F$ in $\mathfrak{A}$, we define a *kernel* of f to be a morphism $E' \to E$ such that for all objects X in the category, the following sequence is exact:

$$0 \to \mathrm{Mor}(X, E') \to \mathrm{Mor}(X, E) \to \mathrm{Mor}(X, F).$$

We define a *cokernel* for f to be a morphism $F \to F''$ such that for all objects X in the category, the following sequence is exact:

$$0 \leftarrow \mathrm{Mor}(E, X) \leftarrow \mathrm{Mor}(F, X) \leftarrow \mathrm{Mor}(F'', X).$$

It is immediately verified that kernels and cokernels are universal in a suitable category, and hence uniquely determined up to a unique isomorphism if they exist.

AB 3. Kernels and cokernels exist.

AB 4. If $f : E \to F$ is a morphism whose kernel is 0, then f is the kernel of its cokernel. If $f : E \to F$ is a morphism whose cokernel is 0, then f is the cokernel of its kernel. A morphism whose kernel and cokernel are 0 is an isomorphism.

A category $\mathfrak{A}$ satisfying the above four axioms is called an *abelian category*.

For instance, complexes of modules form an abelian category, since it is clear how to define, say, the kernel of a morphism. In topology, vector bundles form an abelian category.

§4. The Jordan-Hölder theorem

We begin by some purely group-theoretic results. As with the elementary isomorphism theorems, these have analogues for modules, which will be stated afterwards.

Butterfly Lemma. (*Zassenhaus*) *Let U, V be subgroups of a group. Let u, v be normal subgroups of U and V respectively. Then*

$$\begin{aligned} u(U \cap v) &\quad \textit{is normal in} \quad u(U \cap V), \\ (u \cap V)v &\quad \textit{is normal in} \quad (U \cap V)v, \end{aligned}$$

and the factor groups are isomorphic, i.e.

$$u(U \cap V)/u(U \cap v) \approx (U \cap V)v/(u \cap V)v.$$

Proof. The combination of groups and factor groups becomes clear if one visualizes the following diagram of subgroups (which gives its name to the lemma):

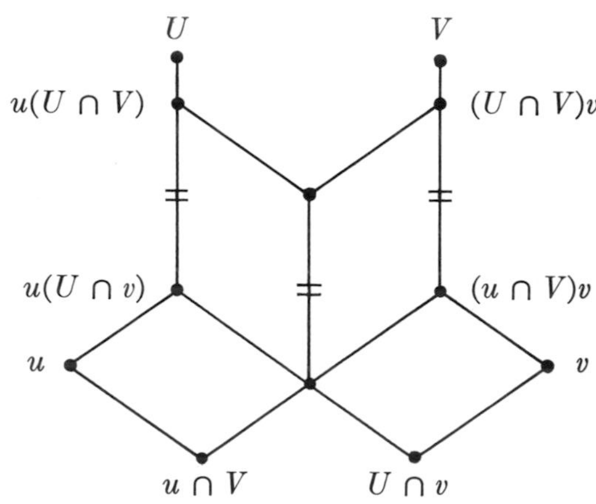

In this diagram, we are given U, u, V, v. All the other points in the diagram correspond to certain groups which can be determined as follows. The intersection of two line segments going downwards represents the intersection of groups. Two lines going upwards meet in a point which represents the product of two subgroups (i.e. the smallest subgroup containing both of them).

We consider the two parallelograms representing the wings of the butterfly, and we shall prove that the opposite sides of the parallelograms are equal.

In fact, the vertical side common to both parallelograms has $U \cap V$ as its top end point, and $(u \cap V)(U \cap v)$ as its bottom end point. We have an isomorphism

$$(U \cap V)/(u \cap V)(U \cap v) \approx u(U \cap V)/u(U \cap v).$$

This is obtained from the isomorphism theorem

$$H/(H \cap N) \approx HN/N$$

by setting $H = U \cap V$ and $N = u(U \cap v)$. Thus the middle vertical side is equal to the vertical side on the left. It is also equal to the vertical side on the right by symmetry, and since equals can be substituted for equals, our lemma is proved.

Let G be a group, and let

$$G = G_1 \supset G_2 \supset \cdots \supset G_r = \{e\},$$
$$G = H_1 \supset H_2 \supset \cdots \supset H_s = \{e\}$$

be normal towers of subgroups, ending with the trivial group. We shall say that these towers are *equivalent* if $r = s$ and if there exists a permutation of the indices $i = 1, \ldots, r - 1$, written $i \mapsto i'$, such that

$$G_i/G_{i+1} \approx H_{i'}/H_{i'+1}.$$

In other words, the sequences of factor groups in our two towers are the same, up to isomorphisms, and a permutation of the indices.

THEOREM 4. (*Schreier*) *Let G be a group. Two normal towers of subgroups ending with the trivial group have equivalent refinements.*

Proof. Let the two towers be as above. For each $i = 1, \ldots, r - 1$ and $j = 1, \ldots, s$ we define

$$G_{ij} = G_{i+1}(H_j \cap G_i).$$

Then $G_{is} = G_{i+1}$, and we have a refinement of the first tower:

$$G = G_{11} \supset G_{12} \supset \cdots \supset G_{1,s-1} \supset G_2$$
$$= G_{21} \supset G_{22} \supset \cdots \supset G_{r-1,1} \supset \cdots \supset G_{r-1,s-1} \supset \{e\}.$$

Similarly, we define

$$H_{ji} = H_{j+1}(G_i \cap H_j),$$

for $j = 1, \ldots, s - 1$ and $i = 1, \ldots, r$. This yields a refinement of the second tower. By the butterfly lemma, for $i = 1, \ldots, r - 1$ and $j = 1, \ldots, s - 1$ we have isomorphisms

$$G_{ij}/G_{i,j+1} \approx H_{ji}/H_{j,i+1}.$$

We view each one of our refined towers as having $(r-1)(s-1)+1$ elements, namely G_{ij} $(i = 1, \ldots, r-1;\ j = 1, \ldots, s-1)$ and $\{e\}$ in the first case, H_{ji} and $\{e\}$ in the second case. The preceding isomorphism for each pair of indices (i, j) shows that our refined towers are equivalent, as was to be proved.

A group G is said to be *simple* if it is non-trivial, and has no normal subgroups other than $\{e\}$ and G itself.

THEOREM 5. (*Jordan-Hölder*) *Let G be a group, and let*

$$G = G_1 \supset G_2 \supset \cdots \supset G_r = \{e\}$$

be a normal tower such that each group G_i/G_{i+1} is simple, and $G_i \neq G_{i+1}$ for $i = 1, \ldots, r-1$. Then any other normal tower of G having the same properties is equivalent to this one.

Proof. Given any refinement $\{G_{ij}\}$ as before for our tower, we observe that for each i, there exists precisely one index j such that $G_i/G_{i+1} = G_{ij}/G_{i,j+1}$. Thus the sequence of non-trivial factors for the original tower, or the refined tower, is the same. This proves our theorem.

Just as with the elementary isomorphism theorems for groups, we have the analogues of Theorems 4 and 5 for modules. Of course in the case of modules, we don't have to worry about the normality of submodules.

If M is a module (over a ring A), then a sequence of submodules

$$M = M_1 \supset M_2 \supset \cdots \supset M_r = 0$$

is also called a *finite filtration*, and we call r the *length* of the filtration. A module M is said to be *simple* if it does not contain any submodule other than 0 and M itself, and if $M \neq 0$. A filtration is said to be *simple* if each M_i/M_{i+1} is simple. *The Jordan-Hölder theorem asserts that two simple filtrations of a module are equivalent.*

A module M is said to be of *finite length* if it is 0, or if it admits a simple (finite) filtration. By the Jordan-Hölder theorem, the length of such a simple filtration is then uniquely determined, and is called the *length of the module.* In the language of Euler characteristics, the Jordan-Hölder theorem can be reformulated as follows:

THEOREM 6. *Let φ be a rule which to each simple module associates an element of a commutative group Γ, and such that if $M \approx M'$ then $\varphi(M) = \varphi(M')$. Then φ has a unique extension to an Euler-Poincaré mapping defined on all modules of finite length.*

Proof. Given a simple filtration

$$M = M_1 \supset M_2 \supset \cdots \supset M_r = 0$$

we define

$$\varphi(M) = \sum_{i=1}^{r-1} \varphi(M_i/M_{i+1}).$$

The Jordan-Hölder theorem shows immediately that this is well-defined, and that this extension of φ is an Euler-Poincaré map.

In particular, we see that the length function is the Euler-Poincaré map taking its values in the additive group of integers, and having the value 1 for any simple module.

Exercises

Take any book on homological algebra, and prove all the theorems without looking at the proofs given in that book.

Homological algebra was invented by Eilenberg-MacLane. General category theory (i.e. the theory of arrow-theoretic results) is generally known as *abstract nonsense* (the terminology is due to Steenrod).

CHAPTER V

Polynomials

§1. Free algebras

Let A be a commutative ring. An *A-algebra* (or an algebra over A) is a module E together with a bilinear map $E \times E \to E$. *In all of this book, unless otherwise specified, we shall only deal with the following special type of algebra.* Let $f : A \to B$ be a ring-homomorphism such that $f(A)$ is contained in the center of B, i.e. $f(a)$ commutes with every element of B for every $a \in A$. Then we may view B as an A-module, defining the operation of A on B by the map

$$(a, b) \mapsto f(a)b$$

for all $a \in A$ and $b \in B$. The axioms for a module are trivially satisfied, and the multiplicative law of composition $B \times B \to B$ is clearly bilinear (i.e. A-bilinear). Thus unless otherwise specified, by an algebra over A, we shall always mean a ring-homomorphism as above. We say that the algebra is *finitely generated* if B is finitely generated as a ring over $f(A)$.

Let G be a *multiplicative* monoid and A a commutative ring. Let $\mathcal{C}$ be the category whose objects are triples (φ, f, B), where $f : A \to B$ is an A-algebra, and $\varphi : G \to B$ is a multiplicative monoid-homomorphism. If (φ', f', B') is another object in $\mathcal{C}$, a morphism in $\mathcal{C}$ from (φ, f, B) to (φ', f', B') is a ring-homomorphism $h : B \to B'$ making the following diagram commutative:

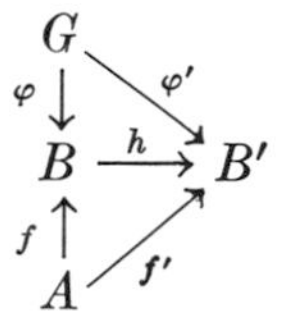

$$\begin{array}{ccc} G & & \\ {\scriptstyle\varphi}\downarrow & \searrow{\scriptstyle\varphi'} & \\ B & \xrightarrow{h} & B' \\ {\scriptstyle f}\uparrow & \nearrow{\scriptstyle f'} & \\ A & & \end{array}$$

A universal (repelling) object in $\mathcal{C}$ is called a *free* (A, G)*-algebra*, or a *free G-algebra over A*. We shall construct one explicitly.

Let $A[G]$ be the set of all maps $\alpha : G \to A$ such that $\alpha(x) = 0$ for almost all $x \in G$. We define addition in $A[G]$ to be the ordinary addition of mappings into an abelian (additive) group. If $\alpha, \beta \in A[G]$, we define their product $\alpha\beta$ by the rule

$$(\alpha\beta)(t) = \sum_{xy=t} \alpha(x)\beta(y).$$

The sum is taken over all pairs (x, y) with $x, y \in G$ such that $xy = t$. This sum is actually finite, because there is only a finite number of pairs of elements $(x, y) \in G \times G$ such that $\alpha(x)\beta(y) \neq 0$. We also see that $(\alpha\beta)(t) = 0$ for almost all t, and thus belongs to our set $A[G]$.

The axioms for a ring are trivially verified. We shall carry out the proof of associativity as an example. Let $\alpha, \beta, \gamma \in A[G]$. Then

$$\begin{aligned}
((\alpha\beta)\gamma)(t) &= \sum_{xy=t} (\alpha\beta)(x)\gamma(y) \\
&= \sum_{xy=t} \left[\sum_{uv=x} \alpha(u)\beta(v)\right]\gamma(y) \\
&= \sum_{xy=t} \left[\sum_{uv=x} \alpha(u)\beta(v)\gamma(y)\right] \\
&= \sum_{\substack{(u,v,y) \\ uvy=t}} \alpha(u)\beta(v)\gamma(y),
\end{aligned}$$

this last sum being taken over all triples (u, v, y) whose product is t. This last sum is now symmetric, and if we had computed $(\alpha(\beta\gamma))(t)$, we would have found this sum also. This proves associativity.

The unit element of $A[G]$ is the function δ such that $\delta(e) = 1$ and $\delta(x) = 0$ for all $x \in G$, $x \neq e$. It is trivial to verify that $\alpha = \delta\alpha = \alpha\delta$ for all $\alpha \in A[G]$.

We shall now adopt a notation which will make the structure of $A[G]$ clearer. Let $a \in A$ and $x \in G$. We denote by $a \cdot x$ (and sometimes also by ax) the function whose value at x is a, and whose value at y is 0 if $y \neq x$. Then an element $\alpha \in A[G]$ can be written as a sum

$$\alpha = \sum_{x \in G} \alpha(x) \cdot x.$$

Indeed, if $\{a_x\}_{x \in G}$ is a set of elements of A almost all of which are 0, and we set

$$\beta = \sum_{x \in G} a_x \cdot x,$$

then for any $y \in G$ we have $\beta(y) = a_y$ (directly from the definitions). This also shows that a given element α admits a unique expression as a sum $\sum a_x \cdot x$.

There is a natural way to make $A[G]$ into an A-module. If $a \in A$ and $\alpha \in A[G]$ is written as a sum $\sum a_x \cdot x$, we define $a\alpha$ to be the element

$\sum (aa_x) \cdot x$. It is then clear that all the axioms of a module are satisfied, and in fact that the set of elements $\{1 \cdot x\}_{x \in G}$ forms a basis of $A[G]$ over A.

With our present notation, multiplication can be written

$$\left(\sum_{x \in G} a_x \cdot x\right)\left(\sum_{y \in G} b_y \cdot y\right) = \sum_{x,y} a_x b_y \cdot xy$$

and addition can be written

$$\sum_{x \in G} a_x \cdot x + \sum_{x \in G} b_x \cdot x = \sum_{x \in G} (a_x + b_x) \cdot x,$$

which looks the way we want it to look. Note that the unit element of $A[G]$ is simply $1 \cdot e$.

Let $\varphi_0 : G \to A[G]$ be the map given by $\varphi_0(x) = 1 \cdot x$. It is immediately verified that φ_0 is a multiplicative monoid-homomorphism, and is in fact injective, i.e. an embedding.

Let $f_0 : A \to A[G]$ be the map given by

$$f_0(a) = a \cdot e.$$

It is immediately verified that f_0 is a ring-homomorphism, and is also an embedding. Thus we have made $A[G]$ into an A-algebra, and one sees at once that the structure of A-module for $A[G]$ as an A-algebra is the same as the structure of A-module which we described previously.

The triple $(\varphi_0, f_0, A[G])$ *is a free* (A, G)*-algebra.* This assertion is a special case of the next proposition.

PROPOSITION 1. *Let $f_0 : A \to B$ be an A-algebra. Let G be a multiplicative submonoid of B, and assume that G is a basis of B as a module over A. Given an A-algebra $f : A \to C$ and a monoid-homomorphism $\varphi : G \to C$, there exists a unique ring-homomorphism $h : B \to C$ such that the diagram is commutative:*

$$\begin{array}{ccc} B & \xrightarrow{h} & C \\ {\scriptstyle f_0}\uparrow & \nearrow{\scriptstyle f} & \\ A & & \end{array}$$

and such that the restriction of h to G is equal to φ.

Proof. For each $x \in G$ and $a \in A$ we write $a \cdot x$ instead of $f_0(a)x$. An element $\alpha \in A[G]$ has a unique expression as a sum

$$\alpha = \sum_{x \in G} a_x \cdot x$$

with $a_x \in A$ because G is a basis of B over A. As we saw in our discussion of bases for modules, there exists a unique module-homomorphism $h : B \to C$ making our diagram commutative, namely the map such that

$$h(\alpha) = \sum_{x \in G} f(a_x)\varphi(x).$$

Furthermore, if

$$\beta = \sum_{y \in G} b_y \cdot y,$$

then

$$\alpha\beta = \sum_{z \in G} \left(\sum_{xy=z} a_x b_y \right) \cdot z$$

and

$$\begin{aligned} h(\alpha\beta) &= \sum_{z \in G} h\left(\sum_{xy=z} a_x b_y \right) \varphi(z) \\ &= \sum_{z \in G} \left(\sum_{xy=z} h(a_x)h(b_y) \right) \varphi(z) \\ &= h(\alpha)h(\beta). \end{aligned}$$

Since h restricted to G is equal to φ, we have $h(1) = 1$. Hence h is also a ring-homomorphism, and our proposition is proved.

To deduce from Proposition 1 that $(\varphi_0, f_0, A[G])$ is a free (A, G)-algebra, we let $B = A[G]$, and we identify G with its image in $A[G]$ under the embedding φ_0.

From now on, we shall write ax instead of $a \cdot x$ without fear of confusion. We shall call $A[G]$ *the monoid algebra of* G *over* A. We call φ_0, f_0 the *canonical maps.*

We shall obtain the polynomial algebra as a special case in the next section. When G is a group, the group algebra $A[G]$ will be discussed at greater length later in this book.

Our monoid algebra satisfies one other universal property:

Proposition 2. *Let* $\varphi : G \to G'$ *be a monoid-homomorphism, and let* $f : A \to A'$ *be a ring-homomorphism such that* A, A' *are both commutative. Then there exists a unique ring-homomorphism*

$$h : A[G] \to A'[G']$$

making the following diagram commutative:

$$\begin{array}{ccc} G & \xrightarrow{\varphi} & G' \\ \downarrow & & \downarrow \varphi_0' \\ A[G] & \xrightarrow{h} & A'[G'] \\ \uparrow & & \uparrow f_0' \\ A & \xrightarrow[f]{} & A' \end{array}$$

(The vertical maps are the canonical maps.)

Proof. This is a direct application of Proposition 1, putting $C = A'[G']$, considering the homomorphism

$$\varphi_0' \circ \varphi \qquad \text{and} \qquad f_0' \circ f$$

and applying Proposition 1 to them.

§2. *Definition of polynomials*

Let S be a set and let $\mathbf{N}$ be the additive monoid of integers $\geqq 0$ (i.e. the monoid of natural numbers). We denote by

$$\mathbf{N}\langle S\rangle$$

the set of functions $S \to \mathbf{N}$ which are 0 for almost all elements of S. (This is essentially the same construction which we performed to get free abelian groups, and in the present case we get the free abelian monoid. However, we shall write it multiplicatively.) If $x \in S$ and $i \in \mathbf{N}$, we denote by x^i the function which takes the value i at x, and 0 at $y \neq x$. If φ, ψ are two functions as above, we define their *product* $\varphi\psi$ by the rule

$$(\varphi\psi)(x) = \varphi(x) + \psi(x).$$

Then $\mathbf{N}\langle S\rangle$ is a multiplicative monoid whose unit element is the zero function. Every $\varphi \in \mathbf{N}\langle S\rangle$ has a unique expression as a product

$$\prod_{x \in S} x^{\nu(x)}$$

where $\nu : S \to \mathbf{N}$ is a map such that $\nu(x) = 0$ for almost all x. Such a product will be called a *primitive monomial*, and will be sometimes denoted by $M_{(\nu)}(S)$, or simply $M_{(\nu)}$.

We have an injection $j_S : S \to \mathbf{N}\langle S\rangle$ given by $x \mapsto x^1$, whose image generates $\mathbf{N}\langle S\rangle$ as a monoid. We note that if n is an integer $\geqq 0$, then

$$(x^1)^n = x^1x^1 \cdots x^1$$

is equal to x^n, i.e. our notation is compatible with the notation used for the product of functions.

We observe that if

$$\prod_{x \in S} x^{\nu(x)} \qquad \text{and} \qquad \prod_{x \in S} x^{\mu(x)}$$

are primitive monomials, then their product is given by

$$\prod_{x \in S} x^{\nu(x)+\mu(x)}.$$

Just as with abelian groups, one has the universal property: Let G be a commutative monoid. Given a map $\lambda : S \to G$ there exists a unique monoid-homomorphism $\mathbf{N}\langle S\rangle \to G$ making the following diagram commutative:

$$\begin{array}{ccc} S & \xrightarrow{\lambda} & G \\ & {\scriptstyle j_S}\searrow \quad \nearrow & \\ & \mathbf{N}\langle S\rangle & \end{array}$$

In particular, given a map $\lambda : S \to S'$ of one set into another, there exists a unique monoid-homomorphism $\lambda_* : \mathbf{N}\langle S\rangle \to \mathbf{N}\langle S'\rangle$ such that the following diagram is commutative:

$$\begin{array}{ccc} S & \xrightarrow{j_S} & \mathbf{N}\langle S\rangle \\ {\scriptstyle\lambda}\downarrow & & \downarrow{\scriptstyle\lambda_*} \\ S' & \xrightarrow[j_{S'}]{} & \mathbf{N}\langle S'\rangle \end{array}$$

namely the map such that

$$\lambda_*\left[\prod_{x\in S} x^{\nu(x)}\right] = \prod_{x\in S} \lambda(x)^{\nu(x)}.$$

The proof of this assertion is trivial, as with abelian groups. One may view $\mathbf{N}\langle S\rangle$ as a functor from the category of sets into the category of commutative monoids.

Let A be a commutative ring. We may then form the monoid algebra $A[\mathbf{N}\langle S\rangle]$ over A, and we shall call it the *polynomial ring* (*or algebra*) *of* S *over* A. For simplicity, we denote it by $A[S]$. By definition, every element of $A[S]$ has a unique expression as a linear combination

$$\sum_{(\nu)} a_{(\nu)} M_{(\nu)}(S) = \sum_{(\nu)} a_{(\nu)} \prod_{x\in S} x^{\nu(x)},$$

where (ν) ranges over maps of S into $\mathbf{N}$ which are 0 for almost all x, and $a_{(\nu)}$ is 0 for almost all (ν). *The primitive monomials form a basis of* $A[S]$ *over* A, as noted previously for a monoid algebra. Elements of $A[S]$ are called *polynomials in* S *over* A. The elements $a_{(\nu)}$ are called the *coefficients* of the polynomial.

Remark on notation. Let T be a subset of a commutative ring B, and let $\nu : T \to \mathbf{N}$ be a map such that $\nu(x) = 0$ for almost all $x \in T$. We shall also denote by $M_{(\nu)}(T)$ the element

$$M_{(\nu)}(T) = \prod_{x\in T} x^{\nu(x)},$$

it being understood that this product is taken over those x such that

$\nu(x) \neq 0$, and that the empty product is the unit element of B. No confusion will arise with the notation for monomials, since the context will always make our meaning clear.

If S is a set of n letters $X_1, \ldots, X_n$, then

$$A[S] = A[X_1, \ldots, X_n],$$

and we call this the polynomial ring (or algebra) in $X_1, \ldots, X_n$ over A. We sometimes use vector notation and write $A[X]$ instead of $A[X_1, \ldots, X_n]$. Every polynomial of $A[X]$ can be written uniquely as a sum

$$\sum a_{(\nu)} M_{(\nu)}(X) = \sum a_{(\nu)} X_1^{\nu_1} \cdots X_n^{\nu_n},$$

the sum being taken over all n-tuples of integers $\nu_1, \ldots, \nu_n \geqq 0$, and almost all coefficients $a_{(\nu)}$ being equal to 0.

Let S be an arbitrary set again. Note that both S and A have canonical injections in $A[S]$, given by

$$x \mapsto 1 \cdot x^1 \qquad \text{and} \qquad a \mapsto a \cdot \prod_{x \in S} x^0.$$

In fact the canonical injection of A in $A[S]$ is a ring-homomorphism, i.e. an embedding. It is harmless to identify S and A with their respective images in $A[S]$. The monomial $\prod_{x \in S} x^0$, which is the unit element in the monoid $\mathbf{N}\langle S \rangle$, is also written 1 because no confusion ever arises from this. Thus if S consists of one letter X, every polynomial can be written in the form

$$a_0 X^0 + a_1 X^1 + \cdots + a_n X^n = a_0 + a_1 X^1 + \cdots + a_n X^n,$$

with $a_\nu \in A$, and some integer $n \geqq 0$.

Let A, B be commutative rings, and let $f_0 : A \to B$ be an A-algebra. Let S be a subset of B. If the family of monomials

$$M_{(\nu)}(S) = \prod_{x \in S} x^{\nu(x)}$$

is linearly independent over A, then we shall say that S, or that the elements of S, are *algebraically independent* over A. We could also deal with an indexed set $S = \{x_i\}_{i \in I}$, form monomials

$$M_{(\nu)}(S) = \prod_{i \in I} x_i^{\nu_i},$$

and define the *family* $\{x_i\}_{i \in I}$ to be algebraically independent if the monomials $M_{(\nu)}(S)$ are linearly independent over A. In particular, when S is finite, and say $S = \{t_1, \dots, t_n\}$, the monomials are of type

$$M_{(\nu)}(t_1, \dots, t_n) = t_1^{\nu_1} \cdots t_n^{\nu_n}$$

where $(\nu_1, \dots, \nu_n)$ ranges over all n-tuples of integers $\geqq 0$.

We have shown by our construction of the polynomial algebra how given a commutative ring A, one can construct an A-algebra having as many algebraically independent elements as one wishes.

The next theorem gives us an essential universal mapping property for algebraically independent elements.

THEOREM 1. *Let A, B be commutative rings, and let $f_0 : A \to B$ be an A-algebra. Let S be a subset of B, which generates B, and assume that the elements of S are algebraically independent over A. Let A' be a commutative ring. Let $f : A \to A'$ be a ring-homomorphism, and $\lambda : S \to A'$ a map. Then there exists a unique ring-homomorphism $h : B \to A'$ such that the diagram is commutative:*

$$\begin{array}{ccc} B & \xrightarrow{h} & A' \\ {\scriptstyle f_0}\uparrow & \nearrow{\scriptstyle f} & \\ A & & \end{array}$$

and such that the restriction of h to S is equal to λ.

Proof. Let G be the multiplicative monoid consisting of all the elements $M_{(\nu)}(S)$ in B. If $\nu \neq \mu$, then $M_{(\nu)}(S) \neq M_{(\mu)}(S)$, otherwise we would have a relation of linear dependence

$$M_{(\nu)}(S) - M_{(\mu)}(S) = 0.$$

Hence the map $\varphi : G \to A'$ such that

$$\varphi\left(\prod_{x \in S} x^{\nu(x)}\right) = \prod_{x \in S} \lambda(x)^{\nu(x)}$$

is a monoid-homomorphism. We apply Proposition 1 to conclude the proof.

We can apply Theorem 1 to the polynomial algebra $A[S]$, identifying S with its canonical image in $A[S]$. Thus if $B = A[S]$ and

$$\alpha = \sum_{(\nu)} a_{(\nu)} \cdot \prod_{x \in S} x^{\nu(x)},$$

the homomorphism h is described by

$$h(\alpha) = \sum_{(\nu)} f(a_{(\nu)}) \prod_{x \in S} \lambda(x)^{\nu(x)}.$$

Consider the special case when S is finite, consisting of distinct elements $t_1, \dots, t_n$, algebraically independent over A. Let $X_1, \dots, X_n$ be n

distinct letters. Then we have a ring-homomorphism

$$A[X_1, \dots, X_n] \to A[t_1, \dots, t_n]$$

mapping X_i on t_i and inducing the identity on A. From the definitions, one sees at once that the kernel must be 0, and that we therefore get an isomorphism. In particular, any two rings generated over A by n algebraically independent elements are isomorphic.

There are several other special cases of Theorem 1, which we write down explicitly.

First let A be fixed, and let S, S' be two sets with a bijection $\lambda : S \to S'$. If we identify S' as a subset of $A[S']$, we obtain an isomorphism

$$A[S] \approx A[S'],$$

inducing the bijection from S onto S'. In case S consists of n letters $X_1, \dots, X_n$ and S' consists of n letters $Y_1, \dots, Y_n$, we see that the polynomial rings are isomorphic, by an isomorphism sending X_i on Y_i for each i.

Suppose that S is contained in S'. Then $A[S]$ has a canonical injection in $A[S']$. If S is the set $\{X_1, \dots, X_n\}$ and S' is the set

$$\{X_1, \dots, X_n, X_{n+1}, \dots, X_N\},$$

then we can view the polynomial ring $A[X_1, \dots, X_n]$ as contained in $A[X_1, \dots, X_N]$. A monomial

$$X_1^{\nu_1} \cdots X_n^{\nu_n}$$

can be regarded as a monomial in $X_1, \dots, X_N$ simply by extending the function ν such that $\nu_i = 0$ for $i > n$.

Next let A be a subring of a ring A', and let S be a set. Then we have a natural embedding of $A[S]$ in $A'[S]$, namely a polynomial

$$\sum a_{(\nu)} \prod_{x \in S} x^{\nu(x)}$$

with coefficients in A may be viewed as having its coefficients in A'. We shall identify $A[S]$ as a subring of $A'[S]$.

More generally, let $\sigma : A \to A'$ be a homomorphism of commutative rings. Then this homomorphism extends uniquely to a ring-homomorphism

$$\bar{\sigma} : A[S] \to A'[S]$$

so as to induce the identity on S. For instance, let S be the set of n letters $X_1, \dots, X_n$. Then

$$\bar{\sigma} : A[X_1, \dots, X_n] \to A'[X_1, \dots, X_n]$$

is the ring-homomorphism given by the map

$$\sum a_{(\nu)} X_1^{\nu_1} \cdots X_n^{\nu_n} \mapsto \sum \sigma(a_{(\nu)}) X_1^{\nu_1} \cdots X_n^{\nu_n}.$$

If α denotes the polynomial on the left of the preceding arrow, then we shall frequently denote the polynomial on the right by the symbols

$$\alpha^\sigma.$$

One may say that α^σ is obtained from α by applying σ to the coefficients of α.

Let A be an entire ring, and let $\mathfrak{p}$ be a prime ideal. Let $\sigma : A \to A'$ be the canonical homomorphism of A onto $A/\mathfrak{p}$. If $\alpha(X)$ is a polynomial in $A[X]$, then α^σ will sometimes be called the *reduction of* α *modulo* $\mathfrak{p}$.

For example, taking $A = \mathbf{Z}$ and $\mathfrak{p} = (p)$ where p is a prime number, we can speak of the polynomial $3X^4 - X + 2$ as a polynomial mod 5, viewing the coefficients 3, -1, 2 as integers mod 5, i.e. elements of $\mathbf{Z}/5\mathbf{Z}$.

§3. *Elementary properties of polynomials*

Let A be a commutative ring and let S be the set of n letters $X_1, \ldots, X_n$. When we identify $X_1, \ldots, X_n$ with their canonical images in the polynomial ring $A[X_1, \ldots, X_n]$, then we call $X_1, \ldots, X_n$ *independent variables* over A, and we call $A[X]$ the polynomial ring in n variables. Every polynomial α in $A[X]$ has a unique expression

$$\alpha = \sum a_{(\nu)} X_1^{\nu_1} \cdots X_n^{\nu_n} = \sum a_{(\nu)} M_{(\nu)}(X).$$

Let $(b_1, \ldots, b_n)$ be an element of $\prod_1^n A$ (the direct product of A with itself n times), which we denote by $A^{(n)}$. By Theorem 1 there exists a unique homomorphism

$$h : A[X_1, \ldots, X_n] \to A$$

such that $h(X_i) = b_i$ for $i = 1, \ldots, n$, and which is the identity on A. We have

$$\sum h(\alpha) = a_{(\nu)} b_1^{\nu_1} \cdots b_n^{\nu_n}.$$

We shall denote this element of A by $\alpha(b_1, \ldots, b_n)$, and say that it is the element obtained by *substituting* $(b_1, \ldots, b_n)$ for $(X_1, \ldots, X_n)$ in α. In this manner we see that α gives rise to a function of $A^{(n)}$ into A.

Similarly, if A is a subring of a (commutative) ring B, and $(b) = (b_1, \ldots, b_n)$ is in $B^{(n)}$, then we can also form $\alpha(b)$ in the same manner as above, and get a function from $B^{(n)}$ into B, given by the map $(b) \mapsto \alpha(b)$.

Writing α as above, we see that

$$\alpha(b_1, \ldots, b_n) = \sum a_{(\nu)} M_{(\nu)}(b_1, \ldots, b_n),$$

or in vector notation,

$$\alpha(b) = \sum a_{(\nu)} M_{(\nu)}(b).$$

With this notation, we see that

$$\alpha = \alpha(X) = \alpha(X_1, \ldots, X_n).$$

We shall see below that when A is an entire ring, then $A[X_1, \ldots, X_n]$ is also entire. If K is the quotient field of A, the quotient field of $A[X_1, \ldots, X_n]$ is denoted by $K(X_1, \ldots, X_n)$. An element of $K(X_1, \ldots, X_n)$ is called a *rational function.* A rational function can be written as a quotient $f(X)/g(X)$ where f, g are polynomials. If $(b_1, \ldots, b_n)$ is in $K^{(n)}$, and a rational function admits an expression as a quotient f/g such that $g(b) \neq 0$, then we say that the rational function is *defined* at (b). From general localization properties, we see that when this is the case, we can substitute (b) in the rational function to get a value $f(b)/g(b)$.

It may happen that a polynomial is not the zero polynomial but gives rise to the zero function.

Example. Let $A = \mathbf{Z}/p\mathbf{Z}$ for some prime number p. If $a \in A$ and $a = 0$, then $a^p = 0$. If $a \neq 0$, then a is an element of the multiplicative group of non-zero elements of A, which has order $p - 1$. Hence $a^{p-1} = 1$, and consequently we obtain

$$a^p = a.$$

This is true for all $a \in A$. Hence the polynomial $X^p - X$ gives rise to the zero function from A into itself, and the polynomials X^p and X give rise to the same function, namely the identity function, on A.

More generally, let F be a finite field, and let q be the number of elements of F. Then X^q and X both give rise to the identity function of F into itself. It can be shown that any function of F into itself is given by a polynomial (in one variable), and similarly every function of $F^{(n)}$ into F is given by a polynomial in n variables. (Cf. the exercises.)

Let again A be a subring of B and let $b_1, \ldots, b_n$ be elements of B. If the homomorphism

$$A[X_1, \ldots, X_n] \to B$$

given by $\alpha(X) \mapsto \alpha(b)$ has a trivial kernel, i.e. if it is an embedding, then we recall that $b_1, \ldots, b_n$ are *algebraically independent over* A. If $n = 1$, and $b = b_1$ is algebraically independent over A, then we also say that b is *transcendental* over A.

Example. It is known (but not trivial to prove) that $e = 2.71\ldots$ and $\pi = 3.14\ldots$ are transcendental over the field of rational numbers $\mathbf{Q}$. It is not known whether they are algebraically independent (or even if $e + \pi$ is rational). Given specific complex numbers, it is usually extremely difficult to determine whether they are transcendental or algebraically independent over the rationals.

Let A be a commutative ring as before and let $S = \{X_1, \ldots, X_n\}$. By the *degree* of a primitive monomial

$$X_1^{\nu_1} \cdots X_n^{\nu_n}$$

we shall mean the integer $\nu_1 + \cdots + \nu_n$ (which is $\geqq 0$).

A polynomial

$$aX_1^{\nu_1} \cdots X_n^{\nu_n} \qquad (a \in A)$$

will be called a *monomial* (not necessarily primitive).

If $\alpha(X)$ is a polynomial in $A[X]$ written as

$$\alpha(X) = \sum a_{(\nu)} X_1^{\nu_1} \cdots X_n^{\nu_n},$$

then either $\alpha = 0$, in which case we say that its degree is $-\infty$, or $\alpha \neq 0$, and then we define the *degree of* α to be the maximum of the degrees of the monomials $M_{(\nu)}(X)$ such that $a_{(\nu)} \neq 0$. (Such monomials are said to *occur* in the polynomial.) We note that the degree of α is 0 if and only if

$$\alpha(X) = a_0 X_1^0 \cdots X_n^0$$

for some $a_0 \in A$, $a_0 \neq 0$. We also write this polynomial simply $\alpha(X) = a_0$, i.e. writing 1 instead of

$$X_1^0 \cdots X_n^0,$$

in other words, we identify the polynomial with the constant a_0.

Note that a polynomial $\alpha(X_1, \ldots, X_n)$ in n variables can be viewed as a polynomial in X_n with coefficients in $A[X_1, \ldots, X_{n-1}]$ (if $n \geqq 2$). Indeed, we have a homomorphism

$$A[X_1, \ldots, X_n] \to A[X_1, \ldots, X_{n-1}][X_n]$$

obtained by substitution, and this homomorphism is obviously an isomorphism. Thus

$$\alpha(X_1, \ldots, X_n) = \sum_{j=0}^{\infty} \alpha_j(X_1, \ldots, X_{n-1}) X_n^j$$

where α_j is an element of $A[X_1, \ldots, X_{n-1}]$. By the *degree of* α *in* X_n we shall mean its degree when viewed as a polynomial in X_n with coefficients in $A[X_1, \ldots, X_{n-1}]$. One sees easily that if this degree is d, then d is the largest integer occurring as an exponent of X_n in a monomial

$$a_{(\nu)}X_1^{\nu_1} \cdots X_n^{\nu_n}$$

with $a_{(\nu)} \neq 0$. Similarly, we define the degree of α in each variable X_i $(i = 1, \ldots, n)$.

The degree of α in each variable is of course usually different from its degree (which is sometimes called the *total degree* if there is need to prevent ambiguity). For instance,

$$X_1^3X_2 + X_2^2$$

has total degree 4, and has degree 3 in X_1 and 2 in X_2.

As a matter of notation, we shall often abbreviate "degree" by "deg".

Let $f(X)$ be a polynomial in one variable in $A[X]$, and write

$$f(X) = a_0 + \cdots + a_nX^n$$

with $a_i \in A$ and some integer $n \geqq 0$. If $f \neq 0$, and $\deg f = n$, then $a_n \neq 0$ by definition, and we call a_n the *leading coefficient* of f. We call a_0 its *constant term*. Observe that $a_0 = f(0)$.

Let

$$g(X) = b_0 + \cdots + b_mX^m$$

be a polynomial in $A[X]$, of degree m, and assume $g \neq 0$. Then

$$f(X)g(X) = a_0b_0 + \cdots + a_nb_mX^{m+n}.$$

If we assume that at least one of the leading coefficients a_n or b_m is not a divisor of 0 in A, then

$$\deg(fg) = \deg f + \deg g$$

and the leading coefficient of fg is a_nb_m. This holds in particular when a_n or b_m is a unit in A, or when A is entire. Consequently, when A is entire, $A[X]$ is also entire.

If f or $g = 0$, then we still have

$$\deg(fg) = \deg f + \deg g$$

if we agree that $-\infty + m = -\infty$ for any integer m.

One verifies trivially that for any polynomial $f, g \in A[X]$ we have

$$\deg(f + g) \leqq \max(\deg f, \deg g),$$

again agreeing that $-\infty < m$ for every integer m.

We leave it as an exercise to show that when A is an entire ring and f, g are polynomials *in several variables* then we still have the same rules:

$$\begin{aligned}\deg(fg) &= \deg f + \deg g,\\ \deg(f+g) &\leqq \max(\deg f, \deg g).\end{aligned}$$

Here degree can be interpreted to mean either total degree or degree in one of the variables. We conclude that $A[X_1, \ldots, X_n]$ is entire.

Let A be an arbitrary commutative ring again, and let d be an integer $\geqq 0$. Let

$$f(X_1, \ldots, X_n) \neq 0$$

be a polynomial in n variables over A. We shall say that f is *homogeneous* of degree d, or is a *form* of degree d, if all the monomials which occur in f are of degree d, i.e. if we write

$$f(X) = \sum a_{(\nu)} X_1^{\nu_1} \cdots X_n^{\nu_n}$$

and $a_{(\nu)} \neq 0$, then

$$\nu_1 + \cdots + \nu_n = d.$$

We shall leave it as an exercise to prove that *a non-zero polynomial f in n variables over A is homogeneous of degree d if and only if, for every set of $n + 1$ algebraically independent elements $u, t_1, \ldots, t_n$ over A we have*

$$f(ut_1, \ldots, ut_n) = u^d f(t_1, \ldots, t_n).$$

If f is homogeneous of degree d, then by Theorem 1, a similar relation holds when we substitute arbitrary elements $b_0, b_1, \ldots, b_n$ for $u, t_1, \ldots, t_n$ (taking the b_i in some commutative ring B containing A as a subring).

We note that if f, g are homogeneous of degree d, e respectively, and $fg \neq 0$, then fg is homogeneous of degree de. If $d = e$ and $f + g \neq 0$, then $f + g$ is homogeneous of degree d.

Finally, we make one remark on terminology. In view of the isomorphism

$$A[X_1, \ldots, X_n] \approx A[t_1, \ldots, t_n]$$

between the polynomial ring in n variables and a ring generated over A by n algebraically independent elements, we can apply all the terminology we have defined for polynomials, to elements of $A[t_1, \ldots, t_n]$. Thus we can speak of the degree of an element in $A[t]$, and the rules for the degree of a product or sum hold. In fact, we shall also call elements of $A[t]$ polynomials in (t). Algebraically independent elements will also be called variables (or independent variables), and any distinction which we make between $A[X]$ and $A[t]$ is more psychological than mathematical.

§4. *The Euclidean algorithm*

THEOREM 2. *Let A be a commutative ring, let $f, g \in A[X]$ be polynomials in one variable, of degrees $\geqq 0$, and assume that the leading coefficient of g is a unit in A. Then there exist unique polynomials q, $r \in A[X]$ such that*

$$f = gq + r$$

and $\deg r < \deg g$.

Proof. Write

$$f(X) = a_n X^n + \cdots + a_0,$$

$$g(X) = b_d X^d + \cdots + b_0,$$

where $n = \deg f$, $d = \deg g$ so that $a_n, b_d \neq 0$ and b_d is a unit in A. We use induction on n.

If $n = 0$, and $\deg g > \deg f$, we let $q = 0$, $r = f$. If $\deg g = \deg f = 0$, then we let $r = 0$ and $q = a_n b_d^{-1}$.

Assume the theorem proved for polynomials of degree $< n$ (with $n > 0$). We may assume $\deg g \leqq \deg f$ (otherwise, take $q = 0$ and $r = f$). Then

$$f(X) = a_n b_d^{-1} X^{n-d} g(X) + f_1(X)$$

where $f_1(X)$ has degree $< n$. By induction, we can find q_1, r such that

$$f(X) = a_n b_d^{-1} X^{n-d} g(X) + q_1(X) g(X) + r(X)$$

and $\deg r < \deg g$. Then we let

$$q(X) = a_n b_d^{-1} X^{n-d} + q_1(X)$$

to conclude the proof of existence for q, r.

As for uniqueness, suppose that

$$f = q_1 g + r_1 = q_2 g + r_2$$

with $\deg r_1 < \deg g$ and $\deg r_2 < \deg g$. Subtracting yields

$$(q_1 - q_2) g = r_2 - r_1.$$

Since the leading coefficient of g is assumed to be a unit, we have

$$\deg(q_1 - q_2) g = \deg(q_1 - q_2) + \deg g.$$

Since $\deg(r_2 - r_1) < \deg g$, this relation can hold only if $q_1 - q_2 = 0$, i.e. $q_1 = q_2$, and hence finally $r_1 = r_2$ as was to be shown.

THEOREM 3. *Let k be a field. Then the polynomial ring in one variable $k[X]$ is entire and principal.*

Proof. Let $\mathfrak{a}$ be an ideal of $k[X]$, and assume $\mathfrak{a} \neq 0$. Let g be an element of $\mathfrak{a}$ of smallest degree $\geqq 0$. Let f be any element of $\mathfrak{a}$ such that $f \neq 0$. By the Euclidean algorithm we can find $q, r \in k[X]$ such that

$$f = qg + r$$

and $\deg r < \deg g$. But $r = f - qg$, whence r is in $\mathfrak{a}$. Since g had minimal degree $\geqq 0$ it follows that $r = 0$, hence that $\mathfrak{a}$ consists of all polynomials qg (with $q \in k[X]$). This proves our theorem.

COROLLARY. *The ring $k[X]$ is factorial.*

If k is a field then every non-zero element of k is a unit in k, and one sees immediately that the units of $k[X]$ are simply the units of k. (No polynomial of degree $\geqq 1$ can be a unit because of the addition formula for the degree of a product.)

Let A be a commutative ring and $f(X)$ a polynomial in $A[X]$. Let A be a subring of B. An element $b \in B$ is called a *root* or a *zero* of f in B if $f(b) = 0$. Similarly, if (X) is an n-tuple of variables, an n-tuple (b) is called a zero of f if $f(b) = 0$.

THEOREM 4. *Let k be a field and f a polynomial in one variable X in $k[X]$, of degree $n \geqq 0$. Then f has at most n roots in k, and if a is a root of f in k, then $X - a$ divides $f(X)$.*

Proof. Suppose $f(a) = 0$. Find q, r such that

$$f(X) = q(X)(X - a) + r(X)$$

and $\deg r < 1$. Then

$$0 = f(a) = r(a).$$

Since $r = 0$ or r is a non-zero constant, we must have $r = 0$, whence $X - a$ divides $f(X)$. If $a_1, \ldots, a_m$ are distinct roots of f in k, then inductively we see that the product

$$(X - a_1) \cdots (X - a_m)$$

divides $f(X)$, whence $m \leqq n$, as contended.

COROLLARY 1. *Let k be a field and T an infinite subset of k. Let $f(X) \in k[X]$ be a polynomial in one variable. If $f(a) = 0$ for all $a \in T$, then $f = 0$, i.e. f induces the zero function.*

COROLLARY 2. *Let k be a field, and let $T_1, \ldots, T_n$ be infinite subsets of k. Let $f(X_1, \ldots, X_n)$ be a polynomial in n variables over k. If $f(a_1, \ldots, a_n) = 0$ for all $a_i \in T_i$ $(i = 1, \ldots, n)$, then $f = 0$.*

Proof. By induction. We have just seen the result is true for one variable. Let $n \geqq 2$, and write

$$f(X_1, \ldots, X_n) = \sum_j f_j(X_1, \ldots, X_{n-1})X_n^j$$

as a polynomial in X_n with coefficients in $k[X_1, \ldots, X_{n-1}]$. If there exists

$$(b_1, \ldots, b_{n-1}) \in T_1 \times \cdots \times T_{n-1}$$

such that for some j we have $f_j(b_1, \ldots, b_{n-1}) \neq 0$, then

$$f(b_1, \ldots, b_{n-1}, X_n)$$

is a non-zero polynomial in $k[X_n]$ which takes on the value 0 for the infinite set of elements T_n. This is impossible. Hence f_j induces the zero function on $T_1 \times \cdots \times T_{n-1}$ for all j, and by induction we have $f_j = 0$ for all j. Hence $f = 0$, as was to be shown.

COROLLARY 3. *Let k be an infinite field and f a polynomial in n variables over k. If f induces the zero function on $k^{(n)}$, then $f = 0$.*

We shall now consider the case of finite fields. Let k be a finite field with q elements. Let $f(X_1, \ldots, X_n)$ be a polynomial in n variables over k. Write

$$f(X_1, \ldots, X_n) = \sum a_{(\nu)} X_1^{\nu_1} \cdots X_n^{\nu_n}.$$

If $a_{(\nu)} \neq 0$, we recall that the monomial $M_{(\nu)}(X)$ *occurs* in f. Suppose this is the case, and that in this monomial $M_{(\nu)}(X)$, some variable X_i occurs with an exponent $\nu_i \geqq q$. We can write

$$X_i^{\nu_i} = X_i^{q+\mu}, \qquad \mu = \text{integer} \geqq 0.$$

If we now replace $X_i^{\nu_i}$ by $X_i^{\mu+1}$ in this monomial, then we obtain a new polynomial which gives rise to the same function as f. The degree of this new polynomial is at most equal to the degree of f.

Performing the above operation a finite number of times, for all the monomials occurring in f and all the variables $X_1, \ldots, X_n$ we obtain some polynomial f^* giving rise to the same function as f, but whose degree in each variable is $< q$.

THEOREM 5. *Let k be a finite field with q elements. Let f be a polynomial in n variables over k such that the degree of f in each variable is $< q$. If f induces the zero function on $k^{(n)}$, then $f = 0$.*

Proof. By induction. If $n = 1$, then the degree of f is $< q$, and hence f cannot have q roots unless it is 0. The inductive step is carried out just as we did for the proof of Corollary 2 above.

Let f be a polynomial in n variables over the finite field k. A polynomial g whose degree in each variable is $< q$ will be said to be *reduced*. We have shown above that there exists a reduced polynomial f^* which gives the same function as f on $k^{(n)}$. Theorem 5 now shows that *this reduced polynomial is unique*. Indeed, if g_1, g_2 are reduced polynomials giving the same function, then $g_1 - g_2$ is reduced and gives the zero function. Hence $g_1 - g_2 = 0$ and $g_1 = g_2$.

We shall give one more application of Theorem 4. Let k be a field. By a multiplicative subgroup of k we shall mean a subgroup of the group k^* (non-zero elements of k).

THEOREM 6. *Let k be a field and let U be a finite multiplicative subgroup of k. Then U is cyclic.*

Proof. Write U as a product of subgroups $U(p)$ for each prime p, where $U(p)$ is a p-group. By Exercise 22 of Chapter I, it will suffice to prove that $U(p)$ is cyclic for each p. Let a be an element of $U(p)$ of maximal period p^r for some integer r. Then $x^{p^r} = 1$ for every element $x \in U(p)$, and hence all elements of $U(p)$ are roots of the polynomial

$$X^{p^r} - 1.$$

The cyclic group generated by a has p^r elements. If this cyclic group is not equal to $U(p)$, then our polynomial has more than p^r roots, which is impossible. Hence a generates $U(p)$, and our theorem is proved.

COROLLARY. *If k is a finite field, then k^* is cyclic.*

An element ζ in a field k such that there exists an integer $n \geqq 1$ such that $\zeta^n = 1$ is called a *root of unity*, or more precisely an n-th root of unity. Thus the set of n-th roots of unity is the set of roots of the polynomial $X^n - 1$. There are at most n such roots, and they obviously form a group, which is cyclic by Theorem 6. We shall study roots of unity in greater detail later. A generator for the group of n-th roots of unity is called a *primitive* n-th root of unity. For example, in the complex numbers, $e^{2\pi i/n}$ is a primitive n-th root of unity, and the n-th roots of unity are of type $e^{2\pi i\nu/n}$ with $1 \leqq \nu \leqq n$.

§5. *Partial fractions*

In this section, we analyze the quotient field of a principal ring, using the factoriality of the ring.

THEOREM 7. *Let A be a principal entire ring, and let P be a set of representatives for its irreducible elements. Let K be the quotient field of A, and let $\alpha \in K$. For each $p \in P$ there exists an element $\alpha_p \in A$ and an* *integer* $j(p) \geqq 0$, *such that* $j(p) = 0$ *for almost all* $p \in P$, α_p *and* $p^{j(p)}$ *are relatively prime, and*

$$\alpha = \sum_{p \in P} \frac{\alpha_p}{p^{j(p)}}.$$

If we have another such expression

$$\alpha = \sum_{p \in P} \frac{\beta_p}{p^{i(p)}},$$

then $j(p) = i(p)$ *for all* p, *and* $\alpha_p \equiv \beta_p \bmod p^{j(p)}$ *for all* p.

Proof. We first prove existence, in a special case. Let a, b be relatively prime non-zero elements of A. Then there exists $x, y \in A$ such that $xa + yb = 1$. Hence

$$\frac{1}{ab} = \frac{x}{b} + \frac{y}{a}.$$

Hence any fraction c/ab with $c \in A$ can be decomposed into a sum of two fractions (namely cx/b and cy/a) whose denominators divide b and a respectively. By induction, it now follows that any $\alpha \in K$ has an expression as stated in the theorem, except possibly for the fact that p may divide α_p. Canceling the greatest common divisor yields an expression satisfying all the desired conditions.

As for uniqueness, suppose that α has two expressions as stated in the theorem. Let q be a fixed prime in P. Then

$$\frac{\alpha_q}{q^{j(q)}} - \frac{\beta_q}{q^{i(q)}} = \sum_{p \neq q} \frac{\beta_p}{p^{i(p)}} - \frac{\alpha_p}{p^{j(p)}}.$$

If $j(q) = i(q) = 0$, our conditions concerning q are satisfied. Suppose one of $j(q)$ or $i(q) > 0$, say $j(q)$, and say $j(q) \geqq i(q)$. Let d be a least common multiple for all powers $p^{j(p)}$ and $p^{i(p)}$ such that $p \neq q$. Multiply the above equation by $dq^{j(q)}$. We get

$$d(\alpha_q - q^{j(q)-i(q)}\beta_q) = q^{j(q)}\beta$$

for some $\beta \in A$. Furthermore, q does not divide d. If $i(q) < j(q)$ then q divides α_q, which is impossible. Hence $i(q) = j(q)$. We now see that $q^{j(q)}$ divides $\alpha_q - \beta_q$, thereby proving the theorem.

We apply Theorem 7 to the polynomial ring $k[X]$ over a field k. We let P be the set of irreducible polynomials, normalized so as to have leading coefficient equal to 1. Then P is a set of representatives for all the irreducible elements of $k[X]$. In the expression given for α in Theorem 7, we can now divide α_p by $p^{j(p)}$, i.e. use the Euclidean algorithm, if $\deg \alpha_p \geqq \deg p^{j(p)}$. We denote the quotient field of $k[X]$ by $k(X)$, and call its elements *rational functions*.

THEOREM 8. *Let $A = k[X]$ be the polynomial ring in one variable over a field k. Let P be the set of irreducible polynomials in $k[X]$ with leading coefficient* 1. *Then any element f of $k(X)$ has a unique expression*

$$f(X) = \sum_{p \in P} \frac{f_p(X)}{p(X)^{j(p)}} + g(X),$$

where f_p, g are polynomials, $f_p = 0$ if $j(p) = 0$, f_p is relatively prime to p if $j(p) > 0$, and $\deg f_p < \deg p^{j(p)}$ *if* $j(p) > 0$.

Proof. The existence follows at once from our previous remarks. The uniqueness follows from the fact that if we have two expressions, with elements f_p and φ_p respectively, and polynomials g, h, then $p^{j(p)}$ divides $f_p - \varphi_p$, whence $f_p - \varphi_p = 0$, and therefore $f_p = \varphi_p$, $g = h$.

One can further decompose the term $f_p/p^{j(p)}$ by expanding f_p according to powers of p. One can in fact do something more general.

THEOREM 9. *Let k be a field and $k[X]$ the polynomial ring in one variable. Let $f, g \in k[X]$, and assume* $\deg g \geqq 1$. *Then there exist unique polynomials*

$$f_0, f_1, \ldots, f_d \in k[X]$$

such that $\deg f_i < \deg g$ *and such that*

$$f = f_0 + f_1 g + \cdots + f_d g^d.$$

Proof. We first prove existence. If $\deg g > \deg f$, then we take $f_0 = f$ and $f_i = 0$ for $i > 0$. Suppose $\deg g \leqq \deg f$. We can find polynomials q, r with $\deg r < \deg g$ such that

$$f = qg + r,$$

and since $\deg g \geqq 1$ we have $\deg q < \deg f$. Inductively, there exist polynomials $h_0, h_1, \ldots, h_s$ such that

$$q = h_0 + h_1 g + \cdots + h_s g^s,$$

and hence

$$f = r + h_0 g + \cdots + h_s g^{s+1},$$

thereby proving existence.

As for uniqueness, let

$$f = f_0 + f_1 g + \cdots + f_d g^d = \varphi_0 + \varphi_1 g + \cdots + \varphi_m g^m$$

be two expressions satisfying the conditions of the theorem. Adding terms equal to 0 to either side, we may assume that $m = d$. Subtracting, we get

$$0 = (f_0 - \varphi_0) + \cdots + (f_d - \varphi_d) g^d.$$

Hence g divides $f_0 - \varphi_0$, and since $\deg(f_0 - \varphi_0) < \deg g$ we see that $f_0 = \varphi_0$. Inductively, take the smallest integer i such that $f_i \neq \varphi_i$ (if such i exists). Dividing the above expression by g^i we find that g divides $f_i - \varphi_i$ and hence that such i cannot exist. This proves uniqueness.

We shall call the expression for f in terms of g in Theorem 9 the *g-adic expansion* of f. If $g(X) = X$, then the g-adic expansion is the usual expression of f as a polynomial.

§6. *Unique factorization in several variables*

Let A be a factorial ring, and K its quotient field. Let $a \in K$, $a \neq 0$. We can write a as a quotient of elements in A, having no prime factor in common. If p is a prime element of A, then we can write

$$a = p^r b,$$

where $b \in K$, r is an integer, and p does not divide the numerator or denominator of b. Using the unique factorization in A, we see at once that r is uniquely determined by a, and we call r the *order of a at p* (and write $r = \operatorname{ord}_p a$). If $a = 0$, we define its order at p to be $-\infty$.

If $a, a' \in K$ and $aa' \neq 0$, then

$$\operatorname{ord}_p(aa') = \operatorname{ord}_p a + \operatorname{ord}_p a'.$$

This is obvious.

Let $f(X) \in K[X]$ be a polynomial in one variable, written

$$f(X) = a_0 + a_1 X + \cdots + a_n X^n.$$

If $f = 0$, we define $\operatorname{ord}_p f$ to be $-\infty$. If $f \neq 0$, we define $\operatorname{ord}_p f$ to be

$$\operatorname{ord}_p f = \min \operatorname{ord}_p a_i,$$

the minimum being taken over all those i such that $a_i \neq 0$.

If $r = \operatorname{ord}_p f$, we call up^r a *p-content* for f, if u is any unit of A. We define the *content* of f to be the product

$$\prod p^{\operatorname{ord}_p f},$$

the product being taken over all p such that $\operatorname{ord}_p f \neq 0$, or any multiple of this product by a unit of A. Thus the content is well-defined up to multiplication by a unit of A. We abbreviate *content* by *cont.*

If $b \in K$, $b \neq 0$, then $\operatorname{cont}(bf) = b \operatorname{cont}(f)$. This is clear. Hence we can write

$$f(X) = c \cdot f_1(X)$$

where $c = \operatorname{cont}(f)$, and $f_1(X)$ has content 1. In particular, all coefficients of f_1 lie in A, and their g.c.d. is 1.

Gauss Lemma. *Let A be a factorial ring, and let K be its quotient field. Let $f, g \in K[X]$ be polynomials in one variable. Then*

$$\operatorname{cont}(fg) = \operatorname{cont}(f)\operatorname{cont}(g).$$

Proof. Writing $f = cf_1$ and $g = dg_1$ where $c = \operatorname{cont}(f)$ and $d = \operatorname{cont}(g)$, we see that it suffices to prove: If f, g have content 1, then fg also has content 1, and for this, it suffices to prove that for each prime p, $\operatorname{ord}_p(fg) = 1$. Let

$$f(X) = a_nX^n + \cdots + a_0, \qquad a_n \neq 0$$
$$g(X) = b_mX^m + \cdots + b_0, \qquad b_m \neq 0$$

be polynomials of content 1. Let p be a prime of A. It will suffice to prove that p does not divide all coefficients of fg. Let r be the largest integer such that $0 \leqq r \leqq n$, $a_r \neq 0$, and p does not divide a_r. Similarly, let b_s be the coefficient of g farthest to the left, $b_s \neq 0$, such that p does not divide b_s. Consider the coefficient of X^{r+s} in $f(X)g(X)$. This coefficient is equal to

$$\begin{aligned} c = a_rb_s &+ a_{r+1}b_{s-1} + \cdots \\ &+ a_{r-1}b_{s+1} + \cdots \end{aligned}$$

and $p \nmid a_rb_s$. However, p divides every other non-zero term in this sum since in each term there will be some coefficient a_i to the left of a_r or some coefficient b_j to the left of b_s. Hence p does not divide c, and our lemma is proved.

Corollary. *Let $f(X) \in A[X]$ have a factorization $f(X) = g(X)h(X)$ in $K[X]$. If $c_g = \operatorname{cont}(g)$, $c_h = \operatorname{cont}(h)$, and $g = c_gg_1$, $h = c_hh_1$, then*

$$f(X) = c_gc_hg_1(X)h_1(X),$$

and c_gc_h is an element of A.

Proof. The only thing to be proved is the last statement. But $\operatorname{cont}(f) = c_gc_h\operatorname{cont}(g_1h_1) = c_gc_h$, whence our assertion follows.

Theorem 10. *Let A be a factorial ring. Then the polynomial ring $A[X]$ in one variable is factorial. Its prime elements are either primes of A, or polynomials in $A[X]$ which are irreducible in $K[X]$ and have content 1.*

Proof. Let $f \in A[X]$, $f \neq 0$. Using the unique factorization in $K[X]$ and the preceding corollary, we can find a factorization

$$f(X) = c \cdot p_1(X) \cdots p_r(X)$$

where $c \in A$, and $p_1, \ldots, p_r$ are polynomials in $A[X]$ which are irreducible in $K[X]$. Extracting their contents, we may assume without loss of generality that the content of p_i is 1 for each i. Then $c = \text{cont}(f)$. This gives us the existence of the factorization. From the Gauss lemma, it follows that each $p_i(X)$ is irreducible in $A[X]$. If we have another such factorization, say

$$f(X) = d \cdot q_1(X) \cdots q_s(X),$$

then from the unique factorization in $K[X]$ we conclude that $r = s$, and after a permutation of the factors we have

$$p_i = a_i q_i$$

with elements $a_i \in K$. Since both p_i, q_i are assumed to have content 1, it follows that a_i in fact lies in A and is a unit. This proves our theorem.

COROLLARY. *Let A be a factorial ring. Then the ring of polynomials in n variables $A[X_1, \ldots, X_n]$ is factorial. Its units are precisely the units of A, and its prime elements are either primes of A or polynomials which are irreducible in $K[X]$ and have content* 1.

Proof. Induction.

In view of Theorem 10, when we deal with polynomials over a factorial ring, and having content 1, it is not necessary to specify whether such polynomials are irreducible over A or over the quotient field K. The two notions are equivalent.

Remark 1. The polynomial ring $K[X_1, \ldots, X_n]$ over a field K is not principal when $n \geqq 2$. For instance, the ideal generated by $X_1, \ldots, X_n$ is not principal (trivial proof).

Remark 2. It is usually not too easy to decide when a given polynomial (say in one variable) is irreducible. For instance, the polynomial $X^4 + 4$ is *reducible* over the rational numbers, because

$$X^4 + 4 = (X^2 - 2X + 2)(X^2 + 2X + 2).$$

Later in this book we shall give a precise criterion when a polynomial $X^n - a$ is irreducible. Other criteria are given in the next section.

§7. *Criteria for irreducibility*

The first criterion is EISENSTEIN'S CRITERION: *Let A be a factorial ring. Let K be its quotient field. Let $f(X) = a_nX^n + \cdots + a_0$ be a polynomial of degree $n \geqq 1$ in $A[X]$. Let p be a prime of A, and assume:*

$$a_n \not\equiv 0 \pmod{p}, \qquad a_i \equiv 0 \pmod{p} \quad \textit{for all } i < n,$$
$$a_0 \not\equiv 0 \pmod{p^2}.$$

Then $f(X)$ is irreducible in $K[X]$.

Proof. Extracting a g.c.d. for the coefficients of f, we may assume without loss of generality that the content of f is 1. If there exists a factorization into factors of degree $\geqq 1$ in $K[X]$, then by the corollary of Gauss' lemma there exists a factorization in $A[X]$, say $f(X) = g(X)h(X)$,

$$g(X) = b_d X^d + \cdots + b_0,$$
$$h(X) = c_m X^m + \cdots + c_0,$$

with $d, m \geqq 1$ and $b_d c_m \neq 0$. Since $b_0 c_0 = a_0$ is divisible by p but not p^2, it follows that one of b_0, c_0 is not divisible by p, say b_0. Then $p \mid c_0$. Since $c_m b_d = a_n$ is not divisible by p, it follows that p does not divide c_m. Let c_r be the coefficient of h furthest to the right such that $c_r \not\equiv 0 \pmod{p}$. Then $r \neq n$ and

$$a_r = b_0 c_r + b_1 c_{r-1} + \cdots$$

Since $p \nmid b_0 c_r$ but p divides every other term in this sum, we conclude that $p \nmid a_r$, a contradiction which proves our theorem.

Example. Let a be a non-zero square-free integer $\neq \pm 1$. Then for any integer $n \geqq 1$, the polynomial $X^n - a$ is irreducible over $\mathbf{Q}$. The polynomials $3X^5 - 15$, $2X^{10} - 21$ are irreducible over $\mathbf{Q}$.

There are some cases in which a polynomial does not satisfy Eisenstein's criterion, but a simple transform of it does.

Example. Let p be a prime number. Then the polynomial

$$f(X) = X^{p-1} + \cdots + 1$$

is irreducible over $\mathbf{Q}$.

Proof. It will suffice to prove that the polynomial $f(X + 1)$ is irreducible over $\mathbf{Q}$. We note that the binomial coefficients

$$\binom{p}{\nu} = \frac{p!}{\nu!(p - \nu)!} \qquad 1 \leqq \nu \leqq p - 1$$

are divisible by p (because the numerator is divisible by p and the denominator is not, and the coefficient is an integer). We have

$$f(X + 1) = \frac{(X + 1)^p - 1}{(X + 1) - 1} = \frac{X^p + pX^{p-1} + \cdots + pX}{X}$$

from which one sees that $f(X + 1)$ satisfies E's criterion.

Example. Let E be a field and t an element of some field containing E such that t is transcendental over E. Let K be the quotient field of $E[t]$. For any integer $n \geqq 1$ the polynomial $X^n - t$ is irreducible in $K[X]$. This comes from the fact that the ring $A = E[t]$ is factorial and that t is a prime in it.

Reduction Criterion. *Let A, B be entire rings, and let*

$$\sigma : A \to B$$

be a homomorphism. Let K, L be the quotient fields of A and B respectively. Let $f \in A[X]$ be such that $f^\sigma \neq 0$ and $\deg f^\sigma = \deg f$. If f^σ is irreducible in $L[X]$, then f does not have a factorization $f(X) = g(X)h(X)$ with

$$g, h \in A[X] \qquad \text{and} \qquad \deg g, \deg h \geqq 1.$$

Proof. Suppose f has such a factorization. Then $f^\sigma = g^\sigma h^\sigma$. Since $\deg g^\sigma \leqq \deg g$ and $\deg h^\sigma \leqq \deg h$, our hypothesis implies that we must have equality in these degree relations. Hence from the irreducibility in $L[X]$ we conclude that g or h is an element of A, as desired.

In the preceding criterion, suppose that A is a local ring, i.e. a ring having a unique maximal ideal $\mathfrak{p}$, and that $\mathfrak{p}$ is the kernel of σ. Then from the irreducibility of f^σ in $L[X]$ we conclude the irreducibility of f in $A[X]$. Indeed, any element of A which does not lie in $\mathfrak{p}$ must be a unit in A, so our last conclusion in the proof can be strengthened to the statement that g or h is a unit in A.

One can also apply the criterion when A is factorial, and in that case deduce the irreducibility of f in $K[X]$.

Example. Let p be a prime number. It will be shown later that $X^p - X - 1$ is irreducible over the field $\mathbf{Z}/p\mathbf{Z}$. Hence $X^p - X - 1$ is irreducible over $\mathbf{Q}$. Similarly,

$$X^5 - 5X^4 - 6X - 1$$

is irreducible over $\mathbf{Q}$.

§8. The derivative and multiple roots

Let A be a commutative ring. We define a map

$$D : A[X] \to A[X]$$

of the polynomial ring into itself. If $f(X) = a_n X^n + \cdots + a_0$ with $a_i \in A$, we define the derivative

$$Df(X) = f'(X) = \sum_{\nu=1}^{n} \nu a_\nu X^{\nu-1} = n a_n X^{n-1} + \cdots + a_1.$$

One verifies easily that if f, g are polynomials in $A[X]$, then

$$(f + g)' = f' + g', \qquad (fg)' = f'g + fg',$$

and if $a \in A$, then

$$(af)' = af'.$$

Let K be a field and f a polynomial in $K[X]$. Let a be a root of f in K. We can write

$$f(X) = (X - a)^m g(X)$$

with some polynomial $g(X)$ relatively prime to $X - a$ (and hence such that $g(a) \neq 0$). We call m the *multiplicity* of a in f, and say that a is a *multiple root* if $m > 1$.

PROPOSITION 1. *Let K, f be as above. The element a of K is a multiple root of f if and only if it is a root and $f'(a) = 0$.*

Proof. Factoring f as above, we get

$$f'(X) = (X - a)^m g'(X) + m(X - a)^{m-1} g(X).$$

If $m > 1$, then obviously $f'(a) = 0$. Conversely, if $m = 1$ then $f'(X) = (X - a)g'(X) + g(X)$, whence $f'(a) = g(a) \neq 0$. Hence if $f'(a) = 0$ we must have $m > 1$, as desired.

PROPOSITION 2. *Let $f \in K[X]$. If K has characteristic 0, and f has degree $\geqq 1$, then $f' \neq 0$. Let K have characteristic $p > 0$ and f have degree $\geqq 1$. Then $f' = 0$ if and only if, in the expression for $f(X)$ given by*

$$f(X) = \sum_{\nu=1}^{n} a_\nu X^\nu,$$

p divides each integer ν such that $a_\nu \neq 0$.

Proof. If K has characteristic 0, then the derivative of a monomial $a_\nu X^\nu$ such that $\nu \geqq 1$ and $a_\nu \neq 0$ is not zero since it is $\nu a_\nu X^{\nu-1}$. If K has characteristic $p > 0$, then the derivative of such a monomial is 0 if and only if $p|\nu$, as contended.

Let K have characteristic $p > 0$, and let f be written as above, and be such that $f'(X) = 0$. Then one can write

$$f(X) = \sum_{\mu=1}^{d} b_\mu X^{p\mu}$$

with $b_\mu \in K$.

Since the binomial coefficients $\binom{p}{\nu}$ are divisible by p for $1 \leqq \nu \leqq p - 1$ we see that if K has characteristic p, then for $a, b \in K$ we have

$$(a + b)^p = a^p + b^p.$$

Since obviously $(ab)^p = a^p b^p$, the map

$$x \mapsto x^p$$

is a homomorphism of K into itself, which has trivial kernel, hence is injective. Iterating, we conclude that for each integer $r \geqq 1$, the map $x \mapsto x^{p^r}$ is an endomorphism of K, called the *Frobenius endomorphism.* Inductively, if $c_1, \ldots, c_n$ are elements of K, then

$$(c_1 + \cdots + c_n)^p = c_1^p + \cdots + c_n^p.$$

Applying these remarks to polynomials, we see that for any element $a \in K$ we have

$$(X - a)^{p^r} = X^{p^r} - a^{p^r}.$$

If $c \in K$ and the polynomial

$$X^{p^r} - c$$

has one root a in K, then $a^{p^r} = c$ and

$$X^{p^r} - c = (X - a)^{p^r}.$$

Hence our polynomial has precisely one root, of multiplicity p^r. For instance, $(X - 1)^{p^r} = X^{p^r} - 1$.

§9. *Symmetric polynomials*

Let A be a commutative ring and let $t_1, \ldots, t_n$ be algebraically independent elements over A. Let X be a variable over $A[t_1, \ldots, t_n]$. We form the polynomial

$$\begin{aligned} \mathbf{F}(X) &= (X - t_1) \cdots (X - t_n) \\ &= X^n - s_1 X^{n-1} + \cdots + (-1)^n s_n \end{aligned}$$

where each $s_i = s_i(t_1, \ldots, t_n)$ is a polynomial in $t_1, \ldots, t_n$. Then for instance,

$$s_1 = t_1 + \cdots + t_n \qquad \text{and} \qquad s_n = t_1 \cdots t_n.$$

The polynomials $s_1, \ldots, s_n$ are called the *elementary symmetric polynomials* of $t_1, \ldots, t_n$.

We leave it as an easy exercise to verify that *s_i is homogeneous of degree i in $t_1, \ldots, t_n$.*

Let σ be a permutation of the integers $(1, \ldots, n)$. Given a polynomial $f(t) \in A[t] = A[t_1, \ldots, t_n]$, we define f^σ to be

$$f^\sigma(t_1, \ldots, t_n) = f(t_{\sigma(1)}, \ldots, t_{\sigma(n)}).$$

If σ, τ are two permutations, then $f^{\sigma\tau} = (f^\sigma)^\tau$ and hence the symmetric group G on n letters operates on the polynomial ring $A[t]$. A polynomial is called *symmetric* if $f^\sigma = f$ for all $\sigma \in G$. It is clear that the set of symmetric polynomials is a subring of $A[t]$, which contains the constant polynomials (i.e. A itself) and also contains the elementary symmetric polynomials $s_1, \ldots, s_n$. We shall see below that it contains nothing else.

Let $X_1, \ldots, X_n$ be variables. We define the *weight* of a monomial

$$X_1^{\nu_1} \cdots X_n^{\nu_n}$$

to be $\nu_1 + 2\nu_2 + \cdots + n\nu_n$. We define the weight of a polynomial $g(X_1, \ldots, X_n)$ to be the maximum of the weights of the monomials occurring in g.

THEOREM 11. *Let $f(t) \in A[t_1, \ldots, t_n]$ be symmetric of degree d. Then there exists a polynomial $g(X_1, \ldots, X_n)$ of weight $\leqq d$ such that*

$$f(t) = g(s_1, \ldots, s_n).$$

Proof. By induction on n. The theorem is obvious if $n = 1$, because $s_1 = t_1$.

Assume the theorem proved for polynomials in $n - 1$ variables.

If we substitute $t_n = 0$ in the expression for $\mathbf{F}(X)$, we find

$$(X - t_1) \cdots (X - t_{n-1})X = X^n - (s_1)_0 X^{n-1} + \cdots + (-1)^{n-1}(s_{n-1})_0 X$$

where $(s_i)_0$ is the expression obtained by substituting $t_n = 0$ in s_i. We see that $(s_1)_0, \ldots, (s_{n-1})_0$ are precisely the elementary symmetric polynomials in $t_1, \ldots, t_{n-1}$.

We now carry out induction on d. If $d = 0$, our assertion is trivial. Assume $d > 0$, and assume our assertion proved for polynomials of degree $< d$. Let $f(t_1, \ldots, t_n)$ have degree d. There exists a polynomial $g_1(X_1, \ldots, X_{n-1})$ of weight $\leqq d$ such that

$$f(t_1, \ldots, t_{n-1}, 0) = g_1((s_1)_0, \ldots, (s_{n-1})_0).$$

We note that $g_1(s_1, \ldots, s_{n-1})$ has degree $\leqq d$ in $t_1, \ldots, t_n$. The polynomial

$$f_1(t_1, \ldots, t_n) = f(t_1, \ldots, t_n) - g_1(s_1, \ldots, s_{n-1})$$

has degree $\leqq d$ (in $t_1, \ldots, t_n$) and is symmetric. We have

$$f_1(t_1, \ldots, t_{n-1}, 0) = 0.$$

Hence f_1 is divisible by t_n, i.e. contains t_n as a factor. Since f_1 is symmetric, it contains $t_1 \cdots t_n$ as a factor. Hence

$$f_1 = s_n f_2(t_1, \ldots, t_n)$$

for some polynomial f_2, which must be symmetric, and whose degree is $\leqq d - n < d$. By induction, there exists a polynomial g_2 in n variables and weight $\leqq d - n$ such that

$$f_2(t_1, \ldots, t_n) = g_2(s_1, \ldots, s_n).$$

We obtain

$$f(t) = g_1(s_1, \ldots, s_{n-1}) + s_n g_2(s_1, \ldots, s_n),$$

and each term on the right has weight $\leqq d$. This proves our theorem.

We shall now prove that the elementary symmetric polynomials $s_1, \ldots, s_n$ *are algebraically independent over* A.

If they are not, take a polynomial $f(X_1, \ldots, X_n) \in A[X]$ of least degree and not equal to 0 such that

$$f(s_1, \ldots, s_n) = 0.$$

Write f as a polynomial in X_n with coefficients in $A[X_1, \ldots, X_{n-1}]$,

$$f(X_1, \ldots, X_n) = f_0(X_1, \ldots, X_{n-1}) + \cdots + f_d(X_1, \ldots, X_{n-1})X_n^d.$$

Then $f_0 \neq 0$. Otherwise, we can write

$$f(X) = X_n \psi(X)$$

with some polynomial ψ, and hence $s_n \psi(s_1, \ldots, s_n) = 0$. From this it follows that $\psi(s_1, \ldots, s_n) = 0$, and ψ has degree smaller than the degree of f.

We substitute s_i for X_i in the above relation, and get

$$0 = f_0(s_1, \ldots, s_{n-1}) + \cdots + f_d(s_1, \ldots, s_{n-1})s_n^d.$$

This is a relation in $A[t_1, \ldots, t_n]$, and we substitute 0 for t_n in this relation. Then all terms become 0 except the first one, which gives

$$0 = f_0\big((s_1)_0, \ldots, (s_{n-1})_0\big),$$

using the same notation as in the proof of Theorem 1. This is a non-trivial relation between the elementary symmetric polynomials in $t_1, \ldots, t_{n-1}$, contradiction.

Example. Consider the product

$$\delta(t) = \prod_{i<j} (t_i - t_j).$$

For any permutation σ of $(1, \ldots, n)$ we see at once that

$$\delta^\sigma(t) = \pm\delta(t).$$

Hence $\delta(t)^2$ is symmetric, and we call it the *discriminant:*

$$D(s_1, \ldots, s_n) = \prod_{i<j} (t_i - t_j)^2.$$

We thus view the discriminant as a polynomial in the elementary symmetric functions.

§10. *The resultant*

In this section, we assume that the reader is familiar with determinants. The theory of determinants will be covered later.

Let A be a commutative ring and let $v_0, \ldots, v_n, w_0, \ldots, w_m$ be algebraically independent over A. We form two polynomials:

$$f_v(X) = v_0X^n + \cdots + v_n,$$
$$g_w(X) = w_0X^m + \cdots + w_m.$$

We define the *resultant* of (v, w), or of f_v, g_w, to be the determinant

$$\underbrace{\begin{array}{c} m \left\{ \begin{array}{l} \\ \\ \\ \\ \end{array} \right. \\ n \left\{ \begin{array}{l} \\ \\ \\ \\ \end{array} \right. \end{array} \left| \begin{array}{llll} v_0 v_1 \ldots v_n & & & \\ & v_0 v_1 \ldots v_n & & \\ & \cdots\cdots\cdots & & \\ & & v_0 v_1 \ldots v_n & \\ w_0 w_1 \ldots w_m & & & \\ & w_0 w_1 \ldots w_m & & \\ & \cdots\cdots\cdots & & \\ & & w_0 w_1 \ldots w_m & \end{array} \right|}_{m+n}$$

The blank spaces are supposed to be filled with zeros.

If we substitute elements $(a) = (a_0, \ldots, a_n)$ and $(b) = (b_0, \ldots, b_m)$ in A for (v) and (w) respectively in the coefficients of f_v and g_w, then we obtain polynomials f_a and g_b with coefficients in A, and *we define their resultant to be the determinant obtained by substituting* (a) *for* (v) *and* (b) *for* (w) *in the determinant.* We shall write the resultant of f_v, g_w in the form

$$R(f_v, g_w) \qquad \text{or} \qquad R(v, w).$$

The resultant $R(f_a, g_b)$ is then obtained by substitution of (a), (b) for (v), (w) respectively.

We observe that $R(v, w)$ is a polynomial with integer coefficients, i.e. we may take $A = \mathbf{Z}$. If z is a variable, then

$$R(zv, w) = z^m R(v, w) \qquad \text{and} \qquad R(v, zw) = z^n R(v, w)$$

as one sees immediately by factoring out z from the first m rows (resp. the last n rows) in the determinant. Thus R is homogeneous of degree m in its first set of variables, and homogeneous of degree n in its second set of variables. Furthermore, $R(v, w)$ contains the monomial

$$v_0^m w_m^n$$

with coefficient 1, when expressed as a sum of monomials.

If we substitute 0 for v_0 and w_0 in the resultant, we obtain 0, because the first column of the determinant vanishes.

Let us work over the integers $\mathbf{Z}$. We consider the linear equations

$$\begin{aligned}
X^{m-1}f_v(X) &= v_0X^{n+m-1} + v_1X^{n+m-2} + \cdots + v_nX^{m-1}, \\
X^{m-2}f_v(X) &= \qquad\qquad v_0X^{n+m-2} + \cdots + v_nX^{m-2}, \\
&\cdots\cdots\cdots\cdots\cdots\cdots\cdots\cdots\cdots\cdots \\
f_v(X) &= \qquad\qquad\qquad v_0X^n + \cdots + v_n, \\
X^{n-1}g_w(X) &= w_0X^{n+m-1} + w_1X^{n+m-2} + \cdots + w_mX^{n-1}, \\
X^{n-2}g_w(X) &= \qquad\qquad w_0X^{n+m-2} + \cdots + w_mX^{n-2}, \\
&\cdots\cdots\cdots\cdots\cdots\cdots\cdots\cdots\cdots\cdots \\
g_w(X) &= \qquad\qquad\qquad w_0X^m + \cdots + w_m.
\end{aligned}$$

Let C be the column vector on the left-hand side, and let

$$C_0, \ldots, C_{m+n}$$

be the column vectors of coefficients. Our equations can be written

$$C = X^{n+m-1}C_0 + \cdots + 1 \cdot C_{m+n}.$$

By Cramer's rule, applied to the last coefficient which is $= 1$,

$$R(v, w) = \det(C_0, \ldots, C_{m+n}) = \det(C_0, \ldots, C_{m+n-1}, C).$$

From this we see that there exist polynomials $\varphi_{v,w}$ and $\psi_{v,w}$ in $\mathbf{Z}[v, w][X]$ such that

$$\varphi_{v,w}f_v + \psi_{v,w}g_w = R(v, w).$$

Note that $R(v, w) \in \mathbf{Z}[v, w]$ but that the polynomials on the left-hand side involve the variable X.

If $\lambda : \mathbf{Z}[v, w] \to A$ is a homomorphism into a commutative ring A and we let $\lambda(v) = (a)$, $\lambda(w) = (b)$, then

$$\varphi_{a,b}f_a + \psi_{a,b}g_b = R(a, b) = R(f_a, f_b).$$

Thus from the universal relation of the resultant over $\mathbf{Z}$ we obtain a similar relation for every pair of polynomials, in any commutative ring A.

PROPOSITION 3. *Let K be a subfield of a field L, and let f_a, g_b be polynomials in $K[X]$ having a common root ξ in L. Then $R(a, b) = 0$.*

Proof. If $f_a(\xi) = g_b(\xi) = 0$, then we substitute ξ for X in the expression obtained for $R(a, b)$ and find $R(a, b) = 0$.

Next, we shall investigate the relationship between the resultant and the roots of our polynomials f_v, g_w. We need a lemma.

LEMMA. *Let $h(X_1, \ldots, X_n)$ be a polynomial in n variables over the integers* **Z**. *If h has the value 0 when we substitute X_1 for X_2 and leave the other X_i fixed ($i \neq 2$), then $h(X_1, \ldots, X_n)$ is divisible by $X_1 - X_2$ in $\mathbf{Z}[X_1, \ldots, X_n]$.*

Proof. Exercise for the reader.

Let $v_0, t_1, \ldots, t_n, w_0, u_1, \ldots, u_m$ be algebraically independent over **Z** and form the polynomials

$$f_v = v_0(X - t_1) \cdots (X - t_n) = v_0X^n + \cdots + v_n,$$
$$g_w = w_0(X - u_1) \cdots (X - u_m) = w_0X^m + \cdots + w_m.$$

Thus we let

$$v_i = (-1)^i v_0 s_i(t) \qquad \text{and} \qquad w_j = (-1)^j w_0 s_j(u).$$

We leave to the reader the easy verification that

$$v_0, v_1, \ldots, v_n, w_0, w_1, \ldots, w_m$$

are algebraically independent over **Z**.

PROPOSITION 4. *Notation being as above, we have*

$$R(f_v, g_w) = v_0^m w_0^n \prod_{i=1}^{n} \prod_{j=1}^{m} (t_i - u_j).$$

Proof. Let S be the expression on the right-hand side of the equality in the statement of the proposition.

Since $R(v, w)$ is homogeneous of degree m in its first variables, and homogeneous of degree n in its second variables, it follows that

$$R = v_0^m w_0^n h(t, u)$$

where $h(t, u) \in \mathbf{Z}[t, u]$. By Proposition 3, the resultant vanishes when we substitute t_i for u_j ($i = 1, \ldots, n$ and $j = 1, \ldots, m$), whence by the lemma, viewing R as an element of $\mathbf{Z}[t, u]$ it follows that R is divisible by $t_i - u_j$ for each pair (i, j). Hence S divides R in $\mathbf{Z}[t, u]$, because $t_i - u_j$ is obviously a prime in that ring, and different pairs (i, j) give rise to different primes.

From the product expression for S, namely

$$S = v_0^m w_0^n \prod_{i=1}^{n} \prod_{j=1}^{m} (t_i - u_j), \tag{1}$$

we obtain

$$\prod_{i=1}^{n} g(t_i) = w_0^n \prod_{i=1}^{n} \prod_{j=1}^{m} (t_i - u_j),$$

whence

$$S = v_0^m \prod_{i=1}^{n} g(t_i). \tag{2}$$

Similarly,

$$S = (-1)^{nm} w_0^n \prod_{j=1}^{m} f(u_j). \tag{3}$$

From (2) we see that S is homogeneous and of degree n in (w), and from (3) we see that S is homogeneous and of degree m in (v). Since R has exactly the same homogeneity properties, and is divisible by S, it follows that $R = cS$ for some integer c. Since both R and S have a monomial $v_0^m w_m^n$ occurring in them with coefficient 1, it follows that $c = 1$, and our proposition is proved.

We also note that the three expressions found for S above now give us a factorization of R. We also get a converse for Proposition 3.

Corollary. *Let f_a, g_b be polynomials with coefficients in a field K, such that $a_0 b_0 \neq 0$, and such that f_a, g_b split in factors of degree 1 in $K[X]$. Then $R(f_a, g_b) = 0$ if and only if f_a and g_b have a root in common.*

Proof. Assume that the resultant is 0. If

$$f_a = a_0(X - \alpha_1) \cdots (X - \alpha_n),$$
$$g_b = b_0(X - \beta_1) \cdots (X - \beta_n)$$

is the factorization of f_a, g_b, then we have a homomorphism

$$\mathbf{Z}[v_0, t, w_0, u] \to K$$

such that $v_0 \mapsto a_0$, $w_0 \mapsto b_0$, $t_i \mapsto \alpha_i$, and $u_j \mapsto \beta_j$ for all i, j. Then

$$0 = R(f_a, g_b) = a_0^m b_0^n \prod_i \prod_j (\alpha_i - \beta_j),$$

whence f_a, f_b have a root in common. The converse has already been proved.

We deduce one more relation for the resultant in a special case. Let f_v be as above,

$$f_v(X) = v_0 X^n + \cdots + v_n = v_0(X - t_1) \cdots (X - t_n).$$

From (2) we know that if f_v' is the derivative of f_v, then

$$R(f_v, f_v') = v_0^{n-1} \prod_i f'(t_i). \tag{4}$$

Using the product rule for differentiation, we find:

$$f_v'(X) = \sum_i v_0(X - t_1) \cdots \widehat{(X - t_i)} \cdots (X - t_n),$$

$$f_v'(t_i) = v_0(t_i - t_1) \cdots \widehat{(t_i - t_i)} \cdots (t_i - t_n),$$

where a roof over a term means that this term is to be omitted.

We define the discriminant of f_v to be

$$D(f_v) = D(v) = v_0^{2n-2} \prod_{i \neq j} (t_i - t_j).$$

PROPOSITION 5. *Let f_v be as above have algebraically independent coefficients over* $\mathbf{Z}$. *Then*

$$R(f_v, f_v') = v_0^{2n-1} \prod_{i \neq j} (t_i - t_j) = v_0 D(f_v). \tag{5}$$

Proof. One substitutes the expression obtained for $f_v'(t_i)$ into the product (4). The result follows at once.

When we substitute 1 for v_0, we find that the discriminant as we defined it in the preceding section coincides with the present definition. In particular, we find an explicit formula for the discriminant. The formulas in the special case of polynomials of degree 2 and 3 will be given as exercises.

EXERCISES

1. (a) State and prove the analogue of Theorem 8 for the rational numbers. (b) State and prove the analogue of Theorem 9 for positive integers.

2. Let f be a polynomial in one variable over a field k. Let X, Y be two variables. Show that in $k[X, Y]$ we have a "Taylor series" expansion

$$f(X + Y) = f(X) + \sum_{i=1}^{n} \varphi_i(X) Y^i,$$

where $\varphi_i(X)$ is a polynomial in X with coefficients in k. If k has characteristic 0, show that

$$\varphi_i(X) = \frac{D^i f(X)}{i!}.$$

3. Generalize the preceding exercise to polynomials in several variables (introduce partial derivatives and show that a finite Taylor expansion exists for a polynomial in several variables).

4. (a) Show that the polynomials $X^4 + 1$ and $X^6 + X^3 + 1$ are irreducible over the rational numbers.

(b) Show that a polynomial of degree 3 over a field is either irreducible or has a root in the field. Is $X^3 - 5X^2 + 1$ irreducible over the rational numbers?

(c) Show that the polynomial in two variables $X^2 + Y^2 - 1$ is irreducible over the rational numbers. Is it irreducible over the complex numbers?

5. Let $f(X) = X^n + a_{n-1}X^{n-1} + \cdots + a_0$ be a polynomial with integer coefficients, $a_0 \neq 0$. If f has a root in the rational numbers, show that this root must be an integer, and that this integer divides a_0. Generalize this statement to a factorial ring and its quotient field.

6. (a) Let k be a finite field with q elements. Let $f(X_1, \ldots, X_n)$ be a polynomial in $k[X]$ of degree d and assume $f(0, \ldots, 0) = 0$. An element $(a_1, \ldots, a_n) \in k^{(n)}$ such that $f(a) = 0$ is called a zero of f. If $n > d$, show that f has at least one other zero in $k^{(n)}$. [*Hint:* Assume the contrary, and compare the degrees of the reduced polynomial belonging to

$$1 - f(X)^{q-1}$$

and $(1 - X_1^{q-1}) \cdots (1 - X_n^{q-1})$. The argument is due to Chevalley.]

(b) Refine the above result by proving that the number N of zeros of f in $k^{(n)}$ is $\equiv 0 \pmod{p}$, arguing as follows. Let i be an integer $\geqq 1$. Show that

$$\sum_{x \in k} x^i = \begin{cases} q - 1 = -1 & \text{if } q - 1 \text{ divides } i \\ 0 & \text{otherwise.} \end{cases}$$

Denote the preceding function of i by $\psi(i)$. Show that

$$N \equiv \sum_{x \in k^{(n)}} (1 - f(x)^{q-1})$$

and for each n-tuple $(i_1, \ldots, i_n)$ of integers $\geqq 0$ that

$$\sum_{x \in k^{(n)}} x_1^{i_1} \cdots x_n^{i_n} = \psi(i_1) \cdots \psi(i_n).$$

Show that both terms in the sum for N above yield 0 mod p. (The above argument is due to Warning.)

(c) Extend Chevalley's theorem to r polynomials $f_1, \ldots, f_r$ of degrees $d_1, \ldots, dr$ respectively, in n variables. If they have no constant term and $n > \sum d_i$, show that they have a non-trivial common zero.

(d) Show that an arbitrary function $f : k^{(n)} \to k$ can be represented by a polynomial. (As before, k is a finite field.)

7. Let A be a commutative entire ring and X a variable over A. Let $a, b \in A$ and assume that a is a unit in A. Show that the map $X \to aX + b$ extends to a unique automorphism of $A[X]$ inducing the identity on A. What is the inverse automorphism?

8. Show that every automorphism of $A[X]$ is of the type described in Exercise 7.

9. Let A be a commutative entire ring, K its quotient field, and $K(X)$ the quotient field of $A[X]$ (or $K[X]$, it is the same thing). Show that every automorphism of $K(X)$ which induces the identity on K is of type

$$X \mapsto \frac{aX + b}{cX + d}$$

with $a, b, c, d \in K$ such that $(aX + b)/(cX + d)$ is not an element of K, or equivalently, $ad - bc \neq 0$.

10. If $f(X) = aX^2 + bX + c$, show that the discriminant of f is $b^2 - 4ac$.

11. If $f(X) = a_0X^3 + a_1X^2 + a_2X + a_3$, show that the discriminant of f is

$$a_1^2a_2^2 - 4a_0a_2^3 - 4a_1^3a_3 - 27a_0^2a_3^2 + 18a_0a_1a_2a_3.$$

In particular, if $f(X) = X^3 + bX + c$, its discriminant is $-4b^3 - 27c^2$.

12. Show that the discriminant of a polynomial vanishes if and only if the polynomial has a multiple root. (You may assume that the polynomial splits into factors of degree 1 in some field.)

13. Let w be a complex number. Show that there exists a constant $c = c(w)$ such that the following is true. Let F, G be non-zero polynomials in one variable with complex coefficients, of degrees d and d' respectively. Let R be their resultant. Then

$$|R| \leqq c^{d+d'}[|F(w)| + |G(w)|]|F|^{d'}|G|^{d}(d + d')^{d+d'}.$$

(We denote by $|F|$ the maximum of the absolute values of the coefficients of F.)

14. Show that one can define partial fractions for positive rational numbers, i.e. get a similar decomposition as in Theorem 8. Show that $\mathbf{Q}/\mathbf{Z}$ is isomorphic to the direct sum of the additive groups $\mathbf{Z}[1/p]/\mathbf{Z}$ taken over all primes p. Generalize to an arbitrary principal ring A. If K is the quotient field of A, what does K/A look like?

15. The following exercise is somewhat harder than the preceding ones. Let m/n be a rational number expressed as a quotient of relatively prime integers m, n. We define its *height* $H(m/n)$ to be the maximum of $|m|$, $|n|$. Let

$$\varphi(X) = \frac{f(X)}{g(X)}$$

be an element of $\mathbf{Q}(X)$ expressed as a quotient of two relatively prime polynomials f, g. We define the *degree* of φ to be the maximum of $\deg f$, $\deg g$. If $a \in \mathbf{Q}$ is such that $g(a) \neq 0$, then we can form $\varphi(a) = f(a)/g(a)$ and we say that φ is defined at a. Let φ be of degree d. Show that there exist two constants $c_1, c_2 > 0$ such that for all rational numbers a such that $\varphi(a)$ is defined, we have

$$c_1H(a)^d \leqq H(\varphi(a)) \leqq c_2H(a)^d.$$

[*Hint:* One equality is trivial. To prove the other, show that the function $H(\varphi(x))/H(x)^d$ is bounded.]

CHAPTER VI

Noetherian Rings and Modules

§1. *Basic criteria*

Let A be a ring and M a module (i.e. a left A-module). We shall say that M is *Noetherian* if it satisfies any one of the following three conditions:

(i) Every submodule of M is finitely generated.

(ii) Every ascending sequence of submodules of M,

$$M_1 \subset M_2 \subset M_3 \subset \cdots,$$

such that $M_i \neq M_{i+1}$ is finite.

(iii) Every non-empty set S of submodules of M has a maximal element (i.e. a submodule M_0 such that for any element N of S which contains M_0 we have $N = M_0$).

We shall now prove that the above three conditions are equivalent.

(i) $\Rightarrow$ (ii) Suppose we have an ascending sequence of submodules of M as above. Let N be the union of all the M_i $(i = 1, 2, \ldots)$. Then N is finitely generated, say by elements $x_1, \ldots, x_r$, and each generator is in some M_i. Hence there exists an index j such that

$$x_1, \ldots, x_r \in M_j.$$

Then

$$\langle x_1, \ldots, x_r \rangle \subset M_j \subset N = \langle x_1, \ldots, x_r \rangle,$$

whence equality holds and our implication is proved.

(ii) $\Rightarrow$ (iii) Let N_0 be an element of S. If N_0 is not maximal, it is properly contained in a submodule N_1. If N_1 is not maximal, it is properly contained in a submodule N_2. Inductively, if we have found N_i which is not maximal, it is contained properly in a submodule N_{i+1}. In this way we could construct an infinite chain, which is impossible.

(iii) $\Rightarrow$ (i) Let N be a submodule of M. Let $a_0 \in N$. If $N \neq \langle a_0 \rangle$, then there exists an element $a_1 \in N$ which does not lie in $\langle a_0 \rangle$. Proceeding inductively, we can find an ascending sequence of submodules of N, namely

$$\langle a_0 \rangle \subset \langle a_0, a_1 \rangle \subset \langle a_0, a_1, a_2 \rangle \subset \cdots$$

where the inclusion each time is proper. The set of these submodules has a maximal element, say a submodule $\langle a_0, a_1, \ldots, a_r \rangle$, and it is then clear that this finitely generated submodule must be equal to N, as was to be shown.

PROPOSITION 1. *Let M be a Noetherian A-module. Then every submodule and every factor module of M is Noetherian.*

Proof. Our assertion is clear for submodules (say from the first condition). For the factor module, let N be a submodule and $f: M \to M/N$ the canonical homomorphism. Let $\overline{M}_1 \subset \overline{M}_2 \subset \cdots$ be an ascending chain of submodules of M/N and let $M_i = f^{-1}(\overline{M}_i)$. Then $M_1 \subset M_2 \subset \cdots$ is an ascending chain of submodules of M, which must have a maximal element, say M_r, so that $M_i = M_r$ for $r \geqq i$. Then $f(M_i) = \overline{M}_i$ and our assertion follows.

PROPOSITION 2. *Let M be a module, N a submodule. Assume that N and M/N are Noetherian. Then M is Noetherian.*

Proof. With every submodule L of M we associate the pair of modules

$$L \mapsto (L \cap N, (L + N)/N).$$

We contend: If $E \subset F$ are two submodules of M such that their associated pairs are equal, then $E = F$. To see this, let $x \in F$. By the hypothesis that $(E + N)/N = (F + N)/N$ there exist elements $u, v \in N$ and $y \in E$ such that $y + u = x + v$. Then

$$x - y = u - v \in F \cap N = E \cap N.$$

Since $x = y + u - v$, it follows that $x \in E$ and our contention is proved. If we have an ascending sequence

$$E_1 \subset E_2 \subset \cdots$$

then the associated pairs form an ascending sequence of submodules of N and M/N respectively, and these sequences must stop. Hence our sequence $E_1 \subset E_2 \cdots$ also stops, by our preceding contention.

Propositions 1 and 2 may be summarized by saying that in an exact sequence $0 \to M' \to M \to M'' \to 0$, M is Noetherian if and only if M' and M'' are Noetherian.

COROLLARY. *Let M be a module, and let N, N' be submodules. If $M = N + N'$ and if both N, N' are Noetherian, then M is Noetherian. A finite direct sum of Noetherian modules is Noetherian.*

Proof. We first observe that the direct product $N \times N'$ is Noetherian since it contains N as a submodule whose factor module is isomorphic to N', and Proposition 2 applies. We have a surjective homomorphism

$$N \times N' \to M$$

such that the pair (x, x') with $x \in N$ and $x' \in N'$ maps on $x + x'$. By Proposition 1, it follows that M is Noetherian. Finite products (or sums) follow by induction.

A ring A is called *Noetherian* if it is Noetherian as a left module over itself. This means that every left ideal is finitely generated.

PROPOSITION 3. *Let A be a Noetherian ring and let M be a finitely generated module. Then M is Noetherian.*

Proof. Let $x_1, \ldots, x_n$ be generators of M. There exists a homomorphism

$$f : A \times A \times \cdots \times A \to M$$

of the product of A with itself n times such that

$$f(a_1, \ldots, a_n) = a_1x_1 + \cdots + a_nx_n.$$

This homomorphism is surjective. By the corollary of the preceding proposition, the product is Noetherian, and hence M is Noetherian by Proposition 1.

PROPOSITION 4. *Let A be a ring which is Noetherian, and let $\varphi : A \to B$ be a surjective ring-homomorphism. Then B is Noetherian.*

Proof. Let $\mathfrak{b}_1 \subset \cdots \subset \mathfrak{b}_n \subset \cdots$ be an ascending chain of left ideals of B and let $\mathfrak{a}_i = \varphi^{-1}(\mathfrak{b}_i)$. Then the $\mathfrak{a}_i$ form an ascending chain of left ideals of A which must stop, say at $\mathfrak{a}_r$. Since $\varphi(\mathfrak{a}_i) = \mathfrak{b}_i$ for all i, our proposition is proved.

PROPOSITION 5. *Let A be a commutative Noetherian ring, and let S be a multiplicative subset of A. Then $S^{-1}A$ is Noetherian.*

Proof. We leave the proof as an exercise.

§2. *Hilbert's theorem*

THEOREM 1. *Let A be a commutative Noetherian ring. Then the polynomial ring $A[X]$ is also Noetherian.*

Proof. Let $\mathfrak{A}$ be an ideal of $A[X]$. Let $\mathfrak{a}_i$ be the set of elements $a \in A$ appearing as leading coefficient in some polynomial

$$a_0 + a_1X + \cdots + aX^i$$

lying in $\mathfrak{A}$. Then it is clear that $\mathfrak{a}_i$ is an ideal. (If a, b are in $\mathfrak{a}_i$, then $a \pm b$ is in $\mathfrak{a}_i$ as one sees by taking the sum and difference of the corresponding polynomials. If $x \in A$, then $xa \in \mathfrak{a}_i$ as one sees by multiplying the corresponding polynomial by x.) Furthermore we have

$$\mathfrak{a}_0 \subset \mathfrak{a}_1 \subset \mathfrak{a}_2 \subset \cdots,$$

in other words, our sequence of ideals $\{\mathfrak{a}_i\}$ is increasing. Indeed, to see this multiply the above polynomial by X to see that $a \in \mathfrak{a}_{i+1}$.

The sequence of ideals $\{\mathfrak{a}_i\}$ stops, say at $\mathfrak{a}_r$:

$$\mathfrak{a}_0 \subset \mathfrak{a}_1 \subset \mathfrak{a}_2 \subset \cdots \subset \mathfrak{a}_r = \mathfrak{a}_{r+1} = \cdots$$

Let

$$a_{01}, \ldots, a_{0n_0} \quad \text{be generators for} \quad \mathfrak{a}_0,$$
$$\cdots\cdots\cdots\cdots\cdots$$
$$a_{r1}, \ldots, a_{rn_r} \quad \text{be generators for} \quad \mathfrak{a}_r.$$

For each $i = 0, \ldots, r$ and $j = 1, \ldots, n_i$ let f_{ij} be a polynomial in $\mathfrak{A}$, of degree i, with leading coefficient a_{ij}. We contend that the polynomials f_{ij} are a set of generators for $\mathfrak{A}$.

Let f be a polynomial of degree d in $\mathfrak{A}$. We shall prove that f is in the ideal generated by the f_{ij}, by induction on d. Say $d \geqq 0$. If $d > r$, then we note that the leading coefficients of

$$X^{d-r}f_{r1}, \ldots, X^{d-r}f_{rn_r}$$

generate $\mathfrak{a}_d$. Hence there exist elements $c_1, \ldots, c_{n_r} \in A$ such that the polynomial

$$f - c_1X^{d-r}f_{r1} - \cdots - c_{n_r}X^{d-r}f_{rn_r}$$

has degree $< d$, and this polynomial also lies in $\mathfrak{A}$. If $d \leqq r$, we can subtract a linear combination

$$f - c_1f_{d1} - \cdots - c_{n_d}f_{dn_d}$$

to get a polynomial of degree $< d$, also lying in $\mathfrak{A}$. We note that the polynomial we have subtracted from f lies in the ideal generated by the f_{ij}. By induction, we can subtract a polynomial g in the ideal generated by the f_{ij} such that $f - g = 0$, thereby proving our theorem.

COROLLARY. *Let A be a Noetherian commutative ring, and let $B = A[x_1, \ldots, x_m]$ be a finitely generated commutative ring containing A as a subring. Then B is Noetherian.*

Proof. Use Theorem 1 and Proposition 4, representing B as a factor ring of a polynomial ring.

§3. *Power series*

Let X be a letter, and let G be the monoid of functions from the set $\{X\}$ to the natural numbers. If $\nu \in \mathbf{N}$, we denote by X^ν the function whose value at X is ν. Then G is a multiplicative monoid, already encountered when we discussed polynomials. Its elements are $X^0, X^1, X^2, \ldots, X^\nu, \ldots$

Let A be a commutative ring, and let $A[[X]]$ be the set of functions from G into A, without any restriction. Then an element of $A[[X]]$ may be viewed as assigning to each monomial X^ν a coefficient $a_\nu \in A$. We denote this element by

$$\sum_{\nu=0}^{\infty} a_\nu X^\nu.$$

The summation symbol is not a sum, of course, but we shall write the above expression also in the form

$$a_0 X^0 + a_1 X^1 + \cdots$$

and we call it a *formal power series* with coefficients in A, in one variable. We call $a_0, a_1, \ldots$ its coefficients.

Given two elements of $A[[X]]$, say

$$\sum_{\nu=0}^{\infty} a_\nu X^\nu \quad \text{and} \quad \sum_{\mu=0}^{\infty} b_\mu X^\mu,$$

we define their product to be

$$\sum_{i=0}^{\infty} c_i X^i$$

where

$$c_i = \sum_{\nu+\mu=i} a_\nu b_\mu.$$

Just as with polynomials, one defines their sum to be

$$\sum_{\nu=0}^{\infty} (a_\nu + b_\nu) X^\nu.$$

Then we see that the power series form a ring, the proof being the same as for polynomials.

One can also construct the power series ring in several variables $A[[X_1, \ldots, X_n]]$ in which every element can be expressed in the form

$$\sum_{(\nu)} a_{(\nu)} X_1^{\nu_1} \cdots X_n^{\nu_n} = \sum a_{(\nu)} M_{(\nu)}(X_1, \ldots, X_n)$$

with unrestricted coefficients $a_{(\nu)}$ in bijection with the n-tuples of integers

$(\nu_1, \ldots, \nu_n)$ such that $\nu_i \geqq 0$ for all i. It is then easy to show that there is an isomorphism between $A[[X_1, \ldots, X_n]]$ and the repeated power series ring $A[[X_1]] \cdots [[X_n]]$. We leave this as an exercise for the reader.

THEOREM 2. *If A is Noetherian, then $A[[X]]$ is also Noetherian.*

Proof. Our argument will be a modification of the argument used in the proof of Hilbert's theorem for polynomials. We shall consider elements of lowest degree instead of elements of highest degree.

Let $\mathfrak{A}$ be an ideal of $A[[X]]$. We let $\mathfrak{a}_i$ be the set of elements $a \in A$ such that a is the coefficient of X^i in a power series

$$aX^i + \text{terms of higher degree}$$

lying in $\mathfrak{A}$. Then $\mathfrak{a}_i$ is an ideal of A, and $\mathfrak{a}_i \subset \mathfrak{a}_{i+1}$ (the proof of this assertion being the same as for polynomials). The ascending chain of ideals stops:

$$\mathfrak{a}_0 \subset \mathfrak{a}_1 \subset \mathfrak{a}_2 \subset \cdots \subset \mathfrak{a}_r = \mathfrak{a}_{r+1} = \cdots$$

As before, let a_{ij} ($i = 0, \ldots, r$ and $j = 1, \ldots, n_i$) be generators for the ideals $\mathfrak{a}_i$, and let f_{ij} be power series in A having a_{ij} as beginning coefficient. Given $f \in \mathfrak{A}$, starting with a term of degree d, say $d \leqq r$, we can find elements

$$c_1, \ldots, c_{n_d} \in A$$

such that

$$f - c_1 f_{d1} - \cdots - c_{n_d} f_{dn_d}$$

starts with a term of degree $\geqq d + 1$. Proceeding inductively, we may assume that $d > r$. We then use a linear combination

$$f - c_1^{(d)} X^{d-r} f_{r1} - \cdots - c_{n_r}^{(d)} X^{d-r} f_{rn_r}$$

to get a power series starting with a term of degree $\geqq d + 1$. In this way, if we start with a power series of degree $d > r$, then it can be expressed as a linear combination of

$$f_{r1}, \ldots, f_{rn_r}$$

by means of the coefficients

$$g_1(X) = \sum_{\nu=d}^{\infty} c_1^{(\nu)} X^{\nu-r}, \ldots, g_{n_r}(X) = \sum_{\nu=d}^{\infty} c_{n_r}^{(\nu)} X^{\nu-r},$$

and we see that the f_{ij} generate our ideal $\mathfrak{A}$, as was to be shown.

COROLLARY. *If A is a Noetherian commutative ring, or a field, then $A[[X_1, \ldots, X_n]]$ is Noetherian.*

§4. *Associated primes*

Throughout this section, we let A be a commutative ring. Modules and homomorphisms are A-modules and A-homomorphisms unless otherwise specified.

PROPOSITION 6. *Let S be a multiplicative subset of A, and assume that S does not contain 0. Then there exists an ideal of A which is maximal in the set of ideals not intersecting S, and any such ideal is prime.*

Proof. The existence of such an ideal $\mathfrak{p}$ follows from Zorn's lemma (the set of ideals not meeting S is not empty, because it contains the zero ideal, and is clearly inductively ordered). Let $\mathfrak{p}$ be maximal in the set. Let $a, b \in A$, $ab \in \mathfrak{p}$, but $a \notin \mathfrak{p}$ and $b \notin \mathfrak{p}$. By hypothesis, the ideals $(a, \mathfrak{p})$ and $(b, \mathfrak{p})$ generated by a and $\mathfrak{p}$ (or b and $\mathfrak{p}$ respectively) meet S, and there exist therefore elements $s, s' \in S$, $c, c', x, x' \in A$, $p, p' \in \mathfrak{p}$ such that

$$s = ca + xp \qquad \text{and} \qquad s' = c'b + x'p'.$$

Multiplying these two expressions, we obtain

$$ss' = cc'ab + p''$$

with some $p'' \in \mathfrak{p}$, whence we see that ss' lies in $\mathfrak{p}$. This contradicts the fact that $\mathfrak{p}$ does not intersect S, and proves that $\mathfrak{p}$ is prime.

An element a of A is said to be *nilpotent* if there exists an integer $n \geqq 1$ such that $a^n = 0$.

COROLLARY 1. *An element a of A is nilpotent if and only if it lies in every prime ideal of A.*

Proof. If $a^n = 0$, then $a^n \in \mathfrak{p}$ for every prime $\mathfrak{p}$, and hence $a \in \mathfrak{p}$. If $a^n \neq 0$ for any positive integer n, we let S be the multiplicative subset of powers of a, namely $\{1, a, a^2, \ldots\}$, and find a prime ideal as in the proposition to prove the converse.

Let $\mathfrak{a}$ be an ideal of A. The *nilradical* of $\mathfrak{a}$ is the set of all $a \in A$ such that $a^n \in \mathfrak{a}$ for some integer $n \geqq 1$, (or equivalently, it is the set of elements $a \in A$ whose image in the factor ring $A/\mathfrak{a}$ is nilpotent). We observe that the nilradical of $\mathfrak{a}$ is an ideal, for if $a^n = 0$ and $b^m = 0$ then $(a + b)^k = 0$ if k is sufficiently large: In the binomial expansion, either a or b will appear with a power at least equal to n or m.

COROLLARY 2. *An element a of A lies in the nilradical of an ideal $\mathfrak{a}$ if and only if it lies in every prime ideal containing $\mathfrak{a}$.*

Proof. Corollary 2 is equivalent to Corollary 1 applied to the ring $A/\mathfrak{a}$.

We shall extend Corollary 1 to modules. We first make some remarks on localization. Let S be a multiplicative subset of A. If M is a module, we can define $S^{-1}M$ in the same way that we defined $S^{-1}A$. We consider equivalence classes of pairs (x, s) with $x \in M$ and $s \in S$, two pairs (x, s) and (x', s') being equivalent if there exists $s_1 \in S$ such that $s_1(s'x - sx') = 0$. We denote the equivalence class of (x, s) by x/s, and verify at once that the set of equivalence classes is an additive group (under the obvious operations). It is in fact an A-module, under the operation

$$(a, x/s) \mapsto ax/s.$$

We shall denote this module of equivalence classes by $S^{-1}M$. (We note that $S^{-1}M$ could also be viewed as an $S^{-1}A$-module.)

If $\mathfrak{p}$ is a prime ideal of A, and S is the complement of $\mathfrak{p}$ in A, then $S^{-1}M$ is also denoted by $M_{\mathfrak{p}}$.

It follows trivially from the definitions that if $N \to M$ is an injective homomorphism, then we have a natural injection $S^{-1}N \to S^{-1}M$. In other words, if N is a submodule of M, then $S^{-1}N$ can be viewed as a submodule of $S^{-1}M$. If $x \in N$ and $s \in S$, then the fraction x/s can be viewed as an element of $S^{-1}N$ or $S^{-1}M$. If $x/s = 0$ in $S^{-1}N$, then there exists $s_1 \in S$ such that $s_1x = 0$, and this means that x/s is also 0 in $S^{-1}M$. Thus if $\mathfrak{p}$ is a prime ideal and N is a submodule of M, we have a natural inclusion of $N_{\mathfrak{p}}$ in $M_{\mathfrak{p}}$. We shall in fact identify $N_{\mathfrak{p}}$ as a submodule of $M_{\mathfrak{p}}$. In particular, we see that $M_{\mathfrak{p}}$ is the sum of its submodules $(Ax)_{\mathfrak{p}}$, for $x \in M$ (but of course not the direct sum).

Let $x \in M$. The *annihilator* $\mathfrak{a}$ of x is the ideal consisting of all elements $a \in A$ such that $ax = 0$. We have an isomorphism (of modules)

$$A/\mathfrak{a} \xrightarrow{\approx} Ax$$

under the map

$$a \mapsto ax.$$

LEMMA. *Let x be an element of a module M, and let $\mathfrak{a}$ be its annihilator. Let $\mathfrak{p}$ be a prime ideal of A. Then $(Ax)_{\mathfrak{p}} \neq 0$ if and only if $\mathfrak{p}$ contains $\mathfrak{a}$.*

Proof. The lemma is an immediate consequence of the definitions, and will be left to the reader.

Let a be an element of A. Let M be a module. The homomorphism

$$x \mapsto ax, \qquad x \in M$$

will be called the *principal homomorphism* associated with a, and will be denoted by a_M. We shall say that a_M is *locally nilpotent* if for each $x \in M$ there exists an integer $n(x) \geqq 1$ such that $a^{n(x)}x = 0$. This condition implies that for every finitely generated submodule N of M, there exists an integer $n \geqq 1$ such that $a^n N = 0$: We take for n the largest power of a annihilating a finite set of generators of N. Therefore, *if M is finitely generated, a_M is locally nilpotent if and only if it is nilpotent.*

PROPOSITION 7. *Let M be a module, $a \in A$. Then a_M is locally nilpotent if and only if a lies in every prime ideal $\mathfrak{p}$ such that $M_{\mathfrak{p}} \neq 0$.*

Proof. Assume that a_M is locally nilpotent. Let $\mathfrak{p}$ be a prime of A such that $M_{\mathfrak{p}} \neq 0$. Then there exists $x \in M$ such that $(Ax)_{\mathfrak{p}} \neq 0$. Let n be a positive integer such that $a^n x = 0$. Let $\mathfrak{a}$ be the annihilator of x. Then $a^n \in \mathfrak{a}$, and hence we can apply the lemma, and Corollary 2 of Proposition 6 to conclude that a lies in every prime $\mathfrak{p}$ such that $M_{\mathfrak{p}} \neq 0$. Conversely, given $x \in M$, $x \neq 0$, we consider the module Ax, and reverse the preceding arguments to prove that $a^n x = 0$ for some $n \geqq 1$, thereby proving that a_M is locally nilpotent.

Let M be a module. A prime ideal $\mathfrak{p}$ of A will be said to be *associated* with M if there exists an element $x \in M$ such that $\mathfrak{p}$ is the annihilator of x. In particular, since $\mathfrak{p} \neq A$, we must have $x \neq 0$.

PROPOSITION 8. *Let M be a module $\neq 0$. Let $\mathfrak{p}$ be a maximal element in the set of ideals which are annihilators of elements $x \in M$, $x \neq 0$. Then $\mathfrak{p}$ is prime.*

Proof. Let $\mathfrak{p}$ be the annihilator of the element $x \neq 0$. Then $\mathfrak{p} \neq A$. Let $a, b \in A$, $ab \in \mathfrak{p}$, $a \notin \mathfrak{p}$. Then $ax \neq 0$. But the ideal $(b, \mathfrak{p})$ annihilates ax, and contains $\mathfrak{p}$. Since $\mathfrak{p}$ is maximal, it follows that $b \in \mathfrak{p}$, and hence $\mathfrak{p}$ is prime.

COROLLARY 1. *If A is Noetherian and M is a module $\neq 0$, then there exists a prime associated with M.*

Proof. The set of ideals as in Proposition 8 is not empty since $M \neq 0$, and has a maximal element because A is Noetherian.

COROLLARY 2. *Assume that both A and M are Noetherian, $M \neq 0$. Then there exists a sequence of submodules*

$$M = M_1 \supset M_2 \supset \cdots \supset M_r = 0$$

such that each factor module M_i/M_{i+1} is isomorphic to $A/\mathfrak{p}_i$ for some prime $\mathfrak{p}_i$.

Proof. Consider the set of submodules having the property described in the Corollary. It is not empty, since there exists an associated prime $\mathfrak{p}$ of M, and if $\mathfrak{p}$ is the annihilator of x, then $Ax \approx A/\mathfrak{p}$. Let N be a maximal element in the set. If $N \neq M$, then by the preceding argument applied to M/N, there exists a submodule N' of M containing N such that N'/N is isomorphic to $A/\mathfrak{p}$ for some $\mathfrak{p}$, and this contradicts the maximality of N.

Proposition 9. *Let A be Noetherian, and $a \in A$. Let M be a module. Then a_M is injective if and only if a does not lie in any associated prime of M.*

Proof. Assume that a_M is not injective, so that $ax = 0$ for some $x \in M$, $x \neq 0$. By Corollary 1 of Proposition 8, there exists an associated prime $\mathfrak{p}$ of Ax, and a is an element of $\mathfrak{p}$. Conversely, if a_M is injective, then a cannot lie in any associated prime because a does not annihilate any non-zero element of M.

Proposition 10. *Let A be Noetherian, and let M be a module. Let $a \in A$. The following conditions are equivalent:*

(1) *a_M is locally nilpotent.*
(2) *a lies in every associated prime of M.*
(3) *a lies in every prime $\mathfrak{p}$ such that $M_{\mathfrak{p}} \neq 0$.*

Proof. The fact that (1) implies (2) is obvious from the definitions, and does not need the hypothesis that A is Noetherian. Neither does the fact that (3) implies (1), which has been proved in Proposition 7. We must therefore prove that (2) implies (3). Let $\mathfrak{p}$ be a prime such that $M_{\mathfrak{p}} \neq 0$. Then there exists $x \in M$ such that $(Ax)_{\mathfrak{p}} \neq 0$. By Proposition 8, there exists an associated prime $\mathfrak{q}$ of $(Ax)_{\mathfrak{p}}$ in A. Hence there exists an element y/s of $(Ax)_{\mathfrak{p}}$, with $y \in Ax$, $s \notin \mathfrak{p}$, and $y/s \neq 0$, such that $\mathfrak{q}$ is the annihilator of y/s. It follows that $\mathfrak{q} \subset \mathfrak{p}$, for otherwise, there exists $b \in \mathfrak{q}$, $b \notin \mathfrak{p}$, and $0 = by/s$, whence $y/s = 0$, contradiction. Let $b_1, \ldots, b_n$ be generators for $\mathfrak{q}$. For each i, there exists $s_i \in A$, $s_i \notin \mathfrak{p}$, such that $s_i b_i y = 0$ because $b_i y/s = 0$. Let $t = s_1 \cdots s_n$. Then it is trivially verified that $\mathfrak{q}$ is the annihilator of ty in A. Hence $a \in \mathfrak{q} \subset \mathfrak{p}$, as desired.

Corollary. *Let A be Noetherian, and let M be a module. The following conditions are equivalent:*

(1) *There exists only one associated prime of M.*
(2) *We have $M \neq 0$, and for every $a \in A$, the homomorphism a_M is injective, or locally nilpotent.*

If these conditions are satisfied, then the set of elements $a \in A$ such that a_M is locally nilpotent is equal to the associated prime of M.

Proof. Immediate consequence of Propositions 9 and 10.

Proposition 11. *Let N be a submodule of M. Every associated prime of N is associated with M also. An associated prime of M is associated with N or with M/N.*

Proof. The first assertion is obvious. Let $\mathfrak{p}$ be an associated prime of M, and say $\mathfrak{p}$ is the annihilator of the element $x \neq 0$. If $Ax \cap N = 0$ then Ax is isomorphic to a submodule of M/N, and hence $\mathfrak{p}$ is associated with M/N. If $Ax \cap N \neq 0$, then we view $Ax \cap N$ as a module over the entire ring $A/\mathfrak{p}$, and it is clear that the annihilator of any non-zero element of $Ax \cap N$ in $A/\mathfrak{p}$ is 0. Hence its annihilator in A is $\mathfrak{p}$, and $\mathfrak{p}$ is associated with N, as was to be shown.

§5. Primary decomposition

We continue to assume that A is a commutative ring, and that modules (resp. homomorphisms) are A-modules (resp. A-homomorphisms), unless otherwise specified.

Let M be a module. A submodule Q of M is said to be *primary* if $Q \neq A$, and if given $a \in A$, the homomorphism $a_{M/Q}$ is either injective or nilpotent. Viewing A as a module over itself, we see that an ideal $\mathfrak{q}$ is primary if and only if it satisfies the following condition:

Given $a, b \in A$, $ab \in \mathfrak{q}$ and $a \notin \mathfrak{q}$, then $b^n \in \mathfrak{q}$ for some $n \geqq 1$.

Let Q be primary. Let $\mathfrak{p}$ be the ideal of elements $a \in A$ such that $a_{M/Q}$ is nilpotent. Then $\mathfrak{p}$ is prime. Indeed, suppose that $a, b \in A$, $ab \in \mathfrak{p}$ and $a \notin \mathfrak{p}$. Then $a_{M/Q}$ is injective, and consequently $a^n_{M/Q}$ is injective for all $n \geqq 1$. Since $(ab)_{M/Q}$ is nilpotent, it follows that $b_{M/Q}$ must be nilpotent, and hence that $b \in \mathfrak{p}$, proving that $\mathfrak{p}$ is prime. We shall call $\mathfrak{p}$ the prime *belonging* to Q, and also say that Q is $\mathfrak{p}$-primary.

PROPOSITION 12. *Let M be a module, and $Q_1, \ldots, Q_r$ submodules which are $\mathfrak{p}$-primary for the same prime $\mathfrak{p}$. Then $Q_1 \cap \cdots \cap Q_r$ is also $\mathfrak{p}$-primary.*

Proof. Let $N = Q_1 \cap \cdots \cap Q_r$. Let $a \in \mathfrak{p}$. Let n_i be such that $(a_{M/Q_i})^{n_i} = 0$ for each $i = 1, \ldots, r$ and let n be the maximum of $n_1, \ldots, n_r$. Then $a^n_{M/Q} = 0$, so that $a_{M/Q}$ is nilpotent. Conversely, suppose $a \notin \mathfrak{p}$. Let $x \in M$, $x \notin Q_j$ for some j. Then $a^n x \notin Q_j$ for all positive integers n, and consequently $a_{M/Q}$ is injective. This proves our proposition.

Let N be a submodule of M. When N is written as a finite intersection of primary submodules, say

$$N = Q_1 \cap \cdots \cap Q_r,$$

we shall call this a *primary decomposition* of N. Using Proposition 11, we see that by grouping the Q_i according to their primes, we can always obtain from a given primary decomposition another one such that the primes belonging to the primary ideals are all distinct. A primary decomposition as above such that the prime ideals $\mathfrak{p}_1, \ldots, \mathfrak{p}_r$ belonging to $Q_1, \ldots, Q_r$ respectively are distinct, and such that N cannot be expressed as an intersection of a proper subfamily of the primary ideals $\{Q_1, \ldots, Q_r\}$ will be said to be *reduced*. By deleting some of the primary modules appearing in a given decomposition, we see that if N admits some primary decomposition, then it admits a reduced one. We shall prove a result giving certain uniqueness properties of a reduced primary decomposition.

Let $Q_1 \cap \cdots \cap Q_r = N$ be a reduced primary decomposition, and let $\mathfrak{p}_i$ belong to Q_i. If $\mathfrak{p}_i$ does not contain $\mathfrak{p}_j$ $(j \neq i)$ then we say that $\mathfrak{p}_i$ is *isolated*. The isolated primes are therefore those primes which are minimal in the set of primes belonging to the primary modules Q_i.

THEOREM 3. *Let N be a submodule of M, and let*

$$N = Q_1 \cap \cdots \cap Q_r = Q'_1 \cap \cdots \cap Q'_s$$

be a reduced primary decomposition of N. Then $r = s$. The set of primes belonging to $Q_1, \ldots, Q_r$ and $Q'_1, \ldots, Q'_r$ is the same. If $\{\mathfrak{p}_1, \ldots, \mathfrak{p}_m\}$ is the set of isolated primes belonging to these decompositions, then $Q_i = Q'_i$ for $i = 1, \ldots, m$, in other words, the primary modules corresponding to isolated primes are uniquely determined.

Proof. After a possible permutation of indices, suppose that $\mathfrak{p}_1$ is maximal among the set of primes belonging to the primary modules Q'_i and Q_i. Suppose that $\mathfrak{p}_1 \neq \mathfrak{p}'_j$ for $j = 1, \ldots, s$. Then there exists $a \in \mathfrak{p}_1$ such that $a \notin \mathfrak{p}_i$ $(i = 2, \ldots, r)$ and $a \notin \mathfrak{p}'_j$ $(j = 1, \ldots, s)$. Let $n \geqq 1$ be an integer such that $a^n M \subset Q_1$. Let N^* be the module of elements $x \in M$ such that $a^n x \subset N$. We contend that $N^* = Q_2 \cap \cdots \cap Q_r$. It is clear that

$$Q_2 \cap \cdots \cap Q_r \subset N^*,$$

and conversely, if $x \in M$, $x \notin Q_i$ for some $i > 1$, then $a^n x \notin Q_i$ because $a^n \notin \mathfrak{p}_i$. Hence $N^* \subset Q_2 \cap \cdots \cap Q_r$. The same argument shows that if $\mathfrak{p}_1 \neq \mathfrak{p}'_j$ for $j = 1, \ldots, s$ then

$$Q_2 \cap \cdots \cap Q_r = Q'_1 \cap \cdots \cap Q'_s,$$

thereby contradicting the hypothesis that our expression of N as an intersection of primary modules is shortest. This proves that $\mathfrak{p}_1$ occurs among the set $\{\mathfrak{p}'_1, \ldots, \mathfrak{p}'_s\}$, say $\mathfrak{p}_1 = \mathfrak{p}'_1$, and also proves that

$$Q_2 \cap \cdots \cap Q_r = Q'_2 \cap \cdots \cap Q'_s.$$

There remains to prove the uniqueness of the primary module belonging to an isolated prime, say $\mathfrak{p}_1$. By definition, for each $j = 2, \ldots, r$ there exists $a_j \in \mathfrak{p}_j$ and $a_j \notin \mathfrak{p}_1$. Let $a = a_2 \cdots a_r$ be the product. Then $a \in \mathfrak{p}_j$ for all $j > 1$, but $a \notin \mathfrak{p}_1$. We can find an integer $n \geqq 1$ such that $a^n_{M/Q_j} = 0$ for $j = 2, \ldots, r$. Let

$$N_1 = \text{set of } x \in M \text{ such that } a^n x \in N.$$

We contend that $Q_1 = N_1$. This will prove the desired uniqueness. Let $x \in Q_1$. Then $a^n x \in Q_1 \cap \cdots \cap Q_r = N$, so $x \in N_1$. Conversely, let $x \in N_1$, so that $a^n x \in N$, and in particular $a^n x \in Q_1$. Since $a \notin \mathfrak{p}_1$, we know by definition that a_{M/Q_1} is injective. Hence $x \in Q_1$, thereby proving our theorem.

THEOREM 4. *Let M be a Noetherian module. Let N be a submodule of M. Then N admits a primary decomposition.*

Proof. We consider the set of submodules of M which do not admit a primary decomposition. If this set is not empty, then it has a maximal element because M is Noetherian. Let N be this maximal element. Then N is not primary, and there exists $a \in A$ such that $a_{M/N}$ is neither injective nor nilpotent. The increasing sequence of modules

$$\operatorname{Ker} a_{M/N} \subset \operatorname{Ker} a_{M/N}^2 \subset \operatorname{Ker} a_{M/N}^3 \subset \cdots$$

stops, say at $a_{M/N}^r$. Let $\varphi : M/N \to M/N$ be the endomorphism $\varphi = a_{M/N}^r$. Then $\operatorname{Ker} \varphi^2 = \operatorname{Ker} \varphi$. Hence $0 = \operatorname{Ker} \varphi \cap \operatorname{Im} \varphi$ in M/N, and neither the kernel nor the image of φ is 0. Taking the inverse image in M, we see that N is the intersection of two submodules of M, unequal to N. We conclude from the maximality of N that each one of these submodules admits a primary decomposition, and therefore that N admits one also, contradiction.

We shall conclude our discussion by relating the primes belonging to a primary decomposition with the associated primes discussed in the previous section.

PROPOSITION 13. *Let A and M be Noetherian. A submodule Q of M is primary if and only if M/Q has exactly one associated prime $\mathfrak{p}$, and in that case, $\mathfrak{p}$ belongs to Q, i.e. Q is $\mathfrak{p}$-primary.*

Proof. Immediate consequence of the definitions, and of the Corollary of Proposition 10.

THEOREM 5. *Let A and M be Noetherian. The associated primes of M are precisely the primes which belong to the primary modules in a reduced primary decomposition of 0 in M. In particular, the set of associated primes of M is finite.*

Proof. Let

$$0 = Q_1 \cap \cdots \cap Q_r$$

be a reduced primary decomposition of 0 in M. We have an injective homomorphism

$$M \to \coprod_{i=1}^{r} M/Q_i.$$

By Proposition 11 of §4, and Proposition 13, we conclude that every associated prime of M belongs to some Q_i. Conversely, let $N = Q_2 \cap \cdots \cap Q_r$. Then $N \neq 0$ because our decomposition is reduced. We have

$$N = N/(N \cap Q_1) \approx (N + Q_1)/Q_1 \subset M/Q_1.$$

Hence N is isomorphic to a submodule of M/Q_1, and consequently has an associated prime which can be none other than the prime $\mathfrak{p}_1$ belonging to Q_1. This proves our theorem.

Exercises

Throughout the exercises, "ring" means "commutative ring".

1. Let A be a ring, and $\mathfrak{a}$ an ideal which is contained in every maximal ideal. Let E be a finitely generated A-module. Suppose that $\mathfrak{a}E = E$. Then $E = 0$. [*Hint:* Induction on the number of generators. Express one generator in terms of the others and use the fact that $1 + a$ is a unit for $a \in \mathfrak{a}$. Cf. NAKAYAMA's LEMMA in Chapter IX.] This statement applies in particular to the case when $A = \mathfrak{o}$ is a local ring, and $\mathfrak{a} = \mathfrak{m}$ is its maximal ideal. Deduce the following two statements as corollaries:

Let E be a finitely generated $\mathfrak{o}$-module, and F a submodule. If $E = F + \mathfrak{m}E$, then $E = F$.

If $x_1, \ldots, x_n$ are generators for $\mathfrak{m}$ mod $\mathfrak{m}^2$, then they are generators for $\mathfrak{m}$ over $\mathfrak{o}$.

2. (Artin-Rees) Let A be Noetherian, $\mathfrak{a}$ an ideal, E a finitely generated module, and F a submodule. Then there exists an integer $s \geqq 1$ such that for all integers $n \geqq 1$ we have $\mathfrak{a}^nE \cap F = \mathfrak{a}^{n-s}(\mathfrak{a}^sE \cap F)$. [*Hint:* Let $A_t = A[t]$ and $A_t' = A[\mathfrak{a}t] = \coprod_{n=0}^{\infty} \mathfrak{a}^n t^n$. If $a_1, \ldots, a_m$ are generators for $\mathfrak{a}$, then $a_1t, \ldots, a_mt$ are generators for A_t' ,which is therefore Noetherian. Define E_t to be the A_t-module $\coprod t^nE$ in the obvious way, and similarly for F_t. Let $E_t' = A_t'E_t = \coprod \mathfrak{a}^n t^n E$. Then E_t' is a finitely generated A_t'-module, and $E_t' \cap F_t$ has a finite number of generators involving only a finite number of powers of t. If t^s is the highest, then

$$\coprod_{n=0}^{\infty} t^n(\mathfrak{a}^nE \cap F) = E_t' \cap F_t = A_t' \coprod_{\nu=0}^{s} t^\nu(\mathfrak{a}^\nu E \cap F).$$

Comparing the coefficients of t^n for $n \geqq s$ gives

$$\mathfrak{a}^nE \cap F = \coprod_{\nu=0}^{s} \mathfrak{a}^{n-\nu}(\mathfrak{a}^\nu E \cap F),$$

whence the result follows at once.

3. (Krull) In the preceding exercise, assume that $\mathfrak{a}$ is contained in every maximal ideal of A. Then $\bigcap_{n=1}^{\infty} \mathfrak{a}^nE = 0$. [*Hint:* Let $F = \bigcap \mathfrak{a}^nE$ and apply

Nakayama.] In particular, let $\mathfrak{o}$ be a Noetherian local ring and $\mathfrak{m}$ its maximal ideal. Then

$$\bigcap_{\nu=1}^{\infty} \mathfrak{m}^{\nu} = 0.$$

4. Let A be a commutative ring. Let M be a module, and N a submodule. Let $N = Q_1 \cap \cdots \cap Q_r$ be a primary decomposition of N. Let $\overline{Q}_i = Q_i/N$. Show that $0 = \overline{Q}_1 \cap \cdots \cap \overline{Q}_r$ is a primary decomposition of 0 in M/N. State and prove the converse.

5. Let $\mathfrak{p}$ be a prime ideal, and $\mathfrak{a}$, $\mathfrak{b}$ ideals of A. If $\mathfrak{a}\mathfrak{b} \subset \mathfrak{p}$, show that $\mathfrak{a} \subset \mathfrak{p}$ or $\mathfrak{b} \subset \mathfrak{p}$.

6. Let $\mathfrak{q}$ be a primary ideal. Let $\mathfrak{a}$, $\mathfrak{b}$ be ideals, and assume $\mathfrak{a}\mathfrak{b} \subset \mathfrak{q}$. Assume that $\mathfrak{b}$ is finitely generated. Show that $\mathfrak{a} \subset \mathfrak{q}$ or there exists some positive integer n such that $\mathfrak{b}^n \subset \mathfrak{q}$.

7. Let A be Noetherian, and let $\mathfrak{q}$ be a $\mathfrak{p}$-primary ideal. Show that there exists some $n \geqq 1$ such that $\mathfrak{p}^n \subset \mathfrak{q}$.

8. Let A be an arbitrary commutative ring and let S be a multiplicative subset. Let $\mathfrak{p}$ be a prime ideal and let $\mathfrak{q}$ be a $\mathfrak{p}$-primary ideal. Then $\mathfrak{p}$ intersects S if and only if $\mathfrak{q}$ intersects S. Furthermore, if $\mathfrak{q}$ does not intersect S, then $S^{-1}\mathfrak{q}$ is $S^{-1}\mathfrak{p}$-primary in $S^{-1}A$.

9. If $\mathfrak{a}$ is an ideal of A, let $\mathfrak{a}_S = S^{-1}\mathfrak{a}$. If $\varphi_S : A \to S^{-1}A$ is the canonical map, abbreviate $\varphi_S^{-1}(\mathfrak{a}_S)$ by $\mathfrak{a}_S \cap A$, even though φ_S is not injective. Show that there is a bijection between the prime ideals of A which do not intersect S and the prime ideals of $S^{-1}A$, given by

$$\mathfrak{p} \mapsto \mathfrak{p}_S \quad \text{and} \quad \mathfrak{p}_S \mapsto \mathfrak{p}_S \cap A = \mathfrak{p}.$$

Prove a similar statement for primary ideals instead of prime ideals.

10. Let $\mathfrak{a} = \mathfrak{q}_1 \cap \cdots \cap \mathfrak{q}_r$ be a reduced primary decomposition of an ideal. Assume that $\mathfrak{q}_1, \ldots, \mathfrak{q}_i$ do not intersect S, but that $\mathfrak{q}_j$ intersects S for $j > i$. Show that

$$\mathfrak{a}_S = \mathfrak{q}_{1S} \cap \cdots \cap \mathfrak{q}_{iS}$$

is a reduced primary decomposition of $\mathfrak{a}_S$.

11. Assume that A is Noetherian. Show that the set of divisors of zero in A is the set-theoretic union of all primes belonging to primary ideals in a reduced primary decomposition of 0.

PART TWO

FIELD THEORY

This part is concerned with the solutions of algebraic equations, in one or several variables. This is the recurrent theme in every chapter of this part, and we lay the foundations for all further studies concerning such equations.

Given a subring A of a ring B, and a finite number of polynomials $f_1, \ldots, f_n$ in $A[X_1, \ldots, X_n]$, we are concerned with the n-tuples $(b_1, \ldots, b_n) \in B^{(n)}$ such that

$$f_i(b_1, \ldots, b_n) = 0$$

for $i = 1, \ldots, r$. For suitable choices of A and B, this includes the general problem of diophantine analysis when A, B have an "arithmetic" structure.

We shall study various cases, first in one variable, over an arbitrary field, taking for B algebraic extensions of this field. Next we discuss the aspects of this question having to do with ring structures (integral extensions). Then we pass to finitely generated ring extensions and polynomials in several variables. Finally we impose additional structures like those of reality, or metric structures given by absolute values. Each one of these structures gives rise to certain theorems describing the structure of the solutions of equations as above.

CHAPTER VII

Algebraic Extensions

§1. *Finite and algebraic extensions*

Let F be a field. If F is a subfield of a field E, then we also say that E is an *extension field* of F. We may view E as a vector space over F, and we say that E is a *finite* or *infinite* extension of F according as the dimension of this vector space is finite or infinite.

Let F be a subfield of a field E. An element α of E is said to be *algebraic* over F if there exist elements $a_0, \ldots, a_n$ ($n \geqq 1$) of F, not all equal to 0, such that

$$a_0 + a_1\alpha + \cdots + a_n\alpha^n = 0.$$

If $\alpha \neq 0$, and α is algebraic, then we can always find elements a_i as above such that $a_0 \neq 0$ (factoring out a suitable power of α).

Let X be a variable over F. We can also say that α is algebraic over F if the homomorphism

$$F[X] \to E$$

which is the identity on F and maps X on α has a non-zero kernel. In that case the kernel is an ideal which is principal, generated by a single polynomial $p(X)$, which we may assume has leading coefficient 1. We then have an isomorphism

$$F[X]/(p(X)) \approx F[\alpha],$$

and since $F[\alpha]$ is entire, it follows that $p(X)$ is irreducible. Having normalized $p(X)$ so that its leading coefficient is 1, we see that $p(X)$ is uniquely determined by α and will be called THE irreducible polynomial of α over F. We sometimes denote it by $\mathrm{Irr}(\alpha, F, X)$.

An extension E of F is said to be *algebraic* if every element of E is algebraic over F.

PROPOSITION 1. *Let E be a finite extension of F. Then E is algebraic over F.*

Proof. Let $\alpha \in E$, $\alpha \neq 0$. The powers of α,

$$1, \alpha, \alpha^2, \ldots, \alpha^n,$$

cannot be linearly independent over F for all positive integers n, otherwise the dimension of E over F would be infinite. A linear relation between these powers shows that α is algebraic over F.

Note that the converse of Proposition 1 is not true; there exist infinite algebraic extensions. We shall see later that the subfield of the complex numbers consisting of all algebraic numbers over $\mathbf{Q}$ is an infinite extension of $\mathbf{Q}$.

If E is an extension of F, we denote by

$$[E : F]$$

the dimension of E as vector space over F. It may be infinite.

PROPOSITION 2. *Let k be a field and $F \subset E$ extension fields of k. Then*

$$[E : k] = [E : F][F : k].$$

If $\{x_i\}_{i\in I}$ is a basis for F over k and $\{y_j\}_{j\in J}$ is a basis for E over F, then $\{x_iy_j\}_{(i,j)\in I\times J}$ is a basis for E over k.

Proof. Let $z \in E$. By hypothesis there exist elements $\alpha_j \in F$, almost all $\alpha_j = 0$, such that

$$z = \sum_{j\in J} \alpha_j y_j.$$

For each $j \in J$ there exist elements $b_{ji} \in k$, almost all of which are equal to 0, such that

$$\alpha_j = \sum_{i\in I} b_{ji}x_i,$$

and hence

$$z = \sum_j \sum_i b_{ji}x_iy_j.$$

This shows that $\{x_iy_j\}$ is a family of generators for E over k. We must show that it is linearly independent. Let $\{c_{ij}\}$ be a family of elements of k, almost all of which are 0, such that

$$\sum_j \sum_i c_{ij}x_iy_j = 0.$$

Then for each j,

$$\sum_i c_{ij}x_i = 0$$

because the elements y_j are linearly independent over F. Finally $c_{ij} = 0$ for each i because $\{x_i\}$ is a basis of F over k, thereby proving our proposition.

COROLLARY. *The extension E of k is finite if and only if E is finite over F and F is finite over k.*

As with groups, we define a *tower* of fields to be a sequence

$$F_1 \subset F_2 \subset \cdots \subset F_n$$

of extension fields. The tower is finite if and only if each step is finite.

Let k be a field, E an extension field, and $\alpha \in E$. We denote by $k(\alpha)$ the smallest subfield of E containing both k and α. It consists of all quotients $f(\alpha)/g(\alpha)$, where f, g are polynomials with coefficients in k and $g(\alpha) \neq 0$.

PROPOSITION 3. *Let α be algebraic over k. Then $k(\alpha) = k[\alpha]$, and $k(\alpha)$ is finite over k. The degree $[k(\alpha) : k]$ is equal to the degree of* $\mathrm{Irr}(\alpha, k, X)$.

Proof. Let $p(X) = \mathrm{Irr}(\alpha, k, X)$. Let $f(X) \in k[X]$ be such that $f(\alpha) \neq 0$. Then $p(X)$ does not divide $f(X)$, and hence there exist polynomials $g(X)$, $h(X) \in k[X]$ such that

$$g(X)p(X) + h(X)f(X) = 1.$$

From this we get $h(\alpha)f(\alpha) = 1$, and we see that $f(\alpha)$ is invertible in $k[\alpha]$. Hence $k[\alpha]$ is not only a ring but a field, and must therefore be equal to $k(\alpha)$. Let $d = \deg p(X)$. The powers

$$1, \alpha, \ldots, \alpha^{d-1}$$

are linearly independent over k, for otherwise suppose

$$a_0 + a_1\alpha + \cdots + a_{d-1}\alpha^{d-1} = 0$$

with $a_i \in k$, not all $a_i = 0$. Let $g(X) = a_0 + \cdots + a_{d-1}X^{d-1}$. Then $g \neq 0$ and $g(\alpha) = 0$. Hence $p(X)$ divides $g(X)$, contradiction. Finally, let $f(\alpha) \in k[\alpha]$, where $f(X) \in k[X]$. There exist polynomials $q(X)$, $r(X) \in k[X]$ such that $\deg r < d$ and

$$f(X) = q(X)p(X) + r(X).$$

Then $f(\alpha) = r(\alpha)$, and we see that $1, \alpha, \ldots, \alpha^{d-1}$ generate $k[\alpha]$ as a vector space over k. This proves our proposition.

Let E, F be extensions of a field k. If E and F are contained in some field L then we denote by EF the smallest subfield of L containing both E and F, and call it the *compositum* of E and F, in L. If E, F are not given as embedded in a common field L, then we cannot define the compositum.

Let k be a subfield of E and let $\alpha_1, \ldots, \alpha_n$ be elements of E. We denote by

$$k(\alpha_1, \ldots, \alpha_n)$$

the smallest subfield of E containing k and $\alpha_1, \ldots, \alpha_n$. Its elements consist of all quotients

$$\frac{f(\alpha_1, \ldots, \alpha_n)}{g(\alpha_1, \ldots, \alpha_n)}$$

where f, g are polynomials in n variables with coefficients in k, and $g(\alpha_1, \ldots, \alpha_n) \neq 0$. Indeed, the set of such quotients forms a field containing k and $\alpha_1, \ldots, \alpha_n$. Conversely, any field containing k and

$$\alpha_1, \ldots, \alpha_n$$

must contain these quotients.

We observe that E is the union of all its subfields $k(\alpha_1, \ldots, \alpha_n)$ as $(\alpha_1, \ldots, \alpha_n)$ ranges over finite subfamilies of elements of E. We could define the *compositum of an arbitrary subfamily of subfields of a field L* as the smallest subfield containing all fields in the family. We say that E is *finitely generated* over k if there is a finite family of elements $\alpha_1, \ldots, \alpha_n$ of E such that

$$E = k(\alpha_1, \ldots, \alpha_n).$$

We see that E is the compositum of all its finitely generated subfields over k.

PROPOSITION 4. *Let E be a finite extension of k. Then E is finitely generated.*

Proof. Let $\{\alpha_1, \ldots, \alpha_n\}$ be a basis of E as vector space over k. Then certainly $E = k(\alpha_1, \ldots, \alpha_n)$.

If $E = k(\alpha_1, \ldots, \alpha_n)$ is finitely generated, and F is an extension of k, both F, E contained in L, then

$$EF = F(\alpha_1, \ldots, \alpha_n),$$

and EF is finitely generated over F. We often draw the following picture:

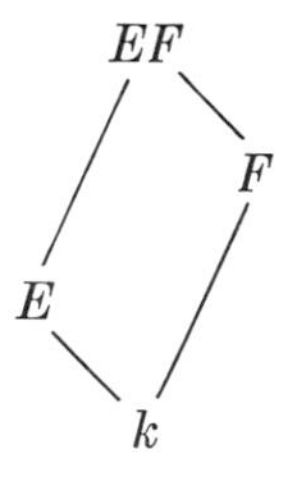

Lines slanting up indicate an inclusion relation between fields. We also call the extension EF of F the *translation* of E to F, or also the *lifting* of E to F.

Let α be algebraic over the field k. Let F be an extension of k, and assume $k(\alpha)$, F both contained in some field L. Then α is algebraic over F. Indeed, the irreducible polynomial for α over k has *a fortiori* coefficients in F, and gives a linear relation for the powers of α over F.

Suppose that we have a tower of fields:

$$k \subset k(\alpha_1) \subset k(\alpha_1, \alpha_2) \subset \cdots \subset k(\alpha_1, \ldots, \alpha_n),$$

each one generated from the preceding field by a single element. Assume that each α_i is algebraic over k, $i = 1, \ldots, n$. As a special case of our preceding remark, we note that α_{i+1} is algebraic over $k(\alpha_1, \ldots, \alpha_i)$. Hence each step of the tower is algebraic.

PROPOSITION 5. *Let $E = k(\alpha_1, \ldots, \alpha_n)$ be a finitely generated extension of a field k, and assume α_i algebraic over k for each $i = 1, \ldots, n$. Then E is finite algebraic over k.*

Proof. From the above remarks, we know that E can be obtained as the end of a tower each of whose steps is generated by one algebraic element, and is therefore finite by Proposition 3. We conclude that E is finite over k by the Corollary of Proposition 2, and that it is algebraic by Proposition 1.

Let $\mathfrak{C}$ be a certain class of extension fields $F \subset E$. We shall say that $\mathfrak{C}$ is *distinguished* if it satisfies the following conditions:

(i) Let $k \subset F \subset E$ be a tower of fields. The extension $k \subset E$ is in $\mathfrak{C}$ if and only if $k \subset F$ is in $\mathfrak{C}$ and $F \subset E$ is in $\mathfrak{C}$.

(ii) If $k \subset E$ is in $\mathfrak{C}$, if F is any extension of k, and E, F are both contained in some field, then $F \subset EF$ is in $\mathfrak{C}$.

(iii) If $k \subset F$ and $k \subset E$ are in $\mathfrak{C}$ and F, E are subfields of a common field, then $k \subset FE$ is in $\mathfrak{C}$.

The diagrams illustrating our properties are as follows:

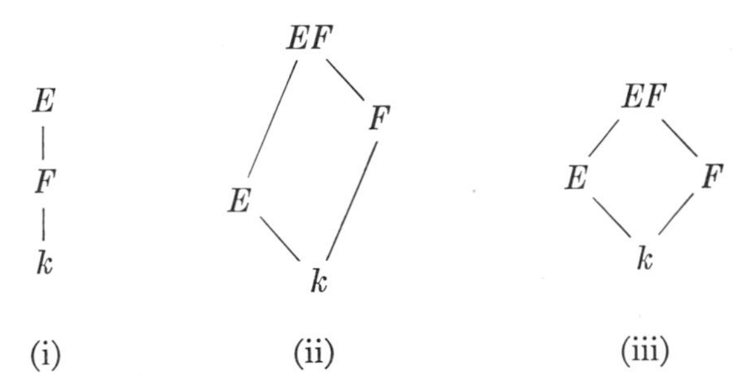

These lattice diagrams of fields are extremely suggestive in handling extension fields.

We observe that (iii) follows formally from the first two conditions. Indeed, one views EF over k as a tower with steps $k \subset F \subset EF$.

As a matter of notation, it is convenient to write E/F instead of $F \subset E$ to denote an extension. There can be no confusion with factor groups since we shall never use the notation E/F to denote such a factor group when E is an extension field of F.

PROPOSITION 6. *The class of algebraic extensions is distinguished, and so is the class of finite extensions.*

Proof. Consider first the class of finite extensions. We have already proved condition (i). As for (ii), assume that E/k is finite, and let F be any extension of k. By Proposition 4 there exist elements $\alpha_1, \ldots, \alpha_n \in E$ such that $E = k(\alpha_1, \ldots, \alpha_n)$. Then $EF = F(\alpha_1, \ldots, \alpha_n)$, and hence EF/F is finitely generated by algebraic elements. Using Proposition 5 we conclude that EF/F is finite.

Consider next the class of algebraic extensions, and let

$$k \subset F \subset E$$

be a tower. Assume that E is algebraic over k. Then *a fortiori*, F is algebraic over k and E is algebraic over F. Conversely, assume each step in the tower to be algebraic. Let $\alpha \in E$. Then α satisfies an equation

$$a_n\alpha^n + \cdots + a_0 = 0$$

with $a_i \in F$, not all $a_i = 0$. Let $F_0 = k(a_n, \ldots, a_0)$. Then F_0 is finite over k by Proposition 5, and α is algebraic over F_0. From the tower

$$k \subset F_0 = k(a_n, \ldots, a_0) \subset F_0(\alpha)$$

and the fact that each step in this tower is finite, we conclude that $F_0(\alpha)$ is finite over k, whence α is algebraic over k, thereby proving that E is algebraic over k and proving condition (i) for algebraic extensions. Condition (ii) has already been observed to hold, i.e. an element remains algebraic under lifting, and hence so does an extension.

Remark. It is true that finitely generated extensions form a distinguished class, but one argument needed to prove part of (i) can be carried out only with more machinery than we have at present. Cf. the chapter on transcendental extensions.

§2. *Algebraic closure*

In this and the next section we shall deal with embeddings of a field into another. We therefore define some terminology.

Let E be an extension of a field F and let

$$\sigma : F \to L$$

be an embedding (i.e. an injective homomorphism) of F into L. Then σ induces an isomorphism of F with its image σF, which is sometimes written F^σ. An embedding τ of E in L will be said to be *over* σ if the restriction of τ to F is equal to σ. We also say that τ *extends* σ. If σ is the identity then we say that τ is an embedding of E *over* F.

These definitions could be made in more general categories, since they depend only on diagrams to make sense:

$$\begin{array}{ccc} E & \xrightarrow{\tau} & L \\ {\scriptstyle \text{inc}}\uparrow & & \uparrow{\scriptstyle id} \\ F & \xrightarrow[\sigma]{} & L \end{array} \qquad\qquad \begin{array}{ccc} E & \xrightarrow{\tau} & L \\ {\scriptstyle \text{inc}}\nwarrow & & \nearrow{\scriptstyle \text{inc}} \\ & F & \end{array}$$

Remark. Let $f(X) \in F[X]$ be a polynomial, and let α be a root of f in E. Say $f(X) = a_0 + \cdots + a_nX^n$. Then

$$0 = f(\alpha) = a_0 + a_1\alpha + \cdots + a_n\alpha^n.$$

If τ extends σ as above, then we see that $\tau\alpha$ is a root of f^σ because

$$0 = \tau(f(\alpha)) = a_0^\sigma + a_1^\sigma(\tau\alpha) + \cdots + a_n^\sigma(\tau\alpha)^n.$$

Here we have written a^σ instead of $\sigma(a)$. This exponential notation is frequently convenient and will be used again in the sequel. Similarly, we write F^σ instead of $\sigma(F)$ or σF.

In our study of embeddings it will also be useful to have a lemma concerning embeddings of algebraic extensions into themselves. For this we note that if $\sigma : E \to L$ is an embedding over k (i.e. inducing the identity on k), then σ can be viewed as a k-homomorphism of vector spaces, because both E, L can be viewed as vector spaces over k. Furthermore σ is injective.

LEMMA 1. *Let E be an algebraic extension of k, and let $\sigma : E \to E$ be an embedding of E into itself over k. Then σ is an automorphism.*

Proof. Since σ is injective, it will suffice to prove that σ is surjective. Let α be an element of E, let $p(X)$ be its irreducible polynomial over k, and let E' be the subfield of E generated by all the roots of $p(X)$ which lie in E. Then E' is finitely generated, hence is a finite extension of k. Furthermore, σ must map a root of $p(X)$ on a root of $p(X)$, and hence σ maps E' into itself. We can view σ as a k-homomorphism of vector spaces because σ induces the identity on k. Since σ is injective, its image $\sigma(E')$ is a subspace of E' having the same dimension as $[E' : k]$. Hence $\sigma(E') = E'$. Since $\alpha \in E'$, it follows that α is in the image of σ, and our lemma is proved.

Let E, F be extensions of a field k, contained in some bigger field L. We can form the ring $E[F]$ generated by the elements of F over E. Then

EF is the quotient field of this ring, and also the quotient field of $F[E]$. It is clear that the elements of $E[F]$ can be written in the form

$$a_1b_1 + \cdots + a_nb_n$$

with $a_i \in E$ and $b_i \in F$. Hence EF is the field of quotients of these elements.

LEMMA 2. *Let E_1, E_2 be extensions of a field k, contained in some bigger field E, and let σ be an embedding of E in some field L. Then $\sigma(E_1E_2) = \sigma(E_1)\sigma(E_2)$.*

Proof. We apply σ to a quotient of elements of the above type, say

$$\sigma\left(\frac{a_1b_1 + \cdots + a_nb_n}{a_1'b_1' + \cdots + a_m'b_m'}\right) = \frac{a_1^\sigma b_1^\sigma + \cdots + a_n^\sigma b_n^\sigma}{a_1'^\sigma b_1'^\sigma + \cdots + a_m'^\sigma b_m'^\sigma},$$

and see that the image is an element of $\sigma(E_1)\sigma(E_2)$. It is clear that the image $\sigma(E_1E_2)$ is $\sigma(E_1)\sigma(E_2)$.

Let k be a field, $f(X)$ a polynomial of degree $\geqq 1$ in $k[X]$. We consider the problem of finding an extension E of k in which f has a root. If $p(X)$ is an irreducible polynomial in $k[X]$ which divides $f(X)$, then any root of $p(X)$ will also be a root of $f(X)$, so we may restrict ourselves to irreducible polynomials.

Let $p(X)$ be irreducible, and consider the canonical homomorphism

$$\sigma : k[X] \to k[X]/(p(X)).$$

Then σ induces a homomorphism on k, whose kernel is 0, because every non-zero element of k is invertible in k, generates the unit ideal, and 1 does not lie in the kernel. Let ξ be the image of X under σ, i.e. $\xi = \sigma(X)$ is the residue class of X mod $p(X)$. Then

$$p^\sigma(\xi) = p^\sigma(X^\sigma) = (p(X))^\sigma = 0.$$

Hence ξ is a root of p^σ, and as such is algebraic over σk. We have now found an extension of σk, namely $\sigma k(\xi)$ in which p^σ has a root.

With a minor set-theoretic argument, we shall have:

PROPOSITION 7. *Let k be a field and f a polynomial in $k[X]$ of degree $\geqq 1$. Then there exists an extension E of k in which f has a root.*

Proof. We may assume that $f = p$ is irreducible. We have shown that there exists a field F and an embedding

$$\sigma : k \to F$$

such that p^σ has a root ξ in F. Let S be a set whose cardinality is the same as that of $F - \sigma k$ (= the complement of σk in F) and which is disjoint from k. Let $E = k \cup S$. We can extend $\sigma : k \to F$ to a bijection of E on F. We now define a field structure on E. If $x, y \in E$ we define

$$xy = \sigma^{-1}(\sigma(x)\sigma(y)),$$
$$x + y = \sigma^{-1}(\sigma(x) + \sigma(y)).$$

Restricted to k, our addition and multiplication coincide with the given addition and multiplication of our original field k, and it is clear that k is a subfield of E. We let $\alpha = \sigma^{-1}(\xi)$. Then it is also clear that $p(\alpha) = 0$, as desired.

COROLLARY. *Let k be a field and let $f_1, \ldots, f_n$ be polynomials in $k[X]$ of degrees $\geqq 1$. Then there exists an extension E of k in which each f_i has a root, $i = 1, \ldots, n$.*

Proof. Let E_1 be an extension in which f_1 has a root. We may view f_2 as a polynomial over E_1. Let E_2 be an extension of E_1 in which f_2 has a root. Proceeding inductively, our corollary follows at once.

We define a field L to be *algebraically closed* if every polynomial in $L[X]$ of degree $\geqq 1$ has a root in L.

THEOREM 1. *Let k be a field. Then there exists an algebraically closed field L containing k as a subfield.*

Proof. We first construct an extension E_1 of k in which every polynomial in $k[X]$ of degree $\geqq 1$ has a root. One can proceed as follows (Artin). To each polynomial f in $k[X]$ of degree $\geqq 1$ we associate a letter X_f and we let S be the set of all such letters X_f (so that S is in bijection with the set of polynomials in $k[X]$ of degree $\geqq 1$). We form the polynomial ring $k[S]$, and contend that the ideal generated by all the polynomials $f(X_f)$ in $k[S]$ is not the unit ideal. If it is, then there is a finite combination of elements in our ideal which is equal to 1:

$$g_1 f_1(X_{f_1}) + \cdots + g_n f_n(X_{f_n}) = 1$$

with $g_i \in k[S]$. For simplicity, write X_i instead of X_{f_i}. The polynomials g_i will involve actually only a finite number of variables, say $X_1, \ldots, X_N$ (with $N \geqq n$). Our relation then reads

$$\sum_{i=1}^{n} g_i(X_1, \ldots, X_N) f_i(X_i) = 1.$$

Let F be a finite extension in which each polynomial $f_1, \ldots, f_n$ has a root, say α_i is a root of f_i in F, for $i = 1, \ldots, n$. Let $\alpha_i = 0$ for $i > n$. Substitute α_i for X_i in our relation. We get $0 = 1$, contradiction.

Let $\mathfrak{m}$ be a maximal ideal containing the ideal generated by all polynomials $f(X_f)$ in $k[S]$. Then $k[S]/\mathfrak{m}$ is a field, and we have a canonical map

$$\sigma : k[S] \to k[S]/\mathfrak{m}.$$

For any polynomial $f \in k[X]$ of degree $\geqq 1$, the polynomial f^σ has a root in $k[S]/\mathfrak{m}$, which is an extension of σk. Using the same type of set-theoretic argument as in Proposition 7, we conclude that there exists an extension E_1 of k in which every polynomial $f \in k[X]$ of degree $\geqq 1$ has a root in E_1.

Inductively, we can form a sequence of fields

$$E_1 \subset E_2 \subset E_3 \subset \cdots \subset E_n \cdots$$

such that every polynomial in $E_n[X]$ of degree $\geqq 1$ has a root in E_{n+1}. Let E be the union of all fields E_n, $n = 1, 2, \ldots$ Then E is naturally a field, for if $x, y \in E$ then there exists some n such that $x, y \in E_n$, and we can take the product or sum xy or $x + y$ in E_n. This is obviously independent of the choice of n such that $x, y \in E_n$, and defines a field structure on E. Every polynomial in $E[X]$ has its coefficients in some subfield E_n, hence a root in E_{n+1}, hence a root in E, as desired.

COROLLARY. *Let k be a field. There exists an extension $\bar{k}$ which is algebraic over k and algebraically closed.*

Proof. Let E be an extension of k which is algebraically closed and let $\bar{k}$ be the union of all subextensions of E, which are algebraic over k. Then $\bar{k}$ is algebraic over k. If $\alpha \in E$ and α is algebraic over $\bar{k}$ then α is algebraic over k by Proposition 6. If f is a polynomial of degree $\geqq 1$ in $\bar{k}[X]$ then f has a root α in E, and α is algebraic over $\bar{k}$. Hence α is in $\bar{k}$ and $\bar{k}$ is algebraically closed.

We observe that if L is an algebraically closed field, and $f \in L[X]$ has degree $\geqq 1$, then there exists $c \in L$ and $\alpha_1, \ldots, \alpha_n \in L$ such that

$$f(X) = c(X - \alpha_1) \cdots (X - \alpha_n).$$

Indeed, f has a root α_1 in L, so there exists $g(X) \in L[X]$ such that $f(X) = (X - \alpha_1)g(X)$. If $\deg g \geqq 1$, we can repeat this argument inductively, and express f as a product of terms $(X - \alpha_i)$ $(i = 1, \ldots, n)$ and an element $c \in L$. Note that c is the leading coefficient of f, i.e.

$$f(X) = cX^n + \text{terms of lower degree.}$$

Hence if the coefficients of f lie in a subfield k of L, then $c \in k$.

Let k be a field and $\sigma : k \to L$ an embedding of k into an algebraically closed field L. We are interested in analyzing the extensions of σ to algebraic extensions E of k. We begin by considering the special case when E is generated by one element.

Let $E = k(\alpha)$ where α is algebraic over k. Let

$$p(X) = \operatorname{Irr}(\alpha, k, X).$$

Let β be a root of p^σ in L. Given an element of $k(\alpha) = k[\alpha]$, we can write it in the form $f(\alpha)$ with some polynomial $f(X) \in k[X]$. We define an extension of σ by mapping

$$f(\alpha) \mapsto f^\sigma(\beta).$$

This is in fact well defined, i.e. independent of the choice of polynomial $f(X)$ used to express our element in $k[\alpha]$. Indeed, if $g(X)$ is in $k[X]$ and such that $g(\alpha) = f(\alpha)$, then $(g - f)(\alpha) = 0$, whence $p(X)$ divides $g(X) - f(X)$. Hence $p^\sigma(X)$ divides $g^\sigma(X) - f^\sigma(X)$, and thus $g^\sigma(\beta) = f^\sigma(\beta)$. It is now clear that our map is a homomorphism inducing σ on k, and that it is an extension of σ to $k(\alpha)$. Hence we get:

PROPOSITION 8. *The number of possible extensions of σ to $k(\alpha)$ is $\leqq$ the number of roots of p, and is equal to the number of distinct roots of p.*

This is an important fact, which we shall analyze more closely later. For the moment, we are interested in extensions of σ to arbitrary algebraic extensions of k. We get them by using Zorn's lemma.

THEOREM 2. *Let k be a field, E an algebraic extension of k, and $\sigma : k \to L$ an embedding of k into an algebraically closed field L. Then there exists an extension of σ to an embedding of E in L. If E is algebraically closed and L is algebraic over σk, then any such extension of σ is an isomorphism of E onto L.*

Proof. Let S be the set of all pairs (F, τ) where F is a subfield of E containing k, and τ is an extension of σ to an embedding of F in L. If (F, τ) and (F', τ') are such pairs, we write $(F, \tau) \leqq (F', \tau')$ if $F \subset F'$ and $\tau' \mid F = \tau$. Note that S is not empty [it contains (k, σ)], and is inductively ordered: If $\{(F_i, \tau_i)\}$ is a totally ordered subset, we let $F = \bigcup F_i$ and define τ on F to be equal to τ_i on each F_i. Then (F, τ) is an upper bound for the totally ordered subset. Using Zorn's lemma, let (K, λ) be a maximal element in S. Then λ is an extension of σ, and we contend that $K = E$. Otherwise, there exists $\alpha \in E$, $\alpha \notin K$. By what we saw above, our embedding λ has an extension to $K(\alpha)$, thereby contradicting the maximality of (K, λ). This proves that there exists an extension of σ to E. We denote this extension again by σ.

If E is algebraically closed, and L is algebraic over σk, then σE is algebraically closed and L is algebraic over σE, hence $L = \sigma E$.

As a corollary, we have a certain uniqueness for an "algebraic closure" of a field k.

Corollary. *Let k be a field and let E, E' be algebraic extensions of k. Assume that E, E' are algebraically closed. Then there exists an isomorphism*

$$\tau : E \to E'$$

of E onto E' inducing the identity on k.

Proof. Extend the identity mapping on k to an embedding of E into E' and apply the theorem.

We see that an algebraically closed and algebraic extension of k is determined up to an isomorphism. Such an extension will be called an *algebraic closure* of k, and we denote it frequently by $\bar{k}$. In fact, unless otherwise specified,we use the symbol $\bar{k}$ only to denote algebraic closure.

It is now worth while to discuss the general situation of isomorphisms and automorphisms in general categories.

Let $\mathfrak{A}$ be a category, and A, B objects in $\mathfrak{A}$. We denote by $\operatorname{Iso}(A, B)$ the set of isomorphisms of A on B. Suppose there exists at least one such isomorphism $\sigma : A \to B$, with inverse $\sigma^{-1} : B \to A$. If φ is an automorphism of A, then $\sigma \circ \varphi : A \to B$ is again an isomorphism. If ψ is an automorphism of B, then $\psi \circ \sigma : A \to B$ is again an isomorphism. Furthermore, the groups of automorphisms $\operatorname{Aut}(A)$ and $\operatorname{Aut}(B)$ are isomorphic, under the mappings

$$\varphi \mapsto \sigma \circ \varphi \circ \sigma^{-1}$$
$$\sigma^{-1} \circ \psi \circ \sigma \leftarrow\!\shortmid \psi$$

which are inverse to each other. The isomorphism $\sigma \circ \varphi \circ \sigma^{-1}$ is the one which makes the following diagram commutative:

$$\begin{array}{ccc} A & \xrightarrow{\sigma} & B \\ \varphi \downarrow & & \downarrow \sigma \circ \varphi \circ \sigma^{-1} \\ A & \xrightarrow[\sigma]{} & B \end{array}$$

We have a similar diagram for $\sigma^{-1} \circ \psi \circ \sigma$.

Let $\tau : A \to B$ be another isomorphism. Then $\tau^{-1} \circ \sigma$ is an automorphism of A, and $\tau \circ \sigma^{-1}$ is an automorphism of B. Thus two isomorphisms differ by an automorphism (of A or B). We see that the group $\operatorname{Aut}(B)$ operates on the set $\operatorname{Iso}(A, B)$ on the left, and $\operatorname{Aut}(A)$ operates on the set $\operatorname{Iso}(A, B)$ on the right.

We also see that $\operatorname{Aut}(A)$ is determined up to a mapping analogous to a conjugation. This is quite different from the type of uniqueness given by universal objects in a category. Such objects have only the identity automorphism, and hence are determined up to a unique isomorphism.

This is not the case with the algebraic closure of a field, which usually has a large amount of automorphisms. Most of this chapter and the next is devoted to the study of such automorphisms.

Examples. It will be proved later in this book that the complex numbers are algebraically closed. Complex conjugation is an automorphism of $\mathbf{C}$. There are many more automorphisms, but the other automorphisms are not continuous. We shall discuss other possible automorphisms in the chapter on transcendental extensions. The subfield of $\mathbf{C}$ consisting of all numbers which are algebraic over $\mathbf{Q}$ is an algebraic closure $\overline{\mathbf{Q}}$ of $\mathbf{Q}$. It is easy to see that $\overline{\mathbf{Q}}$ is denumerable. In fact, prove the following as an exercise:

If k is a field which is not finite, then any algebraic extension of k has the same cardinality as k.

If k is denumerable, one can first enumerate all polynomials in k, then enumerate finite extensions by their degree, and finally enumerate the cardinality of an arbitrary algebraic extension. We leave the counting details as exercises.

In particular, $\overline{\mathbf{Q}} \neq \mathbf{C}$. If $\mathbf{R}$ is the field of real numbers, then $\overline{\mathbf{R}} = \mathbf{C}$.

If k is a finite field, then an algebraic closure $\overline{k}$ of k is denumerable. We shall in fact describe in great detail the nature of algebraic extensions of finite fields later in this chapter.

Not all interesting fields are subfields of the complex numbers. For instance, one wants to investigate the algebraic extensions of a field $\mathbf{C}(X)$ where X is a variable over $\mathbf{C}$. The study of these extensions amounts to the study of ramified coverings of the sphere (viewed as a Riemann surface), and in fact one has precise information concerning the nature of such extensions, because one knows the fundamental group of the sphere from which a finite number of points has been deleted. We shall mention this example again later when we discuss Galois groups.

§3. *Splitting fields and normal extensions*

Let k be a field and let f be a polynomial in $k[X]$ of degree $\geqq 1$. By a *splitting field* K of f we shall mean an extension K of k such that f splits into linear factors in K, i.e.

$$f(X) = c(X - \alpha_1) \cdots (X - \alpha_n)$$

with $\alpha_i \in K$, $i = 1, \ldots, n$, and such that $K = k(\alpha_1, \ldots, \alpha_n)$ is generated by all the roots of f.

THEOREM 3. *Let K be a splitting field of the polynomial $f(X) \in k[X]$. If E is another splitting field of f, then there exists an isomorphism $\sigma : E \to K$ inducing the identity on k. If $k \subset K \subset \overline{k}$, where $\overline{k}$ is an algebraic closure of k, then any embedding of E in $\overline{k}$ inducing the identity on k must be an isomorphism of E on K.*

Proof. Let $\overline{K}$ be an algebraic closure of K. Then $\overline{K}$ is algebraic over k, hence is an algebraic closure of k. By Theorem 2 there exists an embedding

$$\sigma : E \to \overline{K}$$

inducing the identity on k. We have a factorization

$$f(X) = c(X - \beta_1) \cdots (X - \beta_n)$$

with $\beta_i \in E$, $i = 1, \ldots, n$. The leading coefficient c lies in k. We obtain

$$f(X) = f^\sigma(X) = c(X - \sigma\beta_1) \cdots (X - \sigma\beta_n).$$

We have unique factorization in $\overline{K}[X]$. Since f has a factorization

$$f(X) = c(X - \alpha_1) \cdots (X - \alpha_n)$$

in $K[X]$, it follows that $(\sigma\beta_1, \ldots, \sigma\beta_n)$ differs from $(\alpha_1, \ldots, \alpha_n)$ by a permutation. From this we conclude that $\sigma\beta_i \in K$ for $i = 1, \ldots, n$ and hence that $\sigma E \subset K$. But $K = k(\alpha_1, \ldots, \alpha_n) = k(\sigma\beta_1, \ldots, \sigma\beta_n)$, and hence $\sigma E = K$, because $E = k(\beta_1, \ldots, \beta_n)$. This proves our theorem.

We note that a polynomial $f(X) \in k[X]$ always has a splitting field, namely the field generated by its roots in a given algebraic closure $\bar{k}$ of k.

Let I be a set of indices and let $\{f_i\}_{i\in I}$ be a family of polynomials in $k[X]$, of degrees $\geqq 1$. By a *splitting field* for this family we shall mean an extension K of k such that every f_i splits in linear factors in $K[X]$, and K is generated by all the roots of all the polynomials f_i, $i \in I$. In most applications we deal with a finite indexing set I, but it is becoming increasingly important to consider infinite algebraic extensions, and so we shall deal with them fairly systematically. One should also observe that the proofs we shall give for various statements would not be simpler if we restricted ourselves to the finite case.

Let $\bar{k}$ be an algebraic closure of k, and let K_i be a splitting field of f_i in $\bar{k}$. Then the compositum of the K_i is a splitting field for our family, since the two conditions defining a splitting field are immediately satisfied. Furthermore Theorem 3 extends at once to the infinite case:

Corollary. *Let K be a splitting field for the family $\{f_i\}_{i\in I}$ and let E be another splitting field. Any embedding of E into $\overline{K}$ inducing the identity on k gives an isomorphism of E onto K.*

Proof. Let the notation be as above. Note that E contains a unique splitting field E_i of f_i and K contains a unique splitting field K_i of f_i. Any embedding σ of E into $\overline{K}$ must map E_i onto K_i by Theorem 3, and hence maps E into K. Since K is the compositum of the fields K_i, our map σ must send E onto K and hence induces an isomorphism of E onto K.

Remark. If I is finite, and our polynomials are $f_1, \ldots, f_n$, then a splitting field for them is a splitting field for the single polynomial $f(X) = f_1(X) \cdots f_n(X)$ obtained by taking the product. However, even when dealing with finite extensions only, it is convenient to deal simultaneously with sets of polynomials rather than a single one.

THEOREM 4. *Let K be an algebraic extension of k, contained in an algebraic closure $\bar{k}$ of k. Then the following conditions are equivalent:*

NOR 1. *Every embedding σ of K in $\bar{k}$ over k is an automorphism of K.*

NOR 2. *K is the splitting field of a family of polynomials in $k[X]$.*

NOR 3. *Every irreducible polynomial of $k[X]$ which has a root in K splits into linear factors in K.*

Proof. Assume NOR 1. Let α be an element of K and let $p_\alpha(X)$ be its irreducible polynomial over k. Let β be a root of p_α in $\bar{k}$. There exists an isomorphism of $k(\alpha)$ on $k(\beta)$ over k, mapping α on β. Extend this isomorphism to an embedding of K in $\bar{k}$. This extension is an automorphism σ of K by hypothesis, hence $\sigma\alpha = \beta$ lies in K. Hence every root of p_α lies in K, and p_α splits in linear factors in $K[X]$. Hence K is the splitting field of the family $\{p_\alpha\}_{\alpha \in K}$ as α ranges over all elements of K, and NOR 2 is satisfied.

Conversely, assume NOR 2, and let $\{f_i\}_{i \in I}$ be the family of polynomials of which K is the splitting field. If α is a root of some f_i in K, then for any embedding σ of K in $\bar{k}$ over k we know that $\sigma\alpha$ is a root of f_i. Since K is generated by the roots of all the polynomials f_i, it follows that σ maps K into itself. We now apply Lemma 1 to conclude that σ is an automorphism.

Our proof that NOR 1 implies NOR 2 also shows that NOR 3 is satisfied. Conversely, assume NOR 3. Let σ be an embedding of K in $\bar{k}$ over k. Let $\alpha \in K$ and let $p(X)$ be its irreducible polynomial over k. If σ is an embedding of K in $\bar{k}$ over k then σ maps α on a root β of $p(X)$, and by hypothesis β lies in K. Hence $\sigma\alpha$ lies in K, and σ maps K into itself. By Lemma 1, it follows that σ is an automorphism.

An extension K of k satisfying the hypotheses NOR 1, NOR 2, NOR 3 will be said to be *normal*. It is not true that the class of normal extensions is distinguished. For instance, it is easily shown that an extension of degree 2 is normal, but the extension $\mathbf{Q}(\sqrt[4]{2})$ of the rational numbers is not normal (the complex roots of $X^4 - 2$ are not in it), and yet this extension is obtained by successive extensions of degree 2, namely

$$E = \mathbf{Q}(\sqrt[4]{2}) \supset F \supset \mathbf{Q},$$

where

$$F = \mathbf{Q}(\alpha), \qquad \alpha = \sqrt{2} \qquad \text{and} \qquad E = F(\sqrt{\alpha}).$$

Thus a tower of normal extensions is not necessarily normal. However, we still have some of the properties:

THEOREM 5. *Normal extensions remain normal under lifting. If $K \supset E \supset k$ and K is normal over k, then K is normal over E. If K_1, K_2 are normal over k and are contained in some field L, then K_1K_2 is normal over k, and so is $K_1 \cap K_2$.*

Proof. For our first assertion, let K be normal over k, let F be any extension of k, and assume K, F are contained in some bigger field. Let σ be an embedding of KF over F (in $\overline{F}$). Then σ induces the identity on F, hence on k, and by hypothesis its restriction to K maps K into itself. We get $(KF)^\sigma = K^\sigma F^\sigma = KF$ whence KF is normal over F.

Assume that $K \supset E \supset k$ and that K is normal over k. Let σ be an embedding of K over E. Then σ is also an embedding of K over k, and our assertion follows by definition.

Finally, if K_1, K_2 are normal over k, then for any embedding σ of K_1K_2 over k we have

$$\sigma(K_1K_2) = \sigma(K_1)\sigma(K_2)$$

and our assertion again follows from the hypothesis. The assertion concerning the intersection is true because

$$\sigma(K_1 \cap K_2) = \sigma(K_1) \cap \sigma(K_2).$$

We observe that if K is a finitely generated normal extension of k, say $K = k(\alpha_1, \ldots, \alpha_n)$, and $p_1, \ldots, p_n$ are the respective irreducible polynomials of $\alpha_1, \ldots, \alpha_n$ over k then K is already the splitting field of the finite family $p_1, \ldots, p_n$. We shall investigate later when K is the splitting field of a single irreducible polynomial.

§4. *Separable extensions*

Let E be an algebraic extension of a field F and let

$$\sigma : F \to L$$

be an embedding of F in an algebraically closed field L. We investigate more closely extensions of σ to E. Any such extension of σ maps E on a subfield of L which is algebraic over σF. Hence for our purposes, we shall assume that L is algebraic over σF, hence is equal to an algebraic closure of σF.

Let S_σ be the set of extensions of σ to an embedding of E in L.

Let L' be another algebraically closed field, and let $\tau : F \to L'$ be an embedding. We assume as before that L' is an algebraic closure of τF. By Theorem 2 of §2, there exists an isomorphism $\lambda : L \to L'$ extending the map $\tau \circ \sigma^{-1}$ applied to the field σF. This is illustrated in the following diagram:

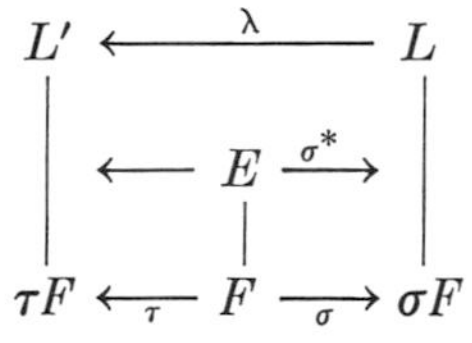

$$\begin{array}{ccccc} L' & & \xleftarrow{\lambda} & & L \\ | & \longleftarrow & E & \xrightarrow{\sigma^*} & | \\ | & & | & & | \\ \tau F & \xleftarrow[\tau]{} & F & \xrightarrow[\sigma]{} & \sigma F \end{array}$$

We let S_τ be the set of embeddings of E in L' extending τ.

If $\sigma^* \in S_\sigma$ is an extension of σ to an embedding of E in L, then $\lambda \circ \sigma^*$ is an extension of τ to an embedding of E into L', because for the restriction to F we have

$$\lambda \circ \sigma^* = \tau \circ \sigma^{-1} \circ \sigma = \tau.$$

Thus λ induces a mapping from S_σ into S_τ. It is clear that the inverse mapping is induced by λ^{-1}, and hence that S_σ, S_τ are in bijection under the mapping

$$\sigma^* \mapsto \lambda \circ \sigma^*.$$

In particular, the cardinality of S_σ, S_τ is the same. Thus this cardinality depends only on the extension E/F, and will be denoted by

$$[E : F]_s.$$

We shall call it the *separable degree* of E over F. It is mostly interesting when E/F is finite.

THEOREM 6. *Let $E \supset F \supset k$ be a tower. Then*

$$[E : k]_s = [E : F]_s [F : k]_s.$$

Furthermore, if E is finite over k, then $[E : k]_s$ is finite and

$$[E : k]_s \leqq [E : k].$$

The separable degree is at most equal to the degree.

Proof. Let $\sigma : k \to L$ be an embedding of k in an algebraically closed field L. Let $\{\sigma_i\}_{i \in I}$ be the family of distinct extensions of σ to F, and for each i, let $\{\tau_{ij}\}$ be the family of distinct extensions of σ_i to E. By what we saw before, each σ_i has precisely $[E : F]_s$ extensions to embeddings of E in L. The set of embeddings $\{\tau_{ij}\}$ contains precisely

$$[E : F]_s [F : k]_s$$

elements. Any embedding of E into L over σ must be one of the τ_{ij}, and thus we see that the first formula holds, i.e. we have multiplicativity in towers.

As to the second, let us assume that E/k is finite. Then we can obtain E as a tower of extensions, each step being generated by one element:

$$k \subset k(\alpha_1) \subset k(\alpha_1, \alpha_2) \subset \cdots \subset k(\alpha_1, \ldots, \alpha_r) = E.$$

If we define inductively $F_{\nu+1} = F_\nu(\alpha_{\nu+1})$ then by Proposition 8, §2,

$$[F_\nu(\alpha_{\nu+1}) : F_\nu]_s \leqq [F_\nu(\alpha_{\nu+1}) : F_\nu].$$

Thus our inequality is true in each step of the tower. By multiplicativity, it follows that the inequality is true for the extension E/k, as was to be shown.

COROLLARY. *Let E be finite over k, and $E \supset F \supset k$. The equality $[E : k]_s = [E : k]$ holds if and only if the corresponding equality holds in each step of the tower, i.e. for E/F and F/k.*

Proof. Clear.

It will be shown later (and it is not difficult to show) that $[E : k]_s$ divides the degree $[E : k]$ when E is finite over k. We define $[E : k]_i$ to be the quotient, so that

$$[E : k]_s[E : k]_i = [E : k].$$

It then follows from the multiplicativity of the separable degree and of the degree in towers that the symbol $[E : k]_i$ is also multiplicative in towers. We shall deal with it in §7.

Let E be a finite extension of k. We shall say that E is *separable* over k if $[E : k]_s = [E : k]$.

An element α algebraic over k is said to be *separable* over k if $k(\alpha)$ is separable over k. We see that this condition is equivalent to saying that the irreducible polynomial $\mathrm{Irr}(\alpha, k, X)$ has no multiple roots.

A polynomial $f(X) \in k[X]$ is called *separable* if it has no multiple roots. If α is a root of a separable polynomial $g(X) \in k[X]$ then the irreducible polynomial of α over k divides g and hence α is separable over k.

We shall now make remarks supplementing those of Proposition 8. The reader can omit these remarks if he is interested only in fields of characteristic 0 or separable extensions.

Let $f(X) = (X - \alpha)^m g(X)$ be a polynomial in $k[X]$, and assume $X - \alpha$ does not divide $g(X)$. We recall that m is called the multiplicity of α in f. We say that α is a *multiple* root of f if $m > 1$. Otherwise, we say that α is a *simple* root.

PROPOSITION 9. *Let α be algebraic over k, $\alpha \in \bar{k}$, and let $f(X) = \mathrm{Irr}(\alpha, k, X)$. If* $\mathrm{char}\, k = 0$, *then all roots of f have multiplicity* 1 *(f is separable). If* $\mathrm{char}\, k = p > 0$, *then there exists an integer $\mu \geqq 0$ such* *that every root of f has multiplicity p^μ. We have*

$$[k(\alpha) : k] = p^\mu [k(\alpha) : k]_s,$$

and α^{p^μ} is separable over k.

Proof. Let $\alpha_1, \dots, \alpha_r$ be the distinct roots of f in $\bar{k}$ and let $\alpha = \alpha_1$. Let m be the multiplicity of α in f. Given $1 \leqq i \leqq r$, there exists an isomorphism

$$\sigma : k(\alpha) \to k(\alpha_i)$$

over k such that $\sigma\alpha = \alpha_i$. Extend σ to an automorphism of $\bar{k}$ and denote this extension also by σ. Since f has coefficients in k we have $f^\sigma = f$. We note that

$$f(X) = \prod_{i=1}^{r} (X - \sigma\alpha_i)^{m_i}$$

if m_i is the multiplicity of α_i in f. By unique factorization, we conclude that $m_i = m_1$ and hence that all m_i are equal to the same integer m.

Consider the derivative $f'(X)$. If f and f' have a root in common, then α is a root of a polynomial of lower degree than $\deg f$. This is impossible unless $\deg f' = -\infty$, in other words, f' is identically 0. If the characteristic is 0, this cannot happen. Hence if f has multiple roots, we are in characteristic p, and $f(X) = g(X^p)$ for some polynomial $g(X) \in k[X]$. Therefore α^p is a root of a polynomial g whose degree is $< \deg f$. Proceeding inductively, we take the smallest integer $\mu \geqq 0$ such that α^{p^μ} is the root of a separable polynomial in $k[X]$, namely the polynomial h such that

$$f(X) = h(X^{p^\mu}).$$

Comparing the degree of f and g, we conclude that

$$[k(\alpha) : k(\alpha^p)] = p.$$

Inductively, we find

$$[k(\alpha) : k(\alpha^{p^\mu})] = p^\mu.$$

Since h has roots of multiplicity 1, we know that

$$[k(\alpha^{p^\mu}) : k]_s = [k(\alpha^{p^\mu}) : k],$$

and comparing the degree of f and the degree of h, we see that the number of distinct roots of f is equal to the number of distinct roots of h. Hence

$$[k(\alpha) : k]_s = [k(\alpha^{p^\mu}) : k]_s.$$

From this our formula for the degree follows by multiplicativity, and our proposition is proved. We note that the roots of h are

$$\alpha_1^{p^\mu}, \ldots, \alpha_r^{p^\mu}.$$

COROLLARY 1. *For any finite extension E of k, the separable degree $[E : k]_s$ divides the degree $[E : k]$. The quotient is 1 if the characteristic is 0, and a power of p if the characteristic is $p > 0$.*

Proof. We decompose E/k into a tower, each step being generated by one element, and apply Proposition 9, together with the multiplicativity of our indices in towers.

If E/K is finite, we call

$$\frac{[E : k]}{[E : k]_s}$$

the *inseparable degree* (or degree of inseparability), and denote it by $[E : k]_i$. We have

$$[E : k]_s[E : k]_i = [E : k].$$

COROLLARY 2. *A finite extension is separable if and only if $[E : k]_i = 1$.*

Proof. By definition.

COROLLARY 3. *If $E \supset F \supset k$ are two finite extensions, then*

$$[E : k]_i = [E : F]_i[F : k]_i.$$

Proof. Obvious.

We note that if α is separable over k, and F is any extension of k then α is separable over F. Indeed, if f is a separable polynomial in $k[X]$ such that $f(\alpha) = 0$, then f also has coefficients in F, and thus α is separable over F. (We may say that a separable element remains separable under lifting.)

THEOREM 7. *Let E be a finite extension of k. Then E is separable over k if and only if each element of E is separable over k.*

Proof. Assume E is separable over k and let $\alpha \in E$. We consider the tower

$$k \subset k(\alpha) \subset E.$$

By Theorem 6, we must have $[k(\alpha) : k] = [k(\alpha) : k]_s$ whence α is separable over k. Conversely, assume that each element of E is separable over k. We can write $E = k(\alpha_1, \ldots, \alpha_n)$ where each α_i is separable over k. We consider the tower

$$k \subset k(\alpha_1) \subset k(\alpha_1, \alpha_2) \subset \cdots \subset k(\alpha_1, \ldots, \alpha_n).$$

Since each α_i is separable over k, each α_i is separable over $k(\alpha_1, \ldots, \alpha_{i-1})$ for $i \geqq 2$. Hence by the tower theorem, it follows that E is separable over k.

We observe that our last argument shows: If E is generated by a finite number of elements, each of which is separable over k, then E is separable over k.

Let E be an arbitrary algebraic extension of k. We define E to be *separable* over k if every finitely generated subextension is separable over k, i.e. if every extension $k(\alpha_1, \ldots, \alpha_n)$ with $\alpha_1, \ldots, \alpha_n \in E$ is separable over k.

THEOREM 8. *Let E be an algebraic extension of k, generated by a family of elements $\{\alpha_i\}_{i \in I}$. If each α_i is separable over k then E is separable over k.*

Proof. Every element of E lies in some finitely generated subfield $k(\alpha_{i_1}, \ldots, \alpha_{i_n})$, and as we remarked above, each such subfield is separable over k. Hence every element of E is separable over k by Theorem 7, and this concludes the proof.

THEOREM 9. *Separable extensions form a distinguished class of extensions.*

Proof. Assume that E is separable over k and let $E \supset F \supset k$. Every element of E is separable over F, and every element of F is an element of E, so separable over k. Hence each step in the tower is separable. Conversely, assume that $E \supset F \supset k$ is some extension such that E/F is separable and F/k is separable. If E is finite over k, then we can use Theorem 6. Namely, we have an equality of the separable degree and the degree in each step of the tower, whence an equality for E over k by multiplicativity.

If E is infinite, let $\alpha \in E$. Then α is a root of a separable polynomial $f(X)$ with coefficients in F. Let these coefficients be $a_n, \ldots, a_0$. Let $F_0 = k(a_n, \ldots, a_0)$. Then F_0 is separable over k, and α is separable over F_0. We now deal with the finite tower

$$k \subset F_0 \subset F_0(\alpha)$$

and we therefore conclude that $F_0(\alpha)$ is separable over k, hence that α is separable over k. This proves condition (i) in the definition of "distinguished".

Let E be separable over k. Let F be any extension of k, and assume that E, F are both subfields of some field. Every element of E is separable over k, whence separable over F. Since EF is generated over F by all the elements of E, it follows that EF is separable over F, by Theorem 8. This proves condition (ii) in the definition of "distinguished", and concludes the proof of our theorem.

Let E be a finite extension of k. The intersection of all normal extensions K of k (in an algebraic closure $\bar{E}$) containing E is a normal extension of k which contains E, and is obviously the smallest normal extension of k containing E. If $\sigma_1, \ldots, \sigma_n$ are the distinct embeddings of E in $\bar{E}$ then the extension

$$K = (\sigma_1 E)(\sigma_2 E) \cdots (\sigma_n E),$$

which is the compositum of all these embeddings, is a normal extension of k, because for any embedding of it, say τ, we can apply τ to each extension $\sigma_i E$. Then $(\tau\sigma_1, \ldots, \tau\sigma_n)$ is a permutation of $(\sigma_1, \ldots, \sigma_n)$ and thus τ maps K into itself. Any normal extension of k containing E must contain $\sigma_i E$ for each i, and thus *the smallest normal extension of k containing E is precisely equal to the compositum*

$$(\sigma_1 E) \cdots (\sigma_n E).$$

If E is separable over k, then from Theorem 9 and induction we conclude that the smallest normal extension of k containing k is also separable over k.

Similar results hold for an infinite algebraic extension E of k, taking an infinite compositum. As a matter of terminology, if E is an algebraic extension of k, and σ any embedding of E in $\bar{k}$ over k, then we call σE a *conjugate* of E in $\bar{k}$. *We can say that the smallest normal extension of k containing E is the compositum of all the conjugates of E in $\bar{E}$.*

Let α be algebraic over k. If $\sigma_1, \ldots, \sigma_r$ are the distinct embeddings of $k(\alpha)$ into $\bar{k}$ over k, then we call $\sigma_1\alpha, \ldots, \sigma_r\alpha$ the *conjugates* of α in $\bar{k}$. These elements are simply the distinct roots of the irreducible polynomial of α over k. The smallest normal extension of k containing one of these conjugates is simply $k(\sigma_1\alpha, \ldots, \sigma_r\alpha)$.

§5. *Finite fields*

We have developed enough general theorems to describe the structure of finite fields. This is interesting for its own sake, and also gives us examples for the general theory.

Let F be a finite field with q elements. As we have noted previously, we have a homomorphism

$$\mathbf{Z} \to F$$

sending 1 on 1, whose kernel cannot be 0, and hence is a principal ideal generated by a prime number p since $\mathbf{Z}/p\mathbf{Z}$ is embedded in F and F has no divisors of zero. Thus F has characteristic p, and contains a field isomorphic to $\mathbf{Z}/p\mathbf{Z}$.

We remark that $\mathbf{Z}/p\mathbf{Z}$ has no automorphisms other than the identity. Indeed, any automorphism must map 1 on 1, hence leaves every element fixed because 1 generates $\mathbf{Z}/p\mathbf{Z}$ additively. We identify $\mathbf{Z}/p\mathbf{Z}$ with its image in F. Then F is a vector space over $\mathbf{Z}/p\mathbf{Z}$, and this vector space must be finite since F is finite. Let its degree be n. Let $\omega_1, \ldots, \omega_n$ be a basis for F over $\mathbf{Z}/p\mathbf{Z}$. Every element of F has a unique expression of the form

$$a_1\omega_1 + \cdots + a_n\omega_n$$

with $a_i \in \mathbf{Z}/p\mathbf{Z}$. Hence $q = p^n$.

The multiplicative group F^* of F has order $q - 1$. Every $\alpha \in F^*$ satisfies the equation $X^{q-1} = 1$. Hence every element of F satisfies the equation

$$f(X) = X^q - X = 0.$$

This implies that the polynomial $f(X)$ has q distinct roots in F, namely all elements of F. Hence f splits into factors of degree 1 in F, namely

$$X^q - X = \prod_{\alpha \in F} (X - \alpha).$$

In particular, F is a splitting field for f. But a splitting field is uniquely determined up to an isomorphism. Hence if a finite field of order p^n exists, it is uniquely determined, up to an isomorphism, as the splitting field of $X^{p^n} - X$ over $\mathbf{Z}/p\mathbf{Z}$.

As a matter of notation, we denote $\mathbf{Z}/p\mathbf{Z}$ by $\mathbf{F}_p$. Let n be an integer $\geqq 1$ and consider the splitting field of

$$X^{p^n} - X = f(X)$$

in an algebraic closure $\overline{\mathbf{F}}_p$. We contend that this splitting field is the set of roots of $f(X)$ in $\overline{\mathbf{F}}_p$. Indeed, let α, β be roots. Then

$$(\alpha + \beta)^{p^n} - (\alpha + \beta) = \alpha^{p^n} + \beta^{p^n} - \alpha - \beta = 0,$$

whence $\alpha + \beta$ is a root. Also,

$$(\alpha\beta)^{p^n} - \alpha\beta = \alpha^{p^n}\beta^{p^n} - \alpha\beta = \alpha\beta - \alpha\beta = 0,$$

and $\alpha\beta$ is a root. Note that 0, 1 are roots of $f(X)$. If $\beta \neq 0$ then

$$(\beta^{-1})^{p^n} - \beta^{-1} = (\beta^{p^n})^{-1} - \beta^{-1} = 0$$

so that β^{-1} is a root. Finally,

$$(-\beta)^{p^n} - (-\beta) = (-1)^{p^n}\beta^{p^n} + \beta.$$

If p is odd, then $(-1)^{p^n} = -1$ and we see that $-\beta$ is a root. If p is even then $-1 = 1$ (in $\mathbf{Z}/2\mathbf{Z}$) and hence $-\beta = \beta$ is a root. This proves our contention.

The derivative of $f(X)$ is

$$f'(X) = p^n X^{p^n - 1} - 1 = -1.$$

Hence $f(X)$ has no multiple roots, and therefore has p^n distinct roots in $\overline{\mathbf{F}}_p$. Hence its splitting field has exactly p^n elements. We summarize our results:

THEOREM 10. *For each prime p and each integer $n \geqq 1$ there exists a finite field of order p^n, denoted by $\mathbf{F}_{p^n}$, uniquely determined as a subfield of an algebraic closure $\overline{\mathbf{F}}_p$. It is the splitting field of the polynomial*

$$X^{p^n} - X,$$

and its elements are the roots of this polynomial. Every finite field is isomorphic to exactly one field $\mathbf{F}_{p^n}$.

We usually write $p^n = q$ and $\mathbf{F}_q$ instead of $\mathbf{F}_{p^n}$.

COROLLARY. *Let $\mathbf{F}_q$ be a finite field. Let n be an integer $\geqq 1$. In a given algebraic closure $\overline{\mathbf{F}}_q$, there exists one and only one extension of $\mathbf{F}_q$ of degree n, and this extension is the field $\mathbf{F}_{q^n}$.*

Proof. Let $q = p^m$. Then $q^n = p^{mn}$. The splitting field of $X^{q^n} - X$ is precisely $\mathbf{F}_{p^{mn}}$ and has degree mn over $\mathbf{Z}/p\mathbf{Z}$. Since $\mathbf{F}_q$ has degree m over $\mathbf{Z}/p\mathbf{Z}$, it follows that $\mathbf{F}_{q^n}$ has degree n over $\mathbf{F}_q$. Conversely, any extension of degree n over $\mathbf{F}_q$ has degree mn over $\mathbf{F}_p$ and hence must be $\mathbf{F}_{p^{mn}}$. This proves our corollary.

THEOREM 11. *The multiplicative group of a finite field is cyclic.*

Proof. This has already been proved in Chapter V, §4, Theorem 6.

We shall determine all automorphisms of a finite field.

Let $q = p^n$ and let $\mathbf{F}_q$ be the finite field with q elements. We consider the Frobenius mapping

$$\varphi : \mathbf{F}_q \to \mathbf{F}_q$$

such that $\varphi(x) = x^p$. Then φ is a homomorphism, and its kernel is 0 since $\mathbf{F}_q$ is a field. Hence φ is injective. Since $\mathbf{F}_q$ is finite, it follows that φ is surjective, and hence that φ is an isomorphism. We note that it leaves $\mathbf{F}_p$ fixed.

THEOREM 12. *The group of automorphisms of $\mathbf{F}_q$ is cyclic of degree n, generated by φ.*

Proof. Let G be the group generated by φ. We note that $\varphi^n = id$ because $\varphi^n(x) = x^{p^n} = x$ for all $x \in \mathbf{F}_q$. Hence n is an exponent for φ.

Let d be the period of φ, so $d \geqq 1$. We have $\varphi^d(x) = x^{p^d}$ for all $x \in \mathbf{F}_q$. Hence each $x \in \mathbf{F}_q$ is a root of the equation

$$X^{p^d} - X = 0.$$

This equation has at most p^d roots. It follows that $d \geqq n$, whence $d = n$.

There remains to be proved that G is the group of all automorphisms of $\mathbf{F}_q$. Any automorphism of $\mathbf{F}_q$ must leave $\mathbf{F}_p$ fixed. Hence it is an automorphism of $\mathbf{F}_q$ over $\mathbf{F}_p$. By Theorem 6 of §4, the number of such automorphisms is $\leqq n$. Hence $\mathbf{F}_q$ cannot have any other automorphisms except for those of G.

THEOREM 13. *Let m, n be integers $\geqq 1$. Then $\mathbf{F}_{p^n}$ is contained in $\mathbf{F}_{p^m}$ if and only if n divides m. If that is the case, let $q = p^n$, and let $m = nd$. Then $\mathbf{F}_{p^m}$ is normal and separable over $\mathbf{F}_q$, and the group of automorphisms of $\mathbf{F}_{p^m}$ over $\mathbf{F}_q$ is cyclic, generated by φ^n.*

Proof. All the statements are trivial consequences of what has already been proved, and will be left to the reader.

§6. Primitive elements

THEOREM 14. *Let E be a finite extension of a field k. There exists an element $\alpha \in E$ such that $E = k(\alpha)$ if and only if there exists only a finite number of fields F such that $k \subset F \subset E$. If E is separable over k, then there exists such an element α.*

Proof. If k is finite, then we know that the multiplicative group of E is generated by one element, which will therefore also generate E over k. We assume that k is infinite.

Assume that there is only a finite number of fields, intermediate between k and E. Let $\alpha, \beta \in E$. As c ranges over elements of k, we can only have a finite number of fields of type $k(\alpha + c\beta)$. Hence there exist elements $c_1, c_2 \in k$ with $c_1 \neq c_2$ such that

$$k(\alpha + c_1\beta) = k(\alpha + c_2\beta).$$

Note that $\alpha + c_1\beta$ and $\alpha + c_2\beta$ are in the same field, whence so is $(c_1 - c_2)\beta$, and hence so is β. Thus α is also in that field, and we see that $k(\alpha, \beta)$ can be generated by one element.

Proceeding inductively, if $E = k(\alpha_1, \ldots, \alpha_n)$ then there will exist elements $c_2, \ldots, c_n \in k$ such that

$$E = k(\xi)$$

where $\xi = \alpha_1 + c_2\alpha_2 + \cdots + c_n\alpha_n$. This proves half of our theorem.

Conversely, assume that $E = k(\alpha)$ for some α, and let $f(X) = \operatorname{Irr}(\alpha, k, X)$. Let $k \subset F \subset E$. Let $g_F(X) = \operatorname{Irr}(\alpha, F, X)$. Then g_F divides f. We have unique factorization in $E[X]$, and any polynomial in $E[X]$ which has leading coefficient 1 and divides $f(X)$ is equal to a product of factors $(X - \alpha_i)$ where $\alpha_1, \ldots, \alpha_n$ are the roots of f. Hence there is only a finite number of such polynomials. Thus we get a mapping

$$F \mapsto g_F$$

from the set of intermediate fields into a finite set of polynomials. Let F_0 be the subfield of F generated over k by the coefficients of $g_F(X)$. Then g_F has coefficients in F_0 and is irreducible over F_0 since it is irreducible over F. Hence the degree of α over F_0 is the same as the degree of α over F. Hence $F = F_0$. Thus our field F is uniquely determined by its associated polynomials g_F, and our mapping is therefore injective. This proves the first assertion of the theorem.

As to the statement concerning separable extensions, using induction, we may assume without loss of generality that $E = k(\alpha, \beta)$ where α, β are separable over k. Let $\sigma_1, \ldots, \sigma_n$ be the distinct embeddings of $k(\alpha, \beta)$ in $\bar{k}$ over k. Let

$$P(X) = \prod_{i \neq j} (\sigma_i\alpha + X\sigma_i\beta - \sigma_j\alpha - X\sigma_j\beta).$$

Then $P(X)$ is not the zero polynomial, and hence there exists $c \in k$ such that $P(c) \neq 0$. Then the elements $\sigma_i(\alpha + c\beta)$ $(i = 1, \ldots, n)$ are distinct, whence $k(\alpha + c\beta)$ has degree at least n over k. But $n = [k(\alpha, \beta) : k]$, and hence $k(\alpha, \beta) = k(\alpha + c\beta)$, as desired.

If $E = k(\alpha)$, then we say that α is a *primitive element* of E (over k).

§7. *Purely inseparable extensions*

This section is of a fairly technical nature, and can be omitted without impairing the understanding of most of the rest of the book.

We assume throughout that k is a field of characteristic $p > 0$.

An element α algebraic over k is said to be *purely inseparable* over k if there exists an integer $n \geqq 0$ such that α^{p^n} lies in k.

Let E be an algebraic extension of k. We contend that the following conditions are equivalent:

P. Ins. 1. We have $[E : k]_s = 1$.

P. Ins. 2. Every element α of E is purely inseparable over k.

P. Ins. 3. For every $\alpha \in E$, the irreducible equation of α over k is of type $X^{p^n} - a = 0$ with some $n \geqq 0$ and $a \in k$.

P. Ins. 4. There exists a set of generators $\{\alpha_i\}_{i \in I}$ of E over k such that each α_i is purely inseparable over k.

To prove the equivalence, assume P. Ins. 1. Let $\alpha \in E$. By Theorem 6, we conclude that $[k(\alpha) : k]_s = 1$. Let $f(X) = \mathrm{Irr}(\alpha, k, X)$. Then f has only one root since

$$[k(\alpha) : k]_s$$

is equal to the number of distinct roots of $f(X)$. Let $m = [k(\alpha) : k]$. Then $\deg f = m$, and the factorization of f over $k(\alpha)$ is $f(X) = (X - \alpha)^m$. Write $m = p^n r$ where r is an integer prime to p. Then

$$\begin{aligned} f(X) &= (X^{p^n} - \alpha^{p^n})^r \\ &= X^{p^n r} - r\alpha^{p^n} X^{p^n(r-1)} + \text{lower terms.} \end{aligned}$$

Since the coefficients of $f(X)$ lie in k, it follows that

$$r\alpha^{p^n}$$

lies in k, and since $r \neq 0$ (in k), then α^{p^n} lies in k. Let $a = \alpha^{p^n}$. Then α is a root of the polynomial $X^{p^n} - a$, which divides $f(X)$. It follows that $f(X) = X^{p^n} - a$.

Essentially the same argument as the preceding one shows that P. Ins. 2 implies P. Ins. 3. It is trivial that the third condition implies the fourth.

Finally, assume P. Ins. 4. Let E be an extension generated by purely inseparable elements α_i $(i \in I)$. Any embedding of E over k maps α_i on a root of

$$f_i(X) = \mathrm{Irr}(\alpha_i, k, X).$$

But $f_i(X)$ divides some polynomial $X^{p^n} - a$, which has only one root. Hence any embedding of E over k is the identity on each α_i, whence the identity on E, and we conclude that $[E : k]_s = 1$, as desired.

An extension satisfying the above four properties will be called *purely inseparable.*

PROPOSITION 10. *Purely inseparable extensions form a distinguished class of extensions.*

Proof. The tower theorem is clear from Theorem 6, and the lifting property is clear from condition P. Ins. 4.

PROPOSITION 11. *Let E be an algebraic extension of k. Let E_0 be the compositum of all subfields F of E such that $F \supset k$ and F is separable over k. Then E_0 is separable over k, and E is purely inseparable over E_0.*

Proof. Since separable extensions form a distinguished class, we know that E_0 is separable over k. In fact, E_0 consists of all elements of E which are separable over k. By Proposition 9, given $\alpha \in E$ there exists a power of p, say p^n such that α^{p^n} is separable over k. Hence E is purely inseparable over E_0, as was to be shown.

COROLLARY 1. *If an algebraic extension E of k is both separable and purely inseparable, then $E = k$.*

Proof. Obvious.

COROLLARY 2. *Let K be normal over k and let K_0 be its maximal separable subextension. Then K_0 is also normal over k.*

Proof. Let σ be an embedding of K_0 in $\overline{K}$ over k and extend σ to an embedding of K. Then σ is an automorphism of K. Furthermore, σK_0 is separable over k, hence is contained in K_0 since K_0 is the maximal separable subfield. Hence $\sigma K_0 = K_0$, as contended.

COROLLARY 3. *Let E, F be two finite extensions of k, and assume that E/k is separable, F/k is purely inseparable. Assume E, F are subfields of a common field. Then*

$$[EF : F] = [E : k] = [EF : k]_s,$$
$$[EF : E] = [F : k] = [EF : k]_i.$$

Proof. The picture is as follows:

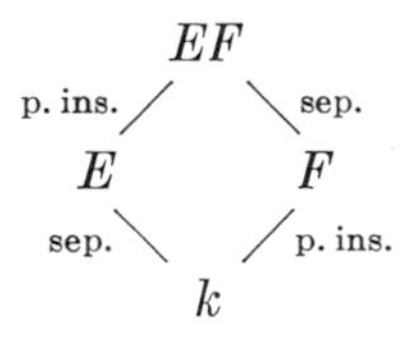

The proof is a trivial juggling of indices, using the corollaries of Proposition 9. We leave it as an exercise.

COROLLARY 4. *Let E^p denote the field of all elements x^p, $x \in E$. Let E be a finite extension of k. If $E^pk = E$, then E is separable over k. If E is separable over k, then $E^{p^n}k = E$ for all $n \geqq 1$.*

Proof. Let E_0 be the maximal separable subfield of E. Assume $E^pk = E$. Let $E = k(\alpha_1, \ldots, \alpha_n)$. Since E is purely inseparable over E_0 there exists m such that $\alpha_i^{p^m} \in E_0$ for each $i = 1, \ldots, n$. Hence $E^{p^m} \subset E_0$. But $E^{p^m}k = E$ whence $E = E_0$ is separable over k. Conversely, assume that E is separable over k. Then E is purely inseparable over E^pk. Since E is also separable over E^pk we conclude that $E = E^pk$. Iterating we get $E = E^{p^n}k$ for $n \geqq 1$, as was to be shown.

Proposition 11 shows that any algebraic extension can be decomposed into a tower consisting of a maximal separable subextension and a purely inseparable step above it. Usually, one cannot reverse the order of the tower. However, there is an important case when it can be done.

PROPOSITION 12. *Let K be a normal extension of k. Let G be its group of automorphisms over k, and let K^G be the fixed field of G. Then K^G is purely inseparable over k, and K is separable over K^G. If K_0 is the maximal separable subextension of K, then $K = K^G K_0$ and $K_0 \cap K^G = k$.*

Proof. Let $\alpha \in K^G$. Let τ be an embedding of $k(\alpha)$ over k in $\overline{K}$ and extend τ to an embedding of K, which we denote also by τ. Then τ is an automorphism of K because K is normal over k. By definition, $\tau\alpha = \alpha$ and hence τ is the identity on $k(\alpha)$. Hence $[k(\alpha) : k]_s = 1$ and α is purely inseparable. Thus K^G is purely inseparable over k. The intersection of K_0 and K^G is both separable and purely inseparable over k, and hence is equal to k.

To prove that K is separable over K^G, assume first that K is finite over k, and hence that G is finite, by Theorem 6. Let $\alpha \in K$. Let $\sigma_1, \ldots, \sigma_r$ be a maximal subset of elements of G such that the elements

$$\sigma_1\alpha, \ldots, \sigma_r\alpha$$

are distinct, and such that σ_i is the identity, and α is a root of the polynomial

$$f(X) = \prod_{i=1}^{r} (X - \sigma_i\alpha).$$

For any $\tau \in G$ we note that $f^\tau = f$ because τ permutes the roots. We note that f is separable, and that its coefficients are in the fixed field K^G. Hence α is separable over K^G. The reduction of the infinite case to the finite case is done by observing that every $\alpha \in K$ is contained in some finite normal subextension of K. We leave the details to the reader.

We now have the following picture:

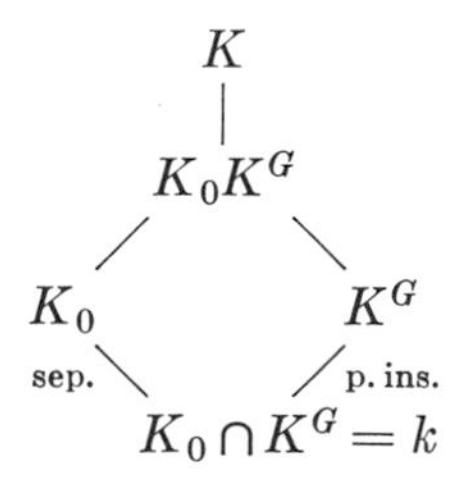

By Proposition 11, K is purely inseparable over K_0, hence purely inseparable over K_0K^G. Furthermore, K is separable over K^G, hence separable over K_0K^G. Hence $K = K_0K^G$, thereby proving our theorem.

We see that every normal extension decomposes into a compositum of a purely inseparable and a separable extension. We shall define a Galois extension in the next chapter to be a normal separable extension. Then K_0

is Galois over k and the normal extension is decomposed into a Galois and a purely inseparable extension. The group G is called the *Galois group* of the extension K/k.

A field k is called *perfect* if $k^p = k$. (Every field of characteristic zero is also called perfect.)

COROLLARY. *If k is perfect, then every algebraic extension of k is separable. Every algebraic extension of k is perfect.*

Proof. Every finite algebraic extension is contained in a normal extension, and we apply Proposition 12 to get what we want.

EXERCISES

1. Let k be a finite field with q elements. Let $f(X) \in k[X]$ be irreducible. Show that $f(X)$ divides $X^{q^n} - X$ if and only if deg f divides n.

2. Show that

$$X^{q^n} - X = \prod_{d|n} \prod_{f_d \text{ irr}} f_d(X),$$

where the inner product is over all irreducible polynomials of degree d with leading coefficient 1. Counting degrees, show that

$$q^n = \sum_{d|n} d\psi(d),$$

where $\psi(d)$ is the number of irreducible polynomials of degree d. Invert by elementary number theory, and find that

$$n\psi(n) = \sum_{d|n} \mu(d)q^{n/d}.$$

3. Let k be a field of characteristic p and let t, u be algebraically independent over k. Prove the following:
(i) $k(t, u)$ has degree p^2 over $k(t^p, u^p)$.
(ii) There exist infinitely many extensions between $k(t, u)$ and $k(t^p, u^p)$.

4. Let E be a finite extension of k and let $p^r = [E : k]_i$. We assume that the characteristic is $p > 0$. Assume that there is no exponent p^s with $s < r$ such that $E^{p^s}k$ is separable over k (i.e. such that α^{p^s} is separable over k for each α in E). Show that E can be generated by one element over k. [*Hint:* Assume first that E is purely inseparable.]

5. Let k be a field, $f(X)$ an irreducible polynomial in $k[X]$, and let K be a finite normal extension of k. If g, h are irreducible factors of $f(X)$ in $K[X]$, show that there exists an automorphism σ of K over k such that $g = h^\sigma$. Give an example when this conclusion is not valid if K is not normal over k.

6. Let $x_1, \ldots, x_n$ be algebraically independent over a field k. Let y be algebraic over $k(x) = k(x_1, \ldots, x_n)$. Let $P(X_{n+1})$ be the irreducible polynomial

of y over $k(x)$. Let $\varphi(x)$ be the least common multiple of the denominators of the coefficients of P. Then the coefficients of $\varphi(x)P$ are elements of $k[x]$. Show that the polynomial

$$f(X_1, \ldots, X_{n+1}) = \varphi(X_1, \ldots, X_n)P(X_{n+1})$$

is irreducible over k, as a polynomial in $n + 1$ variables.

Conversely, let $f(X_1, \ldots, X_{n+1})$ be an irreducible polynomial over k. Let $x_1, \ldots, x_n$ be algebraically independent over k. Show that

$$f(x_1, \ldots, x_n, X_{n+1})$$

is irreducible over $k(x_1, \ldots, x_n)$.

If f is a polynomial in n variables, and $(b) = (b_1, \ldots, b_n)$ is an n-tuple of elements such that $f(b) = 0$, then we say that (b) is a *zero* of f. We say that (b) is *non-trivial* if not all coordinates b_i are equal to 0.

7. Let $f(X_1, \ldots, X_n)$ be a homogeneous polynomial of degree 2 (resp. 3) over a field k. Show that if f has a non-trivial zero in an extension of odd degree (resp. degree 2) over k, then f has a non-trivial zero in k.

8. Let $f(X, Y)$ be an irreducible polynomial in two variables over a field k. Let t be transcendental over k, and assume that there exist integers $m, n \neq 0$ and elements $a, b \in k$, $ab \neq 0$, such that $f(at^n, bt^m) = 0$. Show that after inverting possibly X or Y, and up to a constant factor, f is of type

$$X^m Y^n - c$$

with some $c \in k$.

The answer to the following exercise is not known.

9. (Artin) Let f be a homogeneous polynomial of degree d in n variables, with rational coefficients. If $n > d$, show that there exists a root of unity ζ, and elements $x_1, \ldots, x_n \in \mathbf{Q}[\zeta]$ not all 0 such that $f(x_1, \ldots, x_n) = 0$.

CHAPTER VIII

Galois Theory

§1. *Galois extensions*

Let K be a field and let G be a group of automorphisms of K. We denote by K^G the subset of K consisting of all elements $x \in K$ such that $x^\sigma = x$ for all $\sigma \in G$. It is also called the *fixed field* of G. It is a field because if $x, y \in K^G$ then

$$(x + y)^\sigma = x^\sigma + y^\sigma = x + y$$

for all $\sigma \in G$, and similarly, one verifies that K is closed under multiplication, subtraction, and multiplicative inverse. Furthermore, K^G contains 0 and 1, hence contains the prime field.

An algebraic extension K of a field k is called *Galois* if it is normal and separable. We consider K as embedded in an algebraic closure. The group of automorphisms of K over k is called the *Galois group* of K over k, and is denoted by $G(K/k)$, or simply G. It coincides with the set of embeddings of K in $\overline{K}$ over k.

For the convenience of the reader, we shall now state the main result of the Galois theory for finite Galois extensions.

Let K be a finite Galois extension of k, with Galois group G. There is a bijection between the set of subfields E of K containing k, and the set of subgroups H of G, given by $E = K^H$. The field E is Galois over k if and only if H is normal in G, and if that is the case, then the map $\sigma \mapsto \sigma \mid E$ induces an isomorphism of G/H onto the Galois group of E over k.

We shall give the proofs step by step, and as far as possible, we give them for infinite extensions.

THEOREM 1. *Let K be a Galois extension of k. Let G be its Galois group. Then $k = K^G$. If F is an intermediate field, $k \subset F \subset K$, then K is Galois over F. The map*

$$F \mapsto G(K/F)$$

from the set of intermediate fields into the set of subgroups of G is injective.

Proof. Let $\alpha \in K^G$. Let σ be any embedding of $k(\alpha)$ in $\overline{K}$, inducing the identity on k. Extend σ to an embedding of K into $\overline{K}$, and call this extension σ also. Then σ is an automorphism of K over k, hence is an element of G. By assumption, σ leaves α fixed. Therefore

$$[k(\alpha) : k]_s = 1.$$

Since α is separable over k, we have $k(\alpha) = k$ and α is an element of k. This proves our first assertion.

Let F be an intermediate field. Then K is normal over F by Theorem 5 and is separable over F by Theorem 9 of Chapter VII. Hence K is Galois over F. If $H = G(K/F)$ then by what we proved above we conclude that $F = K^H$. If F, F' are intermediate fields, and $H = G(K/F)$, $H' = G(K/F')$, then

$$F = K^H \quad \text{and} \quad F' = K^{H'}.$$

If $H = H'$ we conclude that $F = F'$, whence our map

$$F \mapsto G(K/F)$$

is injective, thereby proving our theorem.

We shall sometimes call the group $G(K/F)$ of an intermediate field the group *associated* with F. We say that a subgroup H of G *belongs* to an intermediate field F if $H = G(K/F)$.

COROLLARY 1. *Let K/k be Galois with group G. Let F, F' be two intermediate fields, and let H, H' be the subgroups of G belonging to F, F' respectively. Then $H \cap H'$ belongs to FF'.*

Proof. Every element of $H \cap H'$ leaves FF' fixed, and every element of G which leaves FF' fixed also leaves F and F' fixed and hence lies in $H \cap H'$. This proves our assertion.

COROLLARY 2. *Let the notation be as in Corollary 1. The fixed field of the smallest subgroup of G containing H, H' is $F \cap F'$.*

Proof. Obvious.

COROLLARY 3. *Let the notation be as in Corollary 1. Then $F \subset F'$ if and only if $H' \subset H$.*

Proof. If $F \subset F'$ and $\sigma \in H'$ leaves F' fixed then σ leaves F fixed, so σ lies in H. Conversely, if $H' \subset H$ then the fixed field of H is contained in the fixed field of H', so $F \subset F'$.

COROLLARY 4. *Let E be a finite separable extension of a field k. Let K be the smallest normal extension of k containing E. Then K is finite Galois over k. There is only a finite number of intermediate fields F such that $k \subset F \subset E$.*

Proof. We know that K is normal and separable, and K is finite over k since we saw that it is the finite compositum of the finite number of conjugates of E. The Galois group of K/k has only a finite number of subgroups. Hence there is only a finite number of subfields of K containing k, whence *a fortiori* a finite number of subfields of E containing k.

Of course, Corollary 4 has been proved in the preceding chapter, but we get another proof here from another point of view.

LEMMA 1. *Let E be an algebraic separable extension of k. Assume that there is an integer $n \geqq 1$ such that every element α of E is of degree $\leqq n$ over k. Then E is finite over k and $[E : k] \leqq n$.*

Proof. Let α be an element of E such that the degree $[k(\alpha) : k]$ is maximal, say $m \leqq n$. We contend that $k(\alpha) = E$. If this is not true, then there exists an element $\beta \in E$ such that $\beta \notin k(\alpha)$, and by the primitive element theorem, there exists an element $\gamma \in k(\alpha, \beta)$ such that $k(\alpha, \beta) = k(\gamma)$. But from the tower

$$k \subset k(\alpha) \subset k(\alpha, \beta)$$

we see that $[k(\alpha, \beta) : k] > m$ whence γ has degree $> m$ over k, contradiction.

THEOREM 2. (Artin) *Let K be a field and let G be a finite group of automorphisms of K, of order n. Let $k = K^G$ be the fixed field. Then K is a finite Galois extension of k, and its Galois group is G. We have $[K : k] = n$.*

Proof. Let $\alpha \in K$ and let $\sigma_1, \dots, \sigma_r$ be a maximal set of elements of G such that $\sigma_1\alpha, \dots, \sigma_r\alpha$ are distinct. If $\tau \in G$ then $(\tau\sigma_1\alpha, \dots, \tau\sigma_r\alpha)$ differs from $(\sigma_1\alpha, \dots, \sigma_r\alpha)$ by a permutation, because τ is injective, and every $\tau\sigma_i\alpha$ is among the set $\{\sigma_1\alpha, \dots, \sigma_r\alpha\}$; otherwise this set is not maximal. Hence α is a root of the polynomial

$$f(X) = \prod_{i=1}^{r} (X - \sigma_i\alpha),$$

and for any $\tau \in G$, $f^\tau = f$. Hence the coefficients of f lie in $K^G = k$. Furthermore, f is separable. Hence every element α of K is a root of a separable polynomial of degree $\leqq n$ with coefficients in k. Furthermore, this polynomial splits in linear factors in K. Hence K is separable over k, is normal over k, hence Galois over k. By Lemma 1, we have $[K : k] \leqq n$. The Galois group of K over k has order $\leqq [K : k]$ (by Theorem 6 of Chapter VII, §4), and hence G must be the full Galois group. This proves all our assertions.

COROLLARY. *Let K be a finite Galois extension of k and let G be its Galois group. Then every subgroup of G belongs to some subfield F such that $k \subset F \subset K$.*

Proof. Let H be a subgroup of G and let $F = K^H$. By Artin's theorem we know that K is Galois over F with group H.

Remark. When K is an infinite Galois extension of k, then the preceding corollary is not true any more. This shows that some counting argument must be used in the proof of the finite case. In the present treatment, we have used an old-fashioned argument. The reader can look up Artin's own proof in his book *Galois Theory*. In the infinite case, one defines the Krull topology on the Galois group G (cf. Exercises), and G becomes a compact totally disconnected group. The subgroups which belong to the intermediate fields are the *closed* subgroups. If the reader wishes to disregard the infinite case entirely throughout our discussions he may do so without impairing his understanding. The proofs in the infinite case are usually identical with those in the finite case.

The notion of a Galois extension and a Galois group are defined completely algebraically. Hence they behave formally under isomorphisms the way one expects from objects in any category. We describe this behavior more explicitly in the present case.

Let K be a Galois extension of k. Let

$$\lambda : K \to K^\lambda = \lambda K$$

be an isomorphism. Then K^λ is a Galois extension of k^λ.

$$\begin{array}{ccc} K & \xrightarrow{\lambda} & K^\lambda \\ | & & | \\ k & \xrightarrow[\lambda]{} & k^\lambda \end{array}$$

Let G be the Galois group of K over k. Then the map

$$\sigma \mapsto \lambda \circ \sigma \circ \lambda^{-1}$$

gives a homomorphism of G into the Galois group of K^λ over k^λ, whose inverse is given by

$$\lambda^{-1} \circ \tau \circ \lambda \leftarrow\!\shortmid \tau.$$

Hence $G(K^\lambda/k^\lambda)$ is isomorphic to $G(K/k)$ under the above map. We may write

$$G(\lambda K/\lambda k)^\lambda = G(K/k)$$

or

$$G(\lambda K/\lambda k) = \lambda G(K/k)\lambda^{-1},$$

where the exponent λ is "conjugation",

$$\sigma^{\lambda} = \lambda^{-1} \circ \sigma \circ \lambda.$$

There is no avoiding the contravariance if we wish to preserve the rule

$$(\sigma^{\lambda})^{\omega} = \sigma^{\lambda\omega}$$

when we compose mappings λ and ω.

In particular, let F be an intermediate field, $k \subset F \subset K$, and let $\lambda : F \to \lambda F$ be an embedding of F in K, which we assume is extended to an automorphism of K. Then $\lambda K = K$. Hence

$$G(K/\lambda F)^{\lambda} = G(K/F)$$

and

$$G(K/\lambda F) = \lambda G(K/F)\lambda^{-1}.$$

THEOREM 3. *Let K be a Galois extension of k with group G. Let F be a subfield, $k \subset F \subset K$, and let $H = G(K/F)$. Then F is normal over k if and only if H is normal in G. If F is normal over k, then the restriction map $\sigma \mapsto \sigma \mid F$ is a homomorphism of G onto the Galois group of F over k, whose kernel is H. We thus have $G(F/k) \approx G/H$.*

Proof. Assume F is normal over k, and let G' be its Galois group. The restriction map $\sigma \to \sigma \mid F$ maps G into G', and by definition, its kernel is H. Hence H is normal in G. Furthermore, any element $\tau \in G'$ extends to an embedding of K in $\overline{K}$, which must be an automorphism of K, so the restriction map is surjective. This proves the last statement. Finally, assume that F is not normal over k. Then there exists an embedding λ of F in K over k which is not an automorphism, i.e. $\lambda F \neq F$. Extend λ to an automorphism of K over k. The Galois groups $G(K/\lambda F)$ and $G(K/F)$ are conjugate, and they belong to distinct subfields, hence cannot be equal. Hence H is not normal in G.

A Galois extension K/k is said to be *abelian* (resp. *cyclic*) if its Galois group G is abelian (resp. cyclic).

COROLLARY. *Let K/k be abelian (resp. cyclic). If F is an intermediate field, $k \subset F \subset K$, then F is Galois over k and abelian (resp. cyclic).*

Proof. This follows at once from the fact that a subgroup of an abelian group is abelian, and a factor group of an abelian (resp. cyclic) group is abelian (resp. cyclic).

THEOREM 4. *Let K be a Galois extension of k, let F be an arbitrary extension and assume that K, F are subfields of some other field. Then KF is Galois over F, and K is Galois over $K \cap F$. Let H be the Galois* *group of KF over F, and G the Galois group of K over k. If $\sigma \in H$ then the restriction of σ to K is in G, and the map*

$$\sigma \mapsto \sigma \mid K$$

gives an isomorphism of H on the Galois group of K over $K \cap F$.

Proof. Let $\sigma \in H$. The restriction of σ to K is an embedding of K over k, whence an element of G since K is normal over k. The map $\sigma \to \sigma \mid K$ is clearly a homomorphism. If $\sigma \mid K$ is the identity, then σ must be the identity of KF (since every element of KF can be expressed as a combination of sums, products, and quotients of elements in K and F). Hence our homomorphism $\sigma \to \sigma \mid K$ is injective. Let H' be its image. Then H' leaves $K \cap F$ fixed, and conversely, if an element $\alpha \in K$ is fixed under H', we see that α is also fixed under H, whence $\alpha \in F$ and $\alpha \in K \cap F$. Therefore $K \cap F$ is the fixed field. If K is finite over k, or even KF finite over F, then by Theorem 2, we know that H' is the Galois group of K over $K \cap F$, and the theorem is proved in that case.

(In the infinite case, one must add the remark that for the Krull topology, our map $\sigma \to \sigma \mid K$ is continuous, whence its image is closed since H is compact.)

The diagram illustrating Theorem 4 is as follows:

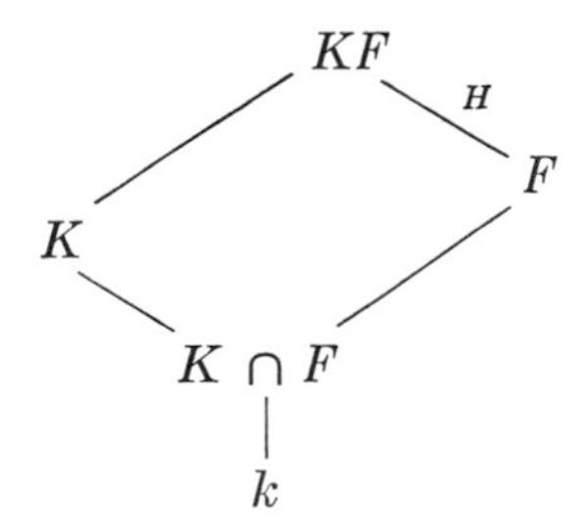

It is suggestive to think of the opposite sides of a parallelogram as being equal.

COROLLARY. *Let K be a finite Galois extension of k. Let F be an arbitrary extension of k. Then $[KF : F]$ divides $[K : k]$.*

Proof. Notation being as above, we know that the order of H divides the order of G, so our assertion follows.

Warning. The assertion of the corollary is not usually valid if K is not Galois over k. For instance, let $\alpha = \sqrt[3]{2}$ be the real cube root of 2, let ζ be a cube root of 1, $\zeta \neq 1$, say

$$\zeta = \frac{1 + \sqrt{-3}}{2},$$

and let $\beta = \zeta\alpha$. Let $E = \mathbf{Q}(\beta)$. Since β is complex and α real, we have $\mathbf{Q}(\beta) \neq \mathbf{Q}(\alpha)$. Let $F = \mathbf{Q}(\alpha)$. Then $E \cap F$ is a subfield of E whose degree over $\mathbf{Q}$ divides 3. Hence this degree is 3 or 1, and must be 1 since $E \neq F$. But

$$EF = \mathbf{Q}(\alpha, \beta) = \mathbf{Q}(\alpha, \zeta) = \mathbf{Q}(\alpha, \sqrt{-3}).$$

Hence EF has degree 2 over F.

THEOREM 5. *Let K_1 and K_2 be Galois extensions of a field k, with Galois groups G_1 and G_2 respectively. Assume K_1, K_2 are subfields of some field. Then K_1K_2 is Galois over k. Let G be its Galois group. Map $G \to G_1 \times G_2$ by restriction, namely*

$$\sigma \mapsto (\sigma \mid K_1, \sigma \mid K_2).$$

This map is injective. If $K_1 \cap K_2 = k$ then the map is an isomorphism.

Proof. Normality and separability are preserved in taking the compositum of two fields, so K_1K_2 is Galois over k. Our map is obviously a homomorphism of G into $G_1 \times G_2$. If an element $\sigma \in G$ induces the identity on K_1 and K_2 then it induces the identity on their compositum, so our map is injective. Assume that $K_1 \cap K_2 = k$. According to Theorem 4, given an element $\sigma_1 \in G_1$ there exists an element σ of the Galois group of K_1K_2 over K_2 which induces σ_1 on K_1. This σ is *a fortiori* in G, and induces the identity on K_2. Hence $G_1 \times \{e_2\}$ is contained in the image of our homomorphism (where e_2 is the unit element of G_2). Similarly, $\{e_1\} \times G_2$ is contained in this image. Hence their product is contained in the image, and their product is precisely $G_1 \times G_2$. This proves Theorem 5.

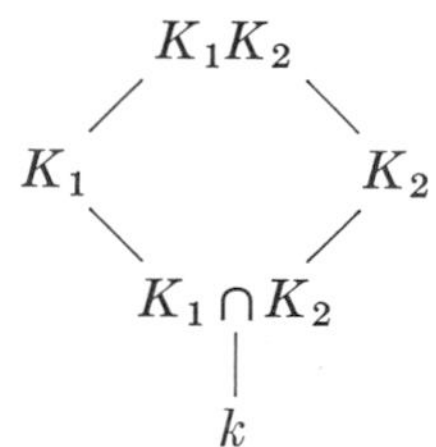

COROLLARY 1. *Let $K_1, \ldots, K_n$ be Galois extensions of k with Galois groups $G_1, \ldots, G_n$. Assume that $K_{i+1} \cap (K_1 \cdots K_i) = k$ for each $i = 1, \ldots, n-1$. Then the Galois group of $K_1 \cdots K_n$ is isomorphic to the product $G_1 \times \cdots \times G_n$ in the natural way.*

Proof. Induction.

COROLLARY 2. *Let K be a finite Galois extension of k with group G, and assume that G can be written as a direct product $G = G_1 \times \cdots \times G_n$.*

Let K_i be the fixed field of

$$G_1 \times \cdots \times \{1\} \times \cdots \times G_n$$

where the group with 1 *element occurs in the i-th place. Then K_i is Galois over k, and $K_{i+1} \cap (K_1 \cdots K_i) = k$. Furthermore $K = K_1 \cdots K_n$.*

Proof. By Corollary 1 of Theorem 1, the compositum of all K_i belongs to the intersection of their corresponding groups, which is clearly the identity. Hence the compositum is equal to K. Each factor of G is normal in G, so K_i is Galois over k. By Corollary 2 of Theorem 1, the intersection of normal extensions belongs to the product of their Galois groups, and it is then clear that $K_{i+1} \cap (K_1 \cdots K_i) = k$.

§2. *Examples and applications*

Let k be a field and $f(X)$ a polynomial of degree $\geqq 1$ in $k[X]$. Let

$$f(X) = (X - \alpha_1) \cdots (X - \alpha_n)$$

be its factorization in a splitting field K over k. Let G be the Galois group of K over k. We call G the *Galois group of f* over k. Then the elements of G permute the roots of f. Thus we have an injective homomorphism of G into the symmetric group S_n on n elements. Not every permutation need be given by an element of G. We shall discuss examples below.

Example 1. Let k be a field and $a \in k$. If a is not a square in k, then the polynomial $X^2 - a$ has no root in k and is therefore irreducible. Assume char $k \neq 2$. Then the polynomial is separable (because $a \neq 0$), and if α is a root, then $k(\alpha)$ is the splitting field, is Galois, and its Galois group is cyclic of order 2. Completing the square shows that every quadratic extension is described as above (for char $\neq 2$).

Example 2. Let k be a field of characteristic $\neq 2$ or 3. Let $f(X) = X^3 + bX + c$. (Any polynomial of degree 3 can be brought into this form by completing the square.) If f has no root in k, then f is irreducible (any factorization must have a factor of degree 1). Let α be a root of $f(X)$. Then $[k(\alpha) : k] = 3$. Let K be the splitting field, and assume f is separable. Let G be the Galois group. Then G has order 3 or 6 since G is a subgroup of the symmetric group S_3. In the second case, $k(\alpha)$ is not normal over k.

There is an easy way to test whether the Galois group is the full symmetric group. We consider the discriminant. If $\alpha_1, \alpha_2, \alpha_3$ are the distinct roots of $f(X)$, we let

$$\delta = (\alpha_1 - \alpha_2)(\alpha_2 - \alpha_3)(\alpha_1 - \alpha_3) \quad \text{and} \quad \Delta = \delta^2.$$

If G is the Galois group and $\sigma \in G$ then $\sigma(\delta) = \pm\delta$. Hence σ leaves Δ fixed. Thus Δ is in the ground field k, and we have seen that

$$\Delta = -4b^3 - 27c^2.$$

The set of σ in G which leave δ fixed is precisely the set of even permutations. Thus G is the symmetric group if and only if Δ is not a square in k.

For instance, consider

$$f(X) = X^3 - X + 1$$

over the rational numbers. Any rational root must be 1 or -1, and so $f(X)$ is irreducible over $\mathbf{Q}$. The discriminant is -23, and is not a square. Hence the Galois group is the symmetric group. The splitting field contains a subfield of order 2, namely $k(\delta) = k(\sqrt{\Delta})$.

Example 3. We consider the polynomial $f(X) = X^4 - 2$ over the rationals $\mathbf{Q}$. It is irreducible by Eisenstein's criterion. Let α be a real root. Let $i = \sqrt{-1}$. Then $\pm\alpha$ and $\pm i\alpha$ are the four roots of $f(X)$, and

$$[\mathbf{Q}(\alpha) : Q] = 4.$$

Hence the splitting field of $f(X)$ is

$$K = \mathbf{Q}(\alpha, i).$$

The field $\mathbf{Q}(\alpha) \cap \mathbf{Q}(i)$ has degree 1 or 2 over $\mathbf{Q}$. The degree cannot be 2 otherwise $i \in \mathbf{Q}(\alpha)$, which is impossible since α is real. Hence the degree is 1. Hence i has degree 2 over $\mathbf{Q}(\alpha)$ and therefore $[K : \mathbf{Q}] = 8$. The Galois group of $f(X)$ has order 8.

There exists an automorphism τ of K leaving $\mathbf{Q}(\alpha)$ fixed, sending i on $-i$, because K is Galois over $\mathbf{Q}(\alpha)$, of degree 2. Then $\tau^2 = id$.

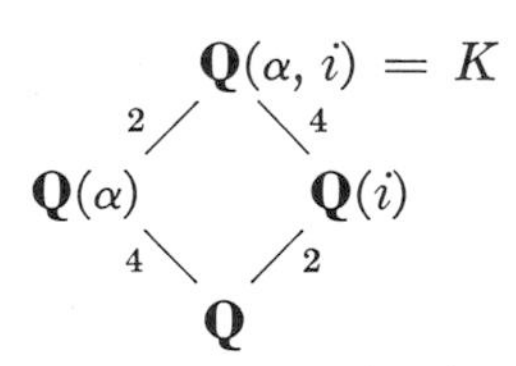

By the multiplicativity of degrees in towers, we see that the degrees are as indicated in the diagram. Thus $X^4 - 2$ is irreducible over $\mathbf{Q}(i)$. Also, K is normal over $\mathbf{Q}(i)$. There exists an automorphism σ of K over $\mathbf{Q}(i)$ mapping the root α of $X^4 - 2$ on the root $i\alpha$. Then one verifies at once that $1, \sigma, \sigma^2, \sigma^3$ are distinct and $\sigma^4 = id$. Thus σ generates a cyclic group of order 4. We denote it by $\langle\sigma\rangle$. Since $\tau \notin \langle\sigma\rangle$ it follows that $G = \langle\sigma, \tau\rangle$ is generated by σ and τ because $\langle\sigma\rangle$ has index 2. Furthermore, one verifies directly that

$$\tau\sigma = \sigma^3\tau,$$

because this relation is true when applied to α and i which generate K over $\mathbf{Q}$. This gives us the structure of G. It is then easy to verify that the lattice of subgroups is as follows:

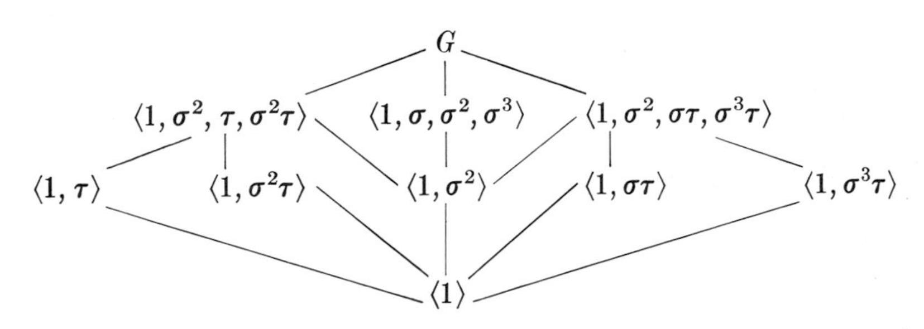

Example 4. Let k be a field and let $t_1, \ldots, t_n$ be algebraically independent over k. Let $K = k(t_1, \ldots, t_n)$. The symmetric group G on n letters operates on K by permuting $(t_1, \ldots, t_n)$ and its fixed field is the field of symmetric functions, by definition the field of those elements of K fixed under G. Let $s_1, \ldots, s_n$ be the elementary symmetric polynomials, and let

$$f(X) = \prod_{i=1}^{n} (X - t_i).$$

Up to a sign, the coefficients of f are $s_1, \ldots, s_n$. We let $F = K^G$. We contend that $F = k(s_1, \ldots, s_n)$. Indeed,

$$k(s_1, \ldots, s_n) \subset F.$$

On the other hand, K is the splitting field of $f(X)$, and its degree over F is $n!$. Its degree over $k(s_1, \ldots, s_n)$ is $\leqq n!$ and hence we have equality, $F = k(s_1, \ldots, s_n)$.

The polynomial $f(X)$ above is called the general polynomial of degree n. We have just constructed a Galois extension whose Galois group is the symmetric group.

Using the Hilbert irreducibility theorem, one can construct a Galois extension of $\mathbf{Q}$ whose Galois group is the symmetric group. (Cf. Chapter IX, and DG [9], Chapter VIII.) It is unknown whether given a finite group G, there exists a Galois extension of $\mathbf{Q}$ whose Galois group is G. By specializing parameters, Emmy Noether remarked that one could prove this if one knew that every field E such that

$$\mathbf{Q}(s_1, \ldots, s_n) \subset E \subset \mathbf{Q}(t_1, \ldots, t_n)$$

is isomorphic to a field generated by n algebraically independent elements. The answer is not known as this book is written.

Note that one can ask the more general question: If $t_1, \ldots, t_n$ are algebraically independent over the complex numbers **C**, is every field E such that $\mathbf{C} \subset E \subset \mathbf{C}(t_1, \ldots, t_n)$ isomorphic to a field generated by r algebraically independent elements ($r \leqq n$)? The answer is known affirmatively when $n \leqq 2$ (Luroth when $n = 1$, and Castelnuovo when $n = 2$). It is not known in any other case. (Fano thought he had a counterexample, but critical reappraisals in recent years have shown the question to be still open.)

Example 5. *We shall prove that the complex numbers are algebraically closed.* This will illustrate almost all the theorems we have proved previously.

We use the following properties of the real numbers **R**: It is an ordered field, every positive element is a square, and every polynomial of odd degree in $\mathbf{R}[X]$ has a root in **R**. We shall discuss ordered fields in general later, and our arguments apply to any ordered field having the above properties.

Let $i = \sqrt{-1}$ (in other words a root of $X^2 + 1$). Every element in $\mathbf{R}(i)$ has a square root. If $a + bi \in \mathbf{R}(i)$, $a, b \in \mathbf{R}$, then the square root is given by $c + di$, where

$$c^2 = \frac{a + \sqrt{a^2 + b^2}}{2} \quad \text{and} \quad d^2 = \frac{-a + \sqrt{a^2 + b^2}}{2}.$$

Each element on the right of our equalities is positive and hence has a square root in **R**. It is then trivial to determine the sign of c and d so that $(c + di)^2 = a + bi$.

Since **R** has characteristic 0, every finite extension is separable. Every finite extension of $\mathbf{R}(i)$ is contained in an extension K which is finite and Galois over **R**. We must show that $K = \mathbf{R}(i)$. Let G be the Galois group over **R** and let H be a 2-Sylow subgroup of G. Let F be its fixed field. Counting degrees and orders, we find that the degree of F over **R** is odd. By the primitive element theorem, there exists an element $\alpha \in F$ such that $F = \mathbf{R}(\alpha)$. Then α is the root of an irreducible polynomial in $\mathbf{R}[X]$ of odd degree. This can happen only if this degree is 1. Hence $G = H$ is a 2-group.

We now see that K is Galois over $\mathbf{R}(i)$. Let G_1 be its Galois group. Since G_1 is a p-group (with $p = 2$), if G_1 is not the trivial group, then G_1 has a subgroup G_2 of index 2. Let F be the fixed field of G_2. Then F is of degree 2 over $\mathbf{R}(i)$; it is a quadratic extension. But we saw that every element of $\mathbf{R}(i)$ has a square root, and hence that $\mathbf{R}(i)$ has no extensions of degree 2. It follows that G_1 is the trivial group and $K = \mathbf{R}(i)$, which is what we wanted.

(The basic ideas of the above proof were already in Gauss. The variation of the ideas which we have selected, making a particularly efficient use of the Sylow group, is due to Artin.)

Example 6. This example is addressed to those who know something about Riemann surfaces and coverings. Let t be transcendental over the complex numbers $\mathbf{C}$, and let $k = \mathbf{C}(t)$. The values of t in $\mathbf{C}$, or ∞, correspond to the points of the Gauss sphere S, viewed as a Riemann surface. Let $P_1, \ldots, P_{n+1}$ be distinct points of S. The finite coverings of

$$S - \{P_1, \ldots, P_{n+1}\}$$

are in bijection with certain finite extensions of $C(t)$, those which are unramified outside $P_1, \ldots, P_{n+1}$. Let K be the union of all these extension fields corresponding to such coverings, and let $\pi_1^{(n)}$ be the fundamental group of $S - \{P_1, \ldots, P_{n+1}\}$. Then it is known that $\pi_1^{(n)}$ is a free group on n generators, and has an embedding in the Galois group of K over $\mathbf{C}(t)$, such that the finite subfields of K over $\mathbf{C}(t)$ are in bijection with the subgroups of $\pi_1^{(n)}$ which are of finite index. Given a finite group G generated by n elements $\sigma_1, \ldots, \sigma_n$ we can find a surjective homomorphism

$$\pi_1^{(n)} \to G$$

mapping the generators of $\pi_1^{(n)}$ on $\sigma_1, \ldots, \sigma_n$. Let H be the kernel. Then H belongs to a subfield K^H of K which is normal over $\mathbf{C}(t)$ and whose Galois group is G. In the language of coverings, H belongs to a finite covering of $S - \{P_1, \ldots, P_{n+1}\}$.

§3. *Roots of unity*

Let k be a field. By a *root of unity* (in k) we shall mean an element $\zeta \in k$ such that $\zeta^n = 1$ for some integer $n \geqq 1$. If the characteristic of k is p, then the equation

$$X^{p^m} = 1$$

has only one root, namely 1, and hence there is no p^m-th root of unity except 1.

Let n be an integer > 1 and prime to the characteristic of k. The polynomial

$$X^n - 1$$

is separable because its derivative is $nX^{n-1} \neq 0$, and the only root of the derivative is 0, so there is no common root. Hence in $\bar{k}$ the polynomial $X^n - 1$ has n distinct roots, which are roots of unity. They obviously form a group, and we know that every finite multiplicative group in a field is cyclic (Chapter V, Theorem 6). Thus the group of n-th roots of unity is cyclic. A generator for this group is called a *primitive* n-th root of unity.

If U_n denotes the group of all n-th roots of unity in $\bar{k}$ and m, n are relatively prime integers, then

$$U_{mn} \approx U_m \times U_n.$$

This follows because U_m, U_n cannot have any element in common except 1, and because $U_m U_n$ has consequently mn elements, each of which is an mn-th root of unity. Hence $U_m U_n = U_{mn}$, and the decomposition is that of a direct product.

THEOREM 6. *Let ζ be a primitive n-th root of unity. Then*

$$[\mathbf{Q}(\zeta) : \mathbf{Q}] = \varphi(n).$$

Proof. Let $f(X)$ be the irreducible polynomial of ζ over $\mathbf{Q}$. Then $f(X)$ divides $X^n - 1$, say $X^n - 1 = f(X)h(X)$, where both f, h have leading coefficient 1. By the Gauss lemma, it follows that f, h have integral coefficients. We shall now prove that if p is a prime number not dividing n, then ζ^p is also a root of f. Since ζ^p is also a primitive n-th root of unity, and since any primitive n-th root of unity can be obtained by raising ζ to a succession of prime powers, with primes not dividing n, this will imply that all the primitive n-th roots of unity are roots of f, which must therefore have degree $\geqq \varphi(n)$, and hence precisely $\varphi(n)$.

Suppose ζ^p is not a root of f. Then ζ^p is a root of h, and ζ itself is a root of $h(X^p)$. Hence $f(X)$ divides $h(X^p)$, and we can write

$$h(X^p) = f(X)g(X).$$

Since f has integral coefficients and leading coefficient 1, we see that g has integral coefficients. Since $a^p \equiv a \pmod{p}$ for any integer a, we conclude that

$$h(X^p) \equiv h(X)^p \pmod{p},$$

and hence

$$h(X)^p \equiv f(X)g(X) \pmod{p}.$$

In particular, if we denote by $\bar{f}$ and $\bar{h}$ the polynomials in $\mathbf{Z}/p\mathbf{Z}$ obtained by reducing f and h respectively mod p, we see that $\bar{f}$ and $\bar{h}$ are not relatively prime, i.e. have a factor in common. But $X^n - \bar{1} = \bar{f}(X)\bar{h}(X)$, and hence $X^n - \bar{1}$ has multiple roots. This is impossible, as one sees by taking the derivative, and our theorem is proved.

COROLLARY. *If n, m are relative prime integers $\geqq 1$, then*

$$\mathbf{Q}(\zeta_n) \cap \mathbf{Q}(\zeta_m) = \mathbf{Q}.$$

Proof. We note that ζ_n and ζ_m are both contained in $\mathbf{Q}(\zeta_{mn})$ since ζ_{mn}^n is a primitive m-th root of unity. Furthermore, $\zeta_m\zeta_n$ is a primitive mn-th root of unity. Hence

$$\mathbf{Q}(\zeta_n)\mathbf{Q}(\zeta_m) = \mathbf{Q}(\zeta_{mn}).$$

Our assertion follows from the multiplicativity $\varphi(mn) = \varphi(m)\varphi(n)$.

Suppose that n is a prime number p (having nothing to do with the characteristic). Then

$$X^p - 1 = (X - 1)(X^{p-1} + \cdots + 1).$$

Any primitive p-th root of unity is a root of the second factor on the right of this equation. Since there are exactly $p - 1$ primitive p-th roots of unity, we conclude that these roots are precisely the roots of

$$X^{p-1} + \cdots + 1.$$

We had seen in Chapter V that this polynomial could be transformed into an Eisenstein polynomial over the rationals. This gives another proof that $[\mathbf{Q}(\zeta_p) : \mathbf{Q}] = p - 1$.

Let k be any field. Let n be prime to the characteristic p in what follows. Let $\zeta = \zeta_n$ be a primitive n-th root of unity in $\bar{k}$. Let σ be an embedding of $k(\zeta)$ in $\bar{k}$ over k. Then

$$(\sigma\zeta)^n = \sigma(\zeta^n) = 1$$

so that $\sigma\zeta$ is an n-th root of unity also. Hence $\sigma\zeta = \zeta^i$ for some integer $i = i(\sigma)$, uniquely determined mod n. It follows that σ maps $k(\zeta)$ into itself, and hence that $k(\zeta)$ is normal over k. If τ is another automorphism of $k(\zeta)$ over k then

$$\sigma\tau\zeta = \zeta^{i(\sigma)i(\tau)}.$$

Since σ and τ are automorphisms, it follows that $i(\sigma)$ and $i(\tau)$ are prime to n (otherwise, $\sigma\zeta$ would have a period smaller than n). In this way we get a homomorphism of the Galois group G of $k(\zeta)$ over k into the multiplicative group $(\mathbf{Z}/n\mathbf{Z})^*$ of integers prime to n, mod n. Our homomorphism is clearly injective since $i(\sigma)$ is uniquely determined by σ mod n, and the effect of σ on $k(\zeta)$ is determined by its effect on ζ. *We conclude that $k(\zeta)$ is abelian over k.*

Let φ be the Euler function. We know that the order of $(\mathbf{Z}/n\mathbf{Z})^*$ is $\varphi(n)$. Hence the degree $[k(\zeta) : k]$ divides $\varphi(n)$.

We investigate more closely the factorization of $X^n - 1$, and suppose that we are in characteristic 0 for simplicity.

We have

$$X^n - 1 = \prod_{\omega} (X - \omega),$$

where the product is taken over all n-th roots of unity. Collect together all terms belonging to roots of unity having the same period. Let

$$f_d(X) = \prod_{\text{period } \omega = d} (X - \omega).$$

Then

$$X^n - 1 = \prod_{d|n} f_d(X).$$

We see that $f_1(X) = X - 1$, and that

$$f_n(X) = \frac{X^n - 1}{\prod_{\substack{d|n \\ d<n}} f_d(X)}.$$

From this we can compute $f_n(X)$ recursively, and we see that $f_n(X)$ is a polynomial in $\mathbf{Q}[X]$ because we divide recursively by polynomials having coefficients in $\mathbf{Q}$. All our polynomials have leading coefficient 1, so that in fact $f_n(X)$ has *integer coefficients* by Theorem 2 of Chapter V, §4. Thus our construction is essentially universal, and would hold over any field (whose characteristic does not divide n).

We call $f_n(X)$ the n-th *cyclotomic polynomial.*

The roots of f_n are precisely the primitive n-th roots of unity, and hence

$$\deg f_n = \varphi(n).$$

From Theorem 6 we conclude that f_n is irreducible over $\mathbf{Q}$, and hence

$$f_n(X) = \operatorname{Irr}(\zeta_n, \mathbf{Q}, X).$$

We leave the proofs of the following recursion formulas as exercises:

1. If p is a prime number, then

$$f_p(X) = X^{p-1} + X^{p-2} + \cdots + 1,$$

and for an integer $r \geqq 1$,

$$f_{p^r}(X) = f_p(X^{p^{r-1}}).$$

2. Let $n = p_1^{r_1} \cdots p_s^{r_s}$ be a positive integer with its prime factorization. Then

$$f_n(X) = f_{p_1 \cdots p_s}(X^{p_1^{r_1-1} \cdots p_s^{r_s-1}}).$$

3. If n is odd, then $f_{2n}(X) = f_n(-X)$.

4. If p is a prime number, not dividing n, then

$$f_{pn}(X) = \frac{f_n(X^p)}{f_n(X)}.$$

5. We have

$$f_n(X) = \prod_{d|n} (X^{n/d} - 1)^{\mu(d)}.$$

As usual, μ is the Möbius function:

$$\mu(n) = \begin{cases} 0 & \text{if } n \text{ is divisible by } p^2 \text{ for some prime } p, \\ (-1)^r & \text{if } n = p_1 \cdots p_r \text{ is a product of distinct primes,} \\ 1 & \text{if } n = 1. \end{cases}$$

As an exercise, show that

$$\sum_{d|n} \mu(d) = \begin{cases} 1 & \text{if} \quad n = 1, \\ 0 & \text{if} \quad n > 1. \end{cases}$$

If ζ is an n-th root of unity and $\zeta \neq 1$, then

$$\frac{1 - \zeta^n}{1 - \zeta} = 1 + \zeta + \cdots + \zeta^{n-1} = 0.$$

This is trivial, but useful.

Let $\mathbf{F}_q$ be the finite field with q elements, q equal to a power of the odd prime number p. Then $\mathbf{F}_q^*$ has $q - 1$ elements and is a cyclic group. Hence we have the index

$$(\mathbf{F}_q^* : \mathbf{F}_q^{*2}) = 2.$$

If ν is a non-zero integer, let

$$\left(\frac{\nu}{p}\right) = \begin{cases} 1 & \text{if} \quad \nu \equiv x^2 \pmod{p}, \\ -1 & \text{if} \quad \nu \not\equiv x^2 \pmod{p}. \end{cases}$$

This is known as the quadratic symbol, and depends only on the residue class of ν mod p.

From our preceding remark, we see that there are as many quadratic residues as there are non-residues mod p.

Let ζ be a primitive p-th root of unity, and let

$$S = \sum_{\nu} \left(\frac{\nu}{p}\right) \zeta^{\nu},$$

the sum being taken over non-zero residue classes mod p. Then

$$S^2 = \left(\frac{-1}{p}\right) p.$$

Every quadratic extension of $\mathbf{Q}$ *is contained in a cyclotomic extension.*

Proof. The last statement follows at once from the explicit expression of $\pm p$ as a square in $\mathbf{Q}(\zeta)$, because the square root of an integer is contained in the field obtained by adjoining the square root of the prime factors in its factorization, and also $\sqrt{-1}$. Furthermore, for the prime 2, we have $(1 + i)^2 = 2i$. We now prove our assertion concerning S^2. We have

$$S^2 = \sum_{\nu,\mu} \left(\frac{\nu}{p}\right)\left(\frac{\mu}{p}\right) \zeta^{\nu+\mu} = \sum_{\nu,\mu} \left(\frac{\nu\mu}{p}\right) \zeta^{\nu+\mu}.$$

As ν ranges over non-zero residue classes, so does $\nu\mu$ for any fixed μ, and hence replacing ν by $\nu\mu$ yields

$$\begin{aligned} S^2 &= \sum_{\nu,\mu} \left(\frac{\nu\mu^2}{p}\right) \zeta^{\mu(\nu+1)} = \sum_{\nu,\mu} \left(\frac{\nu}{p}\right) \zeta^{\mu(\nu+1)} \\ &= \sum_{\mu} \left(\frac{-1}{p}\right) \zeta^0 + \sum_{\nu\neq -1} \left(\frac{\nu}{p}\right) \sum_{\mu} \zeta^{\mu(\nu+1)}. \end{aligned}$$

But $1 + \zeta + \cdots + \zeta^{p-1} = 0$, and the sum on the right over μ consequently yields -1. Hence

$$\begin{aligned} S^2 &= \left(\frac{-1}{p}\right)(p-1) + (-1) \sum_{\nu\neq -1} \left(\frac{\nu}{p}\right) \\ &= p\left(\frac{-1}{p}\right) - \sum_{\nu} \left(\frac{\nu}{p}\right) \\ &= p\left(\frac{-1}{p}\right), \end{aligned}$$

as desired.

We see that $\mathbf{Q}(\sqrt{p})$ is contained in $\mathbf{Q}(\zeta, \sqrt{-1})$ or $\mathbf{Q}(\zeta)$, depending on the sign of the quadratic symbol with -1. An extension of a field is said to be *cyclotomic* if it is contained in a field obtained by adjoining roots of unity. We have shown above that quadratic extensions of $\mathbf{Q}$ are cyclotomic. A theorem of Kronecker asserts that every abelian extension of $\mathbf{Q}$ is cyclotomic, but the proof needs techniques which cannot be covered in this book.

§4. *Linear independence of characters*

Let G be a monoid and K a field. By a *character* of G in K (in this chapter), we shall mean a homomorphism

$$\chi : G \to K^*$$

of G into the multiplicative group of K. The *trivial* character is the homomorphism taking the constant value 1. Functions $f_i : G \to K$ are called *linearly independent* over K if whenever we have a relation

$$a_1 f_1 + \cdots + a_n f_n = 0$$

with $a_i \in K$, then all $a_i = 0$.

THEOREM 7. (Artin) *Let $\chi_1, \ldots, \chi_n$ be distinct characters of G in K. Then they are linearly independent over K.*

Proof. One character is obviously linearly independent. Suppose that we have a relation

$$a_1\chi_1 + \cdots + a_n\chi_n = 0$$

with $a_i \in K$, not all 0. Take such a relation with n as small as possible. Then $n \geqq 2$, and no a_i is equal to 0. Since χ_1, χ_2 are distinct, there exists $z \in G$ such that $\chi_1(z) \neq \chi_2(z)$. For all $x \in G$ we have

$$a_1\chi_1(xz) + \cdots + a_n\chi_n(xz) = 0,$$

and since χ_i is a character,

$$a_1\chi_1(z)\chi_1 + \cdots + a_n\chi_n(z)\chi_n = 0.$$

Divide by $\chi_1(z)$ and subtract from our first relation. The term $a_1\chi_1$ cancels, and we get a relation

$$\left(a_2 \frac{\chi_2(z)}{\chi_1(z)} - a_2\right)\chi_2 + \cdots = 0.$$

The first coefficient is not 0, and this is a relation of smaller length than our first relation, contradiction.

As an application of Artin's theorem, one can consider the case when K is a finite normal extension of a field k, and when the characters are distinct automorphisms $\sigma_1, \ldots, \sigma_n$ of K over k, viewed as homomorphisms of K^* into K^*. This special case had already been considered by Dedekind, who, however, expressed the theorem in a somewhat different way, considering the determinant constructed from $\sigma_i\omega_j$ where ω_j is a suitable set of elements of K, and proving in a more complicated way the fact that this determinant is not 0. The formulation given above and its particularly elegant proof are due to Artin.

As another application, we have:

COROLLARY. *Let $\alpha_1, \ldots, \alpha_n$ be distinct non-zero elements of a field K. If $a_1, \ldots, a_n$ are elements of K such that for all integers ν we have*

$$a_1\alpha_1^\nu + \cdots + a_n\alpha_n^\nu = 0$$

then $a_i = 0$ for all i.

Proof. We apply the theorem to the distinct homomorphisms

$$\nu \mapsto \alpha_i^\nu$$

of $\mathbf{Z}$ into K^*.

Another interesting application will be given as an exercise (relative invariants).

§5. *The norm and trace*

Let E be a finite extension of k. Let $[E : k]_s = r$, and let

$$p^\mu = [E : k]_i$$

if the characteristic is $p > 0$, and 1 otherwise. Let $\sigma_1, \ldots, \sigma_r$ be the distinct embeddings of E in an algebraic closure $\bar{k}$ of k. If α is an element of E, we define its *norm* from E to k to be

$$N_k^E(\alpha) = \prod_{\nu=1}^{r} \sigma_\nu \alpha^{p^\mu} = \left(\prod_{\nu=1}^{r} \sigma_\nu \alpha \right)^{[E:k]_i}.$$

Similarly, we define the *trace*

$$\operatorname{Tr}_k^E(\alpha) = [E : k]_i \sum_{\nu=1}^{r} \sigma_\nu \alpha.$$

The trace is equal to 0 if $[E : k]_i > 1$, in other words, if E/k is not separable. Thus if E is separable over k, we have

$$N_k^E(\alpha) = \prod_\sigma \sigma\alpha$$

where the product is taken over the distinct embeddings of E in $\bar{k}$ over k.

Similarly, if E/k is separable, then

$$\operatorname{Tr}_k^E(\alpha) = \sum_\sigma \sigma\alpha.$$

THEOREM 8. *Let E/k be a finite extension. Then the norm N_k^E is a multiplicative homomorphism of E^* into k^* and the trace is an additive homomorphism of E into k. If $E \supset F \supset k$ is a tower of fields, then the two maps are transitive, in other words,*

$$N_k^E = N_k^F \circ N_F^E \qquad \textit{and} \qquad \operatorname{Tr}_k^E = \operatorname{Tr}_k^F \circ \operatorname{Tr}_F^E.$$

If $E = k(\alpha)$, and $f(X) = \operatorname{Irr}(\alpha, k, X) = X^n + a_{n-1}X^{n-1} + \cdots + a_0$, then

$$N_k^{k(\alpha)}(\alpha) = (-1)^n a_0 \qquad \textit{and} \qquad \operatorname{Tr}_k^{k(\alpha)}(\alpha) = -a_{n-1}.$$

Proof. For the first assertion, we note that α^{p^μ} is separable over k if $p^\mu = [E : k]_i$. On the other hand, the product

$$\prod_{\nu=1}^{r} \sigma_\nu \alpha^{p^\mu}$$

is left fixed under any isomorphism into $\bar{k}$ because applying such an isomorphism simply permutes the factors. Hence this product must lie in k since α^{p^μ} is separable over k. A similar reasoning applies to the trace.

For the second assertion, let $\{\tau_j\}$ be the family of distinct embeddings of F into $\bar{k}$ over k. Extend each τ_j to an embedding of E in $\bar{k}$, and denote this extension by τ_j also. Let $\{\sigma_i\}$ be the family of embeddings of E in $\bar{k}$ over F. (Without loss of generality, we may assume that $E \subset \bar{k}$.) If σ is an embedding of E over k in $\bar{k}$ then for some j, $\tau_j^{-1}\sigma$ leaves F fixed, and hence $\tau_j^{-1}\sigma = \sigma_i$ for some i. Hence $\sigma = \tau_j\sigma_i$ and consequently the family $\{\tau_j\sigma_i\}$ gives all distinct embeddings of E into $\bar{k}$ over k. Since the inseparability degree is multiplicative in towers, our assertion concerning the transitivity of the norm and trace is obvious, because we have already shown that N_F^E maps E into F, and similarly for the trace.

Suppose now that $E = k(\alpha)$. We have

$$f(X) = \big((X - \alpha_1) \cdots (X - \alpha_r)\big)^{[E:k]_i}$$

if $\alpha_1, \ldots, \alpha_r$ are the distinct roots of f. Looking at the constant term of f gives us the expression for the norm, and looking at the penultimate term gives us the expression for the trace.

We observe that the trace is a k-linear map of E into k, namely

$$\mathrm{Tr}_k^E(c\alpha) = c\,\mathrm{Tr}_k^E(\alpha)$$

for all $\alpha \in E$ and $c \in k$. This is clear since c is fixed under every embedding of E over k. Thus the trace is a k-linear functional of E into k. For simplicity, we write $\mathrm{Tr} = \mathrm{Tr}_k^E$.

THEOREM 9. *Let E be a finite separable extension of k. Then $\mathrm{Tr} : E \to k$ is a non-zero functional. The map*

$$(x, y) \mapsto \mathrm{Tr}(xy)$$

of $E \times E \to k$ is bilinear, and identifies E with its dual space.

Proof. That Tr is non-zero follows from the theorem on linear independence of characters. For each $x \in E$, the map

$$\mathrm{Tr}_x : E \to k$$

such that $\mathrm{Tr}_x(y) = \mathrm{Tr}(xy)$ is obviously a k-linear map, and the map

$$x \mapsto \mathrm{Tr}_x$$

is a k-homomorphism of E into its dual space $\hat{E}$. (We don't write E^* for the dual space because we use the star to denote the multiplicative group of E.) If Tr_x is the zero map, then $\mathrm{Tr}(xE) = 0$. If $x \neq 0$ then $xE = E$. Hence the kernel of $x \mapsto \mathrm{Tr}_x$ is 0. Hence we get an injective homomorphism of E into the dual space $\hat{E}$. Since these spaces have the same finite dimension, it follows that we get an isomorphism. This proves our theorem.

COROLLARY 1. *Let $\omega_1, \ldots, \omega_n$ be a basis of E over k. Then there exists a basis $\omega_1', \ldots, \omega_n'$ of E over k such that* $\mathrm{Tr}(\omega_i\omega_j') = \delta_{ij}$.

Proof. The basis $\omega_1', \ldots, \omega_n'$ is none other than the dual basis which we defined when we considered the dual space of an arbitrary vector space.

COROLLARY 2. *Let E be a finite separable extension of k, and let $\sigma_1, \ldots, \sigma_n$ be the distinct set of embeddings of E into $\bar{k}$ over k. Let $w_1, \ldots, w_n$ be elements of E. Then the vectors*

$$\xi_1 = (\sigma_1 w_1, \ldots, \sigma_1 w_n),$$
$$\cdots$$
$$\xi_n = (\sigma_n w_1, \ldots, \sigma_n w_n)$$

are linearly independent over E if and only if $w_1, \ldots, w_n$ form a basis of E over k.

Proof. Assume that $w_1, \ldots, w_n$ form a basis of E/k. Let $\alpha_1, \ldots, \alpha_n$ be elements of E such that

$$\alpha_1\xi_1 + \cdots + \alpha_n\xi_n = 0.$$

Then we see that

$$\alpha_1\sigma_1 + \cdots + \alpha_n\sigma_n$$

applied to each one of $w_1, \ldots, w_n$ gives the value 0. But $\sigma_1, \ldots, \sigma_n$ are linearly independent as characters of the multiplicative group E^* into $\bar{k}^*$. It follows that $\alpha_i = 0$ for $i = 1, \ldots, n$, and our vectors are linearly independent.

Conversely, it is clear that if $w_1, \ldots, w_n$ are linearly dependent over k then our vectors are linearly dependent over E.

Remark. In characteristic 0, one sees much more trivially that the trace is not identically 0. Indeed, if $c \in k$ and $c \neq 0$, then $\mathrm{Tr}(c) = nc$ where $n = [E : k]$, and $n \neq 0$. This argument also holds in characteristic p when n is prime to p.

PROPOSITION 1. *Let* $E = k(\alpha)$ *be a separable extension. Let* $f(X) = \mathrm{Irr}(\alpha, k, X)$, *and let* $f'(X)$ *be its derivative. Let*

$$\frac{f(X)}{(X-\alpha)} = \beta_0 + \beta_1 X + \cdots + \beta_{n-1}X^{n-1}$$

with $\beta_i \in E$. *Then the dual basis of* $1, \alpha, \ldots, \alpha^{n-1}$ *is*

$$\frac{\beta_0}{f'(\alpha)}, \ldots, \frac{\beta_{n-1}}{f'(\alpha)}.$$

Proof. Let $\alpha_1, \ldots, \alpha_n$ be the distinct roots of f. Then

$$\sum_{i=1}^{n} \frac{f(X)}{(X-\alpha_i)} \frac{\alpha_i^r}{f'(\alpha_i)} = X^r \qquad \text{for} \quad 0 \leqq r \leqq n-1.$$

To see this, let $g(X)$ be the difference of the left- and right-hand side of this equality. Then g has degree $\leqq n - 1$, and has n roots $\alpha_1, \ldots, \alpha_n$. Hence g is identically zero.

The polynomials

$$\frac{f(X)}{(X-\alpha_i)} \frac{\alpha_i^r}{f'(\alpha_i)}$$

are all conjugate to each other. If we define the trace of a polynomial with coefficients in E to be the polynomial obtained by applying the trace to the coefficients, then

$$\mathrm{Tr}\left[\frac{f(X)}{(X-\alpha)} \frac{\alpha^r}{f'(\alpha)}\right] = X^r.$$

Looking at the coefficients of each power of X in this equation, we see that

$$\mathrm{Tr}\left(\alpha^i \frac{\beta_j}{f'(\alpha)}\right) = \delta_{ij},$$

thereby proving our proposition.

§6. *Cyclic extensions*

We recall that a finite extension is said to be cyclic if it is Galois and its Galois group is cyclic.

HILBERT'S THEOREM 90. *Let* K/k *be cyclic of degree* n *with Galois group* G. *Let* σ *be a generator of* G. *Let* $\beta \in K$. *The norm* $N_k^K(\beta) = N(\beta)$ *is equal to* 1 *if and only if there exists an element* $\alpha \neq 0$ *in* K *such that* $\beta = \alpha/\sigma\alpha$.

Proof. Assume such an element α exists. Taking the norm of β we get $N(\alpha)/N(\sigma\alpha)$. But the norm is the product over all automorphisms in G. Inserting σ just permutes these automorphisms. Hence the norm is equal to 1.

It will be convenient to use an exponential notation as follows. If τ, $\tau' \in G$ and $\xi \in K$ we write

$$\xi^{\tau+\tau'} = \xi^{\tau}\xi^{\tau'}.$$

By Artin's theorem on characters, the map given by

$$id + \beta\sigma + \beta^{1+\sigma}\sigma^2 + \cdots + \beta^{1+\sigma+\cdots+\sigma^{n-2}}\sigma^{n-1}$$

on K is not identically zero. Hence there exists $\theta \in K$ such that the element

$$\alpha = \theta + \beta\theta^{\sigma} + \beta^{1+\sigma}\theta^{\sigma^2} + \cdots + \beta^{1+\sigma+\cdots+\sigma^{n-2}}\theta^{\sigma^{n-1}}$$

is not equal to 0. It is then clear that $\beta\alpha^{\sigma} = \alpha$ using the fact that $N(\beta) = 1$, and hence that when we apply σ to the last term in the sum, we obtain θ. We divide by α^{σ} to conclude the proof.

THEOREM 10. *Let k be a field, n an integer > 0 prime to the characteristic of k, and assume that there is a primitive n-th root of unity in k.*

(a) *Let K be a cyclic extension of degree n. Then there exists $\alpha \in K$ such that $K = k(\alpha)$, and α satisfies an equation $X^n - a = 0$ for some $a \in k$.*

(b) *Conversely, let $a \in k$. Let α be a root of $X^n - a$. Then $k(\alpha)$ is cyclic over k, of degree d, $d \mid n$, and α^d is an element of k.*

Proof. Let ζ be a primitive n-th root of unity in k, and let K/k be cyclic with group G. Let σ be a generator of G. We have $N(\zeta^{-1}) = (\zeta^{-1})^n = 1$. By Hilbert's theorem 90, there exists $\alpha \in K$ such that $\sigma\alpha = \zeta\alpha$. Since ζ is in k, we have $\sigma^i\alpha = \zeta^i\alpha$ for $i = 1, \ldots, n$. Hence the elements $\zeta^i\alpha$ are n distinct conjugates of α over k, whence $[k(\alpha) : k]$ is at least equal to n. Since $[K : k] = n$, it follows that $K = k(\alpha)$. Furthermore,

$$\sigma(\alpha^n) = \sigma(\alpha)^n = (\zeta\alpha)^n = \alpha^n.$$

Hence α^n is fixed under σ, hence is fixed under each power of σ, hence is fixed under G. Therefore α^n is an element of k, and we let $a = \alpha^n$. This proves the first part of the theorem.

Conversely, let $a \in k$. Let α be a root of $X^n - a$. Then $\zeta^i\alpha$ is also a root for each $i = 1, \ldots, n$, and hence all roots lie in $k(\alpha)$ which is therefore normal over k. All the roots are distinct so $k(\alpha)$ is Galois over k. Let G be the Galois group.

If σ is an automorphism of $k(\alpha)/k$ then $\sigma\alpha$ is also a root of $X^n - a$. Hence $\sigma\alpha = \omega_\sigma\alpha$ where ω_σ is an n-th root of unity, not necessarily primitive. The map $\sigma \mapsto \omega_\sigma$ is obviously a homomorphism of G into the group of n-th roots of unity, and is injective. Since a subgroup of a cyclic group is cyclic, we conclude that G is cyclic, of order d, and $d \mid n$. The image of G is a cyclic group of order d. If σ is a generator of G, then ω_σ is a primitive d-th root of unity. Now we get

$$\sigma(\alpha^d) = (\sigma\alpha)^d = (\omega_\sigma\alpha)^d = \alpha^d.$$

Hence α^d is fixed under σ, and therefore fixed under G. It is an element of k, and our theorem is proved.

We now pass to the analogue of Hilbert's theorem 90 in characteristic p for cyclic extensions of degree p.

HILBERT'S THEOREM 90. (Additive form) *Let k be a field and K/k a cyclic extension of degree n with group G. Let σ be a generator of G. Let $\beta \in K$. The trace $\mathrm{Tr}_k^K(\beta)$ is equal to 0 if and only if there exists an element $\alpha \in K$ such that $\beta = \alpha - \sigma\alpha$.*

Proof. If such an element α exists, then we see that the trace is 0 because the trace is equal to the sum taken over all elements of G, and applying σ permutes these elements.

Conversely, assume $\mathrm{Tr}(\beta) = 0$. There exists an element $\theta \in K$ such that $\mathrm{Tr}(\theta) \neq 0$. Let

$$\alpha = \frac{1}{\mathrm{Tr}(\theta)}\,[\beta\theta^\sigma + (\beta + \sigma\beta)\theta^{\sigma^2} + \cdots + (\beta + \sigma\beta + \cdots + \sigma^{n-2}\beta)\theta^{\sigma^{n-1}}].$$

From this it follows at once that $\beta = \alpha - \sigma\alpha$.

THEOREM 11. (Artin-Schreier) *Let k be a field of characteristic p.*

(a) *Let K be a cyclic extension of k of degree p. Then there exists $\alpha \in K$ such that $K = k(\alpha)$ and α satisfies an equation $X^p - X - a = 0$ with some $a \in k$.*

(b) *Conversely, given $a \in k$, the polynomial $f(X) = X^p - X - a$ either has one root in k, in which case all its roots are in k, or it is irreducible. In this latter case, if α is a root then $k(\alpha)$ is cyclic of degree p over k.*

Proof. Let K/k be cyclic of degree p. Then $\mathrm{Tr}_k^K(-1) = 0$ (it is just the sum of -1 with itself p times). Let σ be a generator of the Galois group. By the additive form of Hilbert's theorem 90, there exists $\alpha \in K$ such that $\sigma\alpha - \alpha = 1$, or in other words, $\sigma\alpha = \alpha + 1$. Hence $\sigma^i\alpha = \alpha + i$ for all integers $i = 1, \ldots, p$ and α has p distinct conjugates. Hence $[k(\alpha) : k] \geqq p$. It follows that $K = k(\alpha)$. We note that

$$\sigma(\alpha^p - \alpha) = \sigma(\alpha)^p - \sigma(\alpha) = (\alpha + 1)^p - (\alpha + 1) = \alpha^p - \alpha.$$

Hence $\alpha^p - \alpha$ is fixed under σ, hence it is fixed under the powers of σ, and therefore under G. It lies in the fixed field k. If we let $a = \alpha^p - \alpha$ we see that our first assertion is proved.

Conversely, let $a \in k$. If α is a root of $X^p - X - a$ then $\alpha + i$ is also a root for $i = 1, \dots, p$. Thus $f(X)$ has p distinct roots. If one root lies in k then all roots lie in k. Assume that no root lies in k. We contend that the polynomial is irreducible. Suppose that

$$f(X) = g(X)h(X)$$

with $g, h \in k[X]$ and $1 \leqq \deg g < p$. Since

$$f(X) = \prod_{i=1}^{p} (X - \alpha - i)$$

we see that $g(X)$ is a product over certain integers i. Let $d = \deg g$. The coefficient of X^{d-1} is a sum of terms $-(\alpha + i)$ taken over precisely d integers i. Hence it is equal to $-d\alpha + j$ for some integer j. But $d \neq 0$ in k, and hence α lies in k, because the coefficients of g lie in k, contradiction. We know therefore that $f(X)$ is irreducible. All roots lie in $k(\alpha)$, which is therefore normal over k. Since $f(X)$ has no multiple roots, it follows that $k(\alpha)$ is Galois over k. There exists an automorphism σ of $k(\alpha)$ over k such that $\sigma\alpha = \alpha + 1$ (because $\alpha + 1$ is also a root). Hence the powers σ^i of σ give $\sigma^i\alpha = \alpha + i$ for $i = 1, \dots, p$ and are distinct. Hence the Galois group consists of these powers and is cyclic, thereby proving the theorem.

§7. *Solvable and radical extensions*

A finite extension E/k (which we shall assume separable for convenience) is said to be *solvable* if the Galois group of the smallest Galois extension K of k containing E is a solvable group. This is equivalent to saying that there exists a solvable Galois extension L of k such that $k \subset E \subset L$. Indeed, we have $k \subset E \subset K \subset L$ and $G(K/k)$ is a homomorphic image of $G(L/k)$.

PROPOSITION 2. *Solvable extensions form a distinguished class of extensions.*

Proof. Let E/k be solvable. Let F be a field containing k and assume E, F are subfields of some algebraically closed field. Let K be Galois solvable over k, and $E \subset K$. Then KF is Galois over F and $G(KF/F)$ is a subgroup of $G(K/k)$ by Theorem 4 of §1. Hence EF/F is solvable. It is clear that a subextension of a solvable extension is solvable. Let $E \supset F \supset k$ be a tower, and assume that E/F is solvable and F/k is solvable. Let K be a finite solvable Galois extension of k containing F. We just saw that EK/K is solvable. Let L be a solvable Galois extension of K containing EK. If σ is any embedding of L over k in a given algebraic closure, then $\sigma K = K$ and hence σL is a solvable extension of K. We let M be the compositum of all extensions σL for all embeddings σ of L over k. Then M is Galois over k, and is therefore Galois over K. The Galois group of M over K is a subgroup of the product

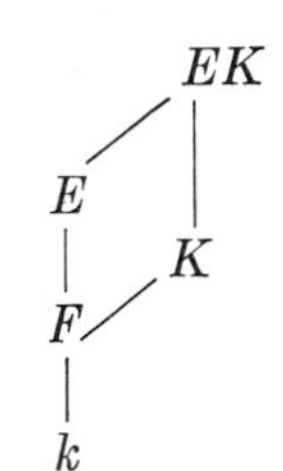

$$\prod_{\sigma} G(\sigma L/K)$$

by Theorem 5 of §1. Hence it is solvable. We have a surjective homomorphism $G(M/k) \to G(K/k)$ by Theorem 3 of §1. Hence the Galois group of M/k has a solvable normal subgroup whose factor group is solvable. It is therefore solvable. Since $E \subset M$ our proof is complete.

A finite extension F of k is said to be *solvable by radicals* if it is separable and if there exists a finite extension E of k containing F, and admitting a tower decomposition

$$k = E_0 \subset E_1 \subset E_2 \subset \cdots \subset E_m = E$$

such that each step E_{i+1}/E_i is one of the following types:

(1) It is obtained by adjoining a root of unity.

(2) It is obtained by adjoining a root of a polynomial $X^n - a$ with $a \in E_i$ and n prime to the characteristic.

(3) It is obtained by adjoining a root of an equation $X^p - X - a$ with $a \in E_i$ if p is the characteristic > 0.

One can see at once that the class of extensions which are solvable by radicals is a distinguished class.

THEOREM 12. *Let E be a separable extension of k. Then E is solvable by radicals if and only if E/k is solvable.*

Proof. Assume that E/k is solvable, and let K be a finite solvable Galois extension of k containing E. Let m be the product of all primes unequal to the characteristic dividing the degree $[K : k]$, and let $F = k(\zeta)$ where ζ is a primitive m-th root of unity. Then F/k is abelian. We lift K over F. Then KF is solvable over F. There is a tower of subfields between F and KF such that each step is cyclic of prime order, because every solvable group admits a tower of subgroups of the same type, and we can use Theorem 3 of §1. By Theorems 10 and 11 we conclude that KF is solvable by radicals over F, and hence is solvable by radicals over k. This proves that E/k is solvable by radicals.

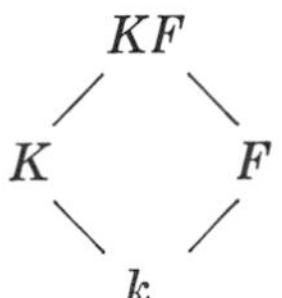

Conversely, assume that E/k is solvable by radicals. For any embedding σ of E in $\overline{E}$ over k, the extension $\sigma E/k$ is also solvable by radicals. Hence the smallest Galois extension K of E containing k, which is a composite of E and its conjugates, is solvable by radicals. Let m be the product of all primes unequal to the characteristic dividing the degree $[K : k]$ and again let $F = k(\zeta)$ where ζ is a primitive m-th root of unity. It will suffice to prove that KF is solvable over F, because it follows then that KF is solvable over k and hence $G(K/k)$ is solvable because it is a homomorphic image of $G(KF/k)$. But KF/F can be decomposed into a tower of extensions, such that each step is prime degree and of the type described in Theorem 10 or Theorem 11, and the corresponding root of unity is in the field F. Hence KF/F is solvable, and our theorem is proved.

Remark. One could modify our preceding discussion by not assuming separability. Then one must deal with normal extensions instead of Galois extensions, and one must allow equations $X^p - a$ in the solvability by radicals, with p equal to the characteristic. Then we still have the theorem corresponding to Theorem 12. The proofs are clear in view of Chapter VII, §7.

§8. *Kummer theory*

In this section we shall carry out a generalization of the theorem concerning cyclic extensions when the ground field contains enough roots of unity.

Let k be a field and m a positive integer. A Galois extension K of k with group G is said to be of *exponent* m if $\sigma^m = 1$ for all $\sigma \in G$.

We shall investigate abelian extensions of exponent m. We first assume that m is prime to the characteristic of k, and that k contains a primitive m-th root of unity. We denote by Z_m the group of m-th roots of unity. We assume that all our algebraic extensions in this section are contained in a fixed algebraic closure $\overline{k}$.

Let $a \in k$. The symbol $a^{1/m}$ (or $\sqrt[m]{a}$) is not well defined. If $\alpha^m = a$ and ζ is an m-th root of unity, then $(\zeta\alpha)^m = a$ also. We shall use the symbol $a^{1/m}$ to denote any such element α, which will be called an m-th root of a. Since the roots of unity are in the ground field, we observe that the field $k(\alpha)$ is the same no matter which m-th root α of a we select. We denote this field by $k(a^{1/m})$.

We denote by k^{*m} the subgroup of k^* consisting of all m-th powers of non-zero elements of k. It is the image of k^* under the homomorphism $x \mapsto x^m$.

Let B be a subgroup of k^* containing k^{*m}. We denote by $k(B^{1/m})$ or K_B the composite of all fields $k(a^{1/m})$ with $a \in B$. It is uniquely determined by B as a subfield of $\overline{k}$.

Let $a \in B$ and let α be an m-th root of a. The polynomial $X^m - a$ splits into linear factors in K_B, and thus K_B is Galois over k, because this holds for all $a \in B$. Let G be the Galois group. Let $\sigma \in G$. Then $\sigma\alpha = \omega_\sigma \alpha$ for some m-th root of unity $\omega_\sigma \in Z_m \subset k^*$. The map

$$\sigma \mapsto \omega_\sigma$$

is obviously a homomorphism of G into Z_m, i.e. for $\tau, \sigma \in G$ we have $\tau\sigma\alpha = \omega_\tau\omega_\sigma\alpha = \omega_\sigma\omega_\tau\alpha$. We may write $\omega_\sigma = \sigma\alpha/\alpha$. This root of unity ω_σ is independent of the choice of m-th root of a, for if α' is another m-th root, then $\alpha' = \zeta\alpha$ for some $\zeta \in Z_m$, whence

$$\sigma\alpha'/\alpha' = \zeta\sigma\alpha/\zeta\alpha = \sigma\alpha/\alpha.$$

We denote ω_σ by $\langle \sigma, a \rangle$. The map

$$(\sigma, a) \mapsto \langle \sigma, a \rangle$$

gives us a map

$$G \times B \to Z_m.$$

If $a, b \in B$ and $\alpha^m = a$, $\beta^m = b$ then $(\alpha\beta)^m = ab$ and hence

$$\sigma(\alpha\beta)/\alpha\beta = (\sigma\alpha/\alpha)(\sigma\beta/\beta).$$

We conclude that the map above is bilinear. Furthermore, if $a \in k^{*m}$ it follows that $\langle \sigma, a \rangle = 1$.

THEOREM 13. *Let k be a field, m an integer > 0 prime to the characteristic of k, and assume that a primitive m-th root of unity lies in k. Let B be a subgroup of k^* containing k^{*m} and let $K_B = k(B^{1/m})$. Then K_B is Galois, and abelian of exponent m. Let G be its Galois group. We have a bilinear map*

$$G \times B \to Z_m \qquad \text{given by} \qquad (\sigma, a) \mapsto \langle \sigma, a \rangle.$$

If $\sigma \in G$ and $a \in B$, and $\alpha^m = a$ then $\langle \sigma, a \rangle = \sigma\alpha/\alpha$. The kernel on the left is 1 and the kernel on the right is k^{*m}. The extension K_B/k is finite if and only if $(B : k^{*m})$ is finite, and if that is the case, then

$$[K_B : k] = (B : k^{*m}).$$

Proof. Let $\sigma \in G$. Suppose $\langle \sigma, a \rangle = 1$ for all $a \in B$. Then for every generator α of K_B such that $\alpha^m = a \in B$ we have $\sigma\alpha = \alpha$. Hence σ induces the identity on K_B and the kernel on the left is 1. Let $a \in B$ and suppose $\langle \sigma, a \rangle = 1$ for all $\sigma \in G$. Consider the subfield $k(a^{1/m})$ of K_B. If $a^{1/m}$ is not in k, there exists an automorphism of $k(a^{1/m})$ over k which is not the identity. Extend this automorphism to K_B, and call this extension σ. Then clearly $\langle \sigma, a \rangle \neq 1$. This proves our contention.

By the duality theorem of Chapter I, §11 we see that G is finite if and only if B/k^{*m} is finite, and in that case the order of G is equal to $(B : k^{*m})$.

THEOREM 14. *Notation being as in Theorem 13, the map $B \mapsto K_B$ gives a bijection of the set of subgroups of k^* containing k^{*m} and the abelian extensions of k of exponent m.*

Proof. Let B_1, B_2 be subgroups of k^* containing k^{*m}. If $B_1 \subset B_2$ then $k(B_1^{1/m}) \subset k(B_2^{1/m})$. Conversely, assume that $k(B_1^{1/m}) \subset k(B_2^{1/m})$. We wish to prove $B_1 \subset B_2$. Let $b \in B_1$. Then $k(b^{1/m}) \subset k(B_2^{1/m})$ and $k(b^{1/m})$ is contained in a finitely generated subextension of $k(B_2^{1/m})$. Thus we may assume without loss of generality that B_2/k^{*m} is finitely generated, hence finite. Let B_3 be the subgroup of k^* generated by B_2 and b. Then $k(B_2^{1/m}) = k(B_3^{1/m})$ and from what we saw above, the degree of this field over k is precisely

$$(B_2 : k^{*m}) \qquad \text{or} \qquad (B_3 : k^{*m}).$$

Thus these two indices are equal, and $B_2 = B_3$. This proves that $B_1 \subset B_2$.

We now have obtained an injection of our set of groups B into the set of abelian extensions of k of exponent m. Assume finally that K is an abelian extension of k of exponent m. Any finite subextension is a composite of cyclic extensions of exponent m because any finite abelian group is a product of cyclic groups, and we can apply Corollary 2 of Theorem 5, §1. By Theorem 10, every cyclic extension can be obtained by adjoining an m-th root. Hence K can be obtained by adjoining a family of m-th roots, say m-th roots of elements $\{b_j\}_{j \in J}$ with $b_j \in k^*$. Let B be the subgroup of k^* generated by all b_j and k^{*m}. If $b' = ba^m$ with $a, b \in k$ then obviously

$$k(b'^{1/m}) = k(b^{1/m}).$$

Hence $k(B^{1/m}) = K$, as desired.

When we deal with abelian extensions of exponent p equal to the characteristic, then we have to develop an additive theory, which bears the same relationship to Theorems 13 and 14 as Theorem 11 bears to Theorem 10.

If k is a field, we define the operator $\wp$ by

$$\wp(x) = x^p - x$$

for $x \in k$. Then $\wp$ is an additive homomorphism of k into itself. The subgroup $\wp(k)$ plays the same role as the subgroup k^{*m} in the multiplicative theory, whenever m is a prime number. The theory concerning a power of p is slightly more elaborate and is due to Witt. We refer the reader to the exercises if he wishes to see it carried out.

We now assume k has characteristic p. A root of the polynomial $X^p - X - a$ with $a \in k$ will be denoted by $\wp^{-1}a$. If B is a subgroup of k containing $\wp k$ we let $K_B = k(\wp^{-1}B)$ be the field obtained by adjoining $\wp^{-1}a$ to k for all $a \in B$. We emphasize the fact that B is an *additive* subgroup of k.

THEOREM 15. *Let k be a field of characteristic p. The map $B \mapsto k(\wp^{-1}B)$ is a bijection between subgroups of k containing $\wp k$ and abelian extensions of k of exponent p. Let $K = K_B = k(\wp^{-1}B)$, and let G be its Galois group. We have a bilinear map*

$$G \times B \to \mathbf{Z}/p\mathbf{Z} \qquad \text{given by} \qquad (\sigma, a) \to \langle \sigma, a \rangle.$$

If $\sigma \in G$ and $a \in B$, and $\wp\alpha = a$, then $\langle \sigma, a \rangle = \sigma\alpha - \alpha$. The kernel on the left is 1 and the kernel on the right is $\wp k$. The extension K_B/k is finite if and only if $(B : \wp k)$ is finite and if that is the case, then

$$[K_B : k] = (B : \wp k).$$

Proof. The proof is entirely similar to the proof of Theorems 13 and 14. It can be obtained by replacing multiplication by addition, and using the "$\wp$-th root" instead of an m-th root. Otherwise, there is no change in the wording of the proof.

The analogous theorem for abelian extensions of exponent p^n requires Witt vectors, and will be developed in the exercises.

§9. *The equation $X^n - a = 0$*

When the roots of unity are not in the ground field, the equation $X^n - a = 0$ is still interesting but a little more subtle to treat.

THEOREM 16. *Let k be a field and n an integer $\geqq 2$. Let $a \in k$, $a \neq 0$. Assume that for all prime numbers p such that $p \mid n$ we have $a \notin k^p$, and if $4 \mid n$ then $a \notin -4k^4$. Then $X^n - a$ is irreducible in $k[X]$.*

Proof. Our first assumption means that a is not a p-th power in k. We shall reduce our theorem to the case when n is a prime power, by induction.

Write $n = p^r m$ with p prime to m, and p odd. Let

$$X^m - a = \prod_{\nu=1}^{m} (X - \alpha_\nu)$$

be the factorization of $X^m - a$ into linear factors, and say $\alpha = \alpha_1$. Substituting X^{p^r} for X we get

$$X^n - a = X^{p^r m} - a = \prod_{\nu=1}^{m} (X^{p^r} - \alpha_\nu).$$

We may assume inductively that $X^m - a$ is irreducible in $k[X]$. We contend that α is not a p-th power in $k(\alpha)$. Otherwise, $\alpha = \beta^p$, $\beta \in k(\alpha)$. Let N be the norm from $k(\alpha)$ to k. Then

$$-a = (-1)^m N(\alpha) = (-1)^m N(\beta^p) = (-1)^m N(\beta)^p.$$

If m is odd, a is a p-th power, which is impossible. Similarly, if m is even and p is odd, we also get a contradiction. This proves our contention, because m is prime to p. If we know our theorem for prime powers, then we conclude that $X^{p^r} - \alpha$ is irreducible over $k(\alpha)$. If A is a root of $X^{p^r} - \alpha$ then $k \subset k(\alpha) \subset k(A)$ gives a tower, of which the bottom step has degree m and the top step has degree p^r. It follows that A has degree n over k and hence that $X^n - a$ is irreducible.

We now suppose that $n = p^r$ is a prime power.

If p is the characteristic, let α be a p-th root of a. Then $X^p - a = (X - \alpha)^p$ and hence $X^{p^r} - a = (X^{p^{r-1}} - \alpha)^p$ if $r \geqq 2$. By an argument even more trivial than before, we see that α is not a p-th power in $k(\alpha)$, hence $X^{p^{r-1}} - \alpha$ is irreducible over $k(\alpha)$. Hence $X^{p^r} - a$ is irreducible over k.

Suppose that p is not the characteristic. We work inductively again, and let α be a root of $X^p - a$. Let $r \geqq 2$. We let $\alpha = \alpha_1$. We have

$$X^p - a = \prod_{\nu=1}^{p} (X - \alpha_\nu)$$

and

$$X^{p^r} - a = \prod_{\nu=1}^{m} (X^{p^{r-1}} - \alpha_\nu).$$

Assume that α is not a p-th power in $k(\alpha)$. Let A be a root of $X^{p^{r-1}} - \alpha$. If p is odd then by induction, A has degree p^{r-1} over $k(\alpha)$, hence has degree p^r over k and we are done. If $p = 2$, suppose $\alpha = -4\beta^4$ with $\beta \in k(\alpha)$. Let N be the norm from $k(\alpha)$ to k. Then $-a = N(\alpha) = 16N(\beta)^4$, contradicting our assumption on a. Hence again by induction, we find that A has degree p^r over k. We therefore assume that $\alpha = \beta^p$ with some $\beta \in k(\alpha)$, and derive the consequences.

Taking the norm from $k(\alpha)$ to k we find

$$-a = (-1)^p N(\alpha) = (-1)^p N(\beta^p) = (-1)^p N(\beta)^p.$$

If p is odd, then a is a p-th power in k, contradiction. Hence $p = 2$, and

$-a = N(\beta)^2$ is a square in k. Write $-a = b^2$ with $b \in k$. Since a is not a square in k we conclude that -1 is not a square in k. Let $i^2 = -1$. Over $k(i)$ we have the factorization

$$X^{2^r} - a = X^{2^r} + b^2 = (X^{2^{r-1}} + ib)(X^{2^{r-1}} - ib).$$

Each factor is of degree 2^{r-1} and we argue inductively. If $X^{2^{r-1}} \pm ib$ is reducible over $k(i)$ then $\pm ib$ is a square in $k(i)$ or lies in $-4(k(i))^4$. In either case, $\pm ib$ is a square in $k(i)$, say

$$\pm ib = (c + di)^2 = c^2 + 2c\,di - d^2$$

with $c, d \in k$. We conclude that $c^2 = d^2$ or $c = \pm d$, and $\pm ib = 2c\,di = \pm c^2 i$. Squaring gives a contradiction, namely

$$a = -b^2 = -4c^4.$$

We now conclude by unique factorization that $X^{2^r} + b^2$ cannot factor in $k[X]$, thereby proving our theorem.

The conditions of our theorem are necessary because

$$X^4 + 4b^4 = (X^2 + 2bX + 2b^2)(X^2 - 2bX + 2b^2).$$

If $n = 4m$ and $a \in -4k^4$ then $X^n - a$ is reducible.

Corollary 1. *Let k be a field and assume that $a \in k$, $a \neq 0$, and that a is not a p-th power for some prime p. If p is equal to the characteristic, or if p is odd, then for every integer $r \geqq 1$ the polynomial $X^{p^r} - a$ is irreducible over k.*

Proof. The assertion is logically weaker than the assertion of the theorem.

Corollary 2. *Let k be a field and assume that the algebraic closure $\bar{k}$ of k is of finite degree > 1 over k. Then $\bar{k} = k(i)$ where $i^2 = -1$, and k has characteristic 0.*

Proof. We note that $\bar{k}$ is normal over k. If $\bar{k}$ is not separable over k, then $\bar{k}$ is purely inseparable over some subfield of degree > 1 (by Chapter VII, §7), and hence there is a subfield E containing k, and an element $a \in E$ such that $X^p - a$ is irreducible over E. By Corollary 1, $\bar{k}$ cannot be of finite degree over E. (The reader may restrict his attention to characteristic 0 if he omitted Chapter VII, §7.)

We may therefore assume that $\bar{k}$ is Galois over k. Let $k_1 = k(i)$. Then $\bar{k}$ is also Galois over k_1. Let G be the Galois group of $\bar{k}/k_1$. Suppose that there is a prime number p dividing the order of G, and let H be a subgroup of order p. Let F be its fixed field. Then $[\bar{k} : F] = p$. If p is the characteristic, then Exercise 5 at the end of the chapter will give the contradiction. We may assume that p is not the characteristic. The p-th roots of unity $\neq 1$ are the roots of a polynomial of degree $\leqq p - 1$ (namely $X^{p-1} + \cdots + 1$), and hence must lie in F. By Theorem 10 of §6, it follows that $\bar{k}$ is the splitting field of some polynomial $X^p - a$ with $a \in F$. The polynomial $X^{p^2} - a$ is necessarily reducible. By the theorem, we must have $p = 2$ and $a = -4b^4$ with $b \in F$. This implies

$$\bar{k} = F(a^{1/2}) = F(i).$$

But we assumed $i \in k_1$, contradiction.

There remains to prove that k has characteristic 0. Assume that k has characteristic > 0 (but don't use a letter for the characteristic since p is already occupied). The field obtained by adjoining a primitive root of unity ζ_{2^r} to the prime field $\mathbf{F}$ is cyclic over the prime field. By Theorem 4 of §1, it follows that the Galois group of $\bar{k}$ over k, which is cyclic of order 2, generated say by σ, corresponds to a subgroup of $\mathbf{F}(\zeta_{2^r})$ over $\mathbf{F}$. However, $\mathbf{F}(\zeta_{2^r})$ being cyclic over $\mathbf{F}$ has a unique subfield of degree 2 over $\mathbf{F}$, and this subfield must contain i, since i has degree 1 or 2 over $\mathbf{F}$. Since $\sigma i \neq i$, the fixed subfield of $\mathbf{F}$ under σ must therefore be equal to $\mathbf{F}$. This implies that $\mathbf{F}(\zeta_{2^r})$ is of degree 2 over $\mathbf{F}$, and yields a contradiction if we take r sufficiently large.

Corollary 2 above is due to Artin.

§10. *Galois cohomology*

Let G be a group and A an abelian group which we write additively for the general remarks which we make, preceding our theorems. Let us assume that G operates on A, by means of a homomorphism $G \to \operatorname{Aut}(A)$. By a 1-*cocycle* of G in A one means a family of elements $\{\alpha_\sigma\}_{\sigma \in G}$ with $\alpha_\sigma \in A$, satisfying the relations

$$\alpha_\sigma + \sigma\alpha_\tau = \alpha_{\sigma\tau}$$

for all $\sigma, \tau \in G$. If $\{\alpha_\sigma\}_{\sigma \in G}$ and $\{\beta_\sigma\}_{\sigma \in G}$ are 1-cocycles, then we can add them to get a 1-cocycle $\{\alpha_\sigma + \beta_\sigma\}_{\sigma \in G}$. It is then clear that 1-cocycles form a group, denoted by $Z^1(G, A)$. By a 1-*coboundary* of G in A one means a family of elements $\{\alpha_\sigma\}_{\sigma \in G}$ such that there exists an element $\beta \in A$ for which $\alpha_\sigma = \sigma\beta - \beta$ for all $\sigma \in G$. It is then clear that a 1-coboundary is a 1-cocycle, and that the 1-coboundaries form a group, denoted by $B^1(G, A)$. The factor group $Z^1(G, A)/B^1(G, A)$ is called the *first cohomology group* of G in A and is denoted by $H^1(G, A)$.

THEOREM 17. *Let K/k be a finite Galois extension with Galois group G. Then for the operation of G on K^* we have $H^1(G, K^*) = 1$ and for the operation of G on the additive group of K we have $H^1(G, K) = 0$. In other words, the first cohomology group is trivial in both cases.*

Proof. Let $\{\alpha_\sigma\}_{\sigma \in G}$ be a 1-cocycle of G in K^*. The multiplicative cocycle relation reads

$$\alpha_\sigma \alpha_\tau^\sigma = \alpha_{\sigma\tau}.$$

By the linear independence of characters, there exists $\theta \in K$ such that the element

$$\beta = \sum_{\tau \in G} \alpha_\tau \tau(\theta)$$

is $\neq 0$. Then

$$\sigma\beta = \sum_{\tau \in G} \alpha_\tau^\sigma \sigma\tau(\theta) = \sum_{\tau \in G} \alpha_{\sigma\tau}\alpha_\sigma^{-1}\sigma\tau(\theta)$$

$$= \alpha_\sigma^{-1} \sum_{\tau \in G} \alpha_{\sigma\tau}\sigma\tau(\theta) = \alpha_\sigma^{-1}\beta.$$

We get $\alpha_\sigma = \beta/\sigma\beta$, and using β^{-1} instead of β gives what we want.

For the additive part of the theorem, we find an element $\theta \in K$ such that the trace $\mathrm{Tr}(\theta)$ is not equal to 0. Given a 1-cocycle $\{\alpha_\sigma\}$ in the additive group of K, we let

$$\beta = \frac{1}{\mathrm{Tr}(\theta)} \sum_{\tau \in G} \alpha_\tau \tau(\theta).$$

It follows at once that $\alpha_\sigma = \beta - \sigma\beta$, as desired.

§11. *Algebraic independence of homomorphisms*

Let A be an additive group, and let K be a field. Let $\lambda_1, \ldots, \lambda_n : A \to K$ be additive homomorphisms. We shall say that $\lambda_1, \ldots, \lambda_n$ are *algebraically dependent* (over K) if there exists a polynomial $f(X_1, \ldots, X_n)$ in $K[X_1, \ldots, X_n]$ such that for all $x \in A$ we have

$$f(\lambda_1(x), \ldots, \lambda_n(x)) = 0,$$

but such that f does not induce the zero function on $K^{(n)}$, i.e. on the direct product of K with itself n times. We know that with each polynomial we can associate a unique reduced polynomial giving the same function. If K is infinite, the reduced polynomial is equal to f itself. In our definition of dependence, we could as well assume that f is reduced.

A polynomial $f(X_1, \ldots, X_n)$ will be called *additive* if it induces an additive homomorphism of $K^{(n)}$ into K. Let $(Y) = (Y_1, \ldots, Y_n)$ be variables independent from (X). Let

$$g(X, Y) = f(X + Y) - f(X) - f(Y)$$

where $X + Y$ is the componentwise vector addition. Then the total degree of g viewed as a polynomial in (X) with coefficients in $K[Y]$ is strictly less than the total degree of f, and similarly, its degree in each X_i is strictly less than the degree of f in each X_i. One sees this easily by considering the difference of monomials,

$$\begin{aligned} M_{(\nu)}(X + Y) &- M_{(\nu)}(X) - M_{(\nu)}(Y) \\ &= (X_1 + Y_1)^{\nu_1} \cdots (X_n + Y_n)^{\nu_n} - X_1^{\nu_1} \cdots X_n^{\nu_n} - Y_1^{\nu_1} \cdots Y_n^{\nu_n}. \end{aligned}$$

A similar assertion holds for g viewed as a polynomial in (Y) with coefficients in $K[X]$.

If f is reduced, it follows that g is reduced. Hence if f is additive, it follows that g is the zero polynomial.

Example. Let K have characteristic p. Then in one variable, the map

$$\xi \mapsto a\xi^{p^m}$$

for $a \in K$ and $m \geqq 1$ is additive, and given by the additive polynomial aX^{p^m}. We shall see later that this is a typical example.

THEOREM 18. (Artin) *Let $\lambda_1, \ldots, \lambda_n : A \to K$ be additive homomorphisms of an additive group into a field. If these homomorphisms are algebraically dependent over K, then there exists an additive polynomial $f(X_1, \ldots, X_n) \neq 0$ in $K[X]$ such that*

$$f(\lambda_1(x), \ldots, \lambda_n(x)) = 0$$

for all $x \in A$.

Proof. Let $f(X) = f(X_1, \ldots, X_n) \in K[X]$ be a reduced polynomial of lowest possible degree such that $f \neq 0$ but for all $x \in A$, $f(\Lambda(x)) = 0$, where $\Lambda(x)$ is the vector $(\lambda_1(x), \ldots, \lambda_n(x))$. We shall prove that f is additive.

Let $g(X, Y) = f(X + Y) - f(X) - f(Y)$. Then

$$g(\Lambda(x), \Lambda(y)) = f(\Lambda(x + y)) - f(\Lambda(x)) - f(\Lambda(y)) = 0$$

for all $x, y \in A$. We shall prove that g induces the zero function on $K^{(n)} \times K^{(n)}$. Assume otherwise. We have two cases.

Case 1. We have $g(\xi, \Lambda(y)) = 0$ for all $\xi \in K^{(n)}$ and all $y \in A$. By hypothesis, there exists $\xi' \in K^{(n)}$ such that $g(\xi', Y)$ is not identically 0. Let $P(Y) = g(\xi', Y)$. Since the degree of g in (Y) is strictly smaller than the degree of f, we have a contradiction.

Case 2. There exist $\xi' \in K^{(n)}$ and $y' \in A$ such that $g(\xi', \Lambda(y')) \neq 0$. Let $P(X) = g(X, \Lambda(y'))$. Then P is not the zero polynomial, but $P(\Lambda(x)) = 0$ for all $x \in A$, again a contradiction.

We conclude that g induces the zero function on $K^{(n)} \times K^{(n)}$, which proves what we wanted, namely that f is additive.

We now consider additive polynomials more closely.

Let f be an additive polynomial in n variables over K, and assume that f is reduced. Let

$$f_i(X_i) = f(0, \dots, X_i, \dots, 0)$$

with X_i in the i-th place, and zeros in the other components. By additivity, it follows that

$$f(X_1, \dots, X_n) = f_1(X_1) + \cdots + f_n(X_n)$$

because the difference of the right-hand side and left-hand side is a reduced polynomial taking the value 0 on $K^{(n)}$. Furthermore, each f_i is an additive polynomial in one variable. We now study such polynomials.

Let $f(X)$ be a reduced polynomial in one variable, which induces a linear map of K into itself. Suppose that there occurs a monomial $a_r X^r$ in f with coefficient $a_r \neq 0$. Then the monomials of degree r in

$$g(X, Y) = f(X + Y) - f(X) - f(Y)$$

are given by

$$a_r(X + Y)^r - a_r X^r - a_r Y^r.$$

We have already seen that g is identically 0. Hence the above expression is identically 0. Hence the polynomial

$$(X + Y)^r - X^r - Y^r$$

is the zero polynomial. It contains the term $rX^{r-1}Y$. Hence if $r > 1$, our field must have characteristic p and r is divisible by p. Write $r = p^m s$ where s is prime to p. Then

$$0 = (X + Y)^r - X^r - Y^r = (X^{p^m} + Y^{p^m})^s - (X^{p^m})^s - (Y^{p^m})^s.$$

Arguing as before, we conclude that $s = 1$.

Hence if f is an additive polynomial in one variable, we have

$$f(X) = \sum_{\nu=0}^{m} a_\nu X^{p^\nu},$$

with $a_\nu \in K$. In characteristic 0, the only additive polynomials in one variable are of type aX with $a \in K$.

As expected, we define $\lambda_1, \dots, \lambda_n$ to be *algebraically independent* if, whenever f is a reduced polynomial such that $f(\Lambda(x)) = 0$ for all $x \in A$, then f is the zero polynomial.

We shall apply Theorem 18 to the case when $\lambda_1, \ldots, \lambda_n$ are automorphisms of a field, and combine Theorem 18 with the theorem on the linear independence of characters.

THEOREM 19. *Let K be an infinite field, and let $\sigma_1, \ldots, \sigma_n$ be the distinct elements of a finite group of automorphisms of K. Then $\sigma_1, \ldots, \sigma_n$ are algebraically independent over K.*

Proof. (Artin) In characteristic 0, Theorem 18 and the linear independence of characters show that our assertion is true. Let the characteristic be $p > 0$, and assume that $\sigma_1, \ldots, \sigma_n$ are algebraically dependent.

There exists an additive polynomial $f(X_1, \ldots, X_n)$ in $K[X]$ which is reduced, $f \neq 0$, and such that

$$f(\sigma_1(x), \ldots, \sigma_n(x)) = 0$$

for all $x \in K$. By what we saw above, we can write this relation in the form

$$\sum_{i=1}^{n} \sum_{r=1}^{m} a_{ir}\sigma_i(x)^{p^r} = 0$$

for all $x \in K$, and with not all coefficients a_{ir} equal to 0. Therefore by the linear independence of characters, the automorphisms

$$\{\sigma_i^{p^r}\} \qquad \text{with} \quad i = 1, \ldots, n \quad \text{and} \quad r = 1, \ldots, m$$

cannot be all distinct. Hence we have

$$\sigma_i^{p^r} = \sigma_j^{p^s}$$

with either $i \neq j$ or $r \neq s$. Say $r \leqq s$. For all $x \in K$ we have

$$\sigma_i(x)^{p^r} = \sigma_j(x)^{p^s}.$$

Extracting p-th roots in characteristic p is unique. Hence

$$\sigma_i(x) = \sigma_j(x)^{p^{s-r}} = \sigma_j(x^{p^{s-r}})$$

for all $x \in K$. Let $\sigma = \sigma_j^{-1}\sigma_i$. Then

$$\sigma(x) = x^{p^{s-r}}$$

for all $x \in K$. Taking $\sigma^n = id$ shows that

$$x = x^{p^{n(s-r)}}$$

for all $x \in K$. Since K is infinite, this can hold only if $s = r$. But in that case, $\sigma_i = \sigma_j$, contradicting the fact that we started with distinct automorphisms.

§12. *The normal basis theorem*

THEOREM 20. *Let K/k be a finite Galois extension of degree n. Let $\sigma_1, \dots, \sigma_n$ be the elements of the Galois group G. Then there exists an element $w \in K$ such that $\sigma_1 w, \dots, \sigma_n w$ form a basis of K over k.*

Proof. We prove this here only when k is infinite. The case when k is finite can be proved later by methods of linear algebra, as an exercise.

For each $\sigma \in G$, let X_σ be a variable, and let $t_{\sigma,\tau} = X_{\sigma^{-1}\tau}$. Let $X_i = X_{\sigma_i}$. Let

$$f(X_1, \dots, X_n) = \det(t_{\sigma_i, \sigma_j}).$$

Then f is not identically 0, as one sees by substituting 1 for X_{id} and 0 for X_σ if $\sigma \neq id$. Since k is infinite, f is reduced. Hence the determinant will not be 0 for all $x \in K$ if we substitute $\sigma_i(x)$ for X_i in f. Hence there exists $w \in K$ such that

$$\det(\sigma_i^{-1}\sigma_j(w)) \neq 0.$$

Suppose $a_1, \dots, a_n \in k$ are such that

$$a_1\sigma_1(w) + \cdots + a_n\sigma_n(w) = 0.$$

Apply σ_i^{-1} to this relation for each $i = 1, \dots, n$. Since $a_j \in k$ we get a system of linear equations, regarding the a_j as unknowns. Since the determinant of the coefficients is $\neq 0$, it follows that

$$a_j = 0 \qquad \text{for } j = 1, \dots, n$$

and hence that w is the desired element.

EXERCISES

1. Let k be a field and X a variable over k. Let

$$\varphi(X) = \frac{f(X)}{g(X)}$$

be a rational function in $k(X)$, expressed as a quotient of two polynomials f, g which are relatively prime. Define the degree of φ to be $\max(\deg f, \deg g)$. Let $Y = \varphi(X)$. (a) Show that the degree of φ is equal to the degree of the field extension $k(X)$ over $k(Y)$ (assuming $Y \notin k$). (b) Show that every automorphism of $k(X)$ over k can be represented by a rational function φ of degree 1, and is therefore induced by a map

$$X \mapsto \frac{aX + b}{cX + d}$$

with $a, b, c, d \in k$ and $ad - bc \neq 0$. (c) Let G be the group of automorphisms of $k(X)$ over k. Show that G is generated by the following automorphisms:

$$\tau_b : X \mapsto X + b, \qquad \sigma_a : X \mapsto aX \quad (a \neq 0), \qquad X \mapsto X^{-1}$$

with $a, b \in k$.

2. Let k be a finite field with q elements. Let $K = k(X)$ be the rational field in one variable. Let G be the group of automorphisms of K obtained by the mappings

$$X \mapsto \frac{aX + b}{cX + d}$$

with a, b, c, d in k and $ad - bc \neq 0$. Prove the following statements:

(i) The order of G is $q^3 - q$.

(ii) The fixed field of G is equal to $k(Y)$ where

$$Y = \frac{(X^{q^2} - X)^{q+1}}{(X^q - X)^{q^2+1}}.$$

(iii) Let H_1 be the subgroup of G consisting of the mappings $X \mapsto aX + b$ with $a \neq 0$. The fixed field of H_1 is $k(T)$ where $T = (X^q - X)^{q-1}$.

(iv) Let H_2 be the subgroup of H_1 consisting of the mappings $X \to X + b$ with $b \in k$. The fixed field of H_2 is equal to $k(Z)$ where $Z = X^q - X$.

3. Let $\overline{\mathbf{Q}}$ be a fixed algebraic closure of $\mathbf{Q}$. Let E be a maximal subfield of $\overline{\mathbf{Q}}$ not containing $\sqrt{2}$ (such a subfield exists by Zorn's lemma). Show that every finite extension of E is cyclic. (Your proof should work taking any algebraic irrational number instead of $\sqrt{2}$.)

4. Let k be a field, $\bar{k}$ an algebraic closure, and σ an automorphism of $\bar{k}$ leaving k fixed. Let F be the fixed field of σ. Show that every finite extension of F is cyclic.

(The above two problems are examples of Artin, showing how to dig holes in an algebraically closed field.)

5. (i) Let K be a cyclic extension of a field F, with Galois group G generated by σ. Assume that the characteristic is p, and that $[K : F] = p^{m-1}$ for some integer $m \geqq 2$. Let β be an element of K such that $\mathrm{Tr}_F^K(\beta) = 1$. Show that there exists an element α in K such that

$$\sigma\alpha - \alpha = \beta^p - \beta.$$

(ii) Prove that the polynomial $X^p - X - \alpha$ is irreducible in $K[X]$.

(iii) If θ is a root of this polynomial, prove that $F(\theta)$ is a Galois, cyclic extension of degree p^m of F, and that its Galois group is generated by an extension σ^* of σ such that

$$\sigma^*(\theta) = \theta + \beta.$$

6. Let E be an algebraic extension of k such that every non-constant polynomial $f(X)$ in $k[X]$ has at least one root in E. Prove that E is algebraically closed. [*Hint:* Discuss the separable and purely inseparable cases separately, and use the primitive element theorem.]

7. *Relative invariants* (Sato). Let k be a field and K an extension of k. Let G be a group of automorphisms of K over k, and assume that k is the fixed field of G. (We do not assume that K is algebraic over k.) By a *relative invariant* of G in K we shall mean an element $P \in K$, $P \neq 0$, such that for each $\sigma \in G$ there exists an element $\chi(\sigma) \in k$ for which $P^\sigma = \chi(\sigma)P$. Since σ is an automorphism, we have $\chi(\sigma) \in k^*$. We say that the map $\chi : G \to k^*$ *belongs* to P, and call it a *character*. Prove the following statements:

(a) The map χ above is a homomorphism.

(b) If the same character χ belongs to relative invariants P and Q then there exists $c \in k^*$ such that $P = cQ$.

(c) The relative invariants form a multiplicative group, which we denote by I.

Elements $P_1, \ldots, P_m$ of I are called multiplicatively independent mod k^* if their images in the factor group I/k^* are multiplicatively independent, i.e. if given integers $\nu_1, \ldots, \nu_m$ such that

$$P_1^{\nu_1} \cdots P_m^{\nu_m} = c \in k^*,$$

then $\nu_1 = \cdots = \nu_m = 0$.

(d) If $P_1, \ldots, P_m$ are multiplicatively independent mod k^* prove that they are algebraically independent over k. [*Hint:* Use Artin's theorem on characters.]

(e) Assume that $K = k(X_1, \ldots, X_n)$ is the quotient field of the polynomial ring $k[X_1, \ldots, X_n] = k[X]$, and assume that G induces an automorphism of the polynomial ring. Prove: If $F_1(X)$ and $F_2(X)$ are relative invariant polynomials, then their g.c.d. is relative invariant. If $P(X) = F_1(X)/F_2(X)$ is a relative invariant, and is the quotient of two relatively prime polynomials, then $F_1(X)$ and $F_2(X)$ are relative invariants. Prove that the relative invariant polynomials generate I/k^*. Let S be the set of relative invariant polynomials which cannot be factored into a product of two relative invariant polynomials of degrees $\geqq 1$. Show that the elements of S/k^* are multiplicatively independent, and hence that I/k^* is a free abelian group. [If you know about transcendence degree, then using (d) you can conclude that this group is finitely generated.]

8. Let E be a finite separable extension of k, of degree n. Let $W = (w_1, \ldots, w_n)$ be elements of E. Let $\sigma_1, \ldots, \sigma_n$ be the distinct embeddings of E in $\bar{k}$ over k. Define the *discriminant* of W to be

$$D_{E/k}(W) = \det(\sigma_i w_j)^2.$$

Prove: (a) If $V = (v_1, \ldots, v_n)$ is another set of elements of E and $X = (x_{ij})$ is a matrix of elements of k such that $W = XV$, then

$$D_{E/k}(W) = \det(X)^2 \, D_{E/k}(V).$$

(b) The discriminant is an element of k.

(c) Let $E = k(\alpha)$ and let $f(X) = \operatorname{Irr}(\alpha, k, X)$. Let $\alpha_1, \ldots, \alpha_n$ be the roots of f and say $\alpha = \alpha_1$. Then

$$f'(\alpha) = \prod_{j=2}^{n} (\alpha - \alpha_j).$$

Show that

$$D_{E/k}(1, \alpha, \ldots, \alpha^{n-1}) = (-1)^{n(n-1)/2} N_k^E(f'(\alpha)).$$

(d) Let the notation be as in (a). Show that $\det(\mathrm{Tr}(w_i w_j)) = (\det(\sigma_i w_j))^2$. [*Hint:* Let A be the matrix $(\sigma_i w_j)$. Show that ${}^t A A$ is the matrix $(\mathrm{Tr}(w_i w_j))$.]

9. Let F be a finite field and K a finite extension of F. Show that the norm N_F^K and the trace Tr_F^K are surjective (as maps from K into F).

10. Let $a \neq 0, \neq \pm 1$ be a square-free integer. For each prime number p, let K_p be the splitting field of the polynomial $X^p - a$ over $\mathbf{Q}$. Show that $[K_p : \mathbf{Q}] = p(p-1)$. For each square-free integer $m > 0$, let

$$K_m = \prod_{p \mid m} K_p$$

be the compositum of all fields K_p for $p \mid m$. Let $d_m = [K_m : \mathbf{Q}]$ be the degree of K_m over $\mathbf{Q}$. Show that if m is odd then $d_m = \prod_{p \mid m} d_p$, and if m is even, $m = 2n$ then $d_{2n} = d_n$ or $2d_n$ according as $\sqrt{a}$ is or is not in the field of m-th roots of unity $\mathbf{Q}(\zeta_m)$.

11. Let A be an abelian group and let G be a finite cyclic group operating on A [by means of a homomorphism $G \to \mathrm{Aut}(A)$]. Let σ be a generator of G. We define the trace $\mathrm{Tr}_G = \mathrm{Tr}$ on A by $\mathrm{Tr}(x) = \sum_{\tau \in G} \tau x$. Let A_{Tr} denote the kernel of the trace, and let $(1 - \sigma)A$ denote the subgroup of A consisting of all elements of type $y - \sigma y$. Show that $H^1(G, A) \approx A_{\mathrm{Tr}}/(1 - \sigma)A$.

12. What is the Galois group of the following polynomials:

(a) $X^3 - X - 1$ over $\mathbf{Q}$.
(b) $X^3 - 10$ over $\mathbf{Q}$.
(c) $X^3 - 10$ over $\mathbf{Q}(\sqrt{2})$.
(d) $X^3 - 10$ over $\mathbf{Q}(\sqrt{-3})$.
(e) $X^3 - X - 1$ over $\mathbf{Q}(\sqrt{-23})$.
(f) $X^4 - 5$ over $\mathbf{Q}$, $\mathbf{Q}(\sqrt{5})$, $\mathbf{Q}(\sqrt{-5})$, $\mathbf{Q}(i)$.
(g) $X^4 - a$ where a is any integer $\neq 0, \neq \pm 1$ and is square free. Over $\mathbf{Q}$.
(h) $X^n - a$ where n is odd > 1, and a is any square-free positive integer. Over $\mathbf{Q}$.
(i) $X^4 + 2$ over $\mathbf{Q}$, $\mathbf{Q}(i)$.
(j) $(X^2 - 2)(X^2 - 3)(X^2 - 5)(X^2 - 7)$ over $\mathbf{Q}$.
(k) Let $p_1, \ldots, p_n$ be distinct prime numbers. What is the Galois group of $(X^2 - p_1) \cdots (X^2 - p_n)$ over $\mathbf{Q}$?
(l) $(X^3 - 2)(X^3 - 3)(X^2 - 2)$ over $\mathbf{Q}(\sqrt{-3})$.
(m) $X^n - t$, where t is transcendental over the complex numbers $\mathbf{C}$ and n is a positive integer. Over $\mathbf{C}(t)$.
(n) $X^4 - t$, where t is as before. Over $\mathbf{R}(t)$.

13. Let k be a field, n an odd integer $\geqq 1$, and let ζ be a primitive n-th root of unity, in k. Show that k also contains a primitive $2n$-th root of unity.

14. Let k be a finite extension of the rationals. Show that there is only a finite number of roots of unity in k.

15. Determine which roots of unity lie in the following fields: $\mathbf{Q}(i)$, $\mathbf{Q}(\sqrt{-2})$, $\mathbf{Q}(\sqrt{2})$, $\mathbf{Q}(\sqrt{-3})$, $\mathbf{Q}(\sqrt{3})$, $\mathbf{Q}(\sqrt{-5})$.

16. For which integers m does a primitive m-th root of unity have degree 2 over $\mathbf{Q}$?

17. Let k be a field of characteristic 0. Assume that for each finite extension E of k, the index $(E^* : E^{*n})$ is finite for every positive integer n. Show that for each positive integer n, there exists only a finite number of abelian extensions of k of degree n.

18. Let $f(z)$ be a rational function with coefficients in a finite extension of the rationals. Assume that there are infinitely many roots of unity ζ such that $f(\zeta)$ is a root of unity. Show that there exists an integer n such that $f(z) = cz^n$ for some constant c (which is in fact a root of unity).

This exercise can be generalized as follows: Let Γ_0 be a finitely generated multiplicative group of complex numbers. Let Γ be the group of all complex numbers γ such that γ^m lies in Γ_0 for some integer $m \neq 0$. Let $f(z)$ be a rational function with complex coefficients such that there exist infinitely many $\gamma \in \Gamma$ for which $f(\gamma)$ lies in Γ. Then again, $f(z) = cz^n$ for some c and n.

If one takes values of γ and f to be in Γ_0 rather than Γ, then I gave a proof for the corresponding statement (cf. *Diophantine Geometry*, Chapter VII, Theorem 7).

19. Let K/k be a Galois extension. We define the Krull topology on the group $G(K/k) = G$ by taking as fundamental system of open neighborhoods of the identity the set of subgroups belonging to finite extensions E of k contained in K. Using the representation on left cosets, one sees that the normal subgroups are cofinal with this family, and hence that the family of normal subgroups belonging to finite normal extensions defines the same topology. Show that G is algebraically and topologically isomorphic to the projective limit of the finite factor groups G/U as U ranges over these normal subgroups. Conclude that G is a compact, totally disconnected group. Such a group is called a *profinite* group. Show that every closed subgroup of finite index is open. Show that the closed subgroups are precisely those subgroups which belong to an intermediate field $k \subset F \subset K$. If H is an arbitrary subgroup of G and F is the fixed field, show that the subgroup of G belonging to F is the closure of H in G.

20. Let k be a field such that every finite extension is cyclic, and having one extension of degree n for each integer n. Show that the Galois group $G = G(\bar{k}/k)$ is the inverse limit $\lim \mathbf{Z}/m\mathbf{Z}$, as $m\mathbf{Z}$ ranges over all ideals of $\mathbf{Z}$, ordered by inclusion. Show that this limit is isomorphic to the direct product of the limits

$$\lim_{n \to \infty} \mathbf{Z}/p^n\mathbf{Z}$$

taken over all prime numbers p, in other words, it is isomorphic to the product of all p-adic integers.

21. *Witt vectors.* Let $x_1, x_2, \ldots$ be a sequence of algebraically independent elements over the integers $\mathbf{Z}$. For each integer $n \geqq 1$ define

$$x^{(n)} = \sum_{d \mid n} d x_d^{n/d}.$$

Show that x_n can be expressed in terms of $x^{(d)}$ for $d \mid n$, with rational coefficients.

Using vector notation, we call $(x_1, x_2, \ldots)$ the Witt components of the vector x, and call $(x^{(1)}, x^{(2)}, \ldots)$ its *ghost* components. We call x a *Witt vector*.

Define the power series

$$f_x(t) = \prod_{n \geqq 1} (1 - x_n t^n).$$

Show that

$$-t \frac{d}{dt} \log f_x(t) = \sum_{n \geqq 1} x^{(n)} t^n.$$

[By $\frac{d}{dt} \log f(t)$ we mean $f'(t)/f(t)$ if $f(t)$ is a power series, and the derivative $f'(t)$ is taken formally.]

If x, y are two Witt vectors, define their sum and product componentwise *with respect to the ghost components*, i.e.

$$(x \dotplus y)^{(n)} = x^{(n)} \dotplus y^{(n)}.$$

What is $(x + y)_n$? Well, show that

$$f_x(t) f_y(t) = \sum (x + y)_n t^n = f_{x+y}(t).$$

Hence $(x + y)_n$ is a polynomial with integer coefficients in $x_1, y_1, \ldots, x_n, y_n$. Also show that

$$f_{xy}(t) = \prod_{d,e \geqq 1} (1 - x_d^{m/d} y_e^{m/e} t^m)^{de/m}$$

where m is the least common multiple of d, e and d, e range over all integers $\geqq 1$. Thus $(xy)_n$ is also a polynomial in $x_1, y_1, \ldots, x_n, y_n$ with integer coefficients. The above arguments are due to Witt (oral communication) and differ from those of his original paper.

If A is a commutative ring, then taking a homomorphic image of the polynomial ring over $\mathbf{Z}$ into A, we see that we can define addition and multiplication of Witt vectors with components in A, and that these Witt vectors form a ring $W(A)$. Show that W is a functor, i.e. that any ring homomorphism φ of A into a commutative ring A' induces a homomorphism $W(\varphi) : W(A) \to W(A')$.

22. Let p be a prime number, and consider Witt vectors whose components are 0 except those indexed by a power of p. Now use the log to the base p to index these components, so that we write x_n instead of x_{p^n}. For instance, x_0 now denotes what was x_1 previously. For a Witt vector $x = (x_0, x_1, \ldots, x_n, \ldots)$ define

$$Vx = (0, x_0, x_1, \ldots) \quad \text{and} \quad Fx = (x_0^p, x_1^p, \ldots).$$

Thus V is a shifting operator. We have $V \circ F = F \circ V$. Show that

$$(Vx)^{(n)} = px^{(n-1)} \quad \text{and} \quad x^{(n)} = (Fx)^{(n-1)} + p^n x_n.$$

Also from the definition, we have

$$x^{(n)} = x_0^{p^n} + px_1^{p^{n-1}} + \cdots + p^n x_0.$$

23. Let k be a field of characteristic p, and consider $W(k)$. Then V is an additive endomorphism of $W(k)$, and F is a ring homomorphism of $W(k)$ into itself. Furthermore, if $x \in W(k)$ then

$$px = VFx.$$

If $x, y \in W(k)$, then $(V^i x)(V^j y) = V^{i+j}(F^{p^j}x \cdot F^{p^i}y)$. For $a \in k$ denote by $\{a\}$ the Witt vector $(a, 0, 0, \ldots)$. Then we can write symbolically

$$x = \sum_{i=0}^{\infty} V^i\{x_i\}.$$

Show that if $x \in W(k)$ and $x_0 \neq 0$ then x is a unit in $W(k)$. [*Hint:* One has $1 - x\{x_0^{-1}\} = Vy$ and then

$$x\{x_0^{-1}\} \sum_0^{\infty} (Vy)^i = (1 - Vy) \sum_0^{\infty} (Vy)^i = 1.]$$

24. Let n be an integer $\geqq 1$ and p a prime number again. Let k be a field of characteristic p. Let $W_n(k)$ be the ring of truncated Witt vectors $(x_0, \ldots, x_{n-1})$ with components in k. We view $W_n(k)$ as an additive group. If $x \in W_n(k)$, define $\wp(x) = Fx - x$. Then $\wp$ is a homomorphism. If K is a Galois extension of K, and $\sigma \in G(K/k)$, and $x \in W_n(K)$ we can define σx to have component $(\sigma x_0, \ldots, \sigma x_{n-1})$. Prove the analogue of Hilbert's Theorem 90 for Witt vectors, and prove that the first cohomology group is trivial. (One takes a vector whose trace is not 0, and finds a coboundary the same way as in the proof of Theorem 17, §10.)

25. If $x \in W_n(k)$, show that there exists $\xi \in W_n(\bar{k})$ such that $\wp(\xi) = x$. Do this inductively, solving first for the first component, and then showing that a vector $(0, \alpha_1, \ldots, \alpha_{n-1})$ is in the image of $\wp$ if and only if $(\alpha_1, \ldots, \alpha_{n-1})$ is in the image of $\wp$. Prove inductively that if $\xi, \xi' \in W_n(k')$ for some extension k' of k and if $\wp\xi = \wp\xi'$ then $\xi - \xi'$ is a vector with components in the prime field. Hence the solutions of $\wp\xi = x$ for given $x \in W_n(k)$ all differ by the vectors with components in the prime field, and there are p^n such vectors. We define

$$k(\xi) = k(\xi_0, \ldots, \xi_{n-1}),$$

or symbolically,

$$k(\wp^{-1}x).$$

Prove that it is a Galois extension of k, and show that the cyclic extensions of k, of degree p^n, are precisely those of type $k(\wp^{-1}x)$ with a vector x such that $x_0 \notin \wp k$.

26. Develop the Kummer theory for abelian extensions of k of exponent p^n by using $W_n(k)$. In other words, show that there is a bijection between subgroups B of $W_n(k)$ containing $\wp W_n(k)$ and abelian extensions as above, given by

$$B \mapsto K_B$$

where $K_B = k(\wp^{-1}B)$. All of this is due to Witt, cf. *Journal für die reine und angewandte Mathematik*, 1935 and 1936. The proofs are the same, *mutatis mutandis*, as those given for the Kummer theory in the text.

27. Give an example of a field K which is of degree 2 over two distinct subfields E and F respectively, but such that K is not algebraic over $E \cap F$.

28. Let $F = \mathbf{F}_p$ be the prime field of characteristic p. Let K be the field obtained from F by adjoining all primitive l-th roots of unity, for all prime numbers $l \neq p$. Prove that K is algebraically closed. [*Hint:* Show that if q is a prime number, and r an integer $\geqq 1$, there exists a prime l such that the period of $p \bmod l$ is q^r, by using the following old trick of Van der Waerden: Let l be a prime dividing the number

$$b = \frac{p^{q^r} - 1}{p^{q^{r-1}} - 1} = (p^{q^{r-1}} - 1)^{q-1} + q(p^{q^{r-1}} - 1)^{q-2} + \cdots + q.$$

If l does not divide $p^{q^{r-1}} - 1$, we are done. Otherwise, $l = q$. But in that case q^2 does not divide b, and hence there exists a prime $l \neq q$ such that l divides b. Then the degree of $F(\zeta_l)$ over F is q^r, so K contains subfields of arbitrary degree over F.]

CHAPTER IX

Extensions of Rings

Throughout this chapter, "ring" will mean "commutative ring".

§1. *Integral ring extensions*

In Chapters VII and VIII we have studied algebraic extensions of fields. For a number of reasons, it is desirable to study algebraic extensions of rings. For instance, given a polynomial with integer coefficients, say $X^5 - X - 1$, one can reduce this polynomial mod p for any prime p, and thus get a polynomial with coefficients in a finite field. As another example, consider the polynomial

$$X^n + s_{n-1}X^{n-1} + \cdots + s_0$$

where $s_{n-1}, \ldots, s_0$ are algebraically independent over a field k. This polynomial has coefficients in $k[s_0, \ldots, s_{n-1}]$ and by substituting elements of k for $s_0, \ldots, s_{n-1}$ one obtains a polynomial with coefficients in k. One can then get information about polynomials by taking a homomorphism of the ring in which they have their coefficients. This chapter is devoted to a brief description of the basic facts concerning polynomials over rings.

Let A be a ring, and M an A-module. We say that M is *faithful* if, whenever $a \in A$ is such that $aM = 0$, then $a = 0$. We note that A is a faithful module over itself since A contains a unit element. Furthermore, if $A \neq 0$, then a faithful module over A cannot be the 0-module.

Let A be a subring of a ring B. Let $\alpha \in B$. The following conditions are equivalent:

INT 1. The element α is a root of a polynomial

$$X^n + a_{n-1}X^{n-1} + \cdots + a_0$$

with coefficients $a_i \in A$, and degree $n \geqq 1$. (The essential thing here is that the leading coefficient is equal to 1.)

INT 2. The subring $A[\alpha]$ is a finitely generated A-module.

INT 3. There exists a faithful module over $A[\alpha]$ which is a finitely generated A-module.

We prove the equivalence. Assume INT 1. Let $g(X)$ be a polynomial in $A[X]$ of degree $\geqq 1$ with leading coefficient 1 such that $g(\alpha) = 0$. If $f(X) \in A[X]$ then

$$f(X) = q(X)g(X) + r(X)$$

with $q, r \in A[X]$ and $\deg r < \deg g$. Hence $f(\alpha) = r(\alpha)$, and we see that if $\deg g = n$, then $1, \alpha, \ldots, \alpha^{n-1}$ are generators of $A[\alpha]$ as a module over A.

An equation $g(X) = 0$ with g as above, such that $g(\alpha) = 0$ is called an *integral equation* for α over A.

Assume INT 2. We let the module be $A[\alpha]$ itself.

Assume INT 3, and let M be the faithful module over $A[\alpha]$ which is finitely generated over A, say by elements $w_1, \ldots, w_n$. Since $\alpha M \subset M$ there exist elements $a_{ij} \in A$ such that

$$\begin{aligned} \alpha w_1 &= a_{11}w_1 + \cdots + a_{1n}w_n, \\ &\cdots \\ \alpha w_n &= a_{n1}w_1 + \cdots + a_{nn}w_n. \end{aligned}$$

Transposing $\alpha w_1, \ldots, \alpha w_n$ to the right-hand side of these equations, we conclude that the determinant

$$d = \begin{vmatrix} \alpha - a_{11} & & & & \\ & \alpha - a_{22} & & -a_{ij} & \\ & & \cdot & & \\ & & & \cdot & \\ & -a_{ij} & & & \cdot \\ & & & & \cdot \\ & & & & \alpha - a_{nn} \end{vmatrix}$$

is such that $dM = 0$. (This will be proved in the chapter when we deal with determinants.) Since M is faithful, we must have $d = 0$. Hence α is a root of the polynomial

$$\det(X\,\delta_{ij} - a_{ij}),$$

which gives an integral equation for α over A.

An element α satisfying the three conditions INT 1, 2, 3 above is called *integral* over A.

PROPOSITION 1. *Let A be an entire ring and K its quotient field. Let α be algebraic over K. Then there exists an element $c \neq 0$ in A such that $c\alpha$ is integral over A.*

Proof. There exists an equation

$$a_n\alpha^n + a_{n-1}\alpha^{n-1} + \cdots + a_0 = 0$$

with $a_i \in A$ and $a_n \neq 0$. Multiply it by a_n^{n-1}. Then

$$(a_n\alpha)^n + \cdots + a_0 a_n^{n-1} = 0$$

is an integral equation for $a_n\alpha$ over A.

Let A, B be subrings of a ring C, and let $\alpha \in B$. If α is integral over A then α is *a fortiori* integral over C. Thus integrality is preserved under lifting. In particular, α is integral over any ring which is intermediate between A and B.

Let B be a ring containing A as a subring. We shall say that B is *integral* over A if every element of B is integral over A.

PROPOSITION 2. *If B is integral over A and finitely generated as an A-algebra, then B is finitely generated as an A-module.*

Proof. We may prove this by induction on the number of ring generators, and thus we may assume that $B = A[\alpha]$ for some element α integral over A, by considering a tower

$$A \subset A[\alpha_1] \subset A[\alpha_1, \alpha_2] \subset \cdots \subset A[\alpha_1, \ldots, \alpha_n] = B.$$

But we have already seen that our assertion is true in that case, this being part of the definition of integrality.

Just as we did for extension fields, one may define a class $\mathfrak{C}$ of extension rings $A \subset B$ to be *distinguished* if it satisfies the analogous properties, namely:

(i) Let $A \subset B \subset C$ be a tower of rings. The extension $A \subset C$ is in $\mathfrak{C}$ if and only if $A \subset B$ is in $\mathfrak{C}$ and $B \subset C$ is in $\mathfrak{C}$.

(ii) If $A \subset B$ is in $\mathfrak{C}$, if C is any extension ring of A, and if B, C are both subrings of some ring, then $B \subset B[C]$ is in $\mathfrak{C}$. (We note that $B[C] = C[B]$ is the smallest ring containing both B and C.)

As with fields, we find formally as a consequence of (i) and (ii) that (iii) holds, namely:

(iii) If $A \subset B$ and $A \subset C$ are in $\mathfrak{C}$, and B, C are subrings of some ring, then $A \subset B[C]$ is in $\mathfrak{C}$.

PROPOSITION 3. *Integral ring extensions form a distinguished class.*

Proof. Let $A \subset B \subset C$ be a tower of rings. If C is integral over A then it is clear that B is integral over A and C is integral over B. Conversely, assume that each step in the tower is integral. Let $\alpha \in C$. Then α satisfies an integral equation

$$\alpha^n + b_{n-1}\alpha^{n-1} + \cdots + b_0 = 0$$

with $b_i \in B$. Let $B_1 = A[b_0, \ldots, b_{n-1}]$. Then B_1 is a finitely generated A-module by Proposition 2, and is obviously faithful. Then $B_1[\alpha]$ is finite over B_1, hence over A, and hence α is integral over A. Hence C is integral over A. Finally let B, C be extension rings of A and assume B integral over A. Assume that B, C are subrings of some ring. Then $C[B]$ is generated by elements of B over C, and each element of B is integral over C. That $C[B]$ is integral over C will follow immediately from our next proposition.

PROPOSITION 4. *Let A be a subring of a ring C. Then the elements of C which are integral over A form a subring of C.*

Proof. Let $\alpha, \beta \in C$ be integral over A. Let $M = A[\alpha]$ and $N = A[\beta]$. Then MN contains 1, and is therefore faithful as an A-module. Furthermore, $\alpha M \subset M$ and $\beta N \subset N$. Hence MN is mapped into itself by multiplication with $\alpha \pm \beta$ and $\alpha\beta$. Furthermore MN is finitely generated over A (if $\{w_i\}$ are generators of M and $\{v_j\}$ are generators of N then $\{w_i v_j\}$ are generators of MN). This proves our proposition.

In Proposition 4, the set of elements of C which are integral over A is called the *integral closure of A in C.*

PROPOSITION 5. *Let $A \subset B$ be an extension ring, and let B be integral over A. Let σ be a homomorphism of B. Then $\sigma(B)$ is integral over $\sigma(A)$.*

Proof. Let $\alpha \in B$, and let

$$\alpha^n + a_{n-1}\alpha^{n-1} + \cdots + a_0 = 0$$

be an integral equation for α over A. Applying σ yields

$$\sigma(\alpha)^n + \sigma(a_{n-1})\sigma(\alpha)^{n-1} + \cdots + \sigma(a_0) = 0,$$

thereby proving our assertion.

COROLLARY. *Let A be an entire ring, k its quotient field, and E a finite extension of k. Let $\alpha \in E$ be integral over A. Then the norm and trace of α (from E to k) are integral over A, and so are the coefficients of the irreducible polynomial satisfied by α over k.*

Proof. For each embedding σ of E over k, $\sigma\alpha$ is integral over A. Since the norm is the product of $\sigma\alpha$ over all such σ (raised to a power of the characteristic), it follows that the norm is integral over A. Similarly for the trace, and similarly for the coefficients of $\mathrm{Irr}(\alpha, k, X)$, which are elementary symmetric functions of the roots.

Let A be an entire ring and k its quotient field. We say that A is *integrally closed* if it is equal to its integral closure in k.

PROPOSITION 6. *Let A be entire and factorial. Then A is integrally closed.*

Proof. Suppose that there exists a quotient a/b with $a, b \in A$ which is integral over A, and a prime element p in A which divides b but not a. We have, for some integer $n \geqq 1$, and $a_i \in A$,

$$(a/b)^n + a_{n-1}(a/b)^{n-1} + \cdots + a_0 = 0$$

whence

$$a^n + a_{n-1}ba^{n-1} + \cdots + a_0b^n = 0.$$

Since p divides b, it must divide a^n, and hence must divide a, contradiction.

Let $f : A \to B$ be a ring-homomorphism (A, B being commutative rings). We recall that such a homomorphism is also called an A-*algebra*. We may view B as an A-module. We say that B is integral over A (for this ring-homomorphism f) if B is integral over $f(A)$. This extension of our definition of integrality is useful because there are applications when certain collapsings take place, and we still wish to speak of integrality. Strictly speaking we should not say that B is integral over A, but that f *is an integral ring-homomorphism*, or simply that f is *integral*. We shall use this terminology frequently.

Some of our preceding propositions have immediate consequences for integral ring-homomorphisms; for instance, if $f : A \to B$ and $g : B \to C$ are integral, then $g \circ f : A \to C$ is integral. However, it is not necessarily true that if $g \circ f$ is integral, so is f.

Let $f : A \to B$ be integral, and let S be a multiplicative subset of A. Then we get a homomorphism

$$S^{-1}f : S^{-1}A \to S^{-1}B,$$

where strictly speaking, $S^{-1}B = (f(S))^{-1}B$, and $S^{-1}f$ is defined by

$$(S^{-1}f)(x/s) = f(x)/f(s).$$

It is trivially verified that this is a homomorphism. We have a commutative diagram

$$\begin{array}{ccc} B & \to & S^{-1}B \\ {\scriptstyle f}\uparrow & & \uparrow{\scriptstyle S^{-1}f} \\ A & \to & S^{-1}A \end{array}$$

the horizontal maps being the canonical ones: $x \to x/1$.

Proposition 7. *Let $f : A \to B$ be integral, and let S be a multiplicative subset of A. Then $S^{-1}f : S^{-1}A \to S^{-1}B$ is integral.*

Proof. If $\alpha \in B$ is integral over $f(A)$, then writing $a\beta$ instead of $f(a)\beta$ for $a \in A$ and $\beta \in B$ we have

$$\alpha^n + a_{n-1}\alpha^{n-1} + \cdots + a_0 = 0$$

with $a_i \in A$. Taking the canonical image in $S^{-1}A$ and $S^{-1}B$ respectively, we see that this relation proves the integrality of $\alpha/1$ over $S^{-1}A$, the coefficients being now $a_i/1$.

PROPOSITION 8. *Let A be entire and integrally closed. Let S be a multiplicative subset of A, $0 \notin S$. Then $S^{-1}A$ is integrally closed.*

Proof. Let α be an element of the quotient field, integral over $S^{-1}A$. We have an equation

$$\alpha^n + \frac{a_{n-1}}{s_{n-1}}\alpha^{n-1} + \cdots + \frac{a_0}{s_0} = 0,$$

$a_i \in A$ and $s_i \in S$. Let s be the product $s_{n-1} \cdots s_0$. Then it is clear that $s\alpha$ is integral over A, whence in A. Hence α lies in $S^{-1}A$, and $S^{-1}A$ is integrally closed.

NAKAYAMA'S LEMMA. *Let A be a ring, and $\mathfrak{a}$ an ideal contained in all maximal ideals of A. Let M be a finitely generated A-module. If $\mathfrak{a}M = M$ then $M = 0$.*

Proof. By induction on the number of generators of M. Say M is generated by $w_1, \ldots, w_n$. There exists an expression

$$w_1 = a_1w_1 + \cdots + a_nw_n$$

with $a_i \in \mathfrak{a}$. Hence

$$(1 - a_1)w_1 = a_2w_2 + \cdots + a_nw_n.$$

If $1 - a_1$ is not a unit in A, then it is contained in a maximal ideal $\mathfrak{p}$. Since $a_1 \in \mathfrak{p}$ by hypothesis, we have a contradiction, $1 \in \mathfrak{p}$. Hence $1 - a_1$ is a unit, and dividing by it shows that M can be generated by $n - 1$ elements, thereby concluding the proof.

Let $\mathfrak{p}$ be a prime ideal of a ring A and let S be the complement of $\mathfrak{p}$ in A. We write $S = A - \mathfrak{p}$. If $f : A \to B$ is an A-algebra (i.e. a ring-homomorphism), we shall write $B_\mathfrak{p}$ instead of $S^{-1}B$. We can view $B_\mathfrak{p}$ as an $A_\mathfrak{p} = S^{-1}A$-module.

Let A be a subring of B. Let $\mathfrak{p}$ be a prime ideal of A and let $\mathfrak{P}$ be a prime ideal of B. We say that $\mathfrak{P}$ *lies above* $\mathfrak{p}$ if $\mathfrak{P} \cap A = \mathfrak{p}$. If that is the case, then the injection $A \to B$ induces an injection of the factor rings

$$A/\mathfrak{p} \to B/\mathfrak{P},$$

and in fact we have a commutative diagram:

$$\begin{array}{ccc} B & \to & B/\mathfrak{P} \\ \uparrow & & \uparrow \\ A & \to & A/\mathfrak{p} \end{array}$$

the horizontal arrows being the canonical homomorphisms, and the vertical arrows being injections.

If B is integral over A, then $B/\mathfrak{P}$ is integral over $A/\mathfrak{p}$ by Proposition 5.

PROPOSITION 9. *Let A be a subring of B, let $\mathfrak{p}$ be a prime ideal of A, and assume B integral over A. Then $\mathfrak{p}B \neq B$ and there exists a prime ideal $\mathfrak{P}$ of B lying above $\mathfrak{p}$.*

Proof. We know that $B_{\mathfrak{p}}$ is integral over $A_{\mathfrak{p}}$ and that $A_{\mathfrak{p}}$ is a local ring with maximal ideal $\mathfrak{m}_{\mathfrak{p}} = S^{-1}\mathfrak{p}$, where $S = A - \mathfrak{p}$. Since we obviously have

$$\mathfrak{p}B_{\mathfrak{p}} = \mathfrak{p}A_{\mathfrak{p}}B_{\mathfrak{p}} = \mathfrak{m}_{\mathfrak{p}}B_{\mathfrak{p}},$$

it will suffice to prove our first assertion when A is a local ring. (Note that the existence of a prime ideal $\mathfrak{p}$ implies that $1 \neq 0$, and $\mathfrak{p}B = B$ if and only if $1 \in \mathfrak{p}B$.) In that case, if $\mathfrak{p}B = B$, then 1 has an expression as a finite linear combination of elements of B with coefficients in $\mathfrak{p}$,

$$1 = a_1b_1 + \cdots + a_nb_n$$

with $a_i \in \mathfrak{p}$ and $b_i \in B$. We shall now use notation as if $A_{\mathfrak{p}} \subset B_{\mathfrak{p}}$. We leave it to the reader as an exercise to verify that our arguments are valid when we deal only with a canonical homomorphism $A_{\mathfrak{p}} \to B_{\mathfrak{p}}$. Let $B_0 = A[b_1, \ldots, b_n]$. Then $\mathfrak{p}B_0 = B_0$ and B_0 is a finite A-module by Proposition 2. Hence $B_0 = 0$ by Nakayama's lemma, contradiction.

To prove our second assertion, note the following commutative diagram:

$$\begin{array}{ccc} B & \to & B_{\mathfrak{p}} \\ \uparrow & & \uparrow \\ A & \to & A_{\mathfrak{p}} \end{array}$$

We have just proved $\mathfrak{m}_{\mathfrak{p}}B_{\mathfrak{p}} \neq B_{\mathfrak{p}}$. Hence $\mathfrak{m}_{\mathfrak{p}}B_{\mathfrak{p}}$ is contained in a maximal ideal $\mathfrak{M}$ of $B_{\mathfrak{p}}$. Taking inverse images, we see that the inverse image of $\mathfrak{M}$ in $A_{\mathfrak{p}}$ is an ideal containing $\mathfrak{m}_{\mathfrak{p}}$ (in the case of an inclusion $A_{\mathfrak{p}} \subset B_{\mathfrak{p}}$ the inverse image is $\mathfrak{M} \cap A_{\mathfrak{p}}$). Since $\mathfrak{m}_{\mathfrak{p}}$ is maximal, we have $\mathfrak{M} \cap A_{\mathfrak{p}} = \mathfrak{m}_{\mathfrak{p}}$. Let $\mathfrak{P}$ be the inverse image of $\mathfrak{M}$ in B (in the case of inclusion, $\mathfrak{P} = \mathfrak{M} \cap B$). Then $\mathfrak{P}$ is a prime ideal of B. The inverse image of $\mathfrak{m}_{\mathfrak{p}}$ in A is simply $\mathfrak{p}$. Taking the inverse image of $\mathfrak{M}$ going around both ways in the diagram, we find that

$$\mathfrak{P} \cap A = \mathfrak{p},$$

as was to be shown.

PROPOSITION 10. *Let A be a subring of B, and assume that B is integral over A. Let $\mathfrak{P}$ be a prime ideal of B lying over a prime ideal $\mathfrak{p}$ of A. Then $\mathfrak{P}$ is maximal if and only if $\mathfrak{p}$ is maximal.*

Proof. Assume $\mathfrak{p}$ maximal in A. Then $A/\mathfrak{p}$ is a field, and $B/\mathfrak{P}$ is an entire ring, integral over $A/\mathfrak{p}$. If $\alpha \in B/\mathfrak{P}$, then α is algebraic over $A/\mathfrak{p}$, and we know that $A/\mathfrak{p}[\alpha]$ is a field. Hence every non-zero element of $B/\mathfrak{P}$ is invertible in $B/\mathfrak{P}$, which is therefore a field. Conversely, assume that $\mathfrak{P}$ is maximal in B. Then $B/\mathfrak{P}$ is a field, which is integral over the entire ring $A/\mathfrak{p}$. If $A/\mathfrak{p}$ is not a field, it has a non-zero maximal ideal $\mathfrak{m}$. By Proposition 9, there exists a prime ideal $\mathfrak{M}$ of $B/\mathfrak{P}$ lying above $\mathfrak{m}$, $\mathfrak{M} \neq 0$, contradiction.

§2. *Integral Galois extensions*

We shall now investigate the relationship between the Galois theory of a polynomial, and the Galois theory of this same polynomial reduced modulo a prime ideal.

PROPOSITION 11. *Let A be an entire ring, integrally closed in its quotient field K. Let L be a finite Galois extension of K with group G. Let $\mathfrak{p}$ be a maximal ideal of A, and let $\mathfrak{P}$, $\mathfrak{Q}$ be prime ideals of the integral closure B of A in L lying above $\mathfrak{p}$. Then there exists $\sigma \in G$ such that $\sigma\mathfrak{P} = \mathfrak{Q}$.*

Proof. Suppose that $\mathfrak{Q} \neq \sigma\mathfrak{P}$ for any $\sigma \in G$. Then $\tau\mathfrak{Q} \neq \sigma\mathfrak{P}$ for any pair of elements $\sigma, \tau \in G$. There exists an element $x \in B$ such that

$$x \equiv 0 \pmod{\sigma\mathfrak{P}}, \qquad \text{all } \sigma \in G$$
$$x \equiv 1 \pmod{\sigma\mathfrak{Q}}, \qquad \text{all } \sigma \in G$$

(use the Chinese remainder theorem). The norm

$$N_K^L(x) = \prod_{\sigma \in G} \sigma x$$

lies in $B \cap K = A$ (because A is integrally closed), and lies in $\mathfrak{P} \cap A = \mathfrak{p}$. But $x \notin \sigma\mathfrak{Q}$ for all $\sigma \in G$, so that $\sigma x \notin \mathfrak{Q}$ for all $\sigma \in G$. This contradicts the fact that the norm of x lies in $\mathfrak{p} = \mathfrak{Q} \cap A$.

If one localizes, one can eliminate the hypothesis that $\mathfrak{p}$ is maximal; just assume that $\mathfrak{p}$ is prime.

COROLLARY. *Let A be a ring, integrally closed in its quotient field K. Let E be a finite separable extension of K, and B the integral closure of A in E. Let $\mathfrak{p}$ be a maximal ideal of A. Then there exists only a finite number of prime ideals of B lying above $\mathfrak{p}$.*

Proof. Let L be the smallest Galois extension of K containing E. If $\mathfrak{Q}_1$, $\mathfrak{Q}_2$ are two distinct prime ideals of B lying above $\mathfrak{p}$, and $\mathfrak{P}_1$, $\mathfrak{P}_2$ are two prime ideals of the integral closure of A in L lying above $\mathfrak{Q}_1$ and $\mathfrak{Q}_2$ respectively, then $\mathfrak{P}_1 \neq \mathfrak{P}_2$. This argument reduces our assertion to the case that E is Galois over K, and it then becomes an immediate consequence of the proposition.

Let A be integrally closed in its quotient field K, and let B be its integral closure in a finite Galois extension L, with group G. Then $\sigma B = B$ for every $\sigma \in G$. Let $\mathfrak{p}$ be a maximal ideal of A, and $\mathfrak{P}$ a maximal ideal of B lying above $\mathfrak{p}$. We denote by $G_{\mathfrak{P}}$ the subgroup of G consisting of those automorphisms such that $\sigma\mathfrak{P} = \mathfrak{P}$. Then $G_{\mathfrak{P}}$ operates in a natural way on the residue class field $B/\mathfrak{P}$, and leaves $A/\mathfrak{p}$ fixed. To each $\sigma \in G_{\mathfrak{P}}$ we can associate an automorphism σ' of $B/\mathfrak{P}$ over $A/\mathfrak{p}$, and the map given by

$$\sigma \mapsto \sigma'$$

induces a homomorphism of G into the group of automorphisms of $B/\mathfrak{P}$ over $A/\mathfrak{p}$.

The group $G_{\mathfrak{P}}$ will be called the *decomposition group* of $\mathfrak{P}$. Its fixed field will be denoted by L^d, and will be called the *decomposition field* of $\mathfrak{P}$. Let B^d be the integral closure of A in L^d, and let $\mathfrak{Q} = \mathfrak{P} \cap B^d$. By Proposition 11, we know that $\mathfrak{P}$ is the only prime of B lying above $\mathfrak{Q}$.

Let $G = \bigcup \sigma_j G_{\mathfrak{P}}$ be a coset decomposition of $G_{\mathfrak{P}}$ in G. Then the prime ideals $\sigma_j\mathfrak{P}$ are precisely the distinct primes of B lying above $\mathfrak{p}$. Indeed, for two elements $\sigma, \tau \in G$ we have $\sigma\mathfrak{P} = \tau\mathfrak{P}$ if and only if $\tau^{-1}\sigma\mathfrak{P} = \mathfrak{P}$, i.e. $\tau^{-1}\sigma$ lies in $G_{\mathfrak{P}}$. Thus τ, σ lie in the same coset mod $G_{\mathfrak{P}}$.

It is then immediately clear that the decomposition group of a prime $\sigma\mathfrak{P}$ is $\sigma G_{\mathfrak{P}} \sigma^{-1}$.

PROPOSITION 12. *The field L^d is the smallest subfield E of L containing K such that $\mathfrak{P}$ is the only prime of B lying above $\mathfrak{P} \cap E$ (which is prime in $B \cap E$).*

Proof. Let E be as above, and let H be the Galois group of L over E. Let $\mathfrak{q} = \mathfrak{P} \cap E$. By Proposition 11, all primes of B lying above $\mathfrak{q}$ are conjugate by elements of H. Since there is only one prime, namely $\mathfrak{P}$, it means that H leaves $\mathfrak{P}$ invariant. Hence $H \subset G_{\mathfrak{P}}$ and $E \supset L^d$. We have already observed that L^d has the required property.

PROPOSITION 13. *Notation being as above, we have $A/\mathfrak{p} = B^d/\mathfrak{Q}$ (under the canonical injection $A/\mathfrak{p} \to B^d/\mathfrak{Q}$).*

Proof. If σ is an element of G, not in $G_{\mathfrak{P}}$, then $\sigma\mathfrak{P} \neq \mathfrak{P}$ and $\sigma^{-1}\mathfrak{P} \neq \mathfrak{P}$. Let

$$\mathfrak{Q}_\sigma = \sigma^{-1}\mathfrak{P} \cap B^d.$$

Then $\mathfrak{Q}_\sigma \neq \mathfrak{Q}$. Let x be an element of B^d. There exists an element y of B^d such that

$$y \equiv x \pmod{\mathfrak{Q}}$$
$$y \equiv 1 \pmod{\mathfrak{Q}_\sigma}$$

for each σ in G, but not in $G_{\mathfrak{P}}$. Hence in particular,

$$\begin{aligned} y &\equiv x \pmod{\mathfrak{P}} \\ y &\equiv 1 \pmod{\sigma^{-1}\,\mathfrak{P}} \end{aligned}$$

for each σ not in $G_{\mathfrak{P}}$. This second congruence yields

$$\sigma y \equiv 1 \pmod{\mathfrak{P}}$$

for all $\sigma \notin G_{\mathfrak{P}}$. The norm of y from L^d to K is a product of y and other factors σy with $\sigma \notin G_{\mathfrak{P}}$. Thus we obtain

$$N_K^{L^d}(y) \equiv x \pmod{\mathfrak{P}}.$$

But the norm lies in K, and even in A, since it is a product of elements integral over A. This last congruence holds mod $\mathfrak{Q}$, since both x and the norm lie in B^d. This is precisely the meaning of the assertion in our proposition.

If x is an element of B, we shall denote by x' its image under the homomorphism $B \to B/\mathfrak{P}$. Then σ' is the automorphism of $B/\mathfrak{P}$ satisfying the relation

$$\sigma' x' = (\sigma x)'.$$

If $f(X)$ is a polynomial with coefficients in B, we denote by $f'(X)$ its natural image under the above homomorphism. Thus, if

$$f(X) = b_n X^n + \cdots + b_0,$$

then

$$f'(X) = b'_n X^n + \cdots + b'_0.$$

Proposition 14. *Let A be integrally closed in its quotient field K, and let B be its integral closure in a finite Galois extension L of K, with group G. Let $\mathfrak{p}$ be a maximal ideal of A, and $\mathfrak{P}$ a maximal ideal of B lying above $\mathfrak{p}$. Then $B/\mathfrak{P}$ is a normal extension of $A/\mathfrak{p}$, and the map $\sigma \mapsto \sigma'$ induces a homomorphism of $G_{\mathfrak{P}}$ onto the Galois group of $B/\mathfrak{P}$ over $A/\mathfrak{p}$.*

Proof. Let $B' = B/\mathfrak{P}$ and $A' = A/\mathfrak{p}$. Any element of B' can be written as x' for some $x \in B$. Let x' generate a separable subextension of B' over A', and let f be the irreducible polynomial for x over K. The coefficients of f lie in A because x is integral over A, and all the roots of f are integral over A. Thus

$$f(X) = \prod_{i=1}^{m} (X - x_i)$$

splits into linear factors in B. Since

$$f'(X) = \prod_{i=1}^{m} (X - x'_i)$$

and all the x_i' lie in B', it follows that f' splits into linear factors in B'. We observe that $f(x) = 0$ implies $f'(x') = 0$. Hence B' is normal over A', and

$$[A'(x') : A'] \leqq [K(x) : K] \leqq [L : K].$$

This implies that the maximal separable subextension of A' in B' is of finite degree over A' (using the primitive element theorem of elementary field theory). This degree is in fact bounded by $[L : K]$.

There remains to prove that the map $\sigma \mapsto \sigma'$ gives a surjective homomorphism of $G_{\mathfrak{P}}$ onto the Galois group of B' over A'. To do this, we shall give an argument which reduces our problem to the case when $\mathfrak{P}$ is the only prime ideal of B lying above $\mathfrak{p}$. Indeed, by Proposition 13, the residue class fields of the ground ring and the ring B^d in the decomposition field are the same. This means that to prove our surjectivity, we may take L^d as ground field. This is the desired reduction, and we can assume $K = L^d$, $G = G_{\mathfrak{P}}$.

This being the case, take a generator of the maximal separable subextension of B' over A', and let it be x', for some element x in B. Let f be the irreducible polynomial of x over K. Any automorphism of B' is determined by its effect on x', and maps x' on some root of f'. Suppose that $x = x_1$. Given any root x_i of f, there exists an element σ of $G = G_{\mathfrak{P}}$ such that $\sigma x = x_i$. Hence $\sigma' x' = x_i'$. Hence the automorphisms of B' over A' induced by elements of G operate transitively on the roots of f'. Hence they give us all automorphisms of the residue class field, as was to be shown.

Corollary 1. *Let A be a ring integrally closed in its quotient field K. Let L be a finite Galois extension of K, and B the integral closure of A in L. Let $\mathfrak{p}$ be a maximal ideal of A. Let $\varphi : A \to A/\mathfrak{p}$ be the canonical homomorphism, and let ψ_1, ψ_2 be two homomorphisms of B extending φ in a given algebraic closure of $A/\mathfrak{p}$. Then there exists an automorphism σ of L over K such that*

$$\psi_1 = \psi_2 \circ \sigma.$$

Proof. The kernels of ψ_1, ψ_2 are prime ideals of B which are conjugate by Proposition 11. Hence there exists an element τ of the Galois group G such that $\psi_1, \psi_2 \circ \tau$ have the same kernel. Without loss of generality, we may therefore assume that ψ_1, ψ_2 have the same kernel $\mathfrak{P}$. Hence there exists an automorphism ω of $\psi_1(B)$ onto $\psi_2(B)$ such that $\omega \circ \psi_1 = \psi_2$. There exists an element σ of $G_{\mathfrak{P}}$ such that $\omega \circ \psi_1 = \psi_1 \circ \sigma$, by the preceding proposition. This proves what we wanted.

Remark. In all the above propositions, we could assume $\mathfrak{p}$ prime instead of maximal. In that case, one has to localize at $\mathfrak{p}$ to be able to apply our proofs.

In the above discussions, the kernel of the map

$$G_{\mathfrak{P}} \to G'_{\mathfrak{P}}$$

is called the *inertia group* of $\mathfrak{P}$. It consists of those automorphisms of $G_{\mathfrak{P}}$ which induce the trivial automorphism on the residue class field. Its fixed field is called the *inertia field*, and is denoted by L^t.

COROLLARY 2. *Let the assumptions be as in Corollary* 1, *and assume that* $\mathfrak{P}$ *is the only prime of* B *lying above* $\mathfrak{p}$. *Let* $f(X)$ *be a polynomial in* $A[X]$ *with leading coefficient* 1. *Assume that* f *is irreducible in* $K[X]$, *and has a root* α *in* B. *Then the reduced polynomial* f' *is a power of an irreducible polynomial in* $A'[X]$.

Proof. By Corollary 1, we know that any two roots of f' are conjugate under some isomorphism of B' over A', and hence that f' cannot split into relative prime polynomials. Therefore, f' is a power of an irreducible polynomial.

PROPOSITION 15. *Let* A *be an entire ring, integrally closed in its quotient field* K. *Let* L *be a finite Galois extension of* K. *Let* $L = K(\alpha)$, *where* α *is integral over* A, *and let*

$$f(X) = X^n + a_{n-1}X^{n-1} + \cdots + a_0$$

be the irreducible polynomial of α *over* k, *with* $a_i \in A$. *Let* $\mathfrak{p}$ *be a maximal ideal in* A, *let* $\mathfrak{P}$ *be a prime ideal of the integral closure* B *of* A *in* L, $\mathfrak{P}$ *lying above* $\mathfrak{p}$. *Let* $f'(X)$ *be the reduced polynomial with coefficients in* $A/\mathfrak{p}$. *Let* $G_{\mathfrak{P}}$ *be the decomposition group. If* f' *has no multiple roots, then the map* $\sigma \mapsto \sigma'$ *has trivial kernel, and is an isomorphism of* $G_{\mathfrak{P}}$ *on the Galois group of* f' *over* $A/\mathfrak{p}$.

Proof. Let

$$f(X) = \prod(X - x_i)$$

be the factorization of f in L. We know that all $x_i \in B$. If $\sigma \in G_{\mathfrak{P}}$, then we denote by σ' the homomorphic image of σ in the group $G'_{\mathfrak{P}}$, as before. We have

$$f'(X) = \prod (X - x'_i).$$

Suppose that $\sigma' x'_i = x'_i$ for all i. Since $(\sigma x_i)' = \sigma' x'_i$, and since f' has no multiple roots, it follows that σ is also the identity. Hence our map is injective, the inertia group is trivial. The field $A'[x'_1, \ldots, x'_n]$ is a subfield of B' and any automorphism of B' over A' which restricts to the identity on this subfield must be the identity, because the map $G_{\mathfrak{P}} \to G'_{\mathfrak{P}}$ is onto the Galois group of B' over A'. Hence B' is purely inseparable over $A'[x'_1, \ldots, x'_n]$ and therefore $G_{\mathfrak{P}}$ is isomorphic to the Galois group of f' over A'.

Proposition 15 gives a very efficient tool for analyzing polynomials over a ring. For instance, consider the "generic" polynomial

$$f_w(X) = X^n + w_{n-1}X^{n-1} + \cdots + w_0$$

where $w_0, \ldots, w_{n-1}$ are algebraically independent over a field k. We know that the Galois group of this polynomial over $k(w_0, \ldots, w_{n-1})$ is the symmetric group. Let $t_1, \ldots, t_n$ be the roots. Let α be a generator of the splitting field. Without loss of generality, we can select α to be integral over the ring $k[w_0, \ldots, w_{n-1}]$ (multiply any given generator by a suitably chosen polynomial and use Proposition 1). Let $g_w(X)$ be the irreducible polynomial of α over $k(w_0, \ldots, w_{n-1})$. The coefficients of g are polynomials in (w). If we can substitute values (a) for (w) with $a_0, \ldots, a_{n-1} \in k$ such that g_a remains irreducible, then by Proposition 15 we conclude at once that the Galois group of g_a is the symmetric group also. Similarly, if every field between $k(w_0, \ldots, w_{n-1})$ and $k(t_1, \ldots, t_n)$ is generated by n algebraically independent elements, then we can do a similar substitution to get extensions with given Galois groups. Whether this can be done is one of the major unsolved problems of Galois theory. It is essentially the parametrization of all Galois extensions by independent elements.

As another example, consider $X^5 - X - 1$ over $\mathbf{Z}$. Reducing mod 5 shows that this polynomial is irreducible. Reducing mod 2 gives the irreducible factors

$$(X^2 + X + 1)(X^3 + X^2 + 1) \pmod 2.$$

Hence the Galois group over the rationals contains a 5-cycle and a product of a 2-cycle and a 3-cycle. It is an easy matter to see that this implies that the group must be the full symmetric group.

§3. *Extension of homomorphisms*

When we first discussed the process of localization, we considered very briefly the extension of a homomorphism to a local ring. In our discussion of field theory, we also described an extension theorem for embeddings of one field into another. We shall now treat the extension question in full generality.

First we recall the case of a local ring. Let A be a ring and $\mathfrak{p}$ a prime ideal. We know that the local ring $A_\mathfrak{p}$ is the set of all fractions x/y, with $x, y \in A$ and $y \notin \mathfrak{p}$. Its maximal ideal consists of those fractions with $x \in \mathfrak{p}$. Let L be a field and let $\varphi : A \to L$ be a homomorphism whose kernel is $\mathfrak{p}$. Then we can extend φ to a homomorphism of $A_\mathfrak{p}$ into L by letting

$$\varphi(x/y) = \varphi(x)/\varphi(y)$$

if x/y is an element of $A_\mathfrak{p}$ as above.

Second, we have integral ring extensions. Let $\mathfrak{o}$ be a local ring with maximal ideal $\mathfrak{m}$, let B be integral over $\mathfrak{o}$, and let $\varphi : \mathfrak{o} \to L$ be a homomorphism of $\mathfrak{o}$ into an algebraically closed field L. We assume that the kernel of φ is $\mathfrak{m}$. By Proposition 9 of §1, we know that there exists a maximal ideal $\mathfrak{M}$ of B lying above $\mathfrak{m}$, i.e. such that $\mathfrak{M} \cap \mathfrak{o} = \mathfrak{m}$. Then $B/\mathfrak{M}$ is a field, which is an algebraic extension of $\mathfrak{o}/\mathfrak{m}$, and $\mathfrak{o}/\mathfrak{m}$ is isomorphic to the subfield $\varphi(\mathfrak{o})$ of L because the kernel of φ is $\mathfrak{m}$.

We can find an isomorphism of $\mathfrak{o}/\mathfrak{m}$ onto $\varphi(\mathfrak{o})$ such that the composite homomorphism

$$\mathfrak{o} \to \mathfrak{o}/\mathfrak{m} \to L$$

is equal to φ. We now embed $B/\mathfrak{M}$ into L so as to make the following diagram commutative:

$$\begin{array}{ccccc} B & \to & B/\mathfrak{M} & & \\ \uparrow & & \uparrow & \searrow & \\ \mathfrak{o} & \to & \mathfrak{o}/\mathfrak{m} & \to & L \end{array}$$

and in this way get a homomorphism of B into L which extends φ.

PROPOSITION 16. *Let A be a subring of B and assume that B is integral over A. Let $\varphi : A \to L$ be a homomorphism into a field L which is algebraically closed. Then φ has an extension to a homomorphism of B into L.*

Proof. Let $\mathfrak{p}$ be the kernel of φ and let S be the complement of $\mathfrak{p}$ in A. Then we have a commutative diagram

$$\begin{array}{ccc} B & \to & S^{-1}B \\ \uparrow & & \uparrow \\ A & \to & S^{-1}A = A_{\mathfrak{p}} \end{array}$$

and φ can be factored through the canonical homomorphism of A into $S^{-1}A$. Furthermore, $S^{-1}B$ is integral over $S^{-1}A$. This reduces the question to the case when we deal with a local ring, which has just been discussed above.

THEOREM 1. *Let A be a subring of a field K and let $x \in K$, $x \neq 0$. Let $\varphi : A \to L$ be a homomorphism of A into an algebraically closed field L. Then φ has an extension to a homomorphism of $A[x]$ or $A[x^{-1}]$ into L.*

Proof. We may first extend φ to a homomorphism of the local ring $A_{\mathfrak{p}}$, where $\mathfrak{p}$ is the kernel of φ. Thus without loss of generality, we may assume that A is a local ring with maximal ideal $\mathfrak{m}$. Suppose that

$$\mathfrak{m}A[x^{-1}] = A[x^{-1}].$$

Then we can write

$$1 = a_0 + a_1x^{-1} + \cdots + a_nx^{-n}$$

with $a_i \in \mathfrak{m}$. Multiplying by x^n we obtain

$$(1 - a_0)x^n + b_{n-1}x^{n-1} + \cdots + b_0 = 0$$

with suitable elements $b_i \in A$. Since $a_0 \in \mathfrak{m}$, it follows that $1 - a_0 \notin \mathfrak{m}$ and hence $1 - a_0$ is a unit in A because A is assumed to be a local ring. Dividing by $1 - a_0$ we see that x is integral over A, and hence that our homomorphism has an extension to $A[x]$.

If on the other hand we have

$$\mathfrak{m}A[x^{-1}] \neq A[x^{-1}]$$

then $\mathfrak{m}A[x^{-1}]$ is contained in some maximal ideal $\mathfrak{P}$ of $A[x^{-1}]$ and $\mathfrak{P} \cap A$ contains $\mathfrak{m}$. Since $\mathfrak{m}$ is maximal, we must have $\mathfrak{P} \cap \mathrm{A} = \mathfrak{m}$. Since φ and the canonical map $A \to A/\mathfrak{m}$ have the same kernel, namely $\mathfrak{m}$, we can find an embedding ψ of $A/\mathfrak{m}$ into L such that the composite map

$$A \to A/\mathfrak{m} \xrightarrow{\psi} L$$

is equal to φ. We note that $A/\mathfrak{m}$ is canonically embedded in $B/\mathfrak{P}$ where $B = A[x^{-1}]$, and extend ψ to a homomorphism of $B/\mathfrak{P}$ into L, which we can do whether the image of x^{-1} in $B/\mathfrak{P}$ is transcendental or algebraic over $A/\mathfrak{m}$. The composite $B \to B/\mathfrak{P} \to L$ gives us what we want.

COROLLARY. *Let A be a subring of a field K and let L be an algebraically closed field. Let $\varphi : A \to L$ be a homomorphism. Let B be a maximal subring of K to which φ has an extension homomorphism into L. Then B is a local ring and if $x \in K$, $x \neq 0$, then $x \in B$ or $x^{-1} \in B$.*

Proof. Let S be the set of pairs (C, ψ) where C is a subring of K and $\psi : C \to L$ is a homomorphism extending φ. Then S is not empty [containing (A, φ)], and is partially ordered by ascending inclusion and restriction. In other words, $(C, \psi) \leqq (C', \psi')$ if $C \subset C'$ and the restriction of ψ' to C is equal to ψ. It is clear that S is inductively ordered, and by Zorn's lemma there exists a maximal element, say (B, ψ_0). Then first B is a local ring, otherwise ψ_0 extends to the local ring arising from the kernel, and second, B has the desired property according to Theorem 1.

EXERCISES

1. What is the Galois group over the rationals of

$$X^4 + 2X^2 + X + 3 = 0?$$

2. Exhibit a polynomial of degree 4 whose Galois group is the symmetric group on 4 elements.

3. Let K be a Galois extension of the rationals $\mathbf{Q}$, with group G. Let B be the integral closure of $\mathbf{Z}$ in K, and let $\alpha \in B$ be such that $K = \mathbf{Q}(\alpha)$. Let $f(X) = \operatorname{Irr}(\alpha, \mathbf{Q}, X)$. Let p be a prime number, and assume that f remains irreducible mod p over $\mathbf{Z}/p\mathbf{Z}$, and that its reduction mod p has no multiple roots. What can you say about the Galois group G?

4. Let A be an entire ring and K its quotient field. Let t be transcendental over K. If A is integrally closed, show that $A[t]$ is integrally closed.

5. Let A be an entire ring, integrally closed in its quotient field K. Let L be a finite separable extension of K, and let B be the integral closure of A in L. If A is Noetherian, show that B is a finite A-module. [*Hint:* Let $\{\omega_1, \ldots, \omega_n\}$ be a basis of L over K. Multiplying all elements of this basis by a suitable element of A, we may assume without loss of generality that all ω_1 are integral over A. Let $\{\omega'_1, \ldots, \omega'_n\}$ be the dual basis relative to the trace, so that $\operatorname{Tr}(\omega_i\omega'_j) = \delta_{ij}$. Write an element α of L integral over A in the form

$$\alpha = b_1\omega'_1 + \cdots + b_n\omega'_n$$

with $b_j \in K$. Taking the trace $\operatorname{Tr}(\alpha\omega_i)$, for $i = 1, \ldots, n$, conclude that B is contained in the finite module $A\omega'_1 + \cdots + A\omega'_n$.]

6. The preceding exercise applies to the case when $A = \mathbf{Z}$ and $K = \mathbf{Q}$. A finite extension of $\mathbf{Q}$ is called a number field, and the integral closure of $\mathbf{Z}$ in such an extension L is called the ring of *algebraic integers* of L. We denote it by I_L. Let $\sigma_1, \ldots, \sigma_n$ be the distinct embeddings of L into the complex numbers. Embed I_L into a Euclidean space by the map

$$\alpha \mapsto (\sigma_1\alpha, \ldots, \sigma_n\alpha).$$

Show that in any bounded region of space, there is only a finite number of elements of I_L. [*Hint:* The coefficients in an integral equation for α are elementary symmetric functions of the conjugates of α and thus are bounded integers.] Use Exercise 10 of Chapter III to conclude that I_L is a free $\mathbf{Z}$-module of dimension $\leqq n$. In fact, show that the dimension is n, a basis of I_L over $\mathbf{Z}$ also being a basis of L over $\mathbf{Q}$.

7. Let E be a finite extension of $\mathbf{Q}$, and let I_E be the ring of algebraic integers of E. Let U be the group of units of I_E. Let $\sigma_1, \ldots, \sigma_n$ be the distinct embeddings of E into $\mathbf{C}$. Map U into a Euclidean space, by the map

$$l : \alpha \mapsto (\log |\sigma_1\alpha|, \ldots, \log |\sigma_n\alpha|).$$

Show that $l(U)$ is a free abelian group, finitely generated, by showing that in any finite region of space, there is only a finite number of elements of $l(U)$. Show that the kernel of l is a finite group, and is therefore the group of roots of unity in E. Thus U itself is a finitely generated abelian group.

8. Generalize the results of §2 to infinite Galois extensions, especially Propositions 11 and 14, using Zorn's lemma.

CHAPTER X

Transcendental Extensions

Throughout this chapter, the word "ring" means "commutative ring".

§1. Transcendence bases

Let K be an extension field of a field k. Let S be a subset of K. We recall that S (or the elements of S) is said to be algebraically independent over k, if whenever we have a relation

$$0 = \sum a_{(\nu)} M_{(\nu)}(S) = \sum a_{(\nu)} \prod_{x \in S} x^{\nu(x)}$$

with coefficients $a_{(\nu)} \in k$, almost all $a_{(\nu)} = 0$, then we must necessarily have all $a_{(\nu)} = 0$.

We can introduce an ordering among algebraically independent subsets of K, by ascending inclusion. These subsets are obviously inductively ordered, and thus there exist maximal elements. If S is a subset of K which is algebraically independent over k, and if the cardinality of S is greatest among all such subsets, then we call this cardinality the *transcendence degree* or *dimension* of K over k. Actually, we shall need to distinguish only between finite transcendence degree or infinite transcendence degree. We observe that the notion of transcendence degree bears to the notion of algebraic independence the same relation as the notion of dimension bears to the notion of linear independence.

We frequently deal with families of elements of K, say a family $\{x_i\}_{i \in I}$, and say that such a family is algebraically independent over k if its elements are distinct (in other words, $x_i \neq x_j$ if $i \neq j$) and if the set consisting of the elements in this family is algebraically independent over k.

A subset S of K which is algebraically independent over k and is maximal with respect to the inclusion ordering will be called a *transcendence base* of K over k. From the maximality, it is clear that if S is a transcendence base of K over k, then K is algebraic over $k(S)$.

THEOREM 1. *Let K be an extension of a field k. Any two transcendence bases of K over k have the same cardinality. If Γ is a set of generators of K over k (i.e. $K = k(\Gamma)$) and S is a subset of Γ which is algebraically independent over k, then there exists a transcendence base $\mathcal{B}$ of K over k such that $S \subset \mathcal{B} \subset \Gamma$.*

Proof. We shall prove that if there exists one finite transcendence base, say $\{x_1, \ldots, x_m\}$, $m \geqq 1$, then any other transcendence base must also have m elements. For this it will suffice to prove: If $w_1, \ldots, w_n$ are elements of K which are algebraically independent over k then $n \leqq m$ (for we can then use symmetry). By assumption, there exists a non-zero polynomial f_1 in $m + 1$ variables with coefficients in k such that

$$f_1(w_1, x_1, \ldots, x_m) = 0.$$

Furthermore, by hypothesis, w_1 occurs in f_1, and some x_i also occurs in f_1, say x_1. Then x_1 is algebraic over $k(w_1, x_2, \ldots, x_m)$. Suppose inductively that after a suitable renumbering of $x_2, \ldots, x_m$ we have found $w_1, \ldots, w_r$ $(r < n)$ such that K is algebraic over

$$k(w_1, \ldots, w_r, x_{r+1}, \ldots, x_m).$$

Then there exists a non-zero polynomial f in $m + 1$ variables with coefficients in k such that

$$f(w_{r+1}, w_1, \ldots, w_r, x_{r+1}, \ldots, x_m) = 0,$$

and such that w_{r+1} actually occurs in f. Since the w's are algebraically independent over k, it follows that some x_j $(j = r + 1, \ldots, m)$ also occurs in f. After renumbering we may assume $j = r + 1$. Then x_{r+1} is algebraic over

$$k(w_1, \ldots, w_{r+1}, x_{r+2}, \ldots, x_m).$$

Since a tower of algebraic extensions is algebraic, it follows that K is algebraic over $k(w_1, \ldots, w_{r+1}, x_{r+2}, \ldots, x_m)$. We can repeat the procedure, and if $n \geqq m$ we can replace all the x's by w's, to see that K is algebraic over $k(w_1, \ldots, w_m)$. This shows that $n \geqq m$ implies $n = m$, as desired.

We have now proved: Either the transcendence degree is finite, and is equal to the cardinality of any transcendence base, or it is infinite, and every transcendence base is infinite. The cardinality statement in the infinite case will be left as an exercise. We shall also leave as an exercise the statement that a set of algebraically independent elements can be completed to a transcendence base, selected from a given set of generators. (The reader will note the complete analogy of our statements with those concerning linear bases.)

§2. *Hilbert's Nullstellensatz*

The Nullstellensatz has to do with a special case of the extension theorem for homomorphisms, applied to finitely generated rings over fields.

THEOREM 2. *Let k be a field, and let $k[x] = k[x_1, \ldots, x_n]$ be a finitely generated ring over k. Let $\varphi : k \to L$ be an embedding of k into an algebraically closed field L. Then there exists an extension of φ to a homomorphism of $k[x]$ into L.*

Proof. Let $\mathfrak{M}$ be a maximal ideal of $k[x]$. Let σ be the canonical homomorphism $\sigma : k[x] \to k[x]/\mathfrak{M}$. Then $\sigma k[\sigma x_1, \ldots, \sigma x_n]$ is a field, and is in fact an extension field of σk. If we can prove our theorem when the finitely generated ring is in fact a field, then we apply $\varphi \circ \sigma^{-1}$ on σk and extend this to a homomorphism of $\sigma k[\sigma x_1, \ldots, \sigma x_n]$ into L to get what we want.

Without loss of generality, we therefore assume that $k[x]$ is a field. If it is algebraic over k, we are done (by the known result for algebraic extensions). Otherwise, let $t_1, \ldots, t_r$ be a transcendence basis, $r \geqq 1$. Without loss of generality, we may assume that φ is the identity on k. Each element $x_1, \ldots, x_n$ is algebraic over $k(t_1, \ldots, t_r)$. If we multiply the irreducible polynomial $\mathrm{Irr}(x_i, k(t), X)$ by a suitable non-zero element of $k[t]$, then we get a polynomial all of whose coefficients lie in $k[t]$. Let $a_1(t), \ldots, a_n(t)$ be the set of the leading coefficients of these polynomials, and let $a(t)$ be their product,

$$a(t) = a_1(t) \cdots a_n(t).$$

Since $a(t) \neq 0$, there exist elements $t'_1, \ldots, t'_r \in \bar{k}$ such that $a(t') \neq 0$, and hence $a_i(t') \neq 0$ for any i. Each x_i is integral over the ring

$$k\left[t_1, \ldots, t_r, \frac{1}{a_1(t)}, \ldots, \frac{1}{a_r(t)}\right].$$

Consider the homomorphism

$$\varphi : k[t_1, \ldots, t_r] \to \bar{k}$$

such that φ is the identity on k, and $\varphi(t_j) = t'_j$. Let $\mathfrak{p}$ be its kernel. Then $a(t) \notin \mathfrak{p}$. Our homomorphism φ extends uniquely to the local ring $k[t]_{\mathfrak{p}}$ and by the preceding remarks, it extends to a homomorphism of

$$k[t]_{\mathfrak{p}}[x_1, \ldots, x_n]$$

into $\bar{k}$, using Proposition 16 of Chapter IX, §3. This proves what we wanted.

COROLLARY 1. *Let k be a field and $k[x_1, \ldots, x_n]$ a finitely generated extension ring of k. If $k[x]$ is a field, then $k[x]$ is algebraic over k.*

Proof. All homomorphisms of a field are isomorphisms (onto the image), and there exists a homomorphism of $k[x]$ over k into the algebraic closure of k.

COROLLARY 2. *Let $k[x_1, \dots, x_n]$ be a finitely generated entire ring over a field k, and let $y_1, \dots, y_m$ be non-zero elements of this ring. Then there exists a homomorphism*

$$\psi : k[x] \to \bar{k}$$

over k such that $\psi(y_j) \neq 0$ for all $j = 1, \dots, m$.

Proof. Consider the ring $k[x_1, \dots, x_n, y_1^{-1}, \dots, y_m^{-1}]$ and apply the theorem to this ring.

Let S be a set of polynomials in the polynomial ring $k[X_1, \dots, X_n]$ in n variables. Let L be an extension field of k. By a *zero* of S in L one means an n-tuple of elements $(c_1, \dots, c_n)$ in L such that

$$f(c_1, \dots, c_n) = 0$$

for all $f \in S$. If S consists of one polynomial f, then we also say that (c) is a zero of f. The set of all zeros of S is called an *algebraic set*, in L (or more accurately in $L^{(n)}$. Let $\mathfrak{a}$ be the ideal generated by all elements of S. Since $S \subset \mathfrak{a}$ it is clear that every zero of $\mathfrak{a}$ is also a zero of S. However, the converse obviously holds, namely every zero of S is also a zero of $\mathfrak{a}$ because every element of $\mathfrak{a}$ is of type

$$g_1(X)f_1(X) + \cdots + g_m(X)f_m(X)$$

with $f_j \in S$ and $g_i \in k[X]$. Thus when considering zeros of a set S, we may just consider zeros of an ideal. We note parenthetically that every ideal is finitely generated (Chapter VI), and so every algebraic set is the set of zeros of a finite number of polynomials. As another corollary of Theorem 2 we get:

HILBERT'S NULLSTELLENSATZ. *Let $\mathfrak{a}$ be an ideal in $k[X] = k[X_1, \dots, X_n]$. Let f be a polynomial such that $f(c) = 0$ for every zero $(c) = (c_1, \dots, c_n)$ of $\mathfrak{a}$ in $\bar{k}$. Then there exists an integer $m \geqq 0$ such that $f^m \in \mathfrak{a}$.*

Proof. If $\mathfrak{a}$ is the polynomial ring itself, our assertion is obvious. If not, suppose that no power f^m of f lies in $\mathfrak{a}$ $(m = 0, 1, \dots)$. Let S be the multiplicative set of such powers of f. Let $\mathfrak{p}$ be a maximal element in the set of ideals containing $\mathfrak{a}$ whose intersection with S is empty. Then $\mathfrak{p}$ is prime by Proposition 6 of Chapter VI, §4. We have an isomorphism

$$k[X_1, \dots, X_n]/\mathfrak{p} \approx k[x_1, \dots, x_n],$$

and since $f \notin \mathfrak{p}$, we have $f(x_1, \dots, x_n) \neq 0$. Let

$$\varphi : k[x] \to \bar{k}$$

be a homomorphism over k such that $\varphi(f(x)) \neq 0$. Then $\varphi(f(x)) = f(\varphi(x))$ where $\varphi(x) = (\varphi(x_1), \dots, \varphi(x_n))$. This contradicts the hypothesis that f vanishes on all algebraic zeros of $\mathfrak{a}$.

§3. *Algebraic sets*

We shall make some very elementary remarks on algebraic sets. Let k be a field, and let A be an algebraic set of zeros in some fixed algebraically closed extension field of k. The set of all ploynomials $f \in k[X_1, \dots, X_n]$ such that $f(x) = 0$ for all $(x) \in A$ is obviously an ideal $\mathfrak{a}$ in $k[X]$, and is determined by A. We shall call it the ideal *belonging* to A, or say that it is *associated* with A. If A is the set of zeros of a set S of polynomials, then $S \subset \mathfrak{a}$, but $\mathfrak{a}$ may be bigger than S. On the other hand, we observe that A is also the set of zeros of $\mathfrak{a}$.

Let A, B be algebraic sets, and $\mathfrak{a}$, $\mathfrak{b}$ their associated ideals. Then it is clear that $A \subset B$ if and only if $\mathfrak{a} \supset \mathfrak{b}$. Hence $A = B$ if and only if $\mathfrak{a} = \mathfrak{b}$. This has an important consequence. Since the polynomial ring $k[X]$ is Noetherian, it follows that algebraic sets satisfy the dual property, namely every descending sequence of algebraic sets

$$A_1 \supset A_2 \supset \cdots$$

must be such that $A_m = A_{m+1} = \cdots$ for some integer m, i.e. all A_ν are equal for $\nu \geqq m$. Furthermore, dually to another property characterizing the Noetherian condition, we conclude that every non-empty set of algebraic sets contains a minimal element.

THEOREM 3. *The finite union and the finite intersection of algebraic sets are algebraic sets. If A, B are the algebraic sets of zeros of ideals $\mathfrak{a}$, $\mathfrak{b}$ respectively, then $A \cup B$ is the set of zeros of $\mathfrak{a} \cap \mathfrak{b}$ and $A \cap B$ is the set of zeros of* $(\mathfrak{a}, \mathfrak{b})$.

Proof. We first consider $A \cup B$. Let $(x) \in A \cup B$. Then (x) is a zero of $\mathfrak{a} \cap \mathfrak{b}$. Conversely, let (x) be a zero of $\mathfrak{a} \cap \mathfrak{b}$, and suppose $(x) \notin A$. There exists a polynomial $f \in \mathfrak{a}$ such that $f(x) \neq 0$. But $\mathfrak{a}\mathfrak{b} \subset \mathfrak{a} \cap \mathfrak{b}$ and hence $(fg)(x) = 0$ for all $g \in \mathfrak{b}$, whence $g(x) = 0$ for all $g \in \mathfrak{b}$. Hence (x) lies in B, and $A \cup B$ is an algebraic set of zeros of $\mathfrak{a} \cap \mathfrak{b}$.

To prove that $A \cap B$ is an algebraic set, let $(x) \in A \cap B$. Then (x) is a zero of $(\mathfrak{a}, \mathfrak{b})$. Conversely, let (x) be a zero of $(\mathfrak{a}, \mathfrak{b})$. Then obviously $(x) \in A \cap B$, as desired. This proves our theorem.

An algebraic set V is called *k-irreducible* if it cannot be expressed as a union $V = A \cup B$ of algebraic sets A, B with A, B distinct from V. We also say irreducible instead of *k-irreducible*.

THEOREM 4. *Let A be an algebraic set. Then A can be expressed as a finite union of irreducible algebraic sets,*

$$A = V_1 \cup \cdots \cup V_r.$$

If there is no inclusion relation among the V_i, i.e. if $V_i \not\subset V_j$ for $i \neq j$, then this representation is unique.

Proof. We first show existence. Suppose the set of algebraic sets which cannot be represented as a finite union of irreducible ones is not empty. Let V be a minimal element in it. Then V cannot be irreducible, and we can write $V = A \cup B$ where A, B are algebraic sets, but $A \neq V$ and $B \neq V$. Since each one of A, B is strictly smaller than V, we can express A, B as finite unions of irreducible algebraic sets, and thus get an expression for V, contradiction.

As for uniqueness, let

$$A = V_1 \cup \cdots \cup V_r = W_1 \cup \cdots \cup W_s$$

be an expression of A as unions of irreducible algebraic sets, without inclusion relations. For each W_j we can write

$$W_j = (W_j \cap V_1) \cup \cdots \cup (W_j \cap V_r).$$

Since each $W_j \cap V_i$ is an algebraic set, we must have $W_j = W_j \cap V_i$ for some i. Hence $W_j \subset V_i$ for some i. Similarly, V_i is contained in some W_ν. Since there is no inclusion relation among the W_j's we must have $W_j = V_i = W_\nu$. This argument can be carried out for each W_j and each V_i. This proves that each W_j appears among the V_i's and each V_i appears among the W_j's, and proves the uniqueness of our representation.

As an exercise, prove that an algebraic set is irreducible if and only if its associated ideal is prime. An irreducible algebraic set is usually called a *variety*.

The notion of algebraic set can be generalized to arbitrary (commutative) rings, as follows:

Let A be a commutative ring. By spec(A) we shall mean the set of prime ideals of A. A subset C of spec(A) is said to be *closed* if there exists an ideal $\mathfrak{a}$ of A such that C consists of those prime ideals $\mathfrak{p}$ such that $\mathfrak{a} \subset \mathfrak{p}$. The complement of a closed subset of spec(A) is called an *open subset* of spec(A). The following statements are then very easy to verify and will be left to the reader.

The union of a finite number of closed sets is closed. The intersection of an arbitrary family of closed sets is closed.

The intersection of a finite number of open sets is open. The union of an arbitrary family of open sets is open.

The empty set, and spec(A) *itself are both open and closed.*

If S is a subset of A, then the set of prime ideals $\mathfrak{p} \in \operatorname{spec}(A)$ such that $S \subset \mathfrak{p}$ coincides with the set of prime ideals $\mathfrak{p}$ containing the ideal generated by S.

If $f \in A$, we view the set of prime ideals $\mathfrak{p}$ of spec(A) containing f as the set of zeros of f. Indeed, it is the set of $\mathfrak{p}$ such that the image of f in the canonical homomorphism

$$A \to A/\mathfrak{p}$$

is 0.

Let A, B be rings and $\varphi : A \to B$ a homomorphism. Then φ induces a map

$$\operatorname{spec}(\varphi) = \varphi^{-1} : \operatorname{spec}(B) \to \operatorname{spec}(A)$$

by

$$\mathfrak{p} \mapsto \varphi^{-1}(\mathfrak{p}).$$

The reader will verify at once that spec(φ) is continuous, in the sense that if U is open in spec(B), then $\varphi^{-1}(U)$ is open in spec(A).

We can view spec as a functor from the category of commutative rings into the category of topological spaces. The topology which we defined above on spec(A) is called the *Zariski topology*.

By a *point* of spec(A) in a field L one means a mapping

$$\operatorname{spec}(\varphi) : \operatorname{spec}(L) \to \operatorname{spec}(A)$$

induced by a homomorphism $\varphi : A \to L$ of A into L.

For example, for each prime number p, we get a point of spec$(\mathbf{Z})$, namely the point arising from the reduction map

$$\mathbf{Z} \to \mathbf{Z}/p\mathbf{Z}.$$

The corresponding point is given by the reversed arrow,

$$\operatorname{spec}(\mathbf{Z}) \leftarrow \operatorname{spec}(\mathbf{Z}/p\mathbf{Z}).$$

As another example, consider the polynomial ring $k[X_1, \ldots, X_n]$ over a field k. For each n-tuple $(c_1, \ldots, c_n)$ in $\bar{k}^{(n)}$ we get a homomorphism

$$\varphi : k[X_1, \ldots, X_n] \to \bar{k}$$

such that φ is the identity on k, and $\varphi(X_i) = c_i$ for all i. The corresponding point is given by the reversed arrow

$$\operatorname{spec} k[X] \leftarrow \operatorname{spec}(\bar{k}).$$

Thus we may identify the points in n-space $\bar{k}^{(n)}$ with the points of spec $k[X]$ (over k) in $\bar{k}$.

The notion of algebraic set which we defined above generalizes to the notion of closed set. We leave it as an exercise to prove:

THEOREM 5. *Let A be a Noetherian ring. Then every closed set C can be expressed as a finite union of irreducible closed sets, and this expression is unique if in the union*

$$C = V_1 \cup \cdots \cup V_r$$

of irreducible closed sets, we have $V_i \not\subset V_j$ if $i \neq j$.

Of course, by an irreducible closed set we mean one which cannot be expressed as a proper union of two closed sets.

§4. Noether normalization theorem

THEOREM 6. *Let $k[x_1, \ldots, x_n] = k[x]$ be a finitely generated entire ring over a field k, and assume that $k(x)$ has transcendence degree r. Then there exist elements $y_1, \ldots, y_r$ in $k[x]$ such that $k[x]$ is integral over*

$$k[y] = k[y_1, \ldots, y_r].$$

Proof. If $(x_1, \ldots, x_n)$ are already algebraically independent over k, we are done. If not, there is a non-trivial relation

$$\sum a_{(j)} x_1^{j_1} \cdots x_n^{j_n} = 0$$

with each coefficient $a_{(j)} \in k$ and $a_{(j)} \neq 0$. The sum is taken over a finite number of distinct n-tuples of integers $(j_1, \ldots, j_n)$, $j_\nu \geqq 0$. Let $m_2, \ldots, m_n$ be positive integers, and put

$$y_2 = x_2 - x_1^{m_2}, \quad \ldots, \quad y_n = x_n - x_1^{m_n}.$$

Substitute $x_i = y_i + x_1^{m_i}$ $(i = 2, \ldots, n)$ in the above equation. Using vector notation, we put $(m) = (1, m_2, \ldots, m_n)$ and use the dot product $(j) \cdot (m)$ to denote $j_1 + m_2 j_2 + \cdots + m_n j_n$. If we expand the relation after making the above substitution, we get

$$\sum c_{(j)} x_1^{(j)\cdot(m)} + f(x_1, y_2, \ldots, y_n) = 0$$

where f is a polynomial in which no pure power of x_1 appears. We now select d to be a large integer [say greater than any component of a vector (j) such that $c_{(j)} \neq 0$] and take

$$(m) = (1, d, d^2, \ldots, d^n).$$

Then all $(j) \cdot (m)$ are distinct for those (j) such that $c_{(j)} \neq 0$. In this way we obtain an integral equation for x_1 over $k[y_2, \ldots, y_n]$. Since each x_i $(i > 1)$ is integral over $k[x_1, y_2, \ldots, y_n]$, it follows that $k[x]$ is integral over $k[y_2, \ldots, y_n]$. We can now proceed inductively, using the transitivity of integral extensions to shrink the number of y's until we reach an algebraically independent set of y's.

§5. *Linearly disjoint extensions*

In this section we discuss the way in which two extensions K and L of a field k behave with respect to each other. We assume that all the fields involved are contained in one field Ω, assumed algebraically closed.

K is said to be *linearly disjoint from L over k* if every finite set of elements of K that is linearly independent over k is still such over L.

The definition is unsymmetric, but we prove right away that the property of being linearly disjoint is actually symmetric for K and L. Assume K linearly disjoint from L over k. Let $y_1, \ldots, y_n$ be elements of L linearly independent over k. Suppose there is a non-trivial relation of linear dependence over K,

$$x_1y_1 + x_2y_2 + \cdots + x_ny_n = 0. \tag{1}$$

Say $x_1, \ldots, x_r$ are linearly independent over k, and $x_{r+1}, \ldots, x_n$ are linear combinations $x_i = \sum_{\mu=1}^{r} a_{i\mu}x_\mu$, $i = r + 1, \ldots, n$. We can write the relation (1) as follows:

$$\sum_{\mu=1}^{r} x_\mu y_\mu + \sum_{i=r+1}^{n} \left(\sum_{\mu=1}^{r} a_{i\mu}x_\mu \right) y_i = 0$$

and collecting terms, after inverting the second sum, we get

$$\sum_{\mu=1}^{r} \left(y_\mu + \sum_{i=r+1}^{n} (a_{i\mu}y_i) \right) x_\mu = 0.$$

The y's are linearly independent over k, so the coefficients of x_μ are $\neq 0$. This contradicts the linear disjointness of K and L over k.

We now give two criteria for linear disjointness.

Criterion 1. Suppose that K is the quotient field of a ring R and L the quotient field of a ring S. To test whether L and K are linearly disjoint, it suffices to show that if elements $y_1, \ldots, y_n$ of S are linearly independent over k, then there is no linear relation among the y's with coefficients in R. Indeed, if elements $y_1, \ldots, y_n$ of L are linearly independent over k, and if there is a relation $x_1y_1 + \cdots + x_ny_n = 0$ with $x_i \in K$, then we can select y in S and x in R such that $xy \neq 0$, $yy_i \in S$ for all i, and $xx_i \in R$

for all i. Multiplying the relation by xy gives a linear dependence between elements of R and S. However, the yy_i are obviously linearly independent over k, and this proves our criterion.

Criterion 2. Again let R be a subring of K such that K is its quotient field and R is a vector space over k. Let $\{u_\alpha\}$ be a basis of R considered as a vector space over k. To prove K and L linearly disjoint over k, it suffices to show that the elements $\{u_\alpha\}$ of this basis remain linearly independent over L. Indeed, suppose this is the case. Let $x_1, \ldots, x_m$ be elements of R linearly independent over k. They lie in a finite dimension vector space generated by some of the u_α, say $u_1, \ldots, u_n$. They can be completed to a basis of this space over k. Lifting this vector space of dimension n over L, it must conserve its dimension because the u's remain linearly independent by hypothesis, and hence the x's must also remain linearly independent.

The next proposition gives a useful criterion which allows us to recognize linear disjointness in a tower of fields.

PROPOSITION 1. *Let K be field containing another field k, and let $L \supset E$ be two other extensions of k. Then K and L are linearly disjoint over k if and only if K and E are linearly disjoint over k and KE, L are linearly disjoint over E.*

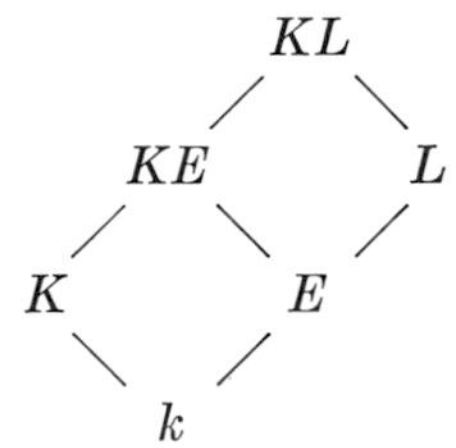

Proof. Assume first that K, E are linearly disjoint over k, and KE, L are linearly disjoint over E. Let $\{\kappa\}$ be a basis of K as vector space over k (we use the elements of this basis as their own indexing set), and let $\{\alpha\}$ be a basis of E over k. Let $\{\lambda\}$ be a basis of L over E. Then $\{\alpha\lambda\}$ is a basis of L over k. If K and L are not linearly disjoint over k, then there exists a relation

$$\sum_{\lambda,\alpha} \left(\sum_{\kappa} c_{\kappa\lambda\alpha}\kappa \right) \lambda\alpha = 0 \qquad \text{with some } c_{\kappa\lambda\alpha} \neq 0, \qquad c_{\kappa\lambda\alpha} \in k.$$

Changing the order of summation gives

$$\sum_{\lambda} \left(\sum_{\kappa,\lambda} c_{\kappa\lambda\alpha}\kappa\alpha \right) \lambda = 0$$

contradicting the linear disjointness of L and KE over E.

Conversely, assume that K and L are linearly disjoint over k. Then *a fortiori*, K and E are also linearly disjoint over k, and the field KE is the quotient field of the ring $E[K]$ generated over E by all elements of K. This ring is a vector space over E, and a basis for K over k is also a basis for this ring $E[K]$ over E. With this remark, and the criteria for linear disjointness, we see that it suffices to prove that the elements of such a basis remain linearly independent over L. At this point we see that the arguments given in the first part of the proof are reversible. We leave the formalism to the reader.

We introduce another notion concerning two extensions K and L of a field k. We shall say that K is *free from L over k* if every finite set of elements of K algebraically independent over k remains such over L. If (x) and (y) are two sets of elements in Ω, we say that they are *free over k* (or *independent over k*) if $k(x)$ and $k(y)$ are free over k.

Just as with linear disjointness, our definition is unsymmetric, and we prove that the relationship expressed therein is actually symmetric. Assume therefore that K is free from L over k. Let $y_1, \ldots, y_n$ be elements of L, algebraically independent over k. Suppose they become dependent over K. They become so in a subfield F of K finitely generated over k, say of transcendence degree r over k. Computing the transcendence degree of $F(y)$ over k in two ways gives a contradiction (cf. Exercise 5).

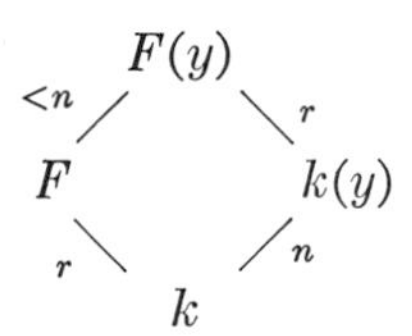

PROPOSITION 2. *If K and L are linearly disjoint over k, then they are free over k.*

Proof. Let $x_1, \ldots, x_n$ be elements of K algebraically independent over k. Suppose they become algebraically dependent over L. We get a relation

$$\sum y_a M_\alpha(x) = 0$$

between monomials $M_\alpha(x)$ with coefficients y_α in L. This gives a linear relation among the $M_\alpha(x)$. But these are linearly independent over k because the x's are assumed algebraically independent over k. This is a contradiction.

PROPOSITION 3. *Let L be an extension of k, and let $(u) = (u_1, \ldots, u_r)$ be a set of quantities algebraically independent over L. Then the field $k(u)$ is linearly disjoint from L over k.*

Proof. According to the criteria for linear disjointness, it suffices to prove that the elements of a basis for the ring $k[u]$ that are linearly independent over k remain so over L. In fact the monomials $M(u)$ give a basis of $k[u]$ over k. They must remain linearly independent over L, because as we have seen, a linear relation gives an algebraic relation. This proves our proposition.

Note finally that the property that two extensions K and L of a field k are linearly disjoint or free is of finite type. To prove that they have either property, it suffices to do it for all subfields K_0 and L_0 of K and L respectively which are finitely generated over k. This comes from the fact that the definitions involve only a finite number of quantities at a time.

§6. *Separable extensions*

Let K be a finitely generated extension of k, $K = k(x)$. We shall say that it is *separably generated* if we can find a transcendence basis $(t_1, \dots, t_r)$ of K/k such that K is separably algebraic over $k(t)$. Such a transcendence base is said to be a *separating transcendence base* for K over k.

We always denote by p the characteristic if it is not 0. The field obtained from k by adjoining all p^m-th roots of all elements of k will be denoted by k^{1/p^m}. The compositum of all such fields for $m = 1, 2, \dots$, is denoted by k^{1/p^∞}.

PROPOSITION 4. *The following conditions concerning an extension field K of k are equivalent:*

(1) *K is linearly disjoint from k^{1/p^∞}.*

(2) *K is linearly disjoint from k^{1/p^m} for some m.*

(3) *Every subfield of K containing k and finitely generated over k is separably generated.*

Proof. It is obvious that (1) implies (2). In order to prove that (2) implies (3), we may clearly assume that K is finitely generated over k, say $K = k(x) = k(x_1, \dots, x_n)$. Let the transcendence degree of this extension be r. If $r = n$, the proof is complete. Otherwise, say $x_1, \dots, x_r$ is a transcendence base. Then x_{r+1} is algebraic over $k(x_1, \dots, x_r)$. Let $f(X_1, \dots, X_{r+1})$ be a polynomial of lowest degree such that

$$f(x_1, \dots, x_{r+1}) = 0.$$

Then f is irreducible. We contend that not all x_i $(i = 1, \dots, r+1)$ appear to the p-th power throughout. If they did, we could write $f(X) = \sum c_\alpha M_\alpha(X)^p$ where $M_\alpha(X)$ are monomials in $X_1, \dots, X_{r+1}$ and $c_\alpha \in k$. This would imply that the $M_\alpha(x)$ are linearly dependent over $k^{1/p}$ (taking the pth root of the equation $\sum c_\alpha M_\alpha(x)^p = 0$). However, the $M_\alpha(x)$ are linearly independent over k (otherwise we would get an equation for

$x_1, \dots, x_{r+1}$ of lower degree) and we thus get a contradiction to the linear disjointness of $k(x)$ and $k^{1/p}$. Say X_1 does not appear to the p-th power throughout, but actually appears in $f(X)$. We know that $f(X)$ is irreducible in $k[X_1, \dots, X_{r+1}]$ and hence $f(x) = 0$ is an irreducible equation for x_1 over $k(x_2, \dots, x_{r+1})$. Since X_1 does not appear to the p-th power throughout, this equation is a separable equation for x_1 over $k(x_2, \dots, x_{r+1})$, in other words, x_1 is separable algebraic over $k(x_2, \dots, x_{r+1})$. From this it follows that it is separable algebraic over $k(x_2, \dots, x_n)$. If $(x_2, \dots, x_n)$ is a transcendence base, the proof is complete. If not, say that x_2 is separable over $k(x_3, \dots, x_n)$. Then $k(x)$ is separable over $k(x_3, \dots, x_n)$. Proceeding inductively, we see that the procedure can be continued until we get down to a transcendence base. This proves that (2) implies (3). It also proves that a separating transcendence base for $k(x)$ over k can be selected from the given set of generators (x).

To prove that (3) implies (1) we may assume that K is finitely generated over k. Let (u) be a transcendence base for K over k. Then K is separably algebraic over $k(u)$. By Proposition 3, $k(u)$ and k^{1/p^∞} are linearly disjoint. Let $L = k^{1/p^\infty}$. Then $k(u)L$ is purely inseparable over $k(u)$, and hence is linearly disjoint from K over $k(u)$ by the elementary theory of finite algebraic extensions. Using Proposition 1, we conclude that K is linearly disjoint from L over k, thereby proving our theorem.

An extension K of k satisfying the conditions of Proposition 4 is called *separable*. This definition is compatible with the use of the word for algebraic extensions.

The first condition of our theorem is known as *MacLane's criterion.* It has the following immediate corollaries.

COROLLARY 1. *If K is separable over k, and E is a subfield of K containing k, then E is separable over k.*

COROLLARY 2. *Let E be a separable extension of k, and K a separable extension of E. Then K is a separable extension of k.*

Proof. Apply Proposition 1 and the definition of separability.

COROLLARY 3. *If k is perfect, every extension of k is separable.*

COROLLARY 4. *Let K be a separable extension of k, and free from an extension L of k. Then KL is a separable extension of L.*

Proof. An element of KL has an expression in terms of a finite number of elements of K and L. Hence any finitely generated subfield of KL containing L is contained in a composite field FL, where F is a subfield of K finitely generated over k. By Corollary 1, we may assume that K is finitely generated over k. Let (t) be a transcendence base of K over k, so K is separable algebraic over $k(t)$. By hypothesis, (t) is a transcendence base of KL over L, and since every element of K is separable algebraic over $k(t)$, it is also separable over $L(t)$. Hence KL is separably generated over L. This proves the corollary.

COROLLARY 5. *Let K and L be two separable extensions of k, free from each other over k. Then KL is separable over k.*

Proof. Use Corollaries 4 and 2.

COROLLARY 6. *Let K, L be two extensions of k, linearly disjoint over k. Then K is separable over k if and only if KL is separable over L.*

Proof. If K is not separable over k, it is not linearly disjoint from $k^{1/p}$ over k, and hence *a fortiori* it is not linearly disjoint from $Lk^{1/p}$ over k. By Proposition 1, this implies that KL is not linearly disjoint from $Lk^{1/p}$ over L, and hence that KL is not separable over L. The converse is a special case of Corollary 4, taking into account that linearly disjoint fields are free.

We conclude our discussion of separability with two results. The first one has already been proved in the first part of Proposition 4, but we state it here explicitly.

PROPOSITION 5. *If K is a separable extension of k, and is finitely generated, then a separating transcendence base can be selected from a given set of generators.*

To state the second result we denote by K^{p^m} the field obtained from K by raising all elements of K to the p^mth power.

PROPOSITION 6. *Let K be a finitely generated extension of a field k. If $K^{p^m}k = K$ for some m, then K is separably algebraic over k. Conversely, if K is separably algebraic over k, then $K^{p^m}k = K$ for all m.*

Proof. If K/k is separably algebraic, then the conclusion follows from the elementary theory of finite algebraic extensions. Conversely, if K/k is finite algebraic but not separable, then the maximal separable extension of k in K cannot be all of K, and hence K^pk cannot be equal to K. Finally, if there exists an element t of K transcendental over k, then $k(t^{1/p^m})$ has degree p^m over $k(t)$, and hence there exists a t such that t^{1/p^m} does not lie in K. This proves our proposition.

§7. *Derivations*

A *derivation* D of a ring R is a mapping $D : R \to R$ of R into itself which is linear and satisfies the ordinary rule for derivatives, i.e. $D(x + y) = Dx + Dy$, and $D(xy) = xDy + yDx$. As an example of derivations, consider the polynomial ring $k[X]$ over a field k. For each variable X_i, the partial derivative $\partial/\partial X_i$ taken in the usual manner is a derivation of

$k[X]$. We also get a derivation of the quotient field in the obvious manner, i.e. by defining $D(u/v) = (vDu - uDv)/v^2$.

We shall work with derivations of a field K. A derivation of K is *trivial* if $Dx = 0$ for all $x \in K$. It is trivial *over a subfield* k of K if $Dx = 0$ for all $x \in k$. A derivation is always trivial over the prime field: one sees that $D(1) = D(1 \cdot 1) = 2D(1)$, whence $D(1) = 0$.

We now consider the problem of extending derivations. Let $L = K(x) = K(x_1, \ldots, x_n)$ be a finitely generated extension. If $f \in K[X]$, we denote by $\partial f/\partial x_i$ the polynomials $\partial f/\partial X_i$ evaluated at (x). Given a derivation D on K, does there exist a derivation D^* on L coinciding with D on K? If $f(X) \in K[X]$ is a polynomial vanishing on (x), then any such D^* must satisfy

$$0 = D^*f(x) = f^D(x) + \sum(\partial f/\partial x_i)D^*x_i, \tag{1}$$

where f^D denotes the polynomial obtained by applying D to all coefficients of f. Note that if relation (1) is satisfied for every element in a finite set of generators of the ideal in $K[X]$ vanishing on (x), then (1) is satisfied by every polynomial of this ideal. This is an immediate consequence of the rules for derivations. The preceding ideal will also be called the ideal determined by (x) in $K[X]$.

The above necessary condition for the existence of a D^* turns out to be sufficient.

THEOREM 7. *Let D be a derivation of a field K. Let (x) be any set of quantities, and let $\{f_\alpha(X)\}$ be a set of generators for the ideal determined by (x) in $K[X]$. Then, if (u) is any set of elements of $K(x)$ satisfying the equations*

$$0 = f_\alpha^D(x) + \sum(\partial f_\alpha/\partial x_i)u_i,$$

there is one and only one derivation D^* of $K(x)$ coinciding with D on K, and such that $D^*x_i = u_i$ for every i.

Proof. The necessity has been shown above. Conversely, if $g(x)$, $h(x)$ are in $K[x]$, and $h(x) \neq 0$, one verifies immediately that the mapping D^* defined by the formulas

$$D^*g(x) = g^D(x) + \sum \frac{\partial g}{\partial x_i} u_i$$

$$D^*(g/h) = \frac{hD^*g - gD^*h}{h^2}$$

is well defined and is a derivation of $K(x)$.

Consider the special case where (x) consists of one element x. Let D be a given derivation on K.

Case 1. x is separable algebraic over K. Let $f(X)$ be the irreducible polynomial satisfied by x over K. Then $f'(x) \neq 0$. We have

$$0 = f^D(x) + f'(x)u,$$

whence $u = -f^D(x)/f'(x)$. Hence D extends to $K(x)$ uniquely. If D is trivial on K, then D is trivial on $K(x)$.

Case 2. x is transcendental over K. Then D extends, and u can be selected arbitrarily in $K(x)$.

Case 3. x is purely inseparable over K, so $x^p - a = 0$, with $a \in K$. Then D extends to $K(x)$ if and only if $Da = 0$. In particular if D is trivial on K, then u can be selected arbitrarily.

PROPOSITION 7. *A finitely generated extension $K(x)$ over K is separably algebraic if and only if every derivation D of $K(x)$ which is trivial on K is trivial on $K(x)$.*

Proof. If $K(x)$ is separable algebraic over K, this is Case 1. Conversely, if it is not, we can make a tower of extensions between K and $K(x)$, such that each step is covered by one of the three above cases. At least one step will be covered by Case 2 or 3. Taking the uppermost step of this latter type, one sees immediately how to construct a derivation trivial on the bottom and non-trivial on top of the tower.

PROPOSITION 8. *Given K and elements $(x) = (x_1, \ldots, x_n)$ in some extension field, assume that there exist n polynomials $f_i \in K[X]$ such that*

1. $f_i(x) = 0$, and
2. $\det(\partial f_i/\partial x_j) \neq 0$.

Then (x) is separably algebraic over K.

Proof. Let D be a derivation on $K(x)$, trivial on K. Having $f_i(x) = 0$ we must have $Df_i(x) = 0$, whence the Dx_i satisfy n linear equations such that the coefficient matrix has non-zero determinant. Hence $Dx_i = 0$, so D is trivial on $K(x)$. Hence $K(x)$ is separable algebraic over K.

The following proposition will follow directly from Cases 1 and 2 above.

PROPOSITION 9. *Let $K = k(x)$ be a finitely generated extension of k. An element z of K is in K^pk if and only if every derivation D of K over k is such that $Dz = 0$.*

Proof. If z is in K^pk, then it is obvious that every derivation D of K over k vanishes on z. Conversely, if $z \notin K^pk$, then z is purely inseparable over K^pk, and by Case 3 of the extension theorem, we can find a derivation D trivial on K^pk such that $Dz = 1$. This derivation is at first defined on the field $K^pk(z)$. One can extend it to K as follows. Suppose there is an element $w \in K$ such that $w \notin K^pk(z)$. Then $w^p \in K^pk$, and D vanishes on w^p. We can then again apply Case 3 to extend D from $K^pk(z)$ to $K^pk(z, w)$. Proceeding stepwise, we finally reach K, thus proving our proposition.

The derivations D of a field K form a vector space over K if we define zD for $z \in K$ by $(zD)(x) = zDx$.

Let K be a finitely generated extension of k, of dimension r over k. We denote by $\mathfrak{D}$ the K-vector space of derivations D of K over k (derivations of K which are trivial on k). For each $z \in K$, we have a pairing

$$(D, z) \mapsto Dz$$

of $(\mathfrak{D}, K)$ into K. Each element z of K gives therefore a K-linear functional of $\mathfrak{D}$. This functional is denoted by dz. We have

$$\begin{aligned} d(yz) &= y\,dz + z\,dy \\ d(y + z) &= dy + dz. \end{aligned}$$

These linear functionals form a subspace $\mathfrak{F}$ of the dual space of $\mathfrak{D}$, if we define ydz by $(D, ydz) \to yDz$.

Proposition 10. *Assume that K is a separably generated and finitely generated extension of k of transcendence degree r. Then the vector space $\mathfrak{D}$ (over K) of derivations of K over k has dimension r. Elements $t_1, \ldots, t_r$ of K form a separating transcendence base of K over k if and only if $dt_1, \ldots, dt_r$ form a basis of the dual space of $\mathfrak{D}$ over K.*

Proof. If $t_1, \ldots, t_r$ is a separating transcendence base for K over k, then we can find derivations $D_1, \ldots, D_r$ of K over k such that $D_it_j = \delta_{ij}$, by Cases 1 and 2 of the extension theorem. Given $D \in \mathfrak{D}$, let $w_i = Dt_i$. Then clearly $D = \sum w_iD_i$, and so the D_i form a basis for $\mathfrak{D}$ over K, and the dt_i form the dual basis. Conversely, if $dt_1, \ldots, dt_r$ is a basis for $\mathfrak{F}$ over K, and if K is not separably generated over $k(t)$, then by Cases 2 and 3 above we can find a derivation D which is trivial on $k(t)$ but non-trivial on K. If $D_1, \ldots, D_r$ is the dual basis of $dt_1, \ldots, dt_r$ (so $D_it_j = \delta_{ij}$) then $D, D_1, \ldots, D_r$ would be linearly independent over K, contradicting the first part of the theorem.

Corollary. *Let K be a finitely generated and separably generated extension of k. Let z be an element of K transcendental over k. Then K is separable over $k(z)$ if and only if there exists a derivation D of K over k such that $Dz \neq 0$.*

Proof. If K is separable over $k(z)$, then z can be completed to a separating base of K over k and we can apply the proposition. If $Dz \neq 0$, then $dz \neq 0$, and we can complete dz to a basis of $\mathfrak{F}$ over K. Again from the proposition, it follows that K will be separable over $k(z)$.

Exercises

1. Prove that the complex numbers have infinitely many automorphisms. [*Hint:* Use transcendence bases.]

2. A subfield k of a field K is said to be algebraically closed in K if every element of K which is algebraic over k is contained in k. Prove: If k is algebraically closed in K, and K, L are free over k, and L is separable over k or K is separable over k, then L is algebraically closed in KL.

3. Prove that the following conditions are equivalent (they define the notion of a *regular extension*).
(i) k is algebraically closed in K and K is separable over k.
(ii) K is linearly disjoint from $\bar{k}$ over k.

4. Prove for regular extensions analogous results to those which were proved for separable extensions before.

5. Let $k \subset E \subset K$ be extension fields. Show that

$$\text{tr. deg. } (K/k) = \text{tr. deg. } (K/E) + \text{tr. deg. } (E/k).$$

If $\{x_i\}$ is a transcendence base of E/k, and $\{y_j\}$ is a transcendence base of K/E, then $\{x_i, y_j\}$ is a transcendence base of K/k.

6. Let K/k be a finitely generated extension, and let $K \supset E \supset k$ be a subextension. Show that E/k is finitely generated.

7. Let k be a field of characteristic 0, and let $z_1, \ldots, z_r$ be algebraically independent over k. Let (e_{ij}), $i = 1, \ldots, m$ and $j = 1, \ldots, r$ be a matrix of integers with $r \geqq m$, and assume that this matrix has rank m. Let

$$w_i = z_1^{e_{i1}} \cdots z_r^{e_{ir}} \qquad \text{for} \quad i = 1, \ldots, m.$$

Show that $w_1, \ldots, w_m$ are algebraically independent over k. [*Hint:* Consider the K-homomorphism mapping the K-space of derivations of K/k into $K^{(r)}$ given by

$$D \mapsto (Dz_1/z_1, \ldots, Dz_r/z_r),$$

and derive a linear condition for those D vanishing on $k(w_1, \ldots, w_m)$.]

8. Let k, (z) be as in Exercise 7. Show that if P is a rational function then

$$d(P(z)) = \text{grad } P(z) \cdot dz,$$

using vector notation, i.e. $dz = (dz_1, \ldots, dz_r)$ and $\text{grad } P = (D_1P, \ldots, D_rP)$. Define $d \log P$ and express it in terms of coordinates. If P, Q are rational functions in $k(z)$ show that

$$d \log(PQ) = d \log P + d \log Q.$$

CHAPTER XI

Real Fields

§1. *Ordered fields*

Let K be a field. An *ordering* of K is a subset P of K having the following properties:

ORD 1. Given $x \in K$, we have either $x \in P$, or $x = 0$, or $-x \in P$, and these three possibilities are mutually exclusive. In other words, K is the disjoint union of P, $\{0\}$, and $-P$.

ORD 2. If $x, y \in P$, then $x + y$ and $xy \in P$.

We shall also say that K is *ordered by* P, and we call P the set of *positive* elements.

Let us assume that K is ordered by P. Since $1 \neq 0$ and $1 = 1^2 = (-1)^2$ we see that $1 \in P$. By ORD 2, it follows that $1 + \cdots + 1 \in P$, whence K has characteristic 0. If $x \in P$, and $x \neq 0$, then $xx^{-1} = 1 \in P$ implies that $x^{-1} \in P$.

Let $x, y \in K$. We define $x < y$ (or $y > x$) to mean that $y - x \in P$. If $x < 0$ we say that x is *negative*. This means that $-x$ is positive. One verifies trivially the usual relations for inequalities, for instance:

$$\begin{aligned} x < y \text{ and } y < z \quad & \text{implies} \quad x < z, \\ x < y \text{ and } z > 0 \quad & \text{implies} \quad xz < yz, \\ x < y \text{ and } x, y > 0 \quad & \text{implies} \quad 1/y < 1/x. \end{aligned}$$

We define $x \leqq y$ to mean $x < y$ or $x = y$. Then $x \leqq y$ and $y \leqq x$ imply $x = y$.

If K is ordered and $x \in K$, $x \neq 0$, then x^2 is positive because $x^2 = (-x)^2$ and either $x \in P$ or $-x \in P$. Thus a sum of squares is positive, or 0.

Let E be a field. Then a product of sums of squares in E is a sum of squares. If $a, b \in E$ are sums of squares and $b \neq 0$ then a/b is a sum of squares.

The first assertion is obvious, and the second also, from the expression $a/b = ab(b^{-1})^2$.

If E has characteristic $\neq 2$, and -1 is a sum of squares in E, then every element $a \in E$ is a sum of squares, because $4a = (1 + a)^2 - (1 - a)^2$.

If K is a field with an ordering P, and F is a subfield, then obviously, $P \cap F$ defines an ordering of F, which is called the *induced* ordering.

We observe that our two axioms ORD 1 and ORD 2 apply to a ring. If A is an ordered ring, with $1 \neq 0$, then clearly A cannot have divisors of 0, and one can extend the ordering of A to the quotient field in the obvious way: A fraction is called positive if it can be written in the form a/b with $a, b \in A$ and $a, b > 0$. One verifies trivially that this defines an ordering on the quotient field.

Example. We define an ordering on the polynomial ring $\mathbf{R}[t]$ over the real numbers. A polynomial

$$f(t) = a_n t^n + \cdots + a_0$$

with $a_n \neq 0$ is defined to be positive if $a_n > 0$. The two axioms are then trivially verified. We note that $t > a$ for all $a \in \mathbf{R}$. Thus t is infinitely large with respect to $\mathbf{R}$. The existence of infinitely large (or infinitely small) elements in an ordered field is the main aspect in which such a field differs from a subfield of the real numbers.

We shall now make some comment on this behavior, i.e. the existence of infinitely large elements.

Let K be an ordered field and let F be a subfield with the induced ordering. As usual, we put $|x| = x$ if $x > 0$ and $|x| = -x$ if $x < 0$. We say that an element α in K is *infinitely large* over F if $|\alpha| \geqq x$ for all $x \in F$. We say that it is *infinitely small* over F if $0 \leqq |\alpha| < |x|$ for all $x \in F$, $x \neq 0$. We see that α is infinitely large if and only if α^{-1} is infinitely small. We say that K is *archimedean* over F if K has no elements which are infinitely large over F. An intermediate field F_1, $K \supset F_1 \supset F$, is maximal archimedean over F in K if it is archimedean over F, and no other intermediate field containing F_1 is archimedean over F. If F_1 is archimedean over F and F_2 is archimedean over F_1 then F_2 is archimedean over F. Hence by Zorn's lemma there always exists a maximal archimedean subfield F_1 of K over F. We say that F is maximal archimedean in K if it is maximal archimedean over itself in K.

Let K be an ordered field and F a subfield. Let $\mathfrak{o}$ be the set of elements of K which are not infinitely large over F. Then it is clear that $\mathfrak{o}$ is a ring, and that for any $\alpha \in K$, we have α or $\alpha^{-1} \in \mathfrak{o}$. Hence $\mathfrak{o}$ is what is called a valuation ring, containing F. Let $\mathfrak{m}$ be the ideal of all $\alpha \in K$ which are infinitely small over F. Then $\mathfrak{m}$ is the unique maximal ideal of $\mathfrak{o}$, because any element in $\mathfrak{o}$ which is not in $\mathfrak{m}$ has an inverse in $\mathfrak{o}$. We call $\mathfrak{o}$ the *valuation ring determined by the ordering of K/F.*

PROPOSITION 1. *Let K be an ordered field and F a subfield. Let $\mathfrak{o}$ be the valuation ring determined by the ordering of K/F, and let $\mathfrak{m}$ be its maximal ideal. Then $\mathfrak{o}/\mathfrak{m}$ is a real field.*

Proof. Otherwise, we could write

$$-1 = \sum \alpha_i^2 + a$$

with $\alpha_i \in \mathfrak{o}$ and $a \in \mathfrak{m}$. Since $\sum \alpha_i^2$ is positive and a is infinitely small, such a relation is clearly impossible.

§2. *Real fields*

A field K is said to be *real* if -1 is not a sum of squares in K. A field K is said to be *real closed* if it is real, and if any algebraic extension of K which is real must be equal to K. In other words, K is maximal with respect to the property of reality in an algebraic closure.

PROPOSITION 2. *Let K be a real field.*

(i) *If $a \in K$ then $K(\sqrt{a})$ or $K(\sqrt{-a})$ is real. If a is a sum of squares in K, then $K(\sqrt{a})$ is real. If $K(\sqrt{a})$ is not real, then $-a$ is a sum of squares in K.*

(ii) *If f is an irreducible polynomial of odd degree n in $K[X]$ and if α is a root of f, then $K(\alpha)$ is real.*

Proof. Let $a \in K$. If a is a square in K, then $K(\sqrt{a}) = K$ and hence is real by assumption. Assume that a is not a square in K. If $K(\sqrt{a})$ is not real, then there exist $b_i, c_i \in K$ such that

$$\begin{aligned}-1 &= \sum(b_i + c_i\sqrt{a})^2 \\ &= \sum(b_i^2 + 2c_ib_i\sqrt{a} + c_i^2a).\end{aligned}$$

Since $\sqrt{a}$ is of degree 2 over K, it follows that

$$-1 = \sum b_i^2 + a\sum c_i^2.$$

If a is a sum of squares in K, this yields a contradiction. In any case, we conclude that

$$-a = \frac{1 + \sum b_i^2}{\sum c_i^2}$$

is a quotient of sums of squares, and by a previous remark, that $-a$ is a sum of squares. Hence $K(\sqrt{-a})$ is real, thereby proving our first assertion.

As to the second, suppose $K(\alpha)$ is not real. Then we can write

$$-1 = \sum g_i(\alpha)^2$$

with polynomials g_i in $K[X]$ of degree $\leqq n - 1$. There exists a polynomial h in $K[X]$ such that

$$-1 = \sum g_i(X)^2 + h(X)f(X).$$

The sum of $g_i(X)^2$ has even degree, and this degree must be > 0, otherwise -1 is a sum of squares in K. This degree is $\leqq 2n - 2$. Since f has odd degree n, it follows that h has odd degree $\leqq n - 2$. If β is a root of h then we see that -1 is a sum of squares in $K(\beta)$. Since $\deg h < \deg f$, our proof is finished by induction.

Let K be a real field. By a *real closure* we shall mean a real closed field L which is algebraic over K.

THEOREM 1. *Let K be a real field. Then there exists a real closure of K. If R is real closed, then R has a unique ordering. The positive elements are the squares of R. Every positive element is a square, and every irreducible polynomial of odd degree in $R[X]$ has a root in R. We have $\overline{R} = R(\sqrt{-1})$.*

Proof. By Zorn's lemma, our field K is contained in some real closed field algebraic over K. Now let R be a real closed field. Let P be the set of non-zero elements of R which are sums of squares. Then P is closed under addition and multiplication. By Proposition 2, every element of P is a square in R, and given $a \in R$, $a \neq 0$, we must have $a \in P$ or $-a \in P$. Thus P defines an ordering. Again by Proposition 2, every irreducible polynomial of odd degree over R has a root in R. Our assertion follows by Example 5 of Chapter VIII, §2.

COROLLARY. *Let K be a real field and a an element of K which is not a sum of squares. Then there exists an ordering of K in which a is negative.*

Proof. The field $K(\sqrt{-a})$ is real by Proposition 1 and hence has an ordering as a subfield of a real closure. In this ordering, $-a > 0$ and hence a is negative.

PROPOSITION 3. *Let R be a field such that $R \neq \overline{R}$ but $\overline{R} = R(\sqrt{-1})$. Then R is real and hence real closed.*

Proof. Let P be the set of elements of R which are squares and $\neq 0$. We contend that P is an ordering of R. Let $a \in R$, $a \neq 0$. Suppose that a is not a square in R. Let α be a root of $X^2 - a = 0$. Then $R(\alpha) = R(\sqrt{-1})$, and hence there exist $c, d \in R$ such that $\alpha = c + d\sqrt{-1}$. Then

$$\alpha^2 = c^2 + 2cd\sqrt{-1} - d^2.$$

Since $1, \sqrt{-1}$ are linearly independent over R, it follows that $c = 0$ (because $a \notin R^2$), and hence $-a$ is a square.

We shall now prove that a sum of squares is a square. For simplicity, write $i = \sqrt{-1}$. Since $R(i)$ is algebraically closed, given $a, b \in R$ we can find $c, d \in R$ such that $(c + di)^2 = a + bi$. Then $a = c^2 - d^2$ and $b = 2cd$. Hence $a^2 + b^2 = (c^2 + d^2)^2$, as was to be shown.

If $a \in R$, $a \neq 0$, then not both a and $-a$ can be squares in R. Hence P is an ordering and our proposition is proved.

THEOREM 2. *Let R be a real closed field, and $f(X)$ a polynomial in $R[X]$. Let $a, b \in R$ and assume that $f(a) < 0$ and $f(b) > 0$. Then there exists c between a and b such that $f(c) = 0$.*

Proof. Since $R(\sqrt{-1})$ is algebraically closed, it follows that f splits into a product of irreducible factors of degree 1 or 2. If $X^2 + \alpha X + \beta$ is irreducible ($\alpha, \beta \in R$) then it is a sum of squares, namely

$$\left(X + \frac{\alpha}{2}\right)^2 + \left(\beta - \frac{\alpha^2}{4}\right),$$

and we must have $4\beta > \alpha^2$ since our factor is assumed irreducible. Hence the change of sign of f must be due to the change of sign of a linear factor, which is trivially verified to be a root lying between a and b.

LEMMA 1. *Let K be a subfield of an ordered field E. Let $\alpha \in E$ be algebraic over K, and a root of the polynomial*

$$f(X) = X^n + a_{n-1}X^{n-1} + \cdots + a_0$$

with coefficients in K. Then $|\alpha| \leqq 1 + |a_{n-1}| + \cdots + |a_0|$.

Proof. If $|\alpha| \leqq 1$, the assertion is obvious. If $|\alpha| > 1$, we express $|\alpha|^n$ in terms of the terms of lower degree, divide by $|\alpha|^{n-1}$, and get a proof for our lemma.

Note that the lemma implies that an element which is algebraic over an ordered field cannot be infinitely large with respect to that field.

Let $f(X)$ be a polynomial with coefficients in a real closed field R, and assume that f has no multiple roots. Let $u < v$ be elements of R. By a *Sturm sequence* for f over the interval $[u, v]$ we shall mean a sequence of polynomials

$$S = \{f = f_0, f' = f_1, \ldots, f_m\}$$

having the following properties:

ST 1. The last polynomial f_m is a non-zero constant.

ST 2. There is no point $x \in [u, v]$ such that $f_j(x) = f_{j+1}(x) = 0$ for any value $0 \leqq j \leqq m - 1$.

ST 3. If $x \in [u, v]$ and $f_j(x) = 0$ for some $j = 1, \ldots, m - 1$, then $f_{j-1}(x)$ and $f_{j+1}(x)$ have opposite signs.

ST 4. We have $f_j(u) \neq 0$ and $f_j(v) \neq 0$ for all $j = 0, \ldots, m$.

For any $x \in [u, v]$ which is not a root of any polynomial f_j we denote by $W_S(x)$ the number of sign changes in the sequence

$$\{f(x), f_1(x), \ldots, f_m(x)\},$$

and call $W_S(x)$ the *variation of signs in the sequence.*

STURM'S THEOREM. *The number of roots of f between u and v is equal to $W_S(u) - W_S(v)$ for any Sturm sequence S.*

Proof. We observe that if $\alpha_1 < \alpha_2 < \cdots < \alpha_r$ is the ordered sequence of roots of the polynomials f_j in $[u, v]$ $(j = 0, \ldots, m - 1)$, then $W_S(x)$ is constant on the open intervals between these roots, by Theorem 2. Hence it will suffice to prove that if there is precisely one element α such that $u < \alpha < v$ and α is a root of some f_j, then $W_S(u) - W_S(v) = 1$ if α is a root of f, and 0 otherwise. Suppose that α is a root of some f_j, for $1 \leqq j \leqq m - 1$. Then $f_{j-1}(\alpha), f_{j+1}(\alpha)$ have opposite signs by ST 3, and these signs do not change when we replace α by u or v. Hence the variation of signs in

$$\{f_{j-1}(u), f_j(u), f_{j+1}(u)\} \qquad \text{and} \qquad \{f_{j-1}(v), f_j(v), f_{j+1}(v)\}$$

is the same, namely equal to 2. If α is not a root of f, we conclude that $W_S(u) = W_S(v)$. If α is a root of f, then $f(u)$ and $f(v)$ have opposite signs, but $f'(u)$ and $f'(v)$ have the same sign, namely the sign of $f'(\alpha)$. Hence in this case, $W_S(u) = W_S(v) + 1$. This proves our theorem.

It is easy to construct a Sturm sequence for a polynomial without multiple roots. We use the Euclidean algorithm, writing

$$\begin{aligned} f &= g_1 f' - f_2, \\ f_2 &= g_2 f_1 - f_3, \\ &\cdots \\ f_{m-2} &= g_{m-1} f_{m-1} - f_m, \end{aligned}$$

using $f' = f_1$. Since f, f' have no common factor, the last term of this sequence is a non-zero constant. The other properties of a Sturm sequence are trivially verified, because if two successive polynomials of the sequence have a common zero, then they must all be 0, contradicting the fact that the last one is not.

COROLLARY. *Let K be an ordered field, f an irreducible polynomial of degree $\geqq 1$ over K. The number of roots of f in two real closures of K inducing the given ordering on K is the same.*

Proof. We can take v sufficiently large positive and u sufficiently large negative in K so that all roots of f and all roots of the polynomials in the

Sturm sequence lie between u and v, using Lemma 1. Then $W_S(u) - W_S(v)$ is the total number of roots of f in any real closure of K inducing the given ordering.

THEOREM 3. *Let K be an ordered field, and let R, R' be real closures of K, whose orderings induce the given ordering on K. Then there exists a unique isomorphism $\sigma : R \to R'$ over K, and this isomorphism is order preserving.*

Proof. We first show that given a finite subextension E of R over K, there exists an embedding of E into R' over K. Let $E = K(\alpha)$, and let $f(X) = \mathrm{Irr}(\alpha, K, X)$. Then $f(\alpha) = 0$ and the Corollary of Sturm's Theorem shows that f has a root β in R'. Thus there exists an isomorphism of $K(\alpha)$ on $K(\beta)$ over K, mapping α on β.

Let $\alpha_1, \ldots, \alpha_n$ be the distinct roots of f in R, and let $\beta_1, \ldots, \beta_m$ be the distinct roots of f in R'. Say

$$\alpha_1 < \cdots < \alpha_n \text{ in the ordering of } R,$$
$$\beta_1 < \cdots < \beta_m \text{ in the ordering of } R'.$$

We contend that $m = n$ and that we can select an embedding σ of $K(\alpha_1, \ldots, \alpha_n)$ into R' such that $\sigma\alpha_i = \beta_i$ for $i = 1, \ldots, n$. Indeed, let γ_i be an element of R such that

$$\gamma_i^2 = \alpha_{i+1} - \alpha_i \qquad \text{for} \quad i = 1, \ldots, n-1$$

and let $E_1 = K(\alpha_1, \ldots, \alpha_n, \gamma_1, \ldots, \gamma_{n-1})$. By what we have seen, there exists an embedding σ of E_1 into R', and then $\sigma\alpha_{i+1} - \sigma\alpha_i$ is a square in R'. Hence

$$\sigma\alpha_1 < \cdots < \sigma\alpha_n.$$

This proves that $m \geqq n$. By symmetry, it follows that $m = n$. Furthermore, the condition that $\sigma\alpha_i = \beta_i$ for $i = 1, \ldots, n$ determines the effect of σ on $K(\alpha_1, \ldots, \alpha_n)$. We contend that σ is order preserving. Let $y \in K(\alpha_1, \ldots, \alpha_n)$ and $0 < y$. Let $\gamma \in R$ be such that $\gamma^2 = y$. There exists an embedding of $K(\alpha_1, \ldots, \alpha_n, \gamma_1, \ldots, \gamma_{n-1}, \gamma)$ into R' over K which must induce σ on $K(\alpha_1, \ldots, \alpha_n)$ and is such that σy is a square, hence > 0, as contended.

Using Zorn's lemma, it is now clear that we get an isomorphism of R onto R' over K. This isomorphism is order preserving because it maps squares on squares, thereby proving our theorem.

PROPOSITION 4. *Let K be an ordered field, K' an extension such that there is no relation*

$$-1 = \sum_{i=1}^{n} a_i\alpha_i^2$$

with $a_i \in K$, $a_i > 0$, and $\alpha_i \in K'$. Let L be the field obtained from K' by adjoining the square roots of all positive elements of K. Then L is real.

Proof. If not, there exists a relation of type

$$-1 = \sum_{i=1}^{n} a_i \alpha_i^2$$

with $a_i \in K$, $a_i > 0$, and $\alpha_i \in L$. (We can take $a_i = 1$.) Let r be the smallest integer such that we can write such a relation with α_i in a subfield of L, of type

$$K'(\sqrt{b_1}, \ldots, \sqrt{b_r})$$

with $b_j \in K$, $b_j > 0$. Write

$$\alpha_i = x_i + y_i\sqrt{b_r}$$

with $x_i, y_i \in K'(\sqrt{b_1}, \ldots, \sqrt{b_{r-1}})$. Then

$$\begin{aligned} -1 &= \sum a_i(x_i + y_i\sqrt{b_r})^2 \\ &= \sum a_i(x_i^2 + 2x_i y_i\sqrt{b_r} + y_i^2 b_r). \end{aligned}$$

By hypothesis, $\sqrt{b_r}$ is not in $K'(\sqrt{b_1}, \ldots, \sqrt{b_{r-1}})$. Hence

$$-1 = \sum a_i x_i^2 + \sum a_i b_r y_i^2,$$

contradicting the minimality of r.

THEOREM 4. *Let K be an ordered field. There exists a real closure R of K inducing the given ordering on K.*

Proof. Take $K' = K$ in Proposition 4. Then L is real, and is contained in a real closure. Our assertion is clear.

COROLLARY. *Let K be an ordered field, and K' an extension field. In order that there exist an ordering on K' inducing the given ordering of K, it is necessary and sufficient that there is no relation of type*

$$-1 = \sum_{i=1}^{n} a_i \alpha_i^2$$

with $a_i \in K$, $a_i > 0$, and $\alpha_i \in K'$.

Proof. If there is no such relation, then Proposition 4 states that L is real, and L is contained in a real closure, whose ordering induces an ordering on K', and the given ordering on K, as desired. The converse is clear.

Example. Let $\overline{\mathbf{Q}}$ be the field of algebraic numbers. One sees at once that $\mathbf{Q}$ admits only one ordering, the ordinary one. Hence any two real closures of $\mathbf{Q}$ in $\overline{\mathbf{Q}}$ are isomorphic, by means of a unique isomorphism.

The real closures of $\mathbf{Q}$ in $\overline{\mathbf{Q}}$ are precisely those subfields of $\overline{\mathbf{Q}}$ which are of finite degree under $\overline{\mathbf{Q}}$. Let K be a finite real extension of $\mathbf{Q}$, contained in $\overline{\mathbf{Q}}$. An element α of K is a sum of squares in K if and only if every conjugate of α in the real numbers is positive, or equivalently, if and only if every conjugate of α in one of the real closures of $\mathbf{Q}$ in $\overline{\mathbf{Q}}$ is positive.

Note. The theory developed in this and the preceding section is due to Artin-Schreier.

§3. *Real zeros and homomorphisms*

Just as we developed a theory of extension of homomorphisms into an algebraically closed field, and Hilbert's Nullstellensatz for zeros in an algebraically closed field, we wish to develop the theory for values in a real closed field. One of the main theorems is the following:

THEOREM 5. *Let k be a field, $K = k(x_1, \ldots, x_n)$ a finitely generated extension. Assume that K is ordered. Let R_k be a real closure of k inducing the same ordering on k as K. Then there exists a homomorphism*

$$\varphi : k[x_1, \ldots, x_n] \to R_k$$

over k.

As applications of Theorem 5, one gets:

COROLLARY 1. *Notation being as in the theorem, let $y_1, \ldots, y_m \in k[x]$ and assume*

$$y_1 < y_2 < \cdots < y_m$$

in the given ordering of K. Then one can choose φ such that

$$\varphi y_1 < \cdots < \varphi y_m.$$

Proof. Let $\gamma_i \in \overline{K}$ be such that $\gamma_i^2 = y_{i+1} - y_i$. Then $K(\gamma_1, \ldots, \gamma_{n-1})$ has an ordering inducing the given ordering on K. We apply the theorem to the ring

$$k[x_1, \ldots, x_n, \gamma_1^{-1}, \ldots, \gamma_{m-1}^{-1}, \gamma_1, \ldots, \gamma_{m-1}].$$

COROLLARY 2. (Artin) *Let k be a real field admitting only one ordering. Let $f(X_1, \ldots, X_n) \in k(X)$ be a rational function having the property that for all $(a) = (a_1, \ldots, a_n) \in k^{(n)}$ such that $f(a)$ is defined, we have $f(a) \geqq 0$. Then $f(X)$ is a sum of squares in $k(X)$.*

Proof. Assume that our conclusion is false. By the Corollary of Theorem 1, §2, there exists an ordering of $k(X)$ in which f is negative. Apply Corollary 1 to the ring

$$k[X_1, \ldots, X_n, h(X)^{-1}]$$

where $h(X)$ is a polynomial denominator for $f(X)$. We can find a homomorphism φ of this ring into R_k (inducing the identity on k) such that $\varphi(f) < 0$. But

$$\varphi(f) = f(\varphi X_1, \ldots, \varphi X_n).$$

By the Lemma of Theorem 3, we can find elements a_i ($i = 1, \ldots, n$) close to φX_i, and $a_i \in k$, so that by continuity, $f(a_1, \ldots, a_n) < 0$, contradiction.

Corollary 2 was a Hilbert problem. The proof which we shall describe for Theorem 5 differs from Artin's proof of the corollary in several technical aspects.

We shall first see how one can reduce Theorem 5 to the case when K has transcendence degree 1 over k, and k is real closed.

LEMMA 1. *Let R be a real closed field and let R_0 be a subfield which is algebraically closed in R (i.e. such that every element of R not in R_0 is transcendental over R_0). Then R_0 is real closed.*

Proof. Let $f(X)$ be an irreducible polynomial over R_0. It splits in R into linear and quadratic factors. Its coefficients in R are algebraic over R_0, and hence must lie in R_0. Hence $f(X)$ is linear itself, or quadratic irreducible already over R_0. By the intermediate value theorem, we may assume that f is positive definite, i.e. $f(a) > 0$ for all $a \in R_0$. Without loss of generality, we may assume that $f(X) = X^2 + b^2$ for some $b \in R_0$. Any root of this polynomial will bring $\sqrt{-1}$ with it and therefore the only algebraic extension of R_0 is $R_0(\sqrt{-1})$. This proves that R_0 is real closed.

Let R_K be a real closure of K inducing the given ordering on K. Let R_0 be the algebraic closure of k in R_K. By the lemma, R_0 is real closed.

We consider the field $R_0(x_1, \ldots, x_n)$. If we can prove our theorem for the ring $R_0[x_1, \ldots, x_n]$, and find a homomorphism

$$\psi : R_0[x_1, \ldots, x_n] \to R_0,$$

then we let $\sigma : R_0 \to R_K$ be an isomorphism over k (it exists by Theorem 3), and we let $\varphi = \sigma \circ \psi$ to solve our problem over k. This reduces our theorem to the case when k is real closed.

Next, let F be an intermediate field, $K \supset F \supset k$, such that K is of transcendence degree 1 over F. Again let R_K be a real closure of K preserving the ordering, and let R_F be the real closure of F contained in R_K. If we know our theorem for extensions of dimension 1, then we can find a homomorphism

$$\psi : R_F[x_1, \ldots, x_n] \to R_F.$$

We note that the field $k(\psi x_1, \ldots, \psi x_n)$ has transcendence degree $\leqq n - 1$, and is real, because it is contained in R_F. Thus we are reduced inductively to the case when K has dimension 1, and as we saw above, when k is real closed.

One can interpret our statement geometrically as follows. We can write $K = R(x, y)$ with x transcendental over R, and (x, y) satisfying some irreducible polynomial $f(X, Y) = 0$ in $R[X, Y]$. What we essentially want to prove is that there are infinitely many points on the curve $f(X, Y) = 0$, with coordinates lying in R, i.e. infinitely many real points.

The main idea is that we find some point $(a, b) \in R^{(2)}$ such that $f(a, b) = 0$ but $D_2 f(a, b) \neq 0$. We can then use the intermediate value theorem. We see that $f(a, b + h)$ changes sign as h changes from a small positive to a small negative element of R. If we take $a' \in R$ close to a, then $f(a', b + h)$ also changes sign for small h, and hence $f(a', Y)$ has a zero in R for all a' sufficiently close to a. In this way we get infinitely many zeros.

To find our point, we consider the polynomial $f(x, Y)$ as a polynomial in one variable Y with coefficients in $R(x)$. Without loss of generality we may assume that this polynomial has leading coefficient 1. We construct a Sturm sequence for this polynomial, say

$$\{f(x, Y), f_1(x, Y), \ldots, f_m(x, Y)\}.$$

Let $d = \deg f$. If we denote by $A(x) = \big(a_{d-1}(x), \ldots, a_0(x)\big)$ the coefficients of $f(x, Y)$, then from the Euclidean algorithm, we see that the coefficients of the polynomials in the Sturm sequence can be expressed as rational functions

$$\{G_v(A(x))\}$$

in terms of $a_{d-1}(x), \ldots, a_0(x)$.

Let

$$v(x) = 1 \pm a_{d-1}(x) \pm \cdots \pm a_0(x) + s,$$

where s is a positive integer, and the signs are selected so that each term in this sum gives a positive contribution. We let $u(x) = -v(x)$, and select s so that neither u nor v is a root of any polynomial in the Sturm sequence for f. Now we need a lemma.

LEMMA 2. *Let R be a real closed field, and $\{h_i(x)\}$ a finite set of rational functions in one variable with coefficients in R. Suppose the rational field $R(x)$ ordered in some way, so that each $h_i(x)$ has a sign attached to it. Then there exist infinitely many special values c of x in R such that $h_i(c)$ is defined and has the same sign as $h_i(x)$, for all i.*

Proof. Considering the numerators and denominators of the rational functions, we may assume without loss of generality that the h_i are polynomials. We then write

$$h_i(x) = \alpha \prod (x - \lambda) \prod p(x),$$

where the first product is extended over all roots λ of h_i in R, and the second product is over positive definite quadratic factors over R. For any $\xi \in R$, $p(\xi)$ is positive. It suffices therefore to show that the signs of

$(x - \lambda)$ can be preserved for all λ by substituting infinitely many values α for x. We order all values of λ and of x and obtain

$$\cdots < \lambda_1 < x < \lambda_2 < \cdots$$

where possibly λ_1 or λ_2 is omitted if x is larger or smaller than any λ. Any value α of x in R selected between λ_1 and λ_2 will then satisfy the requirements of our lemma.

To apply the lemma to the existence of our point, we let the rational functions $\{h_i(x)\}$ consist of all coefficients $a_{\alpha-1}(x), \dots, a_0(x)$, all rational functions $G_v(A(x))$, and all values $f_j(x, u(x)), f_j(x, v(x))$ whose variation in signs satisfied Sturm's theorem. We then find infinitely many special values α of x in R which preserve the signs of these rational functions. Then the polynomials $f(\alpha, Y)$ have roots in R, and for all but a finite number of α, these roots have multiplicity 1.

It is then a matter of simple technique to see that for all but a finite number of points on the curve, the elements $x_1, \dots, x_n$ lie in the local ring of the homomorphism $R[x, y] \to R$ mapping (x, y) on (a, b) such that $f(a, b) = 0$ but $D_2 f(a, b) \neq 0$. (Cf. for instance the example at the end of §4, Chapter XII, and Exercise 12 of that chapter.) One could also give direct proofs here. In this way, we obtain homomorphisms

$$R[x_1, \dots, x_n] \to R,$$

thereby proving Theorem 5.

THEOREM 6. *Let k be a real field, $K = k(x_1, \dots, x_n, y) = k(x, y)$ a finitely generated extension such that $x_1, \dots, x_n$ are algebraically independent over k, and y is algebraic over $k(x)$. Let $f(X, Y)$ be the irreducible polynomial in $k[X, Y]$ such that $f(x, y) = 0$. Let R be a real closed field containing k, and assume that there exists $(a, b) \in R^{(n+1)}$ such that $f(a, b) = 0$ but $D_{n+1} f(a, b) \neq 0$. Then K is real.*

Proof. Let $t_1, \dots, t_n$ be algebraically independent over R. Inductively, we can put an ordering on $R(t_1, \dots, t_n)$ such that each t_i is infinitely small with respect to R, (cf. the example in §1). Let R' be a real closure of $R(t_1, \dots, t_n)$ preserving the ordering. Let $u_i = a_i + t_i$ for each $i = 1, \dots, n$. Then $f(u, b + h)$ changes sign for small h positive and negative in R, and hence $f(u, Y)$ has a root in R', say v. Since f is irreducible, the isomorphism of $k(x)$ on $k(u)$ sending x_i on u_i extends to an embedding of $k(x, y)$ into R', and hence K is real, as was to be shown.

In the language of algebraic geometry, Theorems 5 and 6 state that the function field of a variety over a real field k is real if and only if the variety has a simple point in some real closure of k.

CHAPTER XII

Absolute Values

§1. Definitions, dependence, and independence

Let K be a field. An *absolute value* v on K is a real-valued function $x \mapsto |x|_v$ on K satisfying the following three properties:

AV 1. We have $|x|_v \geqq 0$ for all $x \in K$, and $|x|_v = 0$ if and only if $x = 0$.

AV 2. For all $x, y \in K$, we have $|xy|_v = |x|_v |y|_v$.

AV 3. For all $x, y \in K$, we have $|x + y|_v \leqq |x|_v + |y|_v$.

If instead of AV 3 the absolute value satisfies the stronger condition

AV 4. $|x + y|_v \leqq \max(|x|_v, |y|_v)$

then we shall say that it is a *valuation*, or that it is non-archimedean.

The absolute value which is such that $|x|_v = 1$ for all $x \neq 0$ is called *trivial.*

We shall write $|x|$ instead of $|x|_v$ if we deal with just one fixed absolute value. We also refer to v as the absolute value.

An absolute value on K defines a metric. The distance between two elements x, y of K in this metric is $|x - y|$. Thus an absolute value defines a topology on K. Two absolute values are called *dependent* if they define the same topology. If they do not, they are called independent.

We observe that $|1| = |1^2| = |(-1)^2| = |1|^2$ whence

$$|1| = |-1| = 1.$$

Also, $|-x| = |x|$ for all $x \in K$, and $|x^{-1}| = |x|^{-1}$ for $x \neq 0$.

PROPOSITION 1. *Let $|\ |_1$ and $|\ |_2$ be non-trivial absolute values on a field K. They are dependent if and only if the relation*

$$|x|_1 < 1$$

implies $|x|_2 < 1$. If they are dependent, then there exists a number $\lambda > 0$ such that $|x|_1 = |x|_2^{\lambda}$ for all $x \in K$.

Proof. If the two absolute values are dependent, then our condition is satisfied, because the set of $x \in K$ such that $|x|_1 < 1$ is the same as the set such that $\lim x^n = 0$ for $n \to \infty$. Conversely, assume the condition satisfied. Then $|x|_1 > 1$ implies $|x|_2 > 1$ since $|x^{-1}|_1 < 1$. By hypothesis, there exists an element $x_0 \in K$ such that $|x_0|_1 > 1$. Let $a = |x_0|_1$ and $b = |x_0|_2$. Let

$$\lambda = \frac{\log b}{\log a}.$$

Let $x \in K$, $x \neq 0$. Then $|x|_1 = |x_0|_1^\alpha$ for some number α. If m, n are integers such that $m/n > \alpha$ and $n > 0$, we have

$$|x|_1 < |x_0|_1^{m/n}$$

whence

$$|x^n/x_0^m|_1 < 1,$$

and thus

$$|x^n/x_0^m|_2 < 1.$$

This implies that $|x|_2 < |x_0|_2^{m/n}$. Hence

$$|x|_2 \leqq |x_0|_2^\alpha.$$

Similarly, one proves the reverse inequality, and thus one gets

$$|x|_2 = |x_0|_2^\alpha$$

for all $x \in K$, $x \neq 0$. The assertion of the proposition is now obvious, i.e. $|x|_2 = |x|_1^\lambda$.

We shall give some examples of absolute values.

Consider first the rational numbers. We have the ordinary absolute value such that $|m| = m$ for any positive integer m.

For each prime number p, we have the p-adic absolute value v_p, defined by the formula

$$|p^r m/n|_p = 1/p^r$$

where r is an integer, and m, n are integers $\neq 0$, not divisible by p. One sees at once that the p-adic absolute value is non-archimedean.

One can give a similar definition of a valuation for any field K which is the quotient field of a principal ring. For instance, let $K = k(t)$ where k is a field and t is a variable over k. We have a valuation v_p for each irreducible polynomial $p(t)$ in $k[t]$, defined as for the rational numbers, but there is no way of normalizing it in a natural way. Thus we select a number c with $0 < c < 1$ and for any rational function $p^r f/g$ where f, g are polynomials not divisible by p, we define

$$|p^r f/g|_p = c^r.$$

The various choices of the constant c give rise to dependent valuations.

Any subfield of the complex numbers (or real numbers) has an absolute value, induced by the ordinary absolute value on the complex numbers. We shall see later how to obtain absolute values on certain fields by embedding them into others which are already endowed with natural absolute values.

Suppose that we have an absolute value on a field which is bounded on the prime ring (i.e. the integers $\mathbf{Z}$ *if the characteristic is* 0, *or the integers* mod p *if the characteristic is* p). *Then the absolute value is necessarily non-archimedean.*

Proof. For any elements x, y and any positive integer n, we have

$$|(x + y)^n| \leqq \sum \left| \binom{n}{\nu} x^\nu y^{n-\nu} \right| \leqq nC \max(|x|, |y|)^n.$$

Taking n-th roots and letting n go to infinity proves our assertion. We note that this is always the case in characteristic > 0 because the prime ring is finite!

If the absolute value is archimedean, then we refer the reader to any other book in which there is a discussion of absolute values for a proof of the fact that it is dependent on the ordinary absolute value. This fact is essentially useless (and is never used in the sequel), because we always start with a concretely given set of absolute values on fields which interest us.

In Proposition 1 we derived a strong condition on dependent absolute values. We shall now derive a condition on independent ones.

APPROXIMATION THEOREM. (Artin-Whaples) *Let* K *be a field and* $| \ |_1, \ldots, | \ |_s$ *non-trivial pairwise independent absolute values on* K. *Let* $x_1, \ldots, x_s$ *be elements of* K, *and* $\epsilon > 0$. *Then there exists* $x \in K$ *such that*

$$|x - x_i|_i < \epsilon$$

for all i.

Proof. Consider first two of our absolute values, say v_1 and v_2. By hypothesis we can find $\alpha \in K$ such that $|\alpha|_1 < 1$ and $|\alpha|_s \geqq 1$. Similarly, we can find $\beta \in K$ such that $|\beta|_1 \geqq 1$ and $|\beta|_s < 1$. Put $y = \beta/\alpha$. Then $|y|_1 > 1$ and $|y|_s < 1$.

We shall now prove that there exists $z \in K$ such that $|z|_1 > 1$ and $|z|_j < 1$ for $j = 2, \ldots, s$. We prove this by induction, the case $s = 2$ having just been proved. Suppose we have found $z \in K$ satisfying

$$|z|_1 > 1 \quad \text{and} \quad |z|_j < 1 \qquad \text{for} \quad j = 2, \ldots, s - 1.$$

If $|z|_s \leqq 1$ then the element $z^n y$ for large n will satisfy our requirements.

If $|z|_s > 1$, then the sequence

$$t_n = \frac{z^n}{1 + z^n}$$

tends to 1 at v_1 and v_s, and tends to 0 at v_j ($j = 2, \dots, s - 1$). For large n, it is then clear that $t_n y$ satisfies our requirements.

Using the element z that we have just constructed, we see that the sequence $z^n/(1 + z^n)$ tends to 1 at v_1 and to 0 at v_j for $j = 2, \dots, s$. For each $i = 1, \dots, s$ we can therefore construct an element z_i which is very close to 1 at v_i and very close to 0 at v_j ($j \neq i$). The element

$$x = z_1 x_1 + \cdots + z_s x_s$$

then satisfies the requirement of the theorem.

§2. *Completions*

Let K be a field with a non-trivial absolute value v, which will remain fixed throughout this section. One can then define in the usual manner the notion of a Cauchy sequence. It is a sequence $\{x_n\}$ of elements in K such that, given $\epsilon > 0$, there exists an integer N such that for all $n, m > N$ we have

$$|x_n - x_m| < \epsilon.$$

We say that K is *complete* if every Cauchy sequence converges.

PROPOSITION 2. *There exists a pair (K_v, i) consisting of a field K_v, complete under an absolute value, and an embedding $i : K \to K_v$ such that the absolute value on K is induced by that of K_v (i.e. $|x|_v = |ix|$ for $x \in K$), and such that iK is dense in K_v. If (K'_v, i') is another such pair, then there exists a unique isomorphism $\varphi : K_v \to K'_v$ preserving the absolute values, and making the following diagram commutative:*

$$\begin{array}{ccc} K_v & \xrightarrow{\varphi} & K'_v \\ & {}_i\nwarrow \quad \nearrow_{i'} & \\ & K & \end{array}$$

Proof. The uniqueness is obvious. One proves the existence in the well-known manner, which we shall now recall briefly, leaving the details to the reader.

The Cauchy sequences form a ring, addition and multiplication being taken componentwise.

One defines a null sequence to be a sequence $\{x_n\}$ such that $\lim_{n \to \infty} x_n = 0$. The null sequences form an ideal in the ring of Cauchy sequences, and in fact form a maximal ideal. (If a Cauchy sequence is not a null sequence, then it stays away from 0 for all n sufficiently large, and one can then take the inverse of almost all its terms. Up to a finite number of terms, one then gets again a Cauchy sequence.)

The residue class field of Cauchy sequences modulo null sequences is the field K_v. We embed K in K_v "on the diagonal", i.e. send $x \in K$ on the sequence $(x, x, x, \ldots)$.

We extend the absolute value of K to K_v by continuity. If $\{x_n\}$ is a Cauchy sequence, representing an element ξ in K_v, we define $|\xi| = \lim |x_n|$. It is easily proved that this yields an absolute value (independent of the choice of representative sequence $\{x_n\}$ for ξ), and this absolute value induces the given one on K.

Finally, one proves that K_v is complete. This is done by the usual diagonal process. If $\xi_1, \xi_2, \ldots$ is a Cauchy sequence in K_v, and ξ_j is represented by the Cauchy sequence $\{x_{jn}\}$ in K, then one proves without difficulty that

$$x_{11}, x_{22}, x_{33}, \ldots$$

is a Cauchy sequence in K, and represents an element ξ in K_v such that

$$\lim_{j\to\infty} \xi_j = \xi.$$

A pair (K_v, i) as in Proposition 2 may be called *a* completion of K. The standard pair obtained by the preceding construction could be called *the* completion of K.

Let K have a non-trivial archimedean absolute value v. If one knows that the restriction of v to the rationals is dependent on the ordinary absolute value, then the completion K_v is a complete field, containing the completion of $\mathbf{Q}$ as a closed subfield, i.e. containing the real numbers $\mathbf{R}$ as a closed subfield. It will be worthwhile to state the theorem of Gelfand-Mazur concerning the structure of such fields. First we define the notion of normed vector space.

Let K be a field with a non-trivial absolute value, and let E be a vector space over K. By a *norm* on E (compatible with the absolute value of K) we shall mean a function $\xi \mapsto |\xi|$ of E into the real numbers such that:

NO 1. $|\xi| \geqq 0$ for all $\xi \in E$, and $= 0$ if and only if $\xi = 0$.
NO 2. For all $x \in K$ and $\xi \in E$ we have $|x\xi| \leqq |x|\,|\xi|$.
NO 3. If $\xi, \xi' \in E$ then $|\xi + \xi'| \leqq |\xi| + |\xi'|$.

Two norms $|\ |_1$ and $|\ |_2$ are called *equivalent* if there exist numbers C_1, $C_2 > 0$ such that for all $\xi \in E$ we have

$$C_1|\xi|_1 \leqq |\xi|_2 \leqq C_2|\xi|_1.$$

Suppose that E is finite dimensional, and let $\omega_1, \ldots, \omega_n$ be a basis of E over K. If we write an element

$$\xi = x_1\omega_1 + \cdots + x_n\omega_n$$

in terms of this basis, with $x_i \in K$, then we can define a norm by putting

$$|\xi| = \max_i |x_i|.$$

The three properties defining a norm are trivially satisfied.

Proposition 3. *Let K be a complete field under a non-trivial absolute value, and let E be a finite-dimensional space over K. Then any two norms on E (compatible with the given absolute value on K) are equivalent.*

Proof. We shall first prove that the topology on E is that of a product space, i.e. if $\omega_1, \ldots, \omega_n$ is a basis of E over K, then a sequence

$$\xi^{(\nu)} = x_1^{(\nu)}\omega_1 + \cdots + x_n^{(\nu)}\omega_n, \qquad x_i^{(\nu)} \in K,$$

is a Cauchy sequence in E only if each one of the n sequences $x_i^{(\nu)}$ is a Cauchy sequence in K. We do this by induction on n. It is obvious for $n = 1$. Assume $n \geqq 2$. We consider a sequence as above, and without loss of generality, we may assume that it converges to 0. (If necessary, consider $\xi^{(\nu)} - \xi^{(\mu)}$ for $\nu, \mu \to \infty$.) We must then show that the sequences of the coefficients converge to 0 also. If this is not the case, then there exists a number $a > 0$ such that we have for some j, say $j = 1$,

$$|\xi_1^{(\nu)}| > a$$

for arbitrarily large ν. Thus for a subsequence of (ν), $\xi^{(\nu)}/x_1^{(\nu)}$ converges to 0, and we can write

$$\frac{\xi^{(\nu)}}{x_1^{(\nu)}} - \omega_1 = \frac{x_2^{(\nu)}}{x_1^{(\nu)}}\omega_2 + \cdots + \frac{x_n^{(\nu)}}{x_1^{(\nu)}}\omega_n.$$

We let $\eta^{(\nu)}$ be the right-hand side of this equation. Then the subsequence $\eta^{(\nu)}$ converges (according to the left-hand side of our equation). By induction, we conclude that its coefficients in terms of $\omega_2, \ldots, \omega_n$ also converge in K, say to $y_2, \ldots, y_n$. Taking the limit, we get

$$\omega_1 = y_2\omega_2 + \cdots + y_n\omega_n,$$

contradicting the linear independence of the ω_i.

We must finally see that two norms inducing the same topology are equivalent. Let $| \ |_1$ and $| \ |_2$ be these norms. There exists a number $C > 0$ such that for any $\xi \in E$ we have

$$|\xi|_1 \leqq C \quad \text{implies} \quad |\xi|_2 \leqq 1.$$

Let $a \in K$ be such that $0 < |a| < 1$. For every $\xi \in E$ there exists a unique integer s such that

$$C|a| < |a^s \xi|_1 \leqq C.$$

Hence $|a^s \xi|_2 \leqq 1$ whence we get at once

$$|\xi|_2 \leqq C^{-1}|a|^{-1}|\xi|_1.$$

The other inequality follows by symmetry, with a similar constant.

THEOREM 1. *Let A be a commutative algebra over the real numbers, and assume that A contains an element j such that $j^2 = -1$. Let $\mathbf{C} = \mathbf{R} + \mathbf{R}j$. Assume that A is normed (as a vector space over $\mathbf{R}$), and that $|xy| \leqq |x|\,|y|$ for all $x, y \in A$. Given $x_0 \in A$, $x_0 \neq 0$, there exists an element $c \in \mathbf{C}$ such that $x_0 - c$ is not invertible in A.*

Proof. (Tornheim) Assume that $x_0 - z$ is invertible for all $z \in \mathbf{C}$. Consider the mapping $f : \mathbf{C} \to A$ defined by

$$f(z) = (x_0 - z)^{-1}.$$

It is easily verified (as usual) that taking inverses is a continuous operation. Hence f is continuous, and for $z \neq 0$ we have

$$f(z) = z^{-1}(x_0 z^{-1} - 1)^{-1} = \frac{1}{z}\left(\frac{1}{\dfrac{x_0}{z} - 1}\right).$$

From this we see that $f(z)$ approaches 0 when z goes to infinity (in $\mathbf{C}$). Hence the map $z \mapsto |f(z)|$ is a continuous map of $\mathbf{C}$ into the real numbers $\geqq 0$, is bounded, and is small outside some large circle. Hence it has a maximum, say M. Let D be the set of elements $z \in \mathbf{C}$ such that $|f(z)| = M$. Then D is not empty; D is bounded and closed. We shall prove that D is open, whence a contradiction.

Let c_0 be a point of D, which, after a translation, we may assume to be the origin. We shall see that if r is real > 0 and small, then all points on the circle of radius r lie in D. Indeed, consider the sum

$$S(n) = \frac{1}{n}\sum_{k=1}^{n} \frac{1}{x_0 + \omega^k r}$$

where ω is a primitive n-th root of unity. Taking formally the logarithmic derivative of $X^n - r^n = \prod_{k=1}^{n} (X - \omega^k r)$ shows that

$$\frac{nX^{n-1}}{X^n - r^n} = \sum_{k=1}^{n} \frac{1}{X - \omega^k r},$$

and hence, dividing by n, and by X^{n-1}, and substituting x_0 for X, we obtain

$$S(n) = \frac{1}{x_0 - r(r/x_0)^{n-1}}.$$

If r is small (say $|r/x_0| < 1$), then we see that

$$\lim_{n \to \infty} |S(n)| = \left|\frac{1}{x_0}\right| = M.$$

Suppose that there exists a complex number λ of absolute value 1 such that

$$\left|\frac{1}{x_0 + \lambda r}\right| < M.$$

Then there exists an interval on the unit circle near λ, and there exists $\epsilon > 0$ such that for all roots of unity ζ lying in this interval, we have

$$\left|\frac{1}{x_0 + \zeta r}\right| < M - \epsilon.$$

(This is true by continuity.) Let us take n very large. Let b_n be the number of n-th roots of unity lying in our interval. Then b_n/n is approximately equal to the length of the interval (times 2π). We can express $S(n)$ as a sum

$$S(n) = \frac{1}{n}\left[\sum\nolimits_{\text{I}} \frac{1}{x_0 + \omega^k r} + \sum\nolimits_{\text{II}} \frac{1}{x_0 + \omega^k r}\right],$$

the first sum $\sum_{\text{I}}$ being taken over those roots of unity ω^k lying in our interval, and the second sum being taken over the others. Each term in the second sum has norm $\leqq M$ because M is a maximum. Hence we obtain the estimate

$$\begin{aligned} |S(n)| &\leqq \frac{1}{n}\left[\left|\sum\nolimits_{\text{I}}\right| + \left|\sum\nolimits_{\text{II}}\right|\right] \\ &\leqq \frac{1}{n}\left(b_n(M - \epsilon) + (n - b_n)M\right) \\ &\leqq M - \frac{b_n}{n}\epsilon. \end{aligned}$$

This contradicts the fact that the limit of $|S(n)|$ is equal to M.

COROLLARY. *Let K be a field, which is an extension of* **R**, *and has an absolute value extending the ordinary absolute value on* **R**. *Then $K = \mathbf{R}$ or $K = \mathbf{C}$.*

Proof. Assume first that K contains **C**. Then the assumption that K is a field and Theorem 1 imply that $K = \mathbf{C}$.

If K does not contain **C**, in other words, does not contain a square root of -1, we let $L = K(j)$ where $j^2 = -1$. We define a norm on L (as an **R**-space) by putting

$$|x + yj| = |x| + |y|$$

for $x, y \in K$. This clearly makes L into a normed **R**-space. Furthermore, if $z = x + yj$ and $z' = x' + y'j$ are in L, then

$$\begin{aligned}
|zz'| &= |xx' - yy'| + |xy' + x'y| \\
&\leqq |xx'| + |yy'| + |xy'| + |x'y| \\
&\leqq |x|\,|x'| + |y|\,|y'| + |x|\,|y'| + |x'|\,|y| \\
&\leqq (|x| + |y|)(|x'| + |y'|) \\
&\leqq |z|\,|z'|,
\end{aligned}$$

and we can therefore apply Theorem 1 again to conclude the proof.

As an important application of Proposition 3, we have:

PROPOSITION 4. *Let K be complete with respect to a non-trivial absolute value v. If E is any algebraic extension of K, then v has a unique extension to E. If E is finite over K, then E is complete.*

Proof. In the archimedean case, the existence is obvious since we deal with the real and complex numbers. In the non-archimedean case, we postpone the existence proof to a later section. It uses entirely different ideas from the present ones. As to uniqueness, we may assume that E is finite over K. By Proposition 3, an extension of v to E defines the same topology as the max norm obtained in terms of a basis as above. Given a Cauchy sequence $\xi^{(\nu)}$ in E,

$$\xi^{(\nu)} = x_{\nu 1}\omega_1 + \cdots + x_{\nu n}\omega_n,$$

the n sequences $\{x_{\nu i}\}$ $(i = 1, \ldots, n)$ must be Cauchy sequences in K by the definition of the max norm. If $\{x_{\nu i}\}$ converges to an element z_i in K, then it is clear that the sequence $\xi^{(\nu)}$ converges to $z_1\omega_1 + \cdots + z_n\omega_n$. Hence E is complete. Furthermore, since any two extensions of v to E are equivalent, we can apply Proposition 1, and we see that we must have $\lambda = 1$ since the extensions induce the same absolute value v on K. This proves what we want.

From the uniqueness we can get an explicit determination of the absolute value on an algebraic extension of K. Observe first that if E is a normal extension of K, and σ is an automorphism of E over K, then the function

$$x \mapsto |\sigma x|$$

is an absolute value on E extending that of K. Hence we must have

$$|\sigma x| = |x|$$

for all $x \in E$. If E is algebraic over K, and σ is an embedding of E over K in $\overline{K}$ then the same conclusion remains valid, as one sees immediately by embedding E in a normal extension of K. In particular, if α is algebraic over K, of degree n, and if $\alpha_1, \ldots, \alpha_n$ are its conjugates (counting multiplicities, equal to the degree of inseparability), then all the absolute values $|\alpha_i|$ are equal. Denoting by N the norm from $K(\alpha)$ to K, we see that

$$|N(\alpha)| = |\alpha|^n,$$

and taking the n-th root, we get:

PROPOSITION 5. *Let K be complete with respect to a non-trivial absolute value. Let α be algebraic over K, and let N be the norm from $K(\alpha)$ to K. Let $n = [K(\alpha) : K]$. Then*

$$|\alpha| = |N(\alpha)|^{1/n}.$$

In the special case of the complex numbers over the real numbers, we can write $\alpha = a + bi$ with $a, b \in \mathbf{R}$, and we see that the formula of Proposition 5 is a generalization of the formula for the absolute value of a complex number,

$$\alpha = (a^2 + b^2)^{1/2},$$

since $a^2 + b^2$ is none other than the norm of α from $\mathbf{C}$ to $\mathbf{R}$.

§3. Finite extensions

Throughout this section we shall deal with a field K having a non-trivial absolute value v.

We wish to describe how this absolute value extends to finite extensions of K. If E is an extension of K and w is an absolute value on E extending v, then we shall write $w \mid v$.

If we let K_v be the completion, we know that v can be extended to K_v, and then uniquely to its algebraic closure $\overline{K}_v$. If E is a finite extension of K, or even an algebraic one, then we can extend v to E by embedding E in $\overline{K}_v$ by an isomorphism over K, and taking the induced absolute value on E. We shall now prove that every extension of v can be obtained in this manner.

PROPOSITION 6. *Let E be a finite extension of K. Let w be an absolute value on E extending v, and let E_w be the completion. Let K_w be the closure of K in E_w and identify E in E_w. Then $E_w = EK_w$ (the composite field).*

Proof. We observe that K_w is a completion of K, and that the composite field EK_w is algebraic over K_w and therefore complete by §2, Proposition 4. Since it contains E, it follows that E is dense in it, and hence that $E_w = EK_w$.

If we start with an embedding $\sigma : E \to \overline{K}_v$ (always assumed to be over K), then we know again by §2, Proposition 4 that $\sigma E \cdot K_v$ is complete. Thus this construction and the construction of the proposition are essentially the same, up to an isomorphism. In the future, we take the embedding point of view. We must now determine when two embeddings give us the same absolute value on E.

Given two embeddings $\sigma, \tau : E \to \overline{K}_v$, we shall say that they are *conjugate over* K_v if there exists an automorphism λ of $\overline{K}_v$ over K_v such that $\sigma = \lambda\tau$. We see that actually λ is determined by its effect on τE, or $\tau E \cdot K_v$.

Proposition 7. *Let E be an algebraic extension of K. Two embeddings $\sigma, \tau : E \to \overline{K}_v$ give rise to the same absolute value on E if and only if they are conjugate over K_v.*

Proof. Suppose they are conjugate over K_v. Then the uniqueness of the extension of the absolute value from K_v to $\overline{K}_v$ guarantees that the induced absolute values on E are equal. Conversely, suppose this is the case. Let $\lambda : \tau E \to \sigma E$ be an isomorphism over K. We shall prove that λ extends to an isomorphism of $\tau E \cdot K_v$ onto $\sigma E \cdot K_v$ over K_v. Since τE is dense in $\tau E \cdot K_v$, an element $x \in \tau E \cdot K_v$ can be written

$$x = \lim \tau x_n$$

with $x_n \in E$. Since the absolute values induced by σ and τ on E coincide, it follows that the sequence $\lambda\tau x_n = \sigma x_n$ converges to an element of $\sigma E \cdot K_v$ which we denote by λx. One then verifies immediately that λx is independent of the particular sequence τx_n used, and that the map $\lambda : \tau E \cdot K_v \to \sigma E \cdot K_v$ is an isomorphism, which clearly leaves K_v fixed. This proves our proposition.

In view of the previous two propositions, if w is an extension of v to a finite extension E of K, then we may identify E_w and a composite extension EK_v of E and K_v. If $N = [E : K]$ is finite, then we shall call

$$N_w = [E_w : K_v]$$

the *local degree.*

Proposition 8. *Let E be a finite separable extension of K, of degree N. Then*

$$N = \sum_{w|v} N_w.$$

Proof. We can write $E = K(\alpha)$ for a single element α. Let $f(X)$ be its irreducible polynomial over K. Then over K_v, we have a decomposition

$$f(X) = f_1(X) \cdots f_r(X)$$

into irreducible factors $f_i(X)$. They all appear with multiplicity 1 according to our hypothesis of separability. The embeddings of E into $\overline{K}_v$ correspond to the maps of α onto the roots of the f_i. Two embeddings are conjugate if and only if they map α onto roots of the same polynomial f_i. On the other hand, it is clear that the local degree in each case is precisely the degree of f_i. This proves our proposition.

PROPOSITION 9. *Let E be a finite extension of K. Then*

$$\sum_{w|v} [E_w : K_v] \leqq [E : K].$$

If E is purely inseparable over K, then there exists only one absolute value w on E extending v.

Proof. Let us first prove the second statement. If E is purely inseparable over K, and p^r is its inseparable degree, then $\alpha^{p^r} \in K$ for every α in E. Hence v has a unique extension to E. Consider now the general case of a finite extension, and let $F = E^{p^r}K$. Then F is separable over K and E is purely inseparable over F. By the preceding proposition,

$$\sum_{w|v} [F_w : K_v] = [F : K],$$

and for each w, we have $[E_w : F_w] \leqq [E : F]$. From this our inequality in the statement of the proposition is obvious.

Whenever v is an absolute value on K such that for any finite extension E of K we have $[E : K] = \sum_{w|v} [E_w : K_v]$ we shall say that v is *well behaved*. Suppose we have a tower of finite extensions, $L \supset E \supset K$. Let w range over the absolute values of E extending v, and u over those of L extending v. If $u \mid w$ then L_u contains E_w. Thus we have:

$$\begin{aligned} \sum_{u|v} [L_u : K_v] &= \sum_{w|v} \sum_{u|w} [L_u : E_w][E_w : K_v] \\ &= \sum_{w|v} [E_w : K_v] \sum_{u|w} [L_u : E_w] \\ &\leqq \sum_{w|v} [E_w : K_v][L : E] \\ &\leqq [E : K][L : E]. \end{aligned}$$

From this we immediately see that if v is well behaved, E finite over K, and w extends v on E, then w is well behaved (we must have an equality everywhere).

Let E be a finite extension of K. Let p^r be its inseparable degree. We recall that the norm of an element $\alpha \in K$ is given by the formula

$$N_K^E(\alpha) = \prod_\sigma \sigma\alpha^{p^r}$$

where σ ranges over all distinct isomorphisms of E over K (into a given algebraic closure).

If w is an absolute value extending v on E, then the norm from E_w to K_v will be called the *local norm.*

Replacing the above product by a sum, we get the trace, and the local trace. We abbreviate the trace by Tr.

PROPOSITION 10. *Let E be a finite extension of K, and assume that v is well behaved. Let $\alpha \in E$. Then:*

$$N_K^E(\alpha) = \prod_{w|v} N_{K_v}^{E_w}(\alpha)$$

$$\mathrm{Tr}_K^E(\alpha) = \sum_{w|v} \mathrm{Tr}_{K_v}^{E_w}(\alpha)$$

Proof. Suppose first that $E = K(\alpha)$, and let $f(X)$ be the irreducible polynomial of α over K. If we factor $f(X)$ into irreducible terms over K_v, then

$$f(X) = f_1(X) \cdots f_r(X)$$

where each $f_i(X)$ is irreducible, and the f_i are distinct because of our hypothesis that v is well behaved. The norm $N_K^E(\alpha)$ is equal to $(-1)^{\deg f}$ times the constant term of f, and similarly for each f_i. Since the constant term of f is equal to the product of the constant terms of the f_i, we get the first part of the proposition. The statement for the trace follows by looking at the penultimate coefficient of f and each f_i.

If E is not equal to $K(\alpha)$, then we simply use the transitivity of the norm and trace. We leave the details to the reader.

One can also argue directly on the embeddings. Let $\sigma_1, \ldots, \sigma_m$ be the distinct embeddings of E into $\overline{K}_v$ over K, and let p^r be the inseparable degree of E over K. The inseparable degree of $\sigma E \cdot K_v$ over K_v for any σ is at most equal to p^r. If we separate $\sigma_1, \ldots, \sigma_m$ into distinct conjugacy classes over K_v, then from our hypothesis that v is well behaved, we conclude at once that the inseparable degree of $\sigma_i E \cdot K_v$ over K_v must be equal to p^r also, for each i. Thus the formula giving the norm as a product over conjugates with multiplicity p^r breaks up into a product of factors corresponding to the conjugacy classes over K_v.

Taking into account Proposition 5 of §2, we have:

PROPOSITION 11. *Let K have a well-behaved absolute value v. Let E be a finite extension of K, and $\alpha \in E$. Let*

$$N_w = [E_w : K_v]$$

for each absolute value w on E extending v. Then

$$\prod_{w|v} |\alpha|_w^{N_w} = |N_K^E(\alpha)|_v.$$

§4. *Valuations*

In this section, we shall obtain, among other things, the existence theorem concerning the possibility of extending non-archimedean absolute values to algebraic extensions. We introduce first a generalization of the notion of non-archimedean absolute value.

Let Γ be a multiplicative commutative group. We shall say that an *ordering* is defined in Γ if we are given a subset S of Γ closed under multiplication such that Γ is the disjoint union of S, the unit element 1, and the set S^{-1} consisting of all inverses of elements of S.

If $\alpha, \beta \in \Gamma$ we define $\alpha < \beta$ to mean $\alpha\beta^{-1} \in S$. We have $\alpha < 1$ if and only if $\alpha \in S$. One easily verifies the following properties of the relation $<$:

1. For $\alpha, \beta \in \Gamma$ we have $\alpha < \beta$, or $\alpha = \beta$, or $\beta < \alpha$, and these possibilities are mutually exclusive.
2. $\alpha < \beta$ implies $\alpha\gamma < \beta\gamma$ for any $\gamma \in \Gamma$.
3. $\alpha < \beta$ and $\beta < \gamma$ implies $\alpha < \gamma$.

(Conversely, a relation satisfying the three properties gives rise to a subset S consisting of all elements < 1. However, we don't need this fact in the sequel.)

It is convenient to attach to an ordered group formally an extra element 0, such that $0\alpha = 0$, and $0 < \alpha$ for all $\alpha \in \Gamma$. The ordered group is then analogous to the multiplicative group of positive reals, except that there may be non-archimedean ordering.

If $\alpha \in \Gamma$ and n is an integer $\neq 0$, such that $\alpha^n = 1$, then $\alpha = 1$. This follows at once from the assumption that S is closed under multiplication and does not contain 1. In particular, the map $\alpha \mapsto \alpha^n$ is injective.

Let K be a field. By a *valuation* of K we shall mean a map $x \mapsto |x|$ of K into an ordered group Γ, together with the extra element 0, such that:

VAL 1. $|x| = 0$ if and only if $x = 0$.
VAL 2. $|xy| = |x|\,|y|$ for all $x, y \in K$.
VAL 3. $|x + y| \leqq \max(|x|, |y|)$.

We see that a valuation gives rise to a homomorphism of the multiplicative group K^* into Γ. The valuation is called *trivial* if it maps K^* on 1. If the map giving the valuation is not surjective, then its image is an ordered subgroup of Γ, and by taking its restriction to this image, we obtain a valuation onto an ordered group, called the *value group*.

We shall denote valuations also by v. If v_1, v_2 are two valuations of K, we shall say that they are *equivalent* if there exists an order-preserving isomorphism λ of the image of v_1 onto the image of v_2 such that

$$|x|_2 = \lambda|x|_1$$

for all $x \in K$. (We agree that $\lambda(0) = 0$.)

Valuations have additional properties, like absolute values. For instance, $|1| = 1$ because $|1| = |1|^2$. Furthermore,

$$|\pm x| = |x|$$

for all $x \in K$. Proof obvious. Also, if $|x| < |y|$ then

$$|x + y| = |y|.$$

To see this, note that under our hypothesis, we have

$$|y| = |y + x - x| \leqq \max(|y + x|, |x|) = |x + y| \leqq \max(|x|, |y|) = |y|.$$

Finally, in a sum

$$x_1 + \cdots + x_n = 0,$$

at least two elements of the sum have the same value. This is an immediate consequence of the preceding remark.

Let K be a field. A subring $\mathfrak{o}$ of K is called a *valuation ring* if it has the property that for any $x \in K$ we have $x \in \mathfrak{o}$ or $x^{-1} \in \mathfrak{o}$.

We shall now see that valuation rings give rise to valuations. Let $\mathfrak{o}$ be a valuation ring of K and let U be the group of units of $\mathfrak{o}$. We contend that $\mathfrak{o}$ is a local ring. Indeed suppose that $x, y \in \mathfrak{o}$ are not units. Say $x/y \in \mathfrak{o}$. Then $1 + x/y = (x + y)/y \in \mathfrak{o}$. If $x + y$ were a unit then $1/y \in \mathfrak{o}$, contradicting the assumption that y is not a unit. Hence $x + y$ is not a unit. One sees trivially that for $z \in \mathfrak{o}$, zx is not a unit. Hence the non-units form an ideal, which must therefore be the unique maximal ideal of $\mathfrak{o}$.

Let $\mathfrak{m}$ be the maximal ideal of $\mathfrak{o}$ and let $\mathfrak{m}^*$ be the multiplicative system of non-zero elements of $\mathfrak{m}$. Then

$$K^* = \mathfrak{m}^* \cup U \cup \mathfrak{m}^{*^{-1}}$$

is the disjoint union of $\mathfrak{m}^*$, U, and $\mathfrak{m}^{*^{-1}}$. The factor group K^*/U can now be given an ordering. If $x \in K^*$, we denote the coset xU by $|x|$. We put $|0| = 0$. We define $|x| < 1$ (i.e. $|x| \in S$) if and only if $x \in \mathfrak{m}^*$. Our set S is clearly closed under multiplication, and if we let $\Gamma = K^*/U$ then Γ is the disjoint union of S, 1, S^{-1}. In this way we obtain a valuation of K.

We note that if $x, y \in K$ and $y \neq 0$, then

$$|x| < |y| \Leftrightarrow |x/y| < 1 \Leftrightarrow x/y \in \mathfrak{m}^*.$$

Conversely, given a valuation of K into an ordered group we let $\mathfrak{o}$ be the subset of K consisting of all x such that $|x| < 1$. It follows at once from the axioms of a valuation that $\mathfrak{o}$ is a ring. If $|x| < 1$ then $|x^{-1}| > 1$ so that x^{-1} is not in $\mathfrak{o}$. If $|x| = 1$ then $|x^{-1}| = 1$. We see that $\mathfrak{o}$ is a valuation ring, whose maximal ideal consists of those elements x with $|x| < 1$ and whose units consist of those elements x with $|x| = 1$. The reader will immediately verify that there is a bijection between valuation rings of K and equivalence classes of valuations.

Let F be a field. We let the symbol ∞ satisfy the usual algebraic rules. If $a \in F$, we define

$$a \pm \infty = \infty, \qquad a \cdot \infty = \infty \quad \text{if} \quad a \neq 0,$$
$$\infty \cdot \infty = \infty, \qquad 1/0 = \infty \quad \text{and} \quad 1/\infty = 0.$$

The expressions $\infty \pm \infty$, $0 \cdot \infty$, $0/0$, and ∞/∞ are not defined.

A *place* φ of a field K into a field F is a mapping

$$\varphi : K \to \{F, \infty\}$$

of K into the set consisting of F and ∞ satisfying the usual rules for a homomorphism, namely

$$\varphi(a + b) = \varphi(a) + \varphi(b),$$
$$\varphi(ab) = \varphi(a)\varphi(b)$$

whenever the expressions on the right-hand side of these formulas are defined, and such that $\varphi(1) = 1$. We shall also say that the place is *F-valued*. The elements of K which are not mapped into ∞ will be called *finite* under the place, and the others will be called *infinite*.

The reader will verify at once that the set $\mathfrak{o}$ of elements of K which are finite under a place is a valuation ring of K. The maximal ideal consists of those elements x such that $\varphi(x) = 0$. Conversely, if $\mathfrak{o}$ is a valuation ring of K with maximal ideal $\mathfrak{m}$, we let $\varphi : \mathfrak{o} \to \mathfrak{o}/\mathfrak{m}$ be the canonical homomorphism, and define $\varphi(x) = \infty$ for $x \in K$, $x \notin \mathfrak{o}$. Then it is trivially verified that φ is a place.

If $\varphi_1 : K \to \{F_1, \infty\}$ and $\varphi_2 : K \to \{F_2, \infty\}$ are places of K, we take their restrictions to their images. We may therefore assume that they are surjective. We shall say that they are *equivalent* if there exists an isomorphism $\lambda : F_1 \to F_2$ such that $\varphi_2 = \varphi_1 \circ \lambda$. (We put $\lambda(\infty) = \infty$.) One sees that two places are equivalent if and only if they have the same valuation ring. It is clear that there is a bijection between equivalence classes of places of K, and valuation rings of K. A place is called trivial if it is injective. The valuation ring of the trivial place is simply K itself.

As with homomorphisms, we observe that the composite of two places is also a place (trivial verification).

It is often convenient to deal with places instead of valuation rings, just as it is convenient to deal with homomorphisms and not always with canonical homomorphisms or a ring modulo an ideal. However, in what follows, we use the language of valuation rings, and leave it to the reader to translate into the language of places.

The general theory of valuations and valuation rings is due to Krull (1932). However, the extension theory of homomorphisms of Chapter IX, §3 was realized only around 1945. It gives us the extension theorem for valuations:

THEOREM 1. *Let K be a subfield of a field L. Then a valuation on K has an extension to a valuation on L.*

Proof. Let $\mathfrak{o}$ be the valuation ring on K corresponding to the given valuation. Let $\varphi : \mathfrak{o} \to \mathfrak{o}/\mathfrak{m}$ be the canonical homomorphism on the residue class field, and extend φ to a homomorphism of a valuation ring $\mathfrak{O}$ of L as in §3 of Chapter IX. Let $\mathfrak{M}$ be the maximal ideal of $\mathfrak{O}$. Since $\mathfrak{M} \cap \mathfrak{o}$ contains $\mathfrak{m}$ but does not contain 1, it follows that $\mathfrak{M} \cap \mathfrak{o} = \mathfrak{m}$. Let U' be the group of units of $\mathfrak{O}$. Then $U' \cap K = U$ is the group of units of $\mathfrak{o}$. Hence we have a canonical injection

$$K^*/U \to L^*/U'$$

which is immediately verified to be order-preserving. Identifying K^*/U in L^*/U' we have obtained an extension of our valuation of K to a valuation of L.

Of course, when we deal with absolute values, we require that the value group be a subgroup of the multiplicative reals. Thus we must still prove something about the nature of the value group L^*/U', whenever L is algebraic over K.

PROPOSITION 12. *Let L be a finite extension of K, of degree n. Let w be a valuation of L with value group Γ'. Let Γ be the value group of K. Then $(\Gamma' : \Gamma) \leqq n$.*

Proof. Let $y_1, \ldots, y_r$ be elements of L whose values represent distinct cosets of Γ in Γ'. We shall prove that the y_j are linearly independent over K. In a relation $a_1y_1 + \cdots + a_ry_r = 0$ with $a_j \in K$, $a_j \neq 0$ two terms must have the same value, say $|a_i y_i| = |a_j y_j|$ with $i \neq j$, and hence

$$|y_i| = |a_i^{-1} a_j|\,|y_j|.$$

This contradicts the assumption that the values of y_i, y_j $(i \neq j)$ represent distinct cosets of Γ in Γ', and proves our proposition.

Corollary 1. *There exists an integer $e \geqq 1$ such that the map $\gamma \mapsto \gamma^e$ induces an injective homomorphism of Γ' into Γ.*

Proof. Take e to be the index $(\Gamma' : \Gamma)$.

Corollary 2. *If K is a field with a valuation v whose value group is an ordered subgroup of the ordered group of positive real numbers, and if L is an algebraic extension of K, then there exists an extension of v to L whose value group is also an ordered subgroup of the positive reals.*

Proof. We know that we can extend v to a valuation w of L with some value group Γ', and the value group Γ of v can be identified with a subgroup of $\mathbf{R}^+$. By Corollary 1, every element of Γ' has finite period modulo Γ. Since every element of $\mathbf{R}^+$ has a unique e-th root for every integer $e \geqq 1$, we can find in an obvious way an order-preserving embedding of Γ' into $\mathbf{R}^+$ which induces the identity on Γ. In this way we get our extension of v to an absolute value on L.

Corollary 3. *If L is finite over K, and if Γ is infinite cyclic, then Γ' is also infinite cyclic.*

Proof. Use Corollary 1 and the fact that a subgroup of a cyclic group is cyclic.

We shall now strengthen our preceding proposition to a slightly stronger one. We call $(\Gamma' : \Gamma)$ the *ramification index.*

Proposition 13. *Let L be a finite extension of degree n of a field K, and let $\mathfrak{O}$ be a valuation ring of L. Let $\mathfrak{M}$ be its maximal ideal, let $\mathfrak{o} = \mathfrak{O} \cap K$, and let $\mathfrak{m}$ be the maximal ideal of $\mathfrak{o}$, i.e. $\mathfrak{m} = \mathfrak{M} \cap \mathfrak{o}$. Then the residue class degree $[\mathfrak{O}/\mathfrak{M} : \mathfrak{o}/\mathfrak{m}]$ is finite. If we denote it by f, and if e is the ramification index, then $ef \leqq n$.*

Proof. Let $y_1, \ldots, y_e$ be representatives in L^* of distinct cosets of Γ'/Γ and let $z_1, \ldots, z_s$ be elements of $\mathfrak{O}$ whose residue classes mod $\mathfrak{M}$ are linearly independent over $\mathfrak{o}/\mathfrak{m}$. Consider a relation

$$\sum_{i,j} a_{ij} z_j y_i = 0$$

with $a_{ij} \in K$, not all $a_{ij} = 0$. In an inner sum

$$\sum_{j=1}^{s} a_{ij} z_j,$$

divide by the coefficient $a_{i\nu}$ having the biggest valuation. We obtain a linear combination of $z_1, \ldots, z_s$ with coefficients in $\mathfrak{o}$, and at least one coefficient equal to a unit. Since $z_1, \ldots, z_s$ are linearly independent mod $\mathfrak{M}$ over $\mathfrak{o}/\mathfrak{m}$, it follows that our linear combination is a unit. Hence

$$\left|\sum_{j=1}^{s} a_{ij}z_j\right| = |a_{i\nu}|$$

for some index ν. In the sum

$$\sum_{i=1}^{e}\left(\sum_{j=1}^{s} a_{ij}z_j\right) y_i = 0$$

viewed as a sum on i, at least two terms have the same value. This contradicts the independence of $|y_1|, \ldots, |y_e|$ mod Γ just as in the proof of Proposition 12.

Remark. Our proof also shows that the elements $\{z_j y_i\}$ are linearly independent over K. This will be used again later.

If w is an extension of a valuation v, then the ramification index will be denoted by $e(w \mid v)$ and the residue class degree will be denoted by $f(w \mid v)$.

PROPOSITION 14. *Let K be a field with a valuation v, and let $K \subset E \subset L$ be finite extensions of K. Let w be an extension of v to E and let u be an extension of w to L. Then*

$$e(u \mid w)e(w \mid v) = e(u \mid v),$$
$$f(u \mid w)f(w \mid v) = f(u \mid v).$$

Proof. Obvious.

We can express the above proposition by saying that the ramification index and the residue class degree are multiplicative in towers.

We can obtain a characterization of integral elements by means of valuations (or valuation rings). We shall use the following terminology. If $\mathfrak{o}$, $\mathfrak{O}$ are local rings with maximal ideals $\mathfrak{m}$, $\mathfrak{M}$ respectively, we shall say that $\mathfrak{O}$ *lies above* $\mathfrak{o}$ if $\mathfrak{o} \subset \mathfrak{O}$ and $\mathfrak{M} \cap \mathfrak{o} = \mathfrak{m}$. We then have a canonical injection $\mathfrak{o}/\mathfrak{m} \to \mathfrak{O}/\mathfrak{M}$.

PROPOSITION 15. *Let $\mathfrak{o}$ be a local ring contained in a field L. An element x of L is integral over $\mathfrak{o}$ if and only if x lies in every valuation ring $\mathfrak{O}$ of L lying above $\mathfrak{o}$.*

Proof. Assume that x is not integral over $\mathfrak{o}$. Let $\mathfrak{m}$ be the maximal ideal of $\mathfrak{o}$. Then the ideal $(\mathfrak{m}, 1/x)$ of $\mathfrak{o}[1/x]$ cannot be the entire ring, otherwise we can write

$$-1 = a_n(1/x)^n + \cdots + a_1(1/x) + y$$

with $y \in \mathfrak{m}$ and $a_i \in \mathfrak{o}$. From this we get

$$(1 + y)x^n + \cdots + a_n = 0.$$

But $1 + y$ is not in $\mathfrak{m}$, hence is a unit of $\mathfrak{o}$. We divide the equation by $1 + y$ to conclude that x is integral over $\mathfrak{o}$, contrary to our hypothesis. Thus $(\mathfrak{m}, 1/x)$ is not the entire ring, and is contained in a maximal ideal $\mathfrak{P}$, whose intersection with $\mathfrak{o}$ contains $\mathfrak{m}$ and hence must be equal to $\mathfrak{m}$. Extending the canonical homomorphism $\mathfrak{o}[1/x] \to \mathfrak{o}[1/x]/\mathfrak{P}$ to a homomorphism of a valuation ring $\mathfrak{O}$ of L, we see that the image of $1/x$ is 0 and hence that x cannot be in this valuation ring.

Conversely, assume that x is integral over $\mathfrak{o}$, and let

$$x^n + \cdots + a_0 = 0$$

be an integral equation for x with coefficients in $\mathfrak{o}$. Let $\mathfrak{O}$ be any valuation ring of L lying above $\mathfrak{o}$, and let $|\ |$ be the corresponding valuation. Divide the equation by x^n. If $|x| > 1$ then $|1/x| < 1$, and we get an expression for 1 as a sum of terms each of which has valuation < 1, which is impossible. Hence $|x| \leqq 1$, i.e. $x \in \mathfrak{O}$, as desired.

PROPOSITION 16. *Let A be a ring contained in a field L. An element x of L is integral over A if and only if x lies in every valuation ring $\mathfrak{O}$ of L containing A.*

Proof. The proof is similar to the proof of the preceding proposition, and will be left as an exercise.

We conclude this section by relating valuation rings in a finite extension with the integral closure.

PROPOSITION 17. *Let $\mathfrak{o}$ be a valuation ring in a field K. Let L be a finite extension of K. Let $\mathfrak{O}$ be a valuation ring of L lying above $\mathfrak{o}$, and $\mathfrak{M}$ its maximal ideal. Let B be the integral closure of $\mathfrak{o}$ in L, and let $\mathfrak{P} = \mathfrak{M} \cap B$. Then $\mathfrak{O}$ is equal to the local ring $B_{\mathfrak{P}}$.*

Proof. It is clear that $B_{\mathfrak{P}}$ is contained in $\mathfrak{O}$. Conversely, let x be an element of $\mathfrak{O}$. Then x satisfies an equation with coefficients in K, not all 0, say

$$a_n x^n + \cdots + a_0 = 0, \qquad a_i \in K.$$

Suppose that a_s is the coefficient having the biggest value among the a_i for the valuation associated with the valuation ring $\mathfrak{o}$, and that it is the coefficient farthest to the left having this value. Let $b_i = a_i/a_s$. Then all $b_i \in \mathfrak{o}$ and $b_n, \ldots, b_{s+1} \in \mathfrak{M}$. Divide the equation by x^s. We get

$$(b_n x^{n-s} + \cdots + b_{s+1}x + 1) + \frac{1}{x}\left(b_{s-1} + \cdots + b_0 \frac{1}{x^{s-1}}\right) = 0.$$

Let y and z be the two quantities in parentheses in the preceding equation, so that we can write

$$-y = z/x \qquad \text{and} \qquad -xy = z.$$

To prove our proposition it will suffice to show that y and z lie in B and that y is not in $\mathfrak{P}$.

We use Proposition 15. If a valuation ring of L above $\mathfrak{o}$ contains x, then it contains y because y is a polynomial in x with coefficients in $\mathfrak{o}$. Hence such a valuation ring also contains $z = -xy$. If on the other hand the valuation ring of L above $\mathfrak{o}$ contains $1/x$, then it contains z because z is a polynomial in $1/x$ with coefficients in $\mathfrak{o}$. Hence this valuation ring also contains y. From this we conclude by Proposition 15 that y, z lie in B.

Furthermore, since $x \in \mathfrak{O}$, and $b_n, \ldots, b_{s+1}$ are in $\mathfrak{M}$ by construction, it follows that y cannot be in $\mathfrak{M}$, and hence cannot be in $\mathfrak{P}$. This concludes the proof.

COROLLARY 1. *Let the notation be as in the proposition. Then there is only a finite number of valuation rings of L lying above $\mathfrak{o}$.*

Proof. This comes from the fact that there is only a finite number of maximal ideals $\mathfrak{P}$ of B lying above the maximal ideal of $\mathfrak{o}$ (Corollary of Proposition 11, Chapter IX, §2).

COROLLARY 2. *Let the notation be as in the proposition. Assume in addition that L is Galois over K. If $\mathfrak{O}$ and $\mathfrak{O}'$ are two valuation rings of L lying above $\mathfrak{o}$, with maximal ideals $\mathfrak{M}$, $\mathfrak{M}'$ respectively, then there exists an automorphism σ of L over K such that $\sigma\mathfrak{O} = \mathfrak{O}'$ and $\sigma\mathfrak{M} = \mathfrak{M}'$.*

Proof. Let $\mathfrak{P} = \mathfrak{O} \cap B$ and $\mathfrak{P}' = \mathfrak{O}' \cap B$. By Proposition 11 of Chapter IX, §2, we know that there exists an automorphism σ of L over K such that $\sigma\mathfrak{P} = \mathfrak{P}'$. From this our assertion is obvious.

Example. Let k be a field, and let K be a finitely generated extension of transcendence degree 1. If t is a transcendence base of K over k, then K is finite algebraic over $k(t)$. Let $\mathfrak{O}$ be a valuation ring of K containing k, and assume that $\mathfrak{O}$ is $\neq K$. Let $\mathfrak{o} = \mathfrak{O} \cap k(t)$. Then $\mathfrak{o}$ is obviously a valuation ring of $k(t)$ (the condition about inverses is *a fortiori* satisfied), and the corresponding valuation of $k(t)$ cannot be trivial. Either t or $t^{-1} \in \mathfrak{o}$. Say $t \in \mathfrak{o}$. Then $\mathfrak{o} \cap k[t]$ cannot be the zero ideal, otherwise the canonical homomorphism $\mathfrak{o} \to \mathfrak{o}/\mathfrak{m}$ of $\mathfrak{o}$ modulo its maximal ideal would induce an isomorphism on $k[t]$ and hence an isomorphism on $k(t)$, contrary to hypothesis. Hence $\mathfrak{m} \cap k[t]$ is a prime ideal $\mathfrak{p}$, generated by an irreducible polynomial $p(t)$. The local ring $k[t]_{\mathfrak{p}}$ is obviously a valuation ring, which must be $\mathfrak{o}$ because every element of $k(t)$ has an expression of type $p^r u$ where u is a unit in $k[t]_{\mathfrak{p}}$. Thus we have determined all valuation rings of $k(t)$ containing k, and we see that the value group is cyclic. Such valuations will be called discrete and are studied in greater detail below. In view of Corollary 3 of Proposition 12, it follows that the valuation ring $\mathfrak{O}$ of K is also discrete.

The residue class field $\mathfrak{o}/\mathfrak{m}$ is equal to $k[t]/\mathfrak{p}$ and is therefore a finite extension of k. By Proposition 13, it follows that $\mathfrak{O}/\mathfrak{M}$ is finite over k (if $\mathfrak{M}$ denotes the maximal ideal of $\mathfrak{O}$).

Finally, we observe that there is only a finite number of valuation rings $\mathfrak{O}$ of K containing k such that t lies in the maximal ideal of $\mathfrak{O}$. Indeed, such a valuation ring must lie above $k[t]_{\mathfrak{p}}$ where $\mathfrak{p} = (t)$ is the prime ideal generated by t, and we can apply Corollary 1 above.

§5. *Completions and valuations*

Throughout this section, we deal with a non-archimedean absolute value v on a field K. This absolute value is then a valuation, whose value group Γ_K is a subgroup of the positive reals. We let $\mathfrak{o}$ be its valuation ring, $\mathfrak{m}$ the maximal ideal.

Let us denote by $\hat{K}$ the completion of K at v, and let $\hat{\mathfrak{o}}$ (resp. $\hat{\mathfrak{m}}$) be the closure of $\mathfrak{o}$ (resp. $\mathfrak{m}$) in $\hat{K}$. By continuity, every element of $\hat{\mathfrak{o}}$ has value $\leqq 1$, and every element of $\hat{K}$ which is not in $\hat{\mathfrak{o}}$ has value > 1. If $x \in \hat{K}$ then there exists an element $y \in K$ such that $|x - y|$ is very small, and hence $|x| = |y|$ for such an element y (by the non-archimedean property). Hence $\hat{\mathfrak{o}}$ is a valuation ring in $\hat{K}$, and $\hat{\mathfrak{m}}$ is its maximal ideal. Furthermore,

$$\hat{\mathfrak{o}} \cap K = \mathfrak{o} \qquad \text{and} \qquad \hat{\mathfrak{m}} \cap K = \mathfrak{m},$$

and we have an isomorphism

$$\mathfrak{o}/\mathfrak{m} \xrightarrow{\approx} \hat{\mathfrak{o}}/\hat{\mathfrak{m}}.$$

Thus the residue class field $\mathfrak{o}/\mathfrak{m}$ does not change under completion.

Let E be an extension of K, and let $\mathfrak{o}_E$ be a valuation ring of E lying above $\mathfrak{o}$. Let $\mathfrak{m}_E$ be its maximal ideal. We assume that the valuation corresponding to $\mathfrak{o}_E$ is in fact an absolute value, so that we can form the completion E. We then have a commutative diagram:

$$\begin{array}{ccc} \mathfrak{o}_E/\mathfrak{m}_E & \xrightarrow{\approx} & \hat{\mathfrak{o}}_E/\hat{\mathfrak{m}}_E \\ \uparrow & & \uparrow \\ \mathfrak{o}/\mathfrak{m} & \xrightarrow{\approx} & \hat{\mathfrak{o}}/\hat{\mathfrak{m}} \end{array}$$

the vertical arrows being injections, and the horizontal ones being isomorphisms. Thus the residue class field extension of our valuation can be studied over the completions E of K.

We have a similar remark for the ramification index. Let $\Gamma_v(K)$ and $\Gamma_v(\hat{K})$ denote the value groups of our valuation on K and $\hat{K}$ respectively (i.e. the image of the map $x \mapsto |x|$ for $x \in K^*$ and $x \in \hat{K}^*$ respectively). We saw above that $\Gamma_v(K) = \Gamma_v(\hat{K})$; in other words, the value group is the same under completion, because of the non-archimedean property. (This is of course false in the archimedean case.) If E is again an extension of K and w is an absolute value of E extending v, then we have a commutative diagram

$$\begin{array}{ccc} \Gamma_w(E) & \xrightarrow{=} & \Gamma_w(\hat{E}) \\ \uparrow & & \uparrow \\ \Gamma_v(K) & \xrightarrow{=} & \Gamma_v(\hat{K}) \end{array}$$

from which we see that the ramification index $(\Gamma_w(E) : \Gamma_v(K))$ also does not change under completion.

§6. *Discrete valuations*

A valuation is called *discrete* if its value group is cyclic. In that case, the valuation is an absolute value (if we consider the value group as a subgroup of the positive reals). The p-adic valuation on the rational numbers is discrete for each prime number p. By Corollary 3 of Proposition 12, §4, an extension of a discrete valuation to a finite extension field is also discrete. Aside from the absolute values obtained by embedding a field into the reals or complex numbers, discrete valuations are the most important ones in practice. We shall make some remarks concerning them.

Let v be a discrete valuation on a field K, and let $\mathfrak{o}$ be its valuation ring. Let $\mathfrak{m}$ be the maximal ideal. There exists an element π of $\mathfrak{m}$ which is such that its value $|\pi|$ generates the value group. (The other generator of the value group is $|\pi^{-1}|$.) Such an element π is called a *local parameter* for v (or for $\mathfrak{m}$). Every element x of K can be written in the form

$$x = u\pi^r$$

with some unit u of $\mathfrak{o}$, and some integer r. Indeed, we have $|x| = |\pi|^r = |\pi^r|$ for some $r \in \mathbf{Z}$, whence x/π^r is a unit in $\mathfrak{o}$. We call r the *order* of x at v. It is obviously independent of the choice of parameter selected. We also say that x has a *zero of order* r. (If r is negative, we say that x has a *pole* of order $-r$.)

In particular, we see that $\mathfrak{m}$ is a principal ideal, generated by π. As an exercise, we leave it to the reader to verify that every ideal of $\mathfrak{o}$ is principal, and is a power of $\mathfrak{m}$. Furthermore, we observe that $\mathfrak{o}$ is a factorial ring with exactly one prime element (up to units), namely π.

If $x, y \in K$, we shall write $x \sim y$ if $|x| = |y|$. Let π_i $(i = 1, 2, \ldots)$ be a sequence of elements of $\mathfrak{o}$ such that $\pi_i \sim \pi^i$. Let R be a set of representatives of $\mathfrak{o}/\mathfrak{m}$ in $\mathfrak{o}$. This means that the canonical map $\mathfrak{o} \to \mathfrak{o}/\mathfrak{m}$ induces a bijection of R onto $\mathfrak{o}/\mathfrak{m}$. *Assume that K is complete under our valuation. Then every element x of $\mathfrak{o}$ can be written as a convergent series*

$$x = a_0 + a_1\pi_1 + a_2\pi_2 + \cdots$$

with $a_i \in R$, and the a_i are uniquely determined by x. This is easily proved by a recursive argument. Suppose we have written

$$x \equiv a_0 + \cdots + a_n\pi_n \pmod{\mathfrak{m}^{n+1}},$$

then $x - (a_0 + \cdots + a_n\pi_n) = \pi_{n+1}y$ for some $y \in \mathfrak{o}$. By hypothesis, we can write $y = a_{n+1} + \pi z$ with some $a_{n+1} \in R$. From this we get

$$x \equiv a_0 + \cdots + a_{n+1}\pi_{n+1} \pmod{\mathfrak{m}^{n+2}},$$

and it is clear that the n-th term in our series tends to 0. Therefore our series converges (by the non-archimedean behavior!). The fact that R contains precisely one representative of each residue class mod $\mathfrak{m}$ implies that the a_i are uniquely determined.

Examples. Consider first the case of the rational numbers with the p-adic valuation v_p. The completion is denoted by $\mathbf{Q}_p$. It is the field of p-adic *numbers.* The closure of $\mathbf{Z}$ in $\mathbf{Q}_p$ is the ring of *p-adic integers* $\mathbf{Z}_p$. We note that the prime number p is a prime element in both $\mathbf{Z}$ and its closure $\mathbf{Z}_p$. We can select our set of representatives R to be the set of integers $(0, 1, \ldots, p - 1)$. Thus every p-adic integer can be written uniquely as a convergent sum $\sum a_i p^i$ where a_i is an integer, $0 \leqq a_i \leqq p - 1$. This sum is called its p-adic expansion. Such sums are added and multiplied in the ordinary manner for convergent series.

For instance, we have the usual formalism of geometric series, and if we take $p = 3$, then

$$-1 = \frac{2}{1 - 3} = 2(1 + 3 + 3^2 + \cdots).$$

We note that the representatives $(0, 1, \ldots, p - 1)$ are by no means the only ones which can be used. In fact, it can be shown that $\mathbf{Z}_p$ contains the $(p - 1)$-th roots of unity, and it is often more convenient to select these roots of unity as representatives for the non-zero elements of the residue class field.

Next consider the case of a rational field $k(t)$, where k is any field and t is transcendental over k. We have a valuation determined by the prime element t in the ring $k[t]$. This valuation is discrete, and the completion of $k[t]$ under this valuation is the power series ring $k[[t]]$. In that case, we can take the elements of k itself as representatives of the residue class field, which is canonically isomorphic to k. The maximal ideal of $k[[t]]$ is the ideal generated by t.

This situation amounts to an algebraization of the usual situation arising in the theory of complex variables. For instance, let z_0 be a point in the complex plane. Let $\mathfrak{o}$ be the ring of functions which are holomorphic in some disc around z_0. Then $\mathfrak{o}$ is a discrete valuation ring, whose maximal ideal consists of those functions having a zero at z_0. Every element of $\mathfrak{o}$ has a power series expansion

$$f(z) = \sum_{\nu=m}^{\infty} a_\nu (z - z_0)^\nu.$$

The representatives of the residue class field can be taken to be complex numbers, a_ν. If $a_m \neq 0$, then we say that $f(z)$ has a zero of order m. The order is the same, whether viewed as order with respect to the discrete valuation in the algebraic sense, or the order in the sense of the theory of complex variables. We can select a canonical uniformizing parameter namely $z - z_0$, and

$$f(z) = (z - z_0)^m g(z)$$

where $g(z)$ is a power series beginning with a non-zero constant. Thus $g(z)$ is invertible.

Let K be again complete under a discrete valuation, and let E be a finite extension of K. Let $\mathfrak{o}_E$, $\mathfrak{m}_E$ be the valuation ring and maximal ideal in E lying above $\mathfrak{o}$, $\mathfrak{m}$ in K. Let Π be a prime element in E. If Γ_E and Γ_K are the value groups of the valuations in E and K respectively, and

$$e = (\Gamma_E : \Gamma_K)$$

is the ramification index, then

$$|\Pi^e| = |\pi|,$$

and the elements

$$\Pi^i \pi^j, \qquad 0 \leqq i \leqq e - 1, \quad j = 0, 1, 2, \ldots$$

have order $je + i$ in E.

Let $\omega_1, \ldots, \omega_f$ be elements of E such that their residue classes mod $\mathfrak{m}_E$ form a basis of $\mathfrak{o}_E/\mathfrak{m}_E$. If R is as before a set of representatives of $\mathfrak{o}/\mathfrak{m}$ in $\mathfrak{o}$, then the set consisting of all elements

$$a_1\omega_1 + \cdots + a_f\omega_f$$

with $a_j \in R$ is a set of representatives of $\mathfrak{o}_E/\mathfrak{m}_E$ in $\mathfrak{o}_E$. From this we see that every element of $\mathfrak{o}_E$ admits a convergent expansion

$$\sum_{i=0}^{e-1} \sum_{\nu=1}^{f} \sum_{j=0}^{\infty} a_{\nu,i,j} \pi^j \omega_\nu \Pi^i.$$

Thus the elements $\{\omega_\nu \Pi^i\}$ form a set of generators of $\mathfrak{o}_E$ as a module over $\mathfrak{o}$. On the other hand, we have seen in the proof of Proposition 13 of §4 that these elements are linearly independent over K. Hence we obtain:

PROPOSITION 18. *Let K be complete under a discrete valuation. Let E be a finite extension of K, and let e, f be the ramification index and residue class degree respectively. Then*

$$ef = [E : K].$$

COROLLARY 1. *Let $\alpha \in E$, $\alpha \neq 0$. Let v be the valuation on K and w its extension to E. Then*

$$\operatorname{ord}_v N_K^E(\alpha) = f(w \mid v) \operatorname{ord}_w \alpha.$$

Proof. This is immediate from the formula

$$|N_K^E(\alpha)| = |\alpha|^{ef}$$

and the definitions.

COROLLARY 2. *Let K be any field and v a discrete valuation on K. Let E be a finite extension of K. If v is well behaved in E (for instance if E is separable over K), then*

$$\sum_{w \mid v} e(w \mid v) f(w \mid v) = [E : K].$$

If E is Galois over K, then all e_w are equal to the same number e, all f_w are equal to the same number f, and so

$$efr = [E : K],$$

where r is the number of extensions of v to E.

Proof. Our first assertion comes from our assumption, and Proposition 8 of §3. If E is Galois over K, we know from Corollary 2 of Proposition 17, §4 that any two valuations of E lying above v are conjugate. Hence all ramification indices are equal, and similarly for the residue class degrees. Our relation $efr = [E : K]$ is then obvious.

§7. *Zeros of polynomials in complete fields*

Let K be complete under a non-trivial absolute value.

Let

$$f(X) = \prod (X - \alpha_i)^{r_i}$$

be a polynomial in $K[X]$ having leading coefficient 1, and assume the roots α_i are distinct, with multiplicities r_i. Let d be the degree of f. Let g be another polynomial with coefficients in $\overline{K}$, and assume that the degree of g is also d, and that g has leading coefficient 1. We let $|g|$ be the maximum of the absolute values of the coefficients of g. One sees easily that if $|g|$ is bounded, then the absolute values of the roots of g are also bounded.

Suppose that g comes close to f, in the sense that $|f - g|$ is small. If β is any root of g, then

$$|f(\beta) - g(\beta)| = |f(\beta)| = \prod |\alpha_i - \beta|^{r_i}$$

is small, and hence β must come close to some root of f. As β comes close to say $\alpha = \alpha_1$, its distance from the other roots of f approaches the distance of α_1 from the other roots, and is therefore bounded from below. In that case, we say that β *belongs to* α.

PROPOSITION 19. *If g is sufficiently close to f, and $\beta_1, \dots, \beta_s$ are the roots of g belonging to α (counting multiplicities), then $s = r_1$ is the multiplicity of α in f.*

Proof. Assume the contrary. Then we can find a sequence g_ν of polynomials approaching f with precisely s roots $\beta_1^{(\nu)}, \dots, \beta_s^{(\nu)}$ belonging to α, but with $s \neq r$. (We can take the same multiplicity s since there is only a finite number of choices for such multiplicities.) Furthermore, the other roots of g also belong to roots of f, and we may suppose that these roots are bunched together, according to which root of f they belong to. Since $\lim g_\nu = f$, we conclude that α must have multiplicity s in f, contradiction.

Next we investigate conditions under which a polynomial has a root in a complete field.

We assume that K is complete under a discrete valuation, with valuation ring $\mathfrak{o}$, maximal ideal $\mathfrak{p}$. We let π be a fixed prime element of $\mathfrak{p}$.

We shall deal with n-space over $\mathfrak{o}$. We denote a vector $(a_1, \dots, a_n)$ with $a_i \in \mathfrak{o}$ by A. If $f(X_1, \dots, X_n) \in \mathfrak{o}[X]$ is a polynomial in n variables, with integral coefficients, we shall say that A is a *zero* of f if $f(A) = 0$, and we say that A is a *zero* of f mod $\mathfrak{p}^m$ if $f(A) \equiv 0 \pmod{\mathfrak{p}^m}$.

Let $C = (c_0, \dots, c_n)$ be in $\mathfrak{o}^{(n+1)}$. Let m be an integer $\geqq 1$. We consider the nature of the solutions of a congruence of type

$$(*) \qquad \pi^m(c_0 + c_1x_1 + \cdots + c_nx_n) \equiv 0 \pmod{\mathfrak{p}^{m+1}}.$$

This congruence is equivalent with the linear congruence

$$(**) \qquad c_0 + c_1x_1 + \cdots + c_nx_n \equiv 0 \pmod{\mathfrak{p}}.$$

If some coefficient c_i ($i = 1, \dots, n$) is not $\equiv 0 \pmod{\mathfrak{p}}$, then the set of solutions is not empty, and has the usual structure of a solution of one inhomogeneous linear equation over the field $\mathfrak{o}/\mathfrak{p}$. In particular, it has dimension $n - 1$. A congruence (*) or (**) with some $c_i \not\equiv 0 \pmod{\mathfrak{p}}$ will be called a *proper congruence*.

As a matter of notation, we write $D_i f$ for the formal partial derivative of f with respect to X_i. We write

$$\operatorname{grad} f(X) = \big(D_1 f(X), \ldots, D_n f(X)\big).$$

PROPOSITION 20. *Let $f(X) \in \mathfrak{o}[X]$. Let r be an integer $\geqq 1$ and let $A \in \mathfrak{o}^{(n)}$ be such that*

$$\begin{aligned} f(A) &\equiv 0 \pmod{\mathfrak{p}^{2r-1}}, & & \\ D_i f(A) &\equiv 0 \pmod{\mathfrak{p}^{r-1}}, & \text{for all} \quad & i = 1, \ldots, n, \\ D_i f(A) &\not\equiv 0 \pmod{\mathfrak{p}^{r}}, & \text{for some} \quad & i = 1, \ldots, n. \end{aligned}$$

Let ν be an integer $\geqq 0$ and let $B \in \mathfrak{o}^{(n)}$ be such that

$$B \equiv A \pmod{\mathfrak{p}^r} \qquad \textit{and} \qquad f(B) \equiv 0 \pmod{\mathfrak{p}^{2r-1+\nu}}.$$

A vector $Y \in \mathfrak{o}^{(n)}$ satisfies

$$Y \equiv B \pmod{\mathfrak{p}^{r+\nu}} \qquad \textit{and} \qquad f(Y) \equiv 0 \pmod{\mathfrak{p}^{2r+\nu}}$$

if and only if Y can be written in the form $Y = B + \pi^{r+\nu}C$, with some $C \in \mathfrak{o}^{(n)}$ satisfying the proper congruence

$$f(B) + \pi^{r+\nu} \operatorname{grad} f(B) \cdot C \equiv 0 \pmod{\mathfrak{p}^{2r+\nu}}.$$

Proof. The proof is shorter than the statement of the proposition. Write $Y = B + \pi^{r+\nu}C$. By Taylor's expansion,

$$f(B + \pi^{r+\nu}C) = f(B) + \pi^{r+\nu} \operatorname{grad} f(B) \cdot C \pmod{\mathfrak{p}^{2r+2\nu}}.$$

To solve this last congruence mod $\mathfrak{p}^{2r+\nu}$, we obtain a proper congruence by hypothesis, because $\operatorname{grad} f(B) \equiv \operatorname{grad} f(A) \equiv 0 \pmod{\mathfrak{p}^{r-1}}$.

COROLLARY 1. *Assumptions being as in Proposition 20, there exists a zero of f in $\mathfrak{o}^{(n)}$ which is congruent to A* mod $\mathfrak{p}^r$.

Proof. We can write this zero as a convergent sum

$$A + \pi^{r+1}C_1 + \pi^{r+2}C_2 + \cdots$$

solving for $C_1, C_2, \ldots$ inductively as in the proposition.

COROLLARY 2. *Let f be a polynomial in one variable in $\mathfrak{o}[X]$, and let $a \in \mathfrak{o}$ be such that $f(a) \equiv 0 \pmod{\mathfrak{p}}$ but $f'(a) \not\equiv 0 \pmod{\mathfrak{p}}$. Then there exists $b \in \mathfrak{o}$, $b \equiv a \pmod{\mathfrak{p}}$ such that $f(b) = 0$.*

Proof. Take $n = 1$ and $r = 1$ in the proposition, and apply Corollary 1.

COROLLARY 3. *Let m be a positive integer not divisible by the characteristic of K. There exists an integer r such that for any $a \in \mathfrak{o}$, $a \equiv 1$* (mod $\mathfrak{p}^r$) *the equation $X^m - a = 0$ has a root in K.*

Proof. Apply the proposition.

Example. In the 2-adic field $\mathbf{Q}_2$, there exists a square root of -7, i.e. $\sqrt{-7} \in \mathbf{Q}_2$, because $-7 = 1 - 8$.

(For refinements of the above proposition, cf. Bourbaki, *Algèbre Commutative*, Chapter III, §4, 5.)

When the absolute value is not discrete, it is still possible to formulate a criterion for a polynomial to have a zero.

PROPOSITION 21. *Let K be a complete under a non-archimedean absolute value (non-trivial). Let $\mathfrak{o}$ be the valuation ring and let $f(X) \in \mathfrak{o}[X]$ be a polynomial in one variable. Let $\alpha_0 \in \mathfrak{o}$ be such that*

$$|f(\alpha_0)| < |f'(\alpha_0)^2|$$

(here f' denotes the formal derivative of f). Then the sequence

$$\alpha_{i+1} = \alpha_i - \frac{f(\alpha_i)}{f'(\alpha_i)}$$

converges to a root α of f in $\mathfrak{o}$, and we have

$$|\alpha - \alpha_0| \leqq \left|\frac{f(\alpha_0)}{f'(\alpha_0)^2}\right| < 1.$$

Proof. This is an easy exercise. We leave the details to the reader. We note that here again the exponent 2 gives the precise condition so that we can refine an approximate root to a root. When the absolute value is discrete, Proposition 21 is in fact a special case of Proposition 20.

The technique of the proposition is also useful when dealing with rings, say a local ring $\mathfrak{o}$ with maximal ideal $\mathfrak{m}$ such that $\mathfrak{m}^r = 0$ for some integer $r > 0$. If one has a polynomial f in $\mathfrak{o}[X]$ and an approximate root α_0 such that $f'(\alpha_0) \not\equiv 0 \bmod \mathfrak{m}$, then the Newton approximation sequence shows how to refine α_0 to a root of f.

EXERCISES

1. (a) Let K be a field with a valuation. If

$$f(X) = a_0 + a_1X + \cdots + a_nX^n$$

is a polynomial in $K[X]$, define $|f|$ to be the max of the values $|a_i|$ ($i = 0, \ldots, n$). Show that this defines an extension of the valuation to $K[X]$, and also that the valuation can be extended to the rational field $K(X)$. How is Gauss' lemma a special case of the above statement? Generalize to polynomials in several variables.

(b) Let f be a polynomial with complex coefficients. Define $|f|$ to be the maximum of the absolute values of the coefficients. Let d be an integer $\geqq 1$. Show that there exist constants C_1, C_2 (depending only on d) such that, if f, g are polynomials in $\mathbf{C}[X]$ of degrees $\leqq d$, then

$$C_1|f|\,|g| \leqq |fg| \leqq C_2|f|\,|g|.$$

[*Hint:* Induction on the number of factors of degree 1. Note that the right inequality is trivial.]

2. Let $M_{\mathbf{Q}}$ be the set of absolute values consisting of the ordinary absolute value and all p-adic absolute values v_p on the field of rational numbers $\mathbf{Q}$. Show that for any rational number $a \in \mathbf{Q}$, $a \neq 0$, we have

$$\prod_{v \in M_{\mathbf{Q}}} |a|_v = 1.$$

If K is a finite extension of $\mathbf{Q}$, and M_K denotes the set of absolute values on K extending those of $M_{\mathbf{Q}}$, and for each $w \in M_K$ we let N_w be the local degree $[K_w : \mathbf{Q}_v]$, show that for $\alpha \in K$, $\alpha \neq 0$, we have

$$\prod_{w \in M_K} |\alpha|_w^{N_w} = 1.$$

3. Show that the p-adic numbers $\mathbf{Q}_p$ have no automorphisms other than the identity. [*Hint:* Show that such automorphisms are continuous for the p-adic topology. Use Corollary 3 of Proposition 20 as an algebraic characterization of elements close to 1.]

4. Let A be a principal entire ring, and let K be its quotient field. Let $\mathfrak{o}$ be a valuation ring of K containing A, and assume $\mathfrak{o} \neq K$. Show that $\mathfrak{o}$ is the local ring $A_{(p)}$ for some prime element p. [This applies both to the ring $\mathbf{Z}$ and to a polynomial ring $k[X]$ over a field k.]

5. Let A be an entire ring, and let K be its quotient field. Assume that every finitely generated ideal of A is principal. Let $\mathfrak{o}$ be a discrete valuation ring of K containing A. Show that $\mathfrak{o} = A_{(p)}$ for some element p of A, and that p is a generator of the maximal ideal of $\mathfrak{o}$.

6. (Iss'sa) Let K be the field of meromorphic functions on the complex plane $\mathbf{C}$. Let $\mathfrak{o}$ be a discrete valuation ring of K (containing the constants $\mathbf{C}$). Show that the function z is on $\mathfrak{o}$. [*Hint:* Let $a_1, a_2, \ldots$ be a discrete sequence of complex numbers tending to infinity, for instance the positive integers. Let $\nu_1, \nu_2, \ldots$ be a sequence of integers, $0 \leqq \nu_i \leqq p - 1$, for some prime number p, such that $\sum \nu_i p^i$ is not the p-adic expansion of a rational number. Let f be an entire function having a zero of order $\nu_i p^i$ at a_i for each i and no other zero.

If z is not in $\mathfrak{o}$, consider the quotient

$$g(z) = \frac{f(z)}{\prod_{i=1}^{n} (z - a_i)^{\nu_i p^i}}.$$

From the Weierstrass factorization of an entire function, show that $g(z) = h(z)^{p^{n+1}}$ for some entire function $h(z)$. Now analyze the zero of g at the discrete valuation of $\mathfrak{o}$ in terms of that of f and $\prod(z - a_i)^{\nu_i p^i}$ to get a contradiction.]

If U is a non-compact Riemann surface, and L is the field of meromorphic functions on U, and if $\mathfrak{o}$ is a discrete valuation ring of L containing the constants, show that every holomorphic function φ on U lies in $\mathfrak{o}$. [*Hint:* Map $\varphi : U \to \mathbf{C}$, and get a discrete valuation of K by composing φ with meromorphic functions on $\mathbf{C}$. Apply the first part of the exercise.] Show that the valuation ring is the one associated with a complex number. [Further hint: If you don't know about Riemann surfaces, do it for the complex plane. For each $z \in U$, let f_z be a function holomorphic on U and having only a zero of order 1 at z. If for some z_0 the function f_{z_0} has order $\geqq 1$ at $\mathfrak{o}$, then show that $\mathfrak{o}$ is the valuation ring associated with z_0. Otherwise, every function f_z has order 0 at $\mathfrak{o}$. Conclude that the valuation of $\mathfrak{o}$ is trivial on any holomorphic function by a limit trick analogous to that of the first part of the exercise.]

7. *Witt vectors again.* Let k be a perfect field of characteristic p. We use the Witt vectors as described in the exercises of Chapter VIII. One can define an absolute value on $W(k)$, namely $|x| = p^{-r}$ if x_r is the first non-zero component of x. Show that this is an absolute value, obviously discrete, defined on the ring, and which can be extended at once to the quotient field. Show that this quotient field is complete, and note that $W(k)$ is the valuation ring. The maximal ideal consists of those x such that $x_0 = 0$, i.e. is equal to $pW(k)$.

8. Let F be a complete field with respect to a discrete valuation, let $\mathfrak{o}$ be the valuation ring, π a prime element, and assume that $\mathfrak{o}/(\pi) = k$. Prove that if $a, b \in \mathfrak{o}$ and $a \equiv b \pmod{\pi^r}$ with $r > 0$ then $a^{p^n} \equiv b^{p^n} \pmod{\pi^{r+n}}$ for all integers $n \geqq 0$.

9. Let F be as above. Show that there exists a system of representatives R for $\mathfrak{o}/(\pi)$ in $\mathfrak{o}$ such that $R^p = R$ and that this system is unique (Teichmüller). [*Hint:* Let α be a residue class in k. For each $\nu \geqq 0$ let a_ν be a representative in $\mathfrak{o}$ of $\alpha^{p^{-\nu}}$ and show that the sequence $a_\nu^{p^\nu}$ converges for $\nu \to \infty$, and in fact converges to a representative a of α, independent of the choices of a_ν.] Show that the system of representatives R thus obtained is closed under multiplication, and that if F has characteristic p, then R is closed under addition, and is isomorphic to k.

10. Assume that F has characteristic 0. Map each vector $x \in W(k)$ on the element

$$\sum \xi_i^{p^{-i}} p^i$$

where ξ_i is a representative of x_i in the special system of the preceding exercise. Show that this map is an embedding of $W(k)$ into $\mathfrak{o}$.

11. (Local uniformization) Let k be a field, K a finitely generated extension of transcendence degree 1, and $\mathfrak{o}$ a discrete valuation ring of K over k, with maximal ideal $\mathfrak{m}$. Assume that $\mathfrak{o}/\mathfrak{m} = k$. Let x be a generator of $\mathfrak{m}$, and assume that K is separable over $k(x)$. Show that there exists an element $y \in \mathfrak{o}$ such that $K = k(x, y)$, and also having the following property. Let φ be the place on K determined by $\mathfrak{o}$. Let $a = \varphi(x)$, $b = \varphi(y)$ (of course $a = 0$). Let $f(X, Y)$ be the irreducible polynomial in $k[X, Y]$ such that $f(x, y) = 0$. Then $D_2 f(a, b) \neq 0$. [*Hint:* Write first $K = k(x, z)$ where z is integral over $k[x]$. Let $z = z_1, \ldots, z_n$ $(n \geqq 2)$ be the conjugates of z over $k(x)$, and extend $\mathfrak{o}$ to a valuation ring $\mathfrak{O}$ of $k(x, z_1, \ldots, z_n)$. Let

$$z = a_0 + a_1 x + \cdots + a_r x^r + \cdots$$

be the power series expansion of z with $a_i \in k$, and let $P_r(x) = a_0 + \cdots + a_r x^r$. For $i = 1, \ldots, n$ let

$$y_i = \frac{z_i - P_r(x)}{x^r}.$$

Taking r large enough, show that y_1 has no pole at $\mathfrak{O}$ but $y_2, \ldots, y_n$ have poles at $\mathfrak{O}$. The elements $y_1, \ldots, y_n$ are conjugate over $k(x)$. Let $f(X, Y)$ be the irreducible polynomial of (x, y) over k. Then $f(x, Y) = \psi_n(x) Y^n + \cdots + \psi_0(x)$ with $\psi_i(x) k[x]$. We may also assume $\psi_i(0) \neq 0$ (since f is irreducible). Write $f(x, Y)$ in the form

$$f(x, Y) = \psi_n(x) y_2 \cdots y_n (Y - y_1)(y_2^{-1} Y - 1) \cdots (y_n^{-1} Y - 1).$$

Show that $\psi_n(x) y_2 \cdots y_n = u$ does not have a pole at $\mathfrak{O}$. If $w \in \mathfrak{O}$, let $\overline{w}$ denote its residue class modulo the maximal ideal of $\mathfrak{O}$. Then

$$0 \neq f(\overline{x}, Y) = (-1)^{n-1} \overline{u} (Y - \overline{y}_1).$$

Let $y = y_1$, $\overline{y} = b$. We find that $D_2 f(a, b) = (-1)^{n-1} \overline{u} \neq 0$.]

12. Prove the converse of Exercise 11, i.e. if $K = k(x, y)$, $f(X, Y)$ is the irreducible polynomial of (x, y) over k, and if $a, b \in k$ are such that $f(a, b) = 0$, but $D_2 f(a, b) \neq 0$, then there exists a unique valuation ring $\mathfrak{o}$ of K with maximal ideal $\mathfrak{m}$ such that $x \equiv a$ and $y \equiv b \pmod{\mathfrak{m}}$. Furthermore, $\mathfrak{o}/\mathfrak{m} = k$, and $x - a$ is a generator of $\mathfrak{m}$. [*Hint:* If $g(x, y) \in k[x, y]$ is such that $g(a, b) = 0$, show that $g(x, y) = (x - a) A(x, y)/B(x, y)$ where A, B are polynomials such that $B(a, b) \neq 0$. If $A(a, b) = 0$ repeat the process. Show that the process cannot be repeated indefinitely, and leads to a proof of the desired assertion.]

13. Let K be a field of characteristic 0, complete with respect to a non-archimedean absolute value. Show that the series

$$\exp(x) = 1 + x + \frac{x^2}{2!} + \frac{x^3}{3!} + \cdots$$

$$\log(1 + x) = x - \frac{x^2}{2} + \frac{x^3}{3} - \cdots$$

converge in some neighborhood of 0. (The main problem arises when the characteristic of the residue class field is $p > 0$, so that p divides the denominators $n!$ and n. Get an expression which determines the power of p occurring in $n!$.) Prove that the exp and log give mappings inverse to each other, from a neighborhood of 0 to a neighborhood of 1.

14. Let K be as in the preceding exercise, of characteristic 0, complete with respect to a non-archimedean absolute value. For every integer $n > 0$, show that the usual binomial expansion for $(1 + x)^{1/n}$ converges in some neighborhood of 0. Do this first assuming that the characteristic of the residue class field does not divide n, in which case the assertion is much simpler to prove.

15. Let $\mathbf{Q}_p$ be a p-adic field. Show that $\mathbf{Q}_p$ contains infinitely many quadratic fields of type $\mathbf{Q}(\sqrt{-m})$, where m is a positive integer.

16. Show that the ring of p-adic integers $\mathbf{Z}_p$ is compact. Show that the group of units in $\mathbf{Z}_p$ is compact.

17. If K is a field complete with respect to a discrete valuation, with finite residue class field, and if $\mathfrak{o}$ is the ring of elements of K whose orders are $\geqq 0$, show that $\mathfrak{o}$ is compact. Show that the group of units of $\mathfrak{o}$ is closed in $\mathfrak{o}$ and is compact.

18. Let K be a field complete with respect to a discrete valuation, let $\mathfrak{o}$ be the ring of integers of K, and assume that $\mathfrak{o}$ is compact. Let $f_1, f_2, \ldots$ be a sequence of polynomials in n variables, with coefficients in $\mathfrak{o}$. Assume that all these polynomials have degree $\leqq d$, and that they converge to a polynomial f (i.e. that $|f - f_i| \to 0$ as $i \to \infty$). If each f_i has a zero in $\mathfrak{o}$, show that f has a zero in $\mathfrak{o}$. If the polynomials f_i are homogeneous of degree d, and if each f_i has a non-trivial zero in $\mathfrak{o}$, show that f has a non-trivial zero in $\mathfrak{o}$. [*Hint:* Use the compactness of $\mathfrak{o}$ and of the units of $\mathfrak{o}$ for the homogeneous case.]

(For applications of this exercise, and also of Proposition 21, cf. the paper "On quasi-algebraic closure", *Annals of Math.*, 1951.)

19. Show that if p, p' are two distinct prime numbers, then the fields $\mathbf{Q}_p$ and $\mathbf{Q}_{p'}$ are not isomorphic.

20. Prove that the field $\mathbf{Q}_p$ contains all $(p - 1)$-th roots of unity. [*Hint:* Use Proposition 21, applied to the polynomial $X^{p-1} - 1$ which splits into factors of degree 1 in the residue class field.] Show that two distinct $(p - 1)$-th roots of unity cannot be congruent mod p.

21. Let α be algebraic over $\mathbf{Q}$ and assume that $\mathbf{Q}(\alpha)$ is a real field. Prove that α is a sum of squares in $\mathbf{Q}(\alpha)$ if and only if for every embedding σ of $\mathbf{Q}(\alpha)$ in $\mathbf{R}$ we have $\sigma\alpha > 0$.

22. Let F be a finite extension of $\mathbf{Q}$. Let $\varphi : F \to \mathbf{Q}$ be a $\mathbf{Q}$-linear functional such that $\varphi(x^2) > 0$ for all $x \in F$, $x \neq 0$. Let $\alpha \in F$, $\alpha \neq 0$. If $\varphi(\alpha x^2) \geqq 0$ for all $x \in F$, show that α is a sum of squares in F, and that F is totally real, i.e. every embedding of F in the complex numbers is contained in the real numbers. [*Hint:* Use the fact that the trace gives an identification of F with its dual space over $\mathbf{Q}$, and use the approximation theorem of Chapter XII, §1.]

23. Read the statements of the results in "The theory of real places", *Annals of Math.* (1953), pp. 378–391, and prove these statements without looking at the proofs given in that paper.

24. Let $\alpha \leqq t \leqq \beta$ be a real interval, and let $f(t)$ be a real polynomial which is positive on this interval. Show that $f(t)$ can be written in the form

$$c\left(\sum Q_\nu^2 + \sum (t - \alpha)Q_\mu^2 + \sum (\beta - t)Q_\lambda^2\right)$$

where Q^2 denotes a square, and $c \geqq 0$. [*Hint:* Split the polynomial, and use the identity:

$$(t - \alpha)(\beta - t) = \frac{(t - \alpha)^2(\beta - t) + (t - \alpha)(\beta - t)^2}{\beta - \alpha}.\Bigg]$$

25. Show that the field of real numbers has only the identity automorphism. [*Hint:* Show that an automorphism preserves the ordering.]

PART THREE

LINEAR ALGEBRA and REPRESENTATIONS

We shall be concerned with modules and vector spaces, going into their structure under various points of view. The main theme here is to study a pair, consisting of a module, and an endomorphism, or a ring of endomorphisms, and try to decompose this pair into a direct sum of components whose structure can then be described explicitly. The direct sum theme recurs in every chapter. Sometimes, we use a duality to obtain our direct sum decomposition relative to a pairing, and sometimes we get our decomposition directly. If a module refuses to decompose into a direct sum of simple components, then there is no choice but to apply the Grothendieck construction and see what can be obtained from it.

The extension theme occurs only once, in Witt's theorem, in a brief counterpoint to the decomposition theme.

CHAPTER XIII

Matrices and Linear Maps

Throughout this chapter, we let R be a commutative ring, and we let E, F be R-modules. We suppress the prefix R- in front of linear maps and modules.

§1. Matrices

By an $m \times n$ *matrix* in R one means a doubly indexed family of elements of R, (a_{ij}) ($i = 1, \ldots, m$ and $j = 1, \ldots, n$), usually written in the form

$$\begin{pmatrix} a_{11} & \cdots & a_{1n} \\ & \cdots & \\ a_{m1} & \cdots & a_{mn} \end{pmatrix}.$$

We call the elements a_{ij} the *coefficients* or *components* of the *matrix*. A $1 \times n$ matrix is called a *row vector* (of dimension, or size, n) and a $m \times 1$ matrix is called a *column vector* (of dimension, or size, m). In general, we say that (m, n) is the *size* of the matrix.

We define addition for matrices of the same size by components. If $A = (a_{ij})$ and $B = (b_{ij})$ are matrices of the same size, we define $A + B$ to be the matrix whose ij-component is $a_{ij} + b_{ij}$. Addition is obviously associative. We define the multiplication of a matrix A by an element $c \in R$ to be the matrix (ca_{ij}), whose ij-component is ca_{ij}. Then the set of $m \times n$ matrices in R is a module (i.e. an R-module).

We define the product AB of two matrices only under certain conditions. Namely, when A has size (m, n) and B has size (n, r), i.e. only when the size of the rows of A is the same as the size of the columns of B. If that is the case, let $A = (a_{ij})$ and let $B = (b_{jk})$. We define AB to be the $m \times r$ matrix whose ik-component is

$$\sum_{j=1}^{n} a_{ij}b_{jk}.$$

If A, B, C are matrices such that AB is defined and BC is defined, then so is $(AB)C$ and $A(BC)$ and we have

$$(AB)C = A(BC).$$

This is trivial to prove. If $C = (c_{kl})$, then the reader will see at once that the il-component of either of the above products is equal to

$$\sum_j \sum_k a_{ij} b_{jk} c_{kl}.$$

An $m \times n$ matrix is said to be a *square matrix* if $m = n$. For example, a 1×1 matrix is a square matrix, and will sometimes be identified with the element of R occurring as its single component.

For a given integer $n \geqq 1$ the set of square $n \times n$ matrices forms a ring.

This is again trivially verified and will be left to the reader.

The unit element of the ring of $n \times n$ matrices is the matrix

$$I_n = \begin{pmatrix} 1 & 0 & \cdots & 0 & 0 \\ 0 & 1 & & & 0 \\ \vdots & & \ddots & & \vdots \\ 0 & & & \ddots & 0 \\ 0 & 0 & \cdots & 0 & 1 \end{pmatrix}$$

whose components are equal to 0 except on the diagonal, in which case they are equal to 1. We sometimes write I instead of I_n.

If $A = (a_{ij})$ is a square matrix, we define in general its *diagonal components* to be the elements a_{ii}.

We have a natural ring-homomorphism of R into the ring of $n \times n$ matrices, given by

$$c \mapsto cI_n.$$

Thus cI_n is the square $n \times n$ matrix having all its components equal to 0 except the diagonal components, which are equal to c. Let us denote the ring of $n \times n$ matrices in R by $\mathrm{Mat}_n(R)$. Then $\mathrm{Mat}_n(R)$ is an algebra over R (with respect to the above homomorphism).

Let $A = (a_{ij})$ be an $m \times n$ matrix. We define its *transpose* ${}^tA$ to be the matrix (a_{ji}) ($j = 1, \ldots, n$ and $i = 1, \ldots, m$). Then ${}^tA$ is an $n \times m$ matrix. The reader will verify at once that if A, B are of the same size, then

$${}^t(A + B) = {}^tA + {}^tB.$$

If $c \in R$ then ${}^t(cA) = c\,{}^tA$. If A, B can be multiplied, then ${}^tB\,{}^tA$ is defined and we have

$${}^t(AB) = {}^tB\,{}^tA.$$

We note the operations on matrices commute with homomorphisms. More precisely, let $\varphi : R \to R'$ be a ring-homomorphism. If A, B are matrices in R, we define φA to be the matrix obtained by applying φ to all the components of A. Then

$$\varphi(A + B) = \varphi A + \varphi B, \qquad \varphi(AB) = (\varphi A)(\varphi B), \qquad \varphi(cA) = \varphi(c)\varphi A,$$
$$\varphi({}^tA) = {}^t\varphi(A).$$

A similar remark will hold throughout our discussion of matrices (for instance in the next section).

Let $A = (a_{ij})$ be a square $n \times n$ matrix in a commutative ring R. We define the *trace* of A to be

$$\operatorname{tr}(A) = \sum_{i=1}^{n} a_{ii};$$

in other words, the trace is the sum of the diagonal elements.

If A, B are $n \times n$ matrices, then

$$\operatorname{tr}(AB) = \operatorname{tr}(BA).$$

Indeed, if $A = (a_{ij})$ and $B = (b_{ij})$ then

$$\operatorname{tr}(AB) = \sum_i \sum_\nu a_{i\nu} b_{\nu i} = \operatorname{tr}(BA).$$

As an application, we observe that if B is an invertible $n \times n$ matrix, then

$$\operatorname{tr}(B^{-1}AB) = \operatorname{tr}(A).$$

Indeed, $\operatorname{tr}(B^{-1}AB) = \operatorname{tr}(ABB^{-1}) = \operatorname{tr}(A)$.

§2. The rank of a matrix

Let k be a field and let A be an $m \times n$ matrix in k. By the *row rank* of A we shall mean the maximum number of linearly independent rows of A, and by the *column rank* of A we shall mean the maximum number of linearly independent columns of A. Thus these ranks are the dimensions of the vector spaces generated respectively by the rows of A and the columns of A. We contend that these ranks are equal to the same number, and we define the *rank* of A to be that number.

Let $A^1, \dots, A^n$ be the columns of A, and let $A_1, \dots, A_m$ be the rows of A. Let ${}^tX = (x_1, \dots, x_m)$ have components $x_i \in k$. We have a linear map

$$X \mapsto x_1 A_1 + \cdots + x_m A_m$$

of $k^{(m)}$ onto the space generated by the row vectors. Let W be its kernel. Then W is a subspace of $k^{(m)}$ and

$$\dim W + \text{row rank} = m.$$

If Y is a column vector of dimension m, then the map

$$(X, Y) \mapsto {}^tXY = X \cdot Y$$

is a bilinear map into k, if we view the 1×1 matrix ${}^tXY$ as an element of k. We observe that W is the orthogonal space to the column vectors $A^1, \ldots, A^n$, i.e. it is the space of all X such that $X \cdot A^j = 0$ for all $j = 1, \ldots, n$. By the duality theorem of Chapter III, we know that $k^{(m)}$ is its own dual under the pairing

$$(X, Y) \mapsto X \cdot Y$$

and that $k^{(m)}/W$ is dual to the space generated by $A^1, \ldots, A^n$. Hence

$$\dim k^{(m)}/W = \text{column rank},$$

or

$$\dim W + \text{column rank} = m.$$

From this we conclude that

$$\text{column rank} = \text{row rank},$$

as desired.

We note that W may be viewed as the space of solutions of the system of n linear equations

$$x_1A_1 + \cdots + x_mA_m = 0,$$

in m unknowns $x_1, \ldots, x_m$. Indeed, if we write out the preceding vector equation in terms of all the coordinates, we get the usual system of n linear equations. We let the reader do this if he wishes.

§3. *Matrices and linear maps*

Let E be a module, and assume that there exists a basis $\mathcal{B} = \{\xi_1, \ldots, \xi_n\}$ for E over R. This means that every element of E has a unique expression as a linear combination

$$x = x_1\xi_1 + \cdots + x_n\xi_n$$

with $x_i \in R$. We call $(x_1, \ldots, x_n)$ the *components* of x with respect to the basis. We may view this n-tuple as a row vector. We shall denote by X the transpose of the row vector $(x_1, \ldots, x_n)$. We call X the *column vector of x with respect to the basis.*

We observe that if $\{\xi_1', \ldots, \xi_m'\}$ is another basis of E over R, then $m = n$. Indeed, let $\mathfrak{p}$ be a maximal ideal of R. Then $E/\mathfrak{p}E$ is a vector space over the field $R/\mathfrak{p}R$, and it is immediately clear that if we denote by $\bar{\xi}_i$ the residue class of ξ_i mod $\mathfrak{p}E$, then $\{\bar{\xi}_1, \ldots, \bar{\xi}_n\}$ is a basis for $E/\mathfrak{p}E$ over $R/\mathfrak{p}R$.

Hence n is also the dimension of this vector space, and we know the invariance of the cardinality for bases of vector spaces over fields. Thus $m = n$. We shall call n the *dimension of* the module E over R.

We shall view $R^{(n)}$ as the module of column vectors of size n. It is a free module of dimension n over R. It has a basis consisting of the unit vectors $e^1, \dots, e^n$ such that

$$^t e^i = (0, \dots, 0, 1, 0, \dots, 0)$$

has components 0 except for its i-th component, which is equal to 1.

An $m \times n$ matrix A gives rise to a linear map

$$L_A : R^{(n)} \to R^{(m)}$$

by the rule

$$X \mapsto AX.$$

Namely, we have $A(X + Y) = AX + AY$ and $A(cX) = cAX$ for column vectors X, Y and $c \in R$.

The above considerations can be extended to a slightly more general context, which can be very useful. Let E be an abelian group and assume that R is a commutative subring of

$$\operatorname{End}_{\mathbf{Z}}(E) = \operatorname{Hom}_{\mathbf{Z}}(E, E).$$

Then E is an R-module. Furthermore, if A is an $m \times n$ matrix in R, then we get a linear map

$$L_A : E^{(n)} \to E^{(m)}$$

defined by a rule similar to the above, namely $X \mapsto AX$. However, this has to be interpreted in the obvious way. If $A = (a_{ij})$ and X is a column vector of elements of E, then

$$AX = \begin{pmatrix} a_{11} & \cdots & a_{1n} \\ & \cdots & \\ a_{m1} & \cdots & a_{mn} \end{pmatrix} \begin{pmatrix} x_1 \\ \vdots \\ x_n \end{pmatrix} = \begin{pmatrix} y_1 \\ \vdots \\ y_m \end{pmatrix},$$

where $y_i = \sum_{j=1}^{n} a_{ij} x_j$.

If A, B are matrices in R whose product is defined, then for any $c \in R$ we have

$$L_{AB} = L_A L_B \quad \text{and} \quad L_{cA} = cL_A.$$

Thus we have associativity, namely

$$A(BX) = (AB)X.$$

An arbitrary commutative ring R may be viewed as a module over itself. In this way we recover the special case of our map from $R^{(n)}$ into $R^{(m)}$. Furthermore, if E is a module over R, then R may be viewed as a ring of endomorphisms of E.

PROPOSITION 1. *Let E be a free module over R, and let $\{x_1, \ldots, x_n\}$ be a basis. Let $y_1, \ldots, y_n$ be elements of E. Let A be the matrix in R such that*

$$A\begin{pmatrix} x_1 \\ \vdots \\ x_n \end{pmatrix} = \begin{pmatrix} y_1 \\ \vdots \\ y_n \end{pmatrix}.$$

Then $\{y_1, \ldots, y_n\}$ is a basis of E if and only if A is invertible.

Proof. Let X, Y be the column vectors of our elements. Then $AX = Y$. Suppose Y is a basis. Then there exists a matrix C in R such that $CY = X$. Then $CAX = X$, whence $CA = I$ and A is invertible. Conversely, assume that A is invertible. Then $X = A^{-1}Y$ and hence $x_1, \ldots, x_n$ are in the module generated by $y_1, \ldots, y_n$. Suppose that we have a relation

$$b_1y_1 + \cdots + b_ny_n = 0$$

with $b_i \in R$. Let B be the row vector $(b_1, \ldots, b_n)$. Then

$$BY = 0$$

and hence $BA^{-1}X = 0$. But $\{x_1, \ldots, x_n\}$ is a basis. Hence $BA^{-1} = 0$, and hence $BA^{-1}A = B = 0$. This proves that the components of Y are linearly independent over R, and proves our proposition.

We return to our situation of modules over an arbitrary commutative ring R.

Let E, F be modules. We shall see how we can associate a matrix with a linear map whenever bases of E and F are given. We assume that E, F are free. We let $\mathcal{B} = \{\xi_1, \ldots, \xi_n\}$ and $\mathcal{B}' = \{\xi'_1, \ldots, \xi'_n\}$ be bases of E and F respectively. Let

$$f : E \to F$$

be a linear map. There exist unique elements $a_{ij} \in R$ such that

$$f(\xi_1) = a_{11}\xi'_1 + \cdots + a_{m1}\xi'_m,$$
$$\cdots$$
$$f(\xi_n) = a_{1n}\xi'_1 + \cdots + a_{mn}\xi'_m,$$

or in other words,

$$f(\xi_j) = \sum_{i=1}^{m} a_{ij}\xi'_i$$

(Observe that the sum is over the *first* index.) We define

$$M^{\mathcal{B}}_{\mathcal{B}'}(f) = (a_{ij}).$$

If $x = x_1\xi_1 + \cdots + x_n\xi_n$ is expressed in terms of the basis, let us denote the column vector X of components of x by $M_{\mathcal{B}}(x)$. We see that

$$M_{\mathcal{B}'}(f(x)) = M^{\mathcal{B}}_{\mathcal{B}'}(f)M_{\mathcal{B}}(x).$$

In other words, if X' is the column vector of $f(x)$, and M is the matrix associated with f then $X' = MX$. Thus the operation of the linear map is reflected by the matrix multiplication, and we have $f = L_M$.

PROPOSITION 2. *Let E, F, D be modules, and let $\mathcal{B}$, $\mathcal{B}'$, $\mathcal{B}''$ be finite bases of E, F, D respectively. Let*

$$E \xrightarrow{f} F \xrightarrow{g} D$$

be linear maps. Then

$$M^{\mathcal{B}}_{\mathcal{B}''}(g \circ f) = M^{\mathcal{B}'}_{\mathcal{B}''}(g)M^{\mathcal{B}}_{\mathcal{B}'}(f).$$

Proof. Let A and B be the matrices associated with the maps f, g respectively, with respect to our given bases. If X is the column vector associated with $x \in E$, the vector associated with $g(f(x))$ is $B(AX) = (BA)X$. Hence BA is the matrix associated with $g \circ f$. This proves what we wanted.

COROLLARY 1. *Let $E = F$. Then*

$$M^{\mathcal{B}}_{\mathcal{B}'}(id)M^{\mathcal{B}'}_{\mathcal{B}}(id) = M^{\mathcal{B}'}_{\mathcal{B}'}(id) = I.$$

Each matrix $M^{\mathcal{B}}_{\mathcal{B}'}(id)$ is invertible (i.e. is a unit in the ring of matrices).

Proof. Obvious.

COROLLARY 2. *Let $N = M^{\mathcal{B}}_{\mathcal{B}'}(id)$. Then*

$$M^{\mathcal{B}'}_{\mathcal{B}'}(f) = M^{\mathcal{B}}_{\mathcal{B}'}(id)M^{\mathcal{B}}_{\mathcal{B}}(f)M^{\mathcal{B}'}_{\mathcal{B}}(id) = NM^{\mathcal{B}}_{\mathcal{B}}(f)N^{-1}.$$

Proof. Obvious.

COROLLARY 3. *Let E be a free module of dimension n over R. Let $\mathcal{B}$ be a basis of E over R. The map*

$$f \mapsto M^{\mathcal{B}}_{\mathcal{B}}(f)$$

is a ring-isomorphism of the ring of endomorphisms of E onto the ring of $n \times n$ matrices in R. In fact, the isomorphism is one of algebras over R.

We shall call the matrix $M^{\mathcal{B}}_{\mathcal{B}}(f)$ the *matrix associated with f with respect to the basis $\mathcal{B}$*.

Let E be a free module of dimension n over R. By $GL(E)$ or $\mathrm{Aut}_R(E)$ one means the group of linear automorphisms of E. It is the group of units in $\mathrm{End}_R(E)$. By $GL_n(R)$ one means the group of invertible $n \times n$ matrices in R. Once a basis is selected for E over R, we have a group-isomorphism

$$GL(E) \leftrightarrow GL_n(R)$$

with respect to this basis.

Let E be as above. If

$$f : E \to E$$

is a linear map, we select a basis $\mathcal{B}$ and let M be the matrix associated with f relative to $\mathcal{B}$. We define the *trace* of f to be the trace of M, thus

$$\mathrm{tr}(f) = \mathrm{tr}(M).$$

If M' is the matrix of f with respect to another basis, then there exists an invertible matrix N such that $M' = N^{-1}MN$, and hence the trace is independent of the choice of basis.

§4. Determinants

Let $E_1, \ldots, E_n, F$ be modules. A map

$$f : E_1 \times \cdots \times E_n \to F$$

is said to be *R-multilinear* (or simply multilinear) if it is linear in each variable, i.e. if for every index i and elements $x_1, \ldots, x_{i-1}, x_{i+1}, \ldots, x_n$, $x_j \in E_j$, the map

$$x \mapsto f(x_1, \ldots, x_{i-1}, x, x_{i+1}, \ldots, x_n)$$

is a linear map of E_i into F.

A multilinear map defined on an n-fold product is also called n-multilinear. If $E_1 = \cdots = E_n = E$, we also say that f is a *multilinear map on E*, instead of saying that it is multilinear on $E^{(n)}$.

Let f be an n-multilinear map. If we take two indices i, j and $i \neq j$ then fixing all the variables except the i-th and j-th variable, we can view f as a bilinear map on $E_i \times E_j$.

Assume that $E_1 = \cdots = E_n = E$. We say that the multilinear map f is *alternating* if $f(x_1, \ldots, x_n) = 0$ whenever there exists an index i, $1 \leqq i \leqq n - 1$, such that $x_i = x_{i+1}$ (in other words, when two adjacent elements are equal).

PROPOSITION 3. *Let f be an n-multilinear alternating map on E. Let $x_1, \ldots, x_n \in E$. Then*

$$f(\ldots, x_i, x_{i+1}, \ldots) = -f(\ldots, x_{i+1}, x_i, \ldots).$$

In other words, when we interchange two adjacent arguments of f, the value of f changes by a sign. If $x_i = x_j$ for $i \neq j$ then $f(x_1, \ldots, x_n) = 0$.

Proof. Restricting our attention to the factors in the i-th and j-th place, we may assume f is bilinear for the first statement. Then for all $x, y \in E$ we have

$$0 = f(x + y, x + y) = f(x, y) + f(y, x).$$

This proves what we want, namely $f(y, x) = -f(x, y)$. For the second assertion, we can interchange successively adjacent arguments of f until we obtain an n-tuple of elements of E having two equal adjacent arguments. This shows that when $x_i = x_j$, $i \neq j$, then $f(x_1, \ldots, x_n) = 0$.

COROLLARY. *Let f be an n-multilinear alternating map on E. Let $x_1, \ldots, x_n \in E$. Let $i \neq j$ and let $a \in R$. Then the value of f on $(x_1, \ldots, x_n)$ does not change if we replace x_i by $x_i + ax_j$ and leave all other components fixed.*

Proof. Obvious.

A multilinear alternating map taking its value in R is called a multilinear alternating *form*.

On repeated occasions we shall evaluate multilinear alternating maps on linear combinations of elements of E. Let

$$\begin{aligned} w_1 &= a_{11}v_1 + \cdots + a_{1n}v_n, \\ &\cdots \\ w_n &= a_{n1}v_1 + \cdots + a_{nn}v_n. \end{aligned}$$

Let f be n-multilinear alternating on E. Then

$$f(w_1, \ldots, w_n) = f(a_{11}v_1 + \cdots + a_{1n}v_n, \ldots, a_{n1}v_1 + \cdots + a_{nn}v_n).$$

We expand this by multilinearity, and get a sum of terms of type

$$a_{1,\sigma(1)} \cdots a_{n,\sigma(n)} f(v_{\sigma(1)}, \cdots, v_{\sigma(n)}),$$

where σ ranges over arbitrary maps of $\{1, \ldots, n\}$ into itself. If σ is not a bijection (i.e. a permutation), then two arguments $v_{\sigma(i)}$ and $v_{\sigma(j)}$ are equal for $i \neq j$, and the term is equal to 0. Hence we may restrict our sum to permutations σ. Shuffling back the elements $(v_{\sigma(1)}, \ldots, v_{\sigma(n)})$ to their standard ordering and using Proposition 3, we see that we have obtained the following expansion:

LEMMA. *If $w_1, \ldots, w_n$ are as above, then*

$$f(w_1, \ldots, w_n) = \sum_{\sigma} \epsilon(\sigma) a_{1,\sigma(1)} \cdots a_{n,\sigma(n)} f(v_1, \ldots, v_n)$$

where the sum is taken over all permutations σ *of* $\{1, \dots, n\}$ *and* $\epsilon(\sigma)$ *is the sign of the permutation.*

For determinants, I shall follow Artin's treatment in *Galois Theory*. By an $n \times n$ *determinant* we shall mean a mapping

$$\det : \mathrm{Mat}_n(R) \to R$$

also written

$$D : \mathrm{Mat}_n(R) \to R$$

which, when viewed as a function of the column vectors $A^1, \dots, A^n$ of a matrix A, is multilinear alternating, and such that $D(I) = 1$. In this chapter, we use mostly the letter D to denote determinants.

We shall prove later that determinants exist. For the moment, we derive properties.

CRAMER'S RULE. *Let* $A^1, \dots, A^n$ *be column vectors of dimension* n. *Let* $x_1, \dots, x_n \in R$ *be such that*

$$x_1 A^1 + \cdots + x_n A^n = B$$

for some column vector B. *Then for each* i *we have*

$$x_i D(A^1, \dots, A^n) = D(A^1, \dots, \underset{i}{B}, \dots, A^n),$$

where B *in this last line occurs in the i-th place.*

Proof. Say $i = 1$. We expand

$$D(B, A^2, \dots, A^n) = \sum_{j=1}^{n} x_j D(A^j, A^2, \dots, A^n),$$

and use Proposition 3 to get what we want (all terms on the right are equal to 0 except the one having x_1 in it).

COROLLARY. *Assume that* R *is a field. Then* $A^1, \dots, A^n$ *are linearly dependent if and only if* $D(A^1, \dots, A^n) = 0$.

Proof. Assume we have a relation

$$x_1 A^1 + \cdots + x_n A^n = 0$$

with $x_i \in R$. Then $x_i D(A) = 0$ for all i. If some $x_i \neq 0$ then $D(A) = 0$. Conversely, assume that $A^1, \dots, A^n$ are linearly independent. Then we can express the unit vectors $e^1, \dots, e^n$ as linear combinations

$$e^1 = b_{11} A^1 + \cdots + b_{1n} A^n,$$
$$\cdots$$
$$e^n = b_{n1} A^1 + \cdots + b_{nn} A^n$$

with $b_{ij} \in R$. But

$$1 = D(e^1, \ldots, e^n).$$

Using a previous lemma, we know that this can be expanded into a sum of terms involving $D(A^1, \ldots, A^n)$, and hence $D(A)$ cannot be 0.

PROPOSITION 4. *If determinants exist, they are unique. If $A^1, \ldots, A^n$ are the column vectors of dimension n, of the matrix $A = (a_{ij})$, then*

$$D(A^1, \ldots, A^n) = \sum_\sigma \epsilon(\sigma) a_{\sigma(1),1} \cdots a_{\sigma(n),n},$$

where the sum is taken over all permutations σ of $\{1, \ldots, n\}$, and $\epsilon(\sigma)$ is the sign of the permutation.

Proof. Let $e^1, \ldots, e^n$ be the unit vectors as usual. We can write

$$A^1 = a_{11}e^1 + \cdots + a_{n1}e^n,$$
$$\cdots$$
$$A^n = a_{1n}e^n + \cdots + a_{nn}e^n.$$

Therefore

$$D(A^1, \ldots, A^n) = \sum_\sigma \epsilon(\sigma) a_{\sigma(1),1} \cdots a_{\sigma(n),n}$$

by the lemma. This proves that the value of the determinant is uniquely determined and is given by the expected formula.

COROLLARY. *Let $\varphi : R \to R'$ be a ring-homomorphism into a commutative ring. If A is a square matrix in R, define φA to be the matrix obtained by applying φ to each component of A. Then*

$$\varphi(D(A)) = D(\varphi A).$$

Proof. Apply φ to the expression of Proposition 4.

PROPOSITION 5. *If A is a square matrix in R then*

$$D(A) = D({}^tA).$$

Proof. In a product

$$a_{\sigma(1),1} \cdots a_{\sigma(n),n}$$

each integer k from 1 to n occurs precisely once among the integers $\sigma(1), \ldots, \sigma(n)$. Hence we can rewrite this product in the form

$$a_{1,\sigma^{-1}(1)} \cdots a_{n,\sigma^{-1}(n)}.$$

Since $\epsilon(\sigma) = \epsilon(\sigma^{-1})$, we can rewrite the sum in Proposition 4 in the form

$$\sum_{\sigma} \epsilon(\sigma^{-1}) a_{1,\sigma^{-1}(1)} \cdots a_{n,\sigma^{-1}(n)}.$$

In this sum, each term corresponds to a permutation σ. However, as σ ranges over all permutations, so does σ^{-1}. Hence our sum is equal to

$$\sum_{\sigma} \epsilon(\sigma) a_{1,\sigma(1)} \cdots a_{n,\sigma(n)},$$

which is none other than $D({}^tA)$, as was to be shown.

COROLLARY. *The determinant is multilinear and alternating with respect to the rows of a matrix.*

We shall now prove existence, and prove simultaneously one additional important property of determinants.

When $n = 1$, we define $D(a) = a$ for any $a \in R$.

Assume that we have proved the existence of determinants for all integers $< n$ $(n \geqq 2)$. Let A be an $n \times n$ matrix in R, $A = (a_{ij})$. We let A_{ij} be the $(n-1) \times (n-1)$ matrix obtained from A by deleting the i-th row and j-th column. Let i be a fixed integer, $1 \leqq i \leqq n$. We define inductively

$$D(A) = (-1)^{i+1} a_{i1} D(A_{i1}) + \cdots + (-1)^{i+n} a_{in} D(A_{in}).$$

(This is known as the *expansion of D according to the i-th row.*) We shall prove that D satisfies the definition of a determinant.

Consider D as a function of the k-th column, and consider any term

$$(-1)^{i+j} a_{ij} D(A_{ij}).$$

If $j \neq k$ then a_{ij} does not depend on the k-th column, and $D(A_{ij})$ depends linearly on the k-th column. If $j = k$, then a_{ij} depends linearly on the k-th column, and $D(A_{ij})$ does not depend on the k-th column. In any case our term depends linearly on the k-th column. Since $D(A)$ is a sum of such terms, it depends linearly on the k-th column, and thus D is multilinear.

Next, suppose that two adjacent columns of A are equal, say $A^k = A^{k+1}$. Let j be an index $\neq k$ and $\neq k+1$. Then the matrix A_{ij} has two adjacent equal columns, and hence its determinant is equal to 0. Thus the term corresponding to an index $j \neq k$ or $k+1$ gives a zero contribution to $D(A)$. The other two terms can be written

$$(-1)^{i+k} a_{ik} D(A_{ik}) + (-1)^{i+k+1} a_{i,k+1} D(A_{i,k+1}).$$

The two matrices A_{ik} and $A_{i,k+1}$ are equal because of our assumption that the k-th column of A is equal to the $(k+1)$-th column. Similarly, $a_{ik} = a_{i,k+1}$. Hence these two terms cancel since they occur with opposite signs. This proves that our form is alternating, and gives:

PROPOSITION 6. *Determinants exist and satisfy the rule of expansion according to rows and columns.*

(For columns, we use the fact that $D(A) = D({}^tA)$.)

THEOREM 1. *Let E be a module over R, and let $v_1, \ldots, v_n$ be elements of E. Let $A = (a_{ij})$ be a matrix in R, and let*

$$A\begin{pmatrix} v_1 \\ \vdots \\ v_1 \end{pmatrix} = \begin{pmatrix} w_1 \\ \vdots \\ w_n \end{pmatrix}.$$

Let Δ be an n-multilinear alternating map on E. Then

$$\Delta(w_1, \ldots, w_n) = D(A)\,\Delta(v_1, \ldots, v_n).$$

Proof. We expand

$$\Delta(a_{11}v_1 + \cdots + a_{1n}v_n, \ldots, a_{n1}v_1 + \cdots + a_{nn}v_n),$$

and find precisely what we want, taking into account $D(A) = D({}^tA)$.

Let E, F be modules, and let $L_a^n(E, F)$ denote the set of n-multilinear alternating maps of E into F. If $F = R$, we also write $L_a^n(E, R) = L_a^n(E)$. It is clear that $L_a^n(E, F)$ is a module over R, i.e. is closed under addition and multiplication by elements of R.

COROLLARY 1. *Let E be a free module over R, and let $\{v_1, \ldots, v_n\}$ be a basis. Let F be any module, and let $w \in F$. There exists a unique n-multilinear alternating map*

$$\Delta_w : E \times \cdots \times E \to F$$

such that $\Delta_w(v_1, \ldots, v_n) = w$.

Proof. Without loss of generality, we may assume that $E = R^{(n)}$, and then, if $A^1, \ldots, A^n$ are column vectors, we define $\Delta_w(A^1, \ldots, A^n) = D(A)w$. Then Δ_w obviously has the required properties.

COROLLARY 2. *If E is free over R, and has a basis consisting of n elements, then $L_a^n(E)$ is free over R, and has a basis consisting of 1 element.*

Proof. We let Δ_1 be the multilinear alternating map taking the value 1 on a basis $\{v_1, \ldots, v_n\}$. Any element $\varphi \in L_a^n(E)$ can then be written in a unique way as $c\Delta_1$, with some $c \in R$, namely $c = \varphi(v_1, \ldots, v_n)$. This proves what we wanted.

Any two bases of $L_a^n(E)$ in the preceding corollary differ by a unit in R. In other words, if Δ is a basis of $L_a^n(E)$, then $\Delta = c\Delta_1 = \Delta_c$ for some $c \in R$, and c must be a unit. Our Δ_1 depends of course on the choice of a basis for E. When we consider $R^{(n)}$, our determinant D is precisely Δ_1, relative to the standard basis consisting of the unit vectors $e^1, \ldots, e^n$.

It is sometimes convenient terminology to say that any basis of $L_a^n(E)$ is a *determinant* on E. In that case, the corollary to Cramer's rule can be stated as follows.

COROLLARY 3. *Let R be a field. Let E be a vector space of dimension n. Let Δ be any determinant on E. Let $v_1, \ldots, v_n \in E$. In order that $\{v_1, \ldots, v_n\}$ be a basis of E it is necessary and sufficient that*

$$\Delta(v_1, \ldots, v_n) \neq 0.$$

PROPOSITION 7. *Let A, B be $n \times n$ matrices in R. Then*

$$D(AB) = D(A)D(B).$$

Proof. This is actually a corollary of Theorem 1. We take $v_1, \ldots, v_n$ to be the unit vectors $e^1, \ldots, e^n$, and consider

$$AB\begin{pmatrix} e^1 \\ \vdots \\ e^n \end{pmatrix} = \begin{pmatrix} w_1 \\ \vdots \\ w_n \end{pmatrix}.$$

We obtain

$$D(w_1, \ldots, w_n) = D(AB)D(e^1, \ldots, e^n).$$

On the other hand, by associativity, applying Theorem 1 twice,

$$D(w_1, \ldots, w_n) = D(A)D(B)D(e^1, \ldots, e^n).$$

Since $D(e^1, \ldots, e^n) = 1$, our proposition follows.

Let $A = (a_{ij})$ be an $n \times n$ matrix in R. We let

$$\tilde{A} = (b_{ij})$$

be the matrix such that

$$b_{ij} = (-1)^{i+j}D(A_{ji}).$$

(Note the reversal of indices!)

PROPOSITION 8. *Let $d = D(A)$. Then $A\tilde{A} = \tilde{A}A = dI$. The determinant $D(A)$ is invertible in R if and only if A is invertible, and then*

$$A^{-1} = \frac{1}{d}\tilde{A}.$$

Proof. For any pair of indices i, k the ik-component of $A\tilde{A}$ is

$$a_{i1}b_{1k} + a_{i2}b_{2k} + \cdots + a_{in}b_{nk}$$
$$= a_{i1}(-1)^{k+1}D(A_{k1}) + \cdots + a_{in}(-1)^{k+n}D(A_{kn}).$$

If $i = k$, then this sum is simply the expansion of the determinant according to the i-th row, and hence this sum is equal to d. If $i \neq k$, let $\bar{A}$ be the matrix obtained from A by replacing the k-th row by the i-th row, and leaving all other rows unchanged. If we delete the k-th row and the j-th column from $\bar{A}$, we obtain the same matrix as by deleting the k-th row and j-th column from A. Thus

$$\bar{A}_{kj} = A_{kj},$$

and hence our sum above can be written

$$a_{i1}(-1)^{k+1}D(\bar{A}_{k1}) + \cdots + a_{in}(-1)^{k+n}D(\bar{A}_{kn}).$$

This is the expansion of the determinant of $\bar{A}$ according to the i-th row. Hence $D(\bar{A}) = 0$, and our sum is 0. We have therefore proved that the ik-component of $A\tilde{A}$ is equal to d if $i = k$ (i.e. if it is a diagonal component), and is equal to 0 otherwise. This proves that $A\tilde{A} = dI$. On the other hand, we see at once from the definitions that ${}^t\tilde{A} = \widetilde{{}^tA}$. Then

$${}^t(\tilde{A}A) = {}^tA\,{}^t\tilde{A} = {}^tA\,\widetilde{{}^tA} = dI,$$

and consequently, $\tilde{A}A = dI$ also, since ${}^t(dI) = dI$. When d is a unit in R, then A is invertible, its inverse being $d^{-1}\tilde{A}$. Conversely, if A is invertible, and $AA^{-1} = I$, then $D(A)D(A^{-1}) = 1$, and hence $D(A)$ is invertible, as was to be shown.

Corollary. *Let F be any R-module, and let $w_1, \ldots, w_n$ be elements of F. Let $A = (a_{ij})$ be an $n \times n$ matrix in R. Assume that*

$$a_{11}w_1 + \cdots + a_{1n}w_n = 0,$$
$$\cdots$$
$$a_{n1}w_1 + \cdots + a_{nn}w_n = 0.$$

Then $D(A)w_i = 0$ for all i. In particular, if F is generated by $w_1, \ldots, w_n$, then $D(A)F = 0$.

Proof. This follows from the remarks in §2. We multiply by $\tilde{A}$ to see that

$$\tilde{A}A\begin{pmatrix} w_1 \\ \vdots \\ w_n \end{pmatrix} = d\begin{pmatrix} w_1 \\ \vdots \\ w_n \end{pmatrix},$$

where $d = D(A)$.

Proposition 9. *Let E, F be free modules of dimension n over R. Let $f : E \to F$ be a linear map. Let $\mathcal{B}$, $\mathcal{B}'$ be bases of E, F respectively over R. Then f is an isomorphism if and only if the determinant of its associated matrix $M^{\mathcal{B}}_{\mathcal{B}'}(f)$ is a unit in R.*

Proof. Let $A = M^{\mathcal{B}}_{\mathcal{B}'}(f)$. By definition, f is an isomorphism if and only if there exists a linear map $g : F \to E$ such that $g \circ f = id$ and $f \circ g = id$. If f is an isomorphism, and $B = M^{\mathcal{B}'}_{\mathcal{B}}(g)$, then $AB = BA = 1$. Taking the determinant of the product, we conclude that $D(A)$ is invertible in R. Conversely, if $D(A)$ is a unit, then we can define A^{-1} by Proposition 7. This A^{-1} is the associated matrix of a linear map $g : F \to E$ which is an inverse for f, as desired.

Finally, we shall define the determinant of an endomorphism.

Let E be a free module over R, and let $\mathcal{B}$ be a basis. Let $f : E \to E$ be an endomorphism of E. Let

$$M = M^{\mathcal{B}}_{\mathcal{B}}(f).$$

If $\mathcal{B}'$ is another basis of E, and $M' = M^{\mathcal{B}'}_{\mathcal{B}'}(f)$, then there exists an invertible matrix N such that

$$M' = NMN^{-1}.$$

Taking the determinant, we see that $D(M') = D(M)$. Hence the determinant does not depend on the choice of basis, and will be called the *determinant of the linear map* f. We shall give below a characterization of this determinant which does not depend on the choice of a basis.

Let E be any module. Then we can view $L^n_a(E)$ as a functor in the variable E (contravariant). In fact, we can view $L^n_a(E, F)$ as a functor of two variables, contravariant in the first, and covariant in the second. Indeed, suppose that

$$E' \xrightarrow{f} E$$

is a linear map. To each multilinear map $\varphi : E^{(n)} \to F$ we can associate the composite map $\varphi \circ f^{(n)}$,

$$E' \times \cdots \times E' \xrightarrow{f^{(n)}} E \times \cdots \times E \xrightarrow{\varphi} F$$

where $f^{(n)}$ is the product of f with itself n times. The map

$$L^n_a(f) : L^n_a(E, F) \to L^n_a(E', F)$$

given by

$$\varphi \mapsto \varphi \circ f^{(n)},$$

is obviously a linear map, which defines our functor. We shall sometimes write f^* instead of $L^n_a(f)$.

In particular, consider the case when $E = E'$ and $F = R$. We get an induced map

$$f^* : L_a^n(E) \to L_a^n(E).$$

PROPOSITION 10. *Let E be a free module over R, of dimension n. Let Δ be a basis of $L_a^n(E)$. Let $f : E \to E$ be an endomorphism of E. Then*

$$f^*\Delta = D(f)\Delta.$$

Proof. This is an immediate consequence of Theorem 1. Namely, we let $\{v_1, \ldots, v_n\}$ be a basis of E, and then take A (or ${}^tA$) to be the matrix of f relative to this basis. By definition,

$$f^*\Delta(v_1, \ldots, v_n) = \Delta\big(f(v_1), \ldots, f(v_n)\big),$$

and by Theorem 1, this is equal to

$$D(A)\,\Delta(v_1, \ldots, v_n).$$

By Corollary 1 of Theorem 1, we conclude that $f^*\Delta = D(A)\Delta$ since both of these forms take on the same value on $(v_1, \ldots, v_n)$.

§5. *Duality*

Let R be a commutative ring, and let E, F be modules over R. An *R-bilinear form* on $E \times F$ is a map

$$f : E \times F \to R$$

having the following properties: For each $x \in E$, the map

$$y \mapsto f(x, y)$$

is R-linear, and for each $y \in F$, the map

$$x \mapsto f(x, y)$$

is R-linear. We shall omit the subscript R- in the rest of this section, and write $\langle x, y\rangle_f$ or $\langle x, y\rangle$ instead of $f(x, y)$. If $x \in E$ and $y \in F$, we write $x \perp y$ if $\langle x, y\rangle = 0$. Similarly, if S is a subset of F, we define $x \perp S$ if $x \perp y$ for all $y \in S$. We then say that x is *perpendicular* to S. We let $S^\perp$ consist of all elements of E which are perpendicular to S. It is obviously a submodule of E. We define perpendicularity on the other side in the same way. We define the *kernel* of f on the left to be $F^\perp$ and the kernel on the right to be $E^\perp$. We say that f is *non-degenerate* on the left if its kernel on the left is 0. We say that f is *non-degenerate* on the right if its kernel on the right is 0. If E_0 is the kernel of f on the left, then we get an induced bilinear map

$$E/E_0 \times F \to R$$

which is non-degenerate on the left, as one verifies trivially from the definitions. Similarly, if F_0 is the kernel of f on the right, we get an induced bilinear map

$$E/E_0 \times F/F_0 \to R$$

which is non-degenerate on either side. This map arises from the fact that the value $\langle x, y \rangle$ depends only on the coset of x modulo E_0 and the coset of y modulo F_0.

We shall denote by $L^2(E, F; R)$ the set of all bilinear maps of $E \times F$ into R. It is clear that this set is a module (i.e. an R-module), addition of maps being the usual one, and also multiplication of maps by elements of R.

The form f gives rise to a homomorphism

$$\varphi_f : E \to \operatorname{Hom}_R(F, R)$$

such that

$$\varphi_f(x)(y) = f(x, y) = \langle x, y \rangle,$$

for all $x \in E$ and $y \in F$. We shall call $\operatorname{Hom}_R(F, R)$ the *dual module* of F, and denote it by F^*. We have an *isomorphism*

$$\boxed{L^2(E, F; R) \leftrightarrow \operatorname{Hom}_R(E, \operatorname{Hom}_R(F, R))}$$

given by $f \mapsto \varphi_f$, its inverse being defined in the obvious way: If $\varphi : E \to \operatorname{Hom}_R(F, R)$ is a homomorphism, we let f be such that

$$f(x, y) = \varphi(x)(y).$$

We shall say that f is *non-singular on the left* if φ_f is an isomorphism, in other words if our form can be used to identify E with the dual module of F. We define *non-singular on the right* in a similar way, and say that f is *non-singular* if it is non-singular on the left and on the right. *Warning:* Non-degeneracy does not necessarily imply non-singularity.

We shall now obtain an *isomorphism*

$$\boxed{\operatorname{End}_R(E) \leftrightarrow L^2(E, F; R)}$$

depending on a fixed non-singular bilinear map $f : E \times F \to R$.

Let $A \in \text{End}_R(E)$ be a linear map of E into itself. Then the map

$$(x, y) \mapsto \langle Ax, y \rangle = \langle Ax, y \rangle_f$$

is bilinear, and in this way, we associate linearly with each $A \in \text{End}_R(E)$ a bilinear map in $L^2(E, F; R)$.

Conversely, let $h : E \times F \to R$ be bilinear. Given $x \in E$, the map $h_x : F \to R$ such that $h_x(y) = h(x, y)$ is linear, and is in the dual space F^*. By assumption, there exists a unique element $x' \in E$ such that for all $y \in F$ we have

$$h(x, y) = \langle x', y \rangle.$$

It is clear that the association $x \mapsto x'$ is a linear map of E into itself. Thus with each bilinear map $E \times F \to R$ we have associated a linear map $E \to E$.

It is immediate that the mappings described in the last two paragraphs are inverse isomorphisms between $\text{End}_R(E)$ and $L^2(E, F; R)$. We emphasize of course that they depend on our form f.

Of course, we could also have worked on the right, and thus we have a similar *isomorphism*

$$\boxed{L^2(E, F; R) \leftrightarrow \text{End}_R(F)}$$

depending also on our fixed non-singular form f.

As an application, let $A : E \to E$ be linear, and let $(x, y) \mapsto \langle Ax, y \rangle$ be its associated bilinear map. There exists a unique linear map

$${}^tA : F \to F$$

such that

$$\langle Ax, y \rangle = \langle x, {}^tAy \rangle$$

for all $x \in E$ and $y \in F$. We call ${}^tA$ the *transpose of A with respect to f.*

It is immediately clear that if A, B are linear maps of E into itself, then for $c \in R$,

$${}^t(cA) = c\,{}^tA, \qquad {}^t(A + B) = {}^tA + {}^tB, \qquad \text{and} \qquad {}^t(AB) = {}^tB\,{}^tA.$$

Let us assume that $E = F$. Let $f : E \times E \to R$ be bilinear. By an *automorphism of the pair* (E, f), or simply of f, we shall mean a linear automorphism $A : E \to E$ such that

$$\langle Ax, Ay \rangle = \langle x, y \rangle$$

for all $x, y \in E$. The group of automorphisms of f is denoted by $\text{Aut}(f)$.

PROPOSITION 11. *Let $f : E \times E \to R$ be a non-singular bilinear form. Let $A : E \to E$ be a linear map. Then A is an automorphism of f if and only if ${}^tAA = id$, and A is invertible.*

Proof. From the equality

$$\langle x, y\rangle = \langle Ax, Ay\rangle = \langle x, {}^tAAy\rangle$$

holding for all $x, y \in E$, we conclude that ${}^tAA = id$ if A is an automorphism of f. The converse is equally clear.

Note. If E is free and finite dimensional, then the condition ${}^tAA = id$ implies that A is invertible.

Let $f : E \times E \to R$ be a bilinear form. We say that f is *symmetric* if $f(x, y) = f(y, x)$ for all $x, y \in E$. The set of symmetric bilinear forms on E will be denoted by $L_s^2(E)$. Let us take a fixed symmetric non-singular bilinear form f on E, denoted by $(x, y) \mapsto \langle x, y\rangle$. An endomorphism $A : E \to E$ will be said to be *symmetric with respect to* f if ${}^tA = A$. It is clear that the set of symmetric endomorphisms of E is a module, which we shall denote by $\mathrm{Sym}(E)$. *Depending on our fixed symmetric non-singular f, we have an isomorphism*

$$\boxed{L_s^2(E) \leftrightarrow \mathrm{Sym}(E)}$$

which we describe as follows. If g is symmetric bilinear on E, then there exists a unique linear map A such that

$$g(x, y) = \langle Ax, y\rangle$$

for all $x, y \in E$. Using the fact that both f, g are symmetric, we obtain

$$\langle Ax, y\rangle = \langle Ay, x\rangle = \langle y, {}^tAx\rangle = \langle {}^tAx, y\rangle.$$

Hence $A = {}^tA$. The association $g \mapsto A$ gives us a homomorphism from $L_s^2(E)$ into $\mathrm{Sym}(E)$. Conversely, given a symmetric endomorphism A of E, we can define a symmetric form by the rule $(x, y) \mapsto \langle Ax, y\rangle$, and the association of this form to A clearly gives a homomorphism of $\mathrm{Sym}(E)$ into $L_s^2(E)$ which is inverse to the preceding homomorphism. Hence $\mathrm{Sym}(E)$ and $L_s^2(E)$ are isomorphic.

We recall that a bilinear form $g : E \times E \to R$ is said to be *alternating* if $g(x, x) = 0$ for all $x \in E$, and consequently $g(x, y) = -g(y, x)$ for all $x, y \in E$. The set of bilinear alternating forms on E is a module, denoted by $L_a^2(E)$.

Let f be a fixed *symmetric* non-singular bilinear form on E. An endomorphism $A : E \to E$ will be said to be *skew-symmetric* or *alternating* with respect to f, if ${}^tA = -A$, and also $\langle Ax, x\rangle = 0$ for all $x \in E$. If for all $a \in R$, $2a = 0$ implies $a = 0$, then this second condition $\langle Ax, x\rangle = 0$ is redundant, because $\langle Ax, x\rangle = -\langle Ax, x\rangle$ implies $\langle Ax, x\rangle = 0$. It is clear that the set of alternating endomorphisms of E is a module, denoted by $\mathrm{Alt}(E)$. *Depending on our fixed symmetric non-singular form f, we have an isomorphism*

$$\boxed{L_a^2(E) \leftrightarrow \mathrm{Alt}(E)}$$

described as usual. If g is an alternating bilinear form on E, its corresponding linear map A is the one such that

$$g(x, y) = \langle Ax, y\rangle$$

for all $x, y \in E$. One verifies trivially in a manner similar to the one used in the symmetric case that the correspondence $g \leftrightarrow A$ gives us our desired isomorphism.

Examples. Let k be a field and let E be a finite dimensional vector space over k. Let $f : E \times E \to E$ be a bilinear map, denoted by $(x, y) \mapsto xy$. To each $x \in E$, we associate the linear map $\lambda_x : E \mapsto E$ such that

$$\lambda_x(y) = xy.$$

Then the map obtained by taking the trace, namely

$$(x, y) \mapsto \mathrm{tr}(\lambda_{xy})$$

is a bilinear form on E. If $xy = yx$, then this bilinear form is symmetric.

Next, let E be the space of continuous functions on the interval $[0, 1]$. Let $K(s, t)$ be a continuous function of two real variables defined on the square $0 \leqq s \leqq 1$ and $0 \leqq t \leqq 1$. For $\varphi, \psi \in E$ we define

$$\langle \varphi, \psi\rangle = \iint \varphi(s)K(s, t)\psi(t)\, ds\, dt,$$

the double integral being taken on the square. Then we obtain a bilinear form on E. If $K(s, t) = K(t, s)$, then the bilinear form is symmetric. When we discuss matrices and bilinear forms in the next section, the reader will note the similarity between the preceding formula and the bilinear form defined by a matrix.

Thirdly, let U be an open subset of a real Banach space E (or a finite dimensional Euclidean space, if the reader insists), and let $f : U \to \mathbf{R}$

be a map which is twice continuously differentiable. For each $x \in U$, the derivative $Df(x) : E \to \mathbf{R}$ is a continuous linear map, and the second derivative $D^2f(x)$ can be viewed as a continuous symmetric bilinear map of $E \times E$ into $\mathbf{R}$.

§6. *Matrices and bilinear forms*

We shall investigate the relation between the concepts introduced above and matrices. Let $f : E \times F \to R$ be bilinear. Assume that E, F are free over R. Let $\mathcal{B} = \{v_1, \ldots, v_m\}$ be a basis for E over R, and let $\mathcal{B}' = \{w_1, \ldots, w_n\}$ be a basis for F over R. Let $g_{ij} = \langle v_i, w_j \rangle$. If

$$x = x_1v_1 + \cdots + x_mv_m$$

and

$$y = y_1w_1 + \cdots + y_nw_n$$

are elements of E and F respectively, with coordinates $x_i, y_j \in R$, then

$$\langle x, y \rangle = \sum_{i=1}^{m} \sum_{j=1}^{n} g_{ij}x_iy_j.$$

Let X, Y be the column vectors of coordinates for x, y respectively, with respect to our bases. Then

$$\langle x, y \rangle = {}^tXGY$$

where G is the matrix (g_{ij}). We could write $G = M^{\mathcal{B}}_{\mathcal{B}'}(f)$. We call G the *matrix associated with the form f relative to the bases* $\mathcal{B}$, $\mathcal{B}'$.

Conversely, given a matrix G (of size $m \times n$), we get a bilinear form from the map

$$(X, Y) \mapsto {}^tXGY.$$

In this way, we get a correspondence from bilinear forms to matrices and back, and it is clear that this correspondence induces an *isomorphism* (of R-modules)

$$\boxed{L^2(E, F; R) \leftrightarrow \mathrm{Mat}_{m\times n}(R)}$$

given by

$$f \mapsto M^{\mathcal{B}}_{\mathcal{B}'}(f).$$

The two maps between these two modules which we described above are clearly inverse to each other.

If we have bases $\mathcal{B} = \{v_1, \ldots, v_n\}$ and $\mathcal{B}' = \{w_1, \ldots, w_n\}$ such that $\langle v_i, w_j \rangle = \delta_{ij}$, then we say that these bases are *dual* to each other. In that case, if X is the coordinate vector of an element of E, and Y the coordinate vector of an element of F, then the bilinear map on X, Y has the value

$$X \cdot Y = x_1y_1 + \cdots + x_ny_n$$

given by the usual dot product.

It is easy to derive in general how the matrix G changes when we change bases in E and F. However, we shall write down the explicit formula only when $E = F$ and $\mathcal{B} = \mathcal{B}'$. Thus we have a bilinear form $f : E \times E \to R$. Let $\mathcal{C}$ be another basis of E and write $X_{\mathcal{B}}$ and $X_{\mathcal{C}}$ for the column vectors belonging to an element x of E, relative to the two bases. *Let C be the invertible matrix $M^{\mathcal{C}}_{\mathcal{B}}(id)$*, so that

$$X_{\mathcal{B}} = CX_{\mathcal{C}}.$$

Then our form is given by

$$\langle x, y \rangle = {}^tX_{\mathcal{C}}{}^tCGCY_{\mathcal{C}}.$$

We see that

$$M^{\mathcal{C}}_{\mathcal{C}}(f) = {}^tCM^{\mathcal{B}}_{\mathcal{B}}(f)C. \tag{1}$$

In other words, the matrix of the bilinear form changes by the *transpose*.

If F is free over R, with a basis $\{\eta_1, \ldots, \eta_n\}$, then $\operatorname{Hom}_R(F, R)$ *is also free, and we have a dual basis $\{\eta_1^*, \ldots, \eta_n^*\}$ such that*

$$\eta_i^*(\eta_j) = \delta_{ij}.$$

This is verified just as with vector spaces over fields.

Proposition 12. *Let E, F be free modules of dimension n over R and let $f : E \times F \to R$ be a bilinear form. Then the following conditions are equivalent:*

f is non-singular on the left.

f is non-singular on the right.

f is non-singular.

The determinant of the matrix of f relative to any bases is invertible in R.

Proof. Assume that f is non-singular on the left. Fix a basis of E relative to which we write elements of E as column vectors, and giving rise to the matrix G for f. Then our form is given by

$$(X, Y) \mapsto {}^tXGY$$

where X, Y are column vectors with coefficients in R. By assumption the map

$$X \mapsto {}^tXG$$

gives an isomorphism between the module of column vectors, and the module of row vectors of length n over R. Hence G is invertible, and hence its determinant is a unit in R. The converse is equally clear, and if $\det(G)$ is a unit, we see that the map

$$Y \mapsto GY$$

must also be an isomorphism between the module of column vectors and itself. This proves our assertion.

We shall now investigate how the transpose behaves in terms of matrices. Let E, F be free over R, of dimension n.

Let $f : E \times F \to R$ be a non-singular bilinear form, and assume given a basis $\mathcal{B}$ of E and $\mathcal{B}'$ of F. Let G be the matrix of f relative to these bases. Let $A : E \to E$ be a linear map. If $x, y \in E$, let X, Y be their column vectors relative to $\mathcal{B}$, $\mathcal{B}'$. Let M be the matrix of A relative to $\mathcal{B}$. Then for $x \in E$ and $y \in F$ we have

$$\langle Ax, y \rangle = {}^t(MX)GY = {}^tX{}^tMGY.$$

Let N be the matrix of ${}^tA$ relative to the basis $\mathcal{B}'$. Then NY is the column vector of ${}^tAy$ relative to $\mathcal{B}'$. Hence

$$\langle x, {}^tAy \rangle = {}^tXGNY.$$

From this we conclude that ${}^tMG = GN$, and since G is invertible, we can solve for N in terms of M. We get:

PROPOSITION 13. *Let E, F be free over R, of dimension n. Let $f : E \times F \to R$ be a non-singular bilinear form. Let $\mathcal{B}$, $\mathcal{B}'$ be bases of E and F respectively over R, and let G be the matrix of f relative to these bases. Let $A : E \to E$ be a linear map, and let M be its matrix relative to $\mathcal{B}$. Then the matrix of ${}^tA$ relative to $\mathcal{B}'$ is*

$$(G^{-1}){}^tMG.$$

COROLLARY 1. *If G is the unit matrix, then the matrix of the transpose is equal to the transpose of the matrix.*

In terms of matrices and bases, we obtain the following characterization for a matrix to induce an automorphism of the form.

COROLLARY 2. *Let the notation be as in Proposition 13, and let $E = F$, $\mathcal{B} = \mathcal{B}'$. An $n \times n$ matrix M is the matrix of an automorphism of the form f (relative to our basis) if and only if*

$${}^tMGM = G.$$

If this condition is satisfied, then in particular, M is invertible.

Proof. We use the definitions, together with the formula given in Proposition 13. We note that M is invertible, for instance because its determinant is a unit in R.

A matrix M is said to be *symmetric* (resp. *alternating*) if ${}^tM = M$ (resp. ${}^tM = -M$ and the diagonal elements of M are 0).

Let $f : E \times E \to R$ be a bilinear form. We say that f is *symmetric* if $f(x, y) = f(y, x)$ for all $x, y \in E$. We say that f is *alternating* if $f(x, x) = 0$ for all $x \in E$.

PROPOSITION 14. *Let E be a free module of dimension n over R, and let $\mathcal{B}$ be a fixed basis. The map*

$$f \mapsto M_{\mathcal{B}}^{\mathcal{B}}(f)$$

induces an isomorphism between the module of symmetric bilinear forms on $E \times E$ (resp. the module of alternating forms on $E \times E$) and the module of symmetric $n \times n$ matrices over R (resp. the module of alternating $n \times n$ matrices over R).

Proof. Consider first the symmetric case. Assume that f is symmetric. In terms of coordinates, let $G = M_{\mathcal{B}}^{\mathcal{B}}(f)$. Our form is given by ${}^tXGY$ which must be equal to ${}^tYGX$ by symmetry. However, ${}^tXGY$ may be viewed as a 1×1 matrix, and is equal to its transpose, namely ${}^tY{}^tGX$. Thus

$${}^tYGX = {}^tY{}^tGX$$

for all vectors X, Y. It follows that $G = {}^tG$. Conversely, it is clear that any symmetric matrix defines a symmetric form.

As for the alternating case, replacing x by $x + y$ in the relation $\langle x, x \rangle = 0$ we obtain

$$\langle x, y \rangle + \langle y, x \rangle = 0.$$

In terms of the coordinate vectors X, Y and the matrix G, this yields

$${}^tXGY + {}^tYGX = 0.$$

Taking the transpose of, say, the second of the 1×1 matrices entering in this relation, yields (for all X, Y):

$${}^tXGY + {}^tX{}^tGY = 0.$$

Hence $G + {}^tG = 0$. Furthermore, letting X be any one of the unit vectors

$${}^t(0, \ldots, 0, 1, 0, \ldots, 0)$$

and using the relation ${}^tXGX = 0$, we see that the diagonal elements of G must be equal to 0. Conversely, if G is an $n \times n$ matrix such that

${}^tG + G = 0$, and such that $g_{ii} = 0$ for $i = 1, \ldots, n$ then one verifies immediately that the map

$$(X, Y) \mapsto {}^tXGY$$

defines an alternating form. This proves our proposition.

Of course, if as is usually the case, 2 is invertible in R, then our condition ${}^tM = -M$ implies that the diagonal elements of M must be 0. Thus in that case, showing that $G + {}^tG = 0$ implies that G is alternating.

§7. *Sesquilinear duality*

There exist forms which are not quite bilinear, and for which the results described above hold almost without change, but which must be handled separately for the sake of clarity in the notation involved.

Let R have an automorphism of period 2. We write this automorphism as $a \mapsto \bar{a}$ (and think of complex conjugation).

Following Bourbaki, we say that a map

$$f : E \times F \to R$$

is a *sesquilinear form* if it is $\mathbf{Z}$-bilinear, and if for $x \in E$, $y \in F$, and $a \in R$ we have

$$f(ax, y) = af(x, y)$$

and

$$f(x, ay) = \bar{a}f(x, y).$$

(*Sesquilinear* means $1\frac{1}{2}$ times linear, so the terminology is rather good.)

Let E, E' be modules. A map $\varphi : E \to E'$ is said to be *anti-linear* (or *semi-linear*) if it is $\mathbf{Z}$-linear, and $\varphi(ax) = \bar{a}\varphi(x)$ for all $x \in E$. Thus we may say that a sesquilinear form is linear in its first variable, and anti-linear in its second variable. We let $\overline{\mathrm{Hom}}_R(E, E')$ denote the module of anti-linear maps of E into E'.

We shall now go systematically through the same remarks that we made previously for bilinear forms.

We define perpendicularity as before, and also the kernel on the right and on the left for any sesquilinear form f. These kernels are submodules, say E_0 and F_0, and we get an induced sesquilinear form

$$E/E_0 \times F/F_0 \to R,$$

which is non-degenerate on either side.

Let F be an R-module. We define its *anti-module* $\overline{F}$ to be the module whose additive group is the same as F, and such that the operation $R \times \overline{F} \to \overline{F}$ is given by

$$(a, y) \mapsto \bar{a}y.$$

Then $\overline{F}$ is a module. We have a natural isomorphism

$$\mathrm{Hom}_R(\overline{F}, R) \leftrightarrow \overline{\mathrm{Hom}_R}(F, R),$$

as R-modules.

The sesquilinear form $f : E \times F \to R$ induces a linear map

$$\varphi_f : E \to \mathrm{Hom}_R(\overline{F}, R).$$

We say that f is *non-singular on the left* if φ_f is an isomorphism. Similarly, we have a corresponding linear map

$$\varphi'_f : \overline{F} \to \mathrm{Hom}_R(E, R)$$

from $\overline{F}$ into the dual space of E, and we say that f is *non-singular on the right* if φ'_f is an isomorphism. We say that f is *non-singular* if it is non-singular on the left and on the right.

We observe that our sesquilinear form f can be viewed as a *bilinear* form

$$f : E \times \overline{F} \to R,$$

and that our notions of non-singularity are then compatible with those defined previously for bilinear forms.

If we have a fixed non-singular sesquilinear form on $E \times F$, then depending on this form, we obtain an isomorphism between the module of sesquilinear forms on $E \times F$ and the module of endomorphisms of E. We also obtain an anti-isomorphism between these modules and the module of endomorphisms of F. In particular, we can define the analogue of the transpose, which in the present case we shall call the adjoint. Thus, let $f : E \times F \to R$ be a non-singular sesquilinear form. Let $A : E \to E$ be a linear map. There exists a unique linear map

$$A^* : F \to F$$

such that

$$\langle Ax, y \rangle = \langle x, A^*y \rangle$$

for all $x \in E$ and $y \in F$. Note that A^* is linear, not anti-linear. We call A^* the *adjoint* of A with respect to our form f. We have the rules

$$(cA)^* = \bar{c}A^*, \qquad (A + B)^* = A^* + B^*, \qquad (AB)^* = B^*A^*$$

for all linear maps A, B of E into itself, and $c \in R$.

Let us assume that $E = F$. Let $f : E \times E \to R$ be sesquilinear. By an *automorphism* of f we shall mean a linear automorphism $A : E \to E$ such that

$$\langle Ax, Ay \rangle = \langle x, y \rangle$$

just as we did for bilinear forms.

PROPOSITION 11S. *Let $f : E \times E \to R$ be a non-singular sesquilinear form. Let $A : E \to E$ be a linear map. Then A is an automorphism of f if and only if $A^*A = id$, and A is invertible.*

The proof, and also the proofs of subsequent propositions, which are completely similar to those of the bilinear case, will be omitted.

A sesquilinear form $g : E \times E \to R$ is said to be *hermitian* if

$$g(x, y) = \overline{g(y, x)}$$

for all $x, y \in E$. The set of hermitian forms on E will be denoted by $L_h^2(E)$. Let R_0 be the subring of R consisting of all elements fixed under our automorphism $a \to \bar{a}$ (i.e. consisting of all elements $a \in R$ such that $a = \bar{a}$). Then $L_h^2(E)$ is an R_0-module.

Let us take a fixed hermitian non-singular form f on E, denoted by $(x, y) \mapsto \langle x, y\rangle$. An endomorphism $A : E \to E$ will be said to be *hermitian* with respect to f if $A^* = A$. It is clear that the set of hermitian endomorphisms is an R_0-module, which we shall denote by Herm(E). *Depending on our fixed hermitian non-singular form f, we have an R_0-isomorphism*

$$\boxed{L_h^2(E) \leftrightarrow \mathrm{Herm}(E)}$$

described in the usual way. A hermitian form g corresponds to a hermitian map A if and only if

$$g(x, y) = \langle Ax, y\rangle$$

for all $x, y \in E$.

We can now describe the relation between our concepts and matrices, just as we did with bilinear forms.

We start with a sesquilinear form $f : E \times F \to R$.

If E, F are free, and we have selected bases as before, then we can again associate a matrix G with the form, and in terms of coordinate vectors X, Y our sesquilinear form is given by

$$(X, Y) \mapsto {}^tXG\overline{Y},$$

where $\overline{Y}$ is obtained from Y by applying the automorphism to each component of Y.

If $E = F$ and we use the same basis on the right and on the left, then with the same notation as that used in formula (1), if f is sesquilinear, the formula now reads

$$\text{(1S)} \qquad M_{\mathfrak{C}}^{\mathfrak{C}}(f) = {}^tCM_{\mathfrak{B}}^{\mathfrak{B}}(f)\overline{C}.$$

The automorphism appears.

PROPOSITION 12S. *Let E, F be free modules of dimension n over R, and let $f : E \times F \to R$ be a sesquilinear form. Then the following conditions are equivalent.*

f is non-singular on the left.

f is non-singular on the right.

f is non-singular.

The determinant of the matrix of f relative to any bases is invertible in R.

PROPOSITION 13S. *Let E, F be free over R, of dimension n. Let $f : E \times F \to R$ be a non-singular sesquilinear form. Let $\mathcal{B}$, $\mathcal{B}'$ be bases of E and F respectively over R, and let G be the matrix of f relative to these bases. Let $A : E \to E$ be a linear map, and let M be its matrix relative to $\mathcal{B}$. Then the matrix of A^* relative to $\mathcal{B}'$ is*

$$(\overline{G}^{-1})^t\overline{M}\,\overline{G}.$$

COROLLARY 1. *If G is the unit matrix, then the matrix of A^* is equal to ${}^t\overline{M}$.*

COROLLARY 2. *Let the notation be as in the proposition, and let $\mathcal{B} = \mathcal{B}'$ be a basis of E. An $n \times n$ matrix M is the matrix of an automorphism of f (relative to our basis) if and only if*

$${}^tMG\overline{M} = G.$$

A matrix M is said to be *hermitian* if ${}^tM = \overline{M}$.

Let R_0 be as before the subring of R consisting of all elements fixed under our automorphism $a \mapsto \overline{a}$ (i.e. consisting of all elements $a \in R$ such that $a = \overline{a}$).

PROPOSITION 14S. *Let E be a free module of dimension n over R, and let $\mathcal{B}$ be a basis. The map*

$$f \mapsto M_{\mathcal{B}}^{\mathcal{B}}(f)$$

induces an R_0-isomorphism between the R_0-module of hermitian forms on E and the R_0-module of $n \times n$ hermitian matrices in R.

Remark. If we had assumed at the beginning that our automorphism $a \mapsto \overline{a}$ has period 2 or 1 (i.e. if we allow it to be the identity), then the results on bilinear and symmetric forms become special cases of the results of this section. However, the notational difference is sufficiently disturbing to warrant a repetition of the results as we have done.

Terminology

For some confusing reason, the group of automorphisms of a symmetric (resp. alternating, resp. hermitian) form on a vector space is called the *orthogonal* (resp. *symplectic*, resp. *unitary*) group of the form. The word orthogonal is especially unfortunate, because an orthogonal map preserves more than orthogonality: It also preserves the scalar product, i.e. length. Furthermore, the word symplectic is also unfortunate. It turns out that one can carry out a discussion of hermitian forms over certain division rings (having automorphisms of order 2), and their group of automorphisms have also been called symplectic, thereby creating genuine confusion with the use of the word relative to alternating forms.

In order to unify and improve the terminology, I have discussed the matter with several persons, and it seems that one could adopt the following conventions.

As said in the text, the group of automorphisms of any form f is denoted by $\mathrm{Aut}(f)$.

On the other hand, there is a standard form, described over the real numbers in terms of coordinates by

$$f(x, x) = x_1^2 + \cdots + x_n^2,$$

over the complex numbers by

$$f(x, x) = x_1\bar{x}_1 + \cdots + x_n\bar{x}_n,$$

and over the quaternions by the same formula as in the complex case. The group of automorphisms of this form would be called the *unitary group*, and be denoted by U_n. The points of this group in the reals (resp. complex, resp. quaternions) would be denoted by

$$U_n(\mathbf{R}), \qquad U_n(\mathbf{C}), \qquad U_n(\mathbf{K}),$$

and these three groups would be called the *real unitary group* (resp. *complex unitary group*, resp. *quaternion unitary group*). Similarly, the group of points of U_n in any subfield or subring k of the quaternions would be denoted by $U_n(k)$.

Finally, if f is the standard alternating form, whose matrix is

$$\begin{pmatrix} 0 & I_n \\ -I_n & 0 \end{pmatrix},$$

one would denote its group of automorphisms by A_{2n}, and call it the *alternating form group*, or simply the alternating group, if there is no danger of confusion with the permutation group. The group of points of the alternating form group in a field k would then be denoted by $A_{2n}(k)$.

As usual, the subgroup of $\mathrm{Aut}(f)$ consisting of those elements whose determinant is 1 would be denoted by adding the letter S in front, and would still be called the *special group*. In the four standard cases, this yields $SU_n(\mathbf{R})$, $SU_n(\mathbf{C})$, $SU_n(\mathbf{K})$, $SA_{2n}(k)$.

Exercises

1. Interpret the rank of a matrix A in terms of the dimensions of the image and kernel of the linear map L_A.

2. Let $\mathfrak{g}$ be a module over the commutative ring R. A bilinear map $\mathfrak{g} \times \mathfrak{g} \to \mathfrak{g}$, written $(x, y) \mapsto [x, y]$, is said to make $\mathfrak{g}$ a *Lie algebra* if $[x, x] = 0$ and

$$[[x, y], z] + [[y, z], x] + [[z, x], y] = 0$$

for all $x, y, z \in \mathfrak{g}$.

(a) Let $M_n(R)$ be the ring of matrices over R. If $x, y \in M_n(R)$, show that the product

$$(x, y) \mapsto [x, y] = xy - yx$$

makes $M_n(R)$ into a Lie algebra.

(b) Let $\mathfrak{g}$ be a Lie algebra. Let $x \in \mathfrak{g}$, and let ad x be the linear map given by $(\text{ad}\, x)(y) = [x, y]$. Show that ad x is a derivation of $\mathfrak{g}$ into itself (i.e. satisfies the rule $D([x, y]) = [Dx, y] + [x, Dy]$).

(c) Show that the map $x \mapsto \text{ad}\, x$ is a Lie homomorphism of $\mathfrak{g}$ into the module of derivations of $\mathfrak{g}$ into itself.

3. Given a set of polynomials $\{P_\nu(X_{ij})\}$ in the polynomial ring $R[X_{ij}]$ $(1 \leqq i, j \leqq n)$, a zero of this set in R is a matrix $x = (x_{ij})$ such that $x_{ij} \in R$ and $P_\nu(x_{ij}) = 0$ for all ν. We use vector notation, and write $(X) = (X_{ij})$. We let $G(R)$ denote the set of zeros of our set of polynomials $\{P_\nu\}$. Thus $G(R) \subset M_n(R)$, and if R' is any commutative associative R-algebra we have $G(R') \subset M_n(R')$. We say that the set $\{P_\nu\}$ defines an *algebraic group* over R if $G(R')$ is a subgroup of the group $GL_n(R')$ for all R' (where $GL_n(R')$ is the multiplicative group of invertible matrices in R').

As an example, the group of matrices satisfying the equation ${}^tXX = I_n$ is an algebraic group.

Let R' be the R-algebra which is free, with a basis $\{1, t\}$ such that $t^2 = 0$. Thus $R' = R[t]$. Let $\mathfrak{g}$ be the set of matrices $x \in M_n(R)$ such that $I_n + tx \in G(R[t])$. Show that $\mathfrak{g}$ is a Lie algebra. [*Hint:* Note that

$$P_\nu(I_n + tX) = P_\nu(I_n) + \text{grad}\, P_\nu(I_n)tX.$$

Use the algebra $R[t, u]$ where $t^2 = u^2 = 0$ to show that if $I_n + tx \in G(R[t])$ and $I_n + uy \in G(R[u])$ then $[x, y] \in \mathfrak{g}$.]

(I have taken the above from the first four pages of Serre's notes on Lie groups and Lie algebras, Harvard, 1965. For additional information, besides Serre's notes, c.f. books by Jacobson, Bourbaki, etc.)

4. Let E be a finite extension of a field k. Let $\alpha \in E$. Let $f_\alpha : E \to E$ be the k-linear map such that $f_\alpha(x) = \alpha x$. Show that the trace of this linear map is the trace $\text{Tr}_k^E(\alpha)$ defined in field theory. [*Hint:* First suppose $E = k(\alpha)$, take for a basis the powers of α, and compute the trace of f_α with respect to this basis. In fact, what is the matrix of f_α relative to this basis?]

5. Let E be a finite extension of a field k. Show that the norm $N_k^E(\alpha)$ is equal to the determinant $\det(f_\alpha)$ (notation as in the preceding exercise).

6. Let A be an invertible matrix in a commutative ring R. Show that $({}^tA)^{-1} = {}^t(A^{-1})$.

7. Let f be a non-singular bilinear form on the module E over R. Let A be an R-automorphism of E. Show that $({}^tA)^{-1} = {}^t(A^{-1})$. Prove the same thing in the hermitian case, i.e. $(A^*)^{-1} = (A^{-1})^*$.

8. Let $A_1, \ldots, A_r$ be row vectors of dimension n, over a field k. Let $X = (x_1, \ldots, x_n)$. Let $b_1, \ldots, b_r \in k$. By a system of linear equations in k one means a system of type

$$A_1 \cdot X = b_1, \ldots, A_r \cdot X = b_r.$$

If $b_1 = \cdots = b_r = 0$, one says the system is *homogeneous*. We call n the number of variables, and r the number of equations. A solution X of the homogeneous system is called *trivial* if $x_i = 0$, $i = 1, \ldots, n$.

(a) Show that a homogeneous system of r linear equations in n unknowns with $n > r$ always has a non-trivial solution.

(b) Let L be a system of homogeneous linear equations over a field k. Let k be a subfield of k'. If L has a non-trivial solution in k', show that it has a non-trivial solution in k.

9. Let M be an $n \times n$ matrix over a field k. Assume that $\text{tr}(MX) = 0$ for all $n \times n$ matrices X in k. Show that $M = 0$.

10. Let S be a set of $n \times n$ matrices over a field k. Show that there exists a column vector $X \neq 0$ of dimension n in k, such that $MX = X$ for all $M \in S$ if and only if there exists such a vector in some extension field k' of k.

11. Let $\mathbf{K}$ be the division ring over the reals generated by elements i, j, k such that $i^2 = j^2 = k^2 = -1$, and

$$ij = -ji = k, \qquad jk = -kj = i, \qquad ki = -ik = j.$$

Then $\mathbf{K}$ has an automorphism of order 2, given by

$$a_0 + a_1 i + a_2 j + a_3 k \mapsto a_0 - a_1 i - a_2 j - a_3 k.$$

Denote this automorphism by $\alpha \mapsto \bar{\alpha}$. What is $\alpha\bar{\alpha}$? Show that the theory of hermitian forms can be carried out over $\mathbf{K}$, which is called the division ring of *quaternions* (or by abuse of language, the non-commutative field of quaternions).

12. Let $f_{11}, \ldots, f_{1n}$ be polynomials in n variables over a field k, which may be assumed algebraically closed. Assume that these polynomials generate the unit ideal in the polynomial ring $k[X_1, \ldots, X_n]$. Determine whether there exist polynomials f_{ij} such that the determinant

$$\begin{vmatrix} f_{11} & f_{12} & \cdots & f_{1n} \\ f_{21} & f_{22} & \cdots & f_{2n} \\ & & \cdots & \\ f_{n1} & f_{n2} & \cdots & f_{nn} \end{vmatrix}$$

is equal to 1. (This is a very interesting research problem, which arose first when Serre tried to determine whether a finitely generated projective module over the polynomial ring is free. The answer to this is still unknown at the time this book appears. The analogous problem when $f_{11}, \ldots, f_{1n}$ are taken from a principal entire ring is an easy exercise, however.)

13. Let A, B be square matrices of the same size over a field k. Assume that B is non-singular. If t is a variable, show that $\det(A + tB)$ is a polynomial in t, whose leading coefficient is $\det(B)$, and whose constant term is $\det(A)$.

CHAPTER XIV

Structure of Bilinear Forms

§1. *Preliminaries, orthogonal sums*

The purpose of this chapter is to go somewhat deeper into the structure theory for our three types of forms. To do this we shall assume most of the time that our ground ring is a field, and in fact a field of characteristic $\neq 2$ in the symmetric case.

We recall our three definitions. Let E be a module over a commutative ring R. Let $g : E \times E \to R$ be a map. If g is bilinear, we call g a *symmetric* form if $g(x, y) = g(y, x)$ for all $x, y \in E$. We call g *alternating* if $g(x, x) = 0$, and hence $g(x, y) = -g(y, x)$ for all $x, y \in E$. If R has an automorphism of order 2, written $a \mapsto \bar{a}$, we say that g is a *hermitian* form if it is linear in its first variable, antilinear in its second, and

$$g(x, y) = \overline{g(y, x)}.$$

We shall write $g(x, y) = \langle x, y \rangle$ if the reference to g is clear. We also occasionally write $g(x, y) = x \cdot y$ or $g(x, x) = x^2$. We sometimes call g a *scalar product.*

If $v_1, \ldots, v_m \in E$, we denote by $(v_1, \ldots, v_m)$ the submodule of E generated by $v_1, \ldots, v_m$.

Let g be symmetric, alternating, or hermitian. Then it is clear that the left kernel of g is equal to its right kernel, and it will simply be called the *kernel* of g.

In any one of these cases, we say that g is *non-degenerate* if its kernel is 0. Assume that E is finite dimensional over the field k. The form is non-degenerate if and only if it is non-singular, i.e. induces an isomorphism of E with its dual space (anti-dual in the case of hermitian forms).

Except for the few remarks on the anti-linearity made in the previous chapter, we don't use the results of the duality in that chapter. We need only the duality over fields, given in Chapter III. Furthermore, we don't essentially meet matrices again, except for the remarks on the pfaffian in §10.

We introduce one more notation. In the study of forms on vector spaces, we shall frequently decompose the vector space into direct sums of orthogonal subspaces. If E is a vector space with a form g as above, and F, F' are subspaces, we shall write

$$E = F \perp F'$$

to mean that E is the direct sum of F and F', and that F is orthogonal (or perpendicular) to F', in other words, $x \perp y$ (or $\langle x, y \rangle = 0$) for all $x \in F$ and $y \in F'$. We then say that E is the *orthogonal sum* of F and F'. There will be no confusion with the use of the symbol $\perp$ when we write $F \perp F'$ to mean simply that F is perpendicular to F'. The context always makes our meaning clear.

Most of this chapter is devoted to giving certain orthogonal decompositions of a vector space with one of our three types of forms, so that each factor in the sum is an easily recognizable type.

In the symmetric and hermitian case, we shall be especially concerned with direct sum decompositions into factors which are 1-dimensional. Thus if $\langle \ , \ \rangle$ is symmetric or hermitian, we shall say that $\{v_1, \ldots, v_n\}$ is an *orthogonal basis* (with respect to the form) if $\langle v_i, v_j \rangle = 0$ whenever $i \neq j$. We see that an orthogonal basis gives such a decomposition. If the form is non-degenerate, and if $\{v_1, \ldots, v_n\}$ is an orthogonal basis, then we see at once that $\langle v_i, v_i \rangle \neq 0$ for all i.

PROPOSITION 1. *Let E be a vector space over the field k, and let g be a form of one of the three above types. Suppose that E is expressed as an orthogonal sum,*

$$E = E_1 \perp \cdots \perp E_m.$$

Then g is non-degenerate on E if and only if it is non-degenerate on each E_i. If E_i^0 is the kernel of the restriction of g to E_i, then the kernel of g in E is the orthogonal sum

$$E^0 = E_1^0 \perp \cdots \perp E_m^0.$$

Proof. Elements v, w of E can be written uniquely

$$v = \sum_{i=1}^{m} v_i, \qquad w = \sum_{i=1}^{m} w_i$$

with $v_i, w_i \in E_i$. Then

$$v \cdot w = \sum_{i=1}^{m} v_i \cdot w_i,$$

and $v \cdot w = 0$ if $v_i \cdot w_i = 0$ for each $i = 1, \ldots, m$. From this our assertion is obvious.

Observe that if $E_1, \ldots, E_m$ are vector spaces over k, and $g_1, \ldots, g_m$ are forms on these spaces respectively, then we can define a form

$g = g_1 \oplus \cdots \oplus g_m$ on the direct sum $E = E_1 \oplus \cdots \oplus E_m$; namely if v, w are written as above, then we let

$$g(v, w) = \sum_{i=1}^{m} g_i(v_i, w_i).$$

It is then clear that, in fact, we have $E = E_1 \perp \cdots \perp E_m$. We could also write $g = g_1 \perp \cdots \perp g_m$.

Proposition 2. *Let E be a finite dimensional space over the field k, and let g be a form of the preceding type on E. Assume that g is non-degenerate. Let F be a subspace of E. The form is non-degenerate on F if and only if $F + F^{\perp} = E$, and also if and only if it is non-degenerate on $F^{\perp}$.*

Proof. We have (as a trivial consequence of Chapter III, §5)

$$\dim F + \dim F^{\perp} = \dim E = \dim(F + F^{\perp}) + \dim(F \cap F^{\perp}).$$

Hence $F + F^{\perp} = E$ if and only if $\dim(F \cap F^{\perp}) = 0$. Our first assertion follows at once. Since $F, F^{\perp}$ enter symmetrically in the dimension condition, our second assertion also follows.

Instead of saying that a form is non-degenerate on E, we shall sometimes say, by abuse of language, that E is non-degenerate.

Let E be a finite dimensional space over the field k, and let g be a form of the preceding type. Let E_0 be the kernel of the form. Then we get an induced form of the same type

$$g_0 : E/E_0 \times E/E_0 \to k,$$

because $g(x, y)$ depends only on the coset of x and the coset of y modulo E_0. Furthermore, g_0 is non-degenerate since its kernel on both sides is 0.

Let E, E' be finite dimensional vector spaces, with forms g, g' as above, respectively. A linear map $\sigma : E \to E'$ is said to be *metric* if

$$g'(\sigma x, \sigma y) = g(x, y)$$

or in the dot notation, $\sigma x \cdot \sigma y = x \cdot y$ for all $x, y \in E$. If σ is a linear isomorphism, and is metric, then we say that σ is an *isometry*.

Let E, E_0 be as above. Then we have an induced form on the factor space E/E_0. If W is a complementary subspace of E_0, in other words, $E = E_0 \oplus W$, and if we let $\sigma : E \to E/E_0$ be the canonical map, then σ is metric, and induces an isometry of W on E/E_0. This assertion is obvious, and shows that if $E = E_0 \oplus W'$ is another direct sum decomposition of E, then W' is isometric to W. We know that $W \approx E/E_0$ is non-degenerate. Hence our form determines a unique non-degenerate form, up to isometry, on complementary subspaces of the kernel.

§2. *Quadratic maps*

Let R be a commutative ring and let E, F be R-modules. We suppress the prefix R- as usual. We recall that a bilinear map $f : E \times E \to F$ is said to be symmetric if $f(x, y) = f(y, x)$ for all $x, y \in E$.

We say that F is *without 2-torsion* if for all $y \in F$ such that $2y = 0$ we have $y = 0$. (This holds if 2 is invertible in R.)

Let $f : E \to F$ be a mapping. We shall say that f is *quadratic* (i.e. R-quadratic) if there exists a symmetric bilinear map $g : E \times E \to F$ and a linear map $h : E \to F$ such that for all $x \in E$ we have

$$f(x) = g(x, x) + h(x).$$

PROPOSITION 3. *Assume that F is without 2-torsion. Let $f : E \to F$ be quadratic, expressed as above in terms of a symmetric bilinear map and a linear map. Then g, h are uniquely determined by f. For all $x, y \in E$ we have*

$$2g(x, y) = f(x + y) - f(x) - f(y).$$

Proof. If we compute $f(x + y) - f(x) - f(y)$, then we obtain $2g(x, y)$. If g_1 is symmetric bilinear, h_1 is linear, and $f(x) = g_1(x, x) + h_1(x)$, then $2g(x, y) = 2g_1(x, y)$. Since F is assumed to be without 2-torsion, it follows that $g(x, y) = g_1(x, y)$ for all $x, y \in E$, and thus that g is uniquely determined. But then h is determined by the relation

$$h(x) = f(x) - g(x, x).$$

We call g, h the bilinear and linear maps *associated* with f.

If $f : E \to F$ is a map, we define

$$\Delta f : E \times E \to F$$

by

$$\Delta f(x, y) = f(x + y) - f(x) - f(y).$$

We say that f is *homogeneous quadratic* if it is quadratic, and if its associated linear map is 0. We shall say that F is *uniquely divisible* by 2 if for each $z \in F$ there exists a unique $u \in F$ such that $2u = z$. (Again this holds if 2 is invertible in R.)

PROPOSITION 4. *Let $f : E \to F$ be a map such that Δf is bilinear. Assume that F is uniquely divisible by 2. Then the map $x \mapsto f(x) - \frac{1}{2}\Delta f(x, x)$ is $\mathbf{Z}$-linear. If f satisfies the condition $f(2x) = 4f(x)$, then f is homogeneous quadratic.*

Proof. Obvious.

By a *quadratic form* on E, one means a homogeneous quadratic map $f : E \to R$, with values in R.

In what follows, we are principally concerned with symmetric bilinear forms. The quadratic forms play a secondary role.

For the entire discussion of symmetric forms, §3–§8, we assume that k is a field of characteristic $\neq 2$. Throughout the rest of this chapter, we also assume that all modules and vector spaces are finite dimensional.

§3. Symmetric forms, orthogonal bases

THEOREM 1. *Let E be a vector space over k. Let g be a symmetric form on E. If* $\dim E \geqq 1$ *then there exists an orthogonal basis.*

Proof. We assume first that g is non-degenerate, and prove our assertion by induction in that case. If the dimension n is 1, then our assertion is obvious.

Assume $n > 1$. Let $v_1 \in E$ be such that $v_1^2 \neq 0$ (such an element exists since g is assumed non-degenerate). Let $F = (v_1)$ be the subspace generated by v_1. Then F is non-degenerate, and by Proposition 2, we have

$$V = F + F^{\perp}.$$

Furthermore, $\dim F^{\perp} = n - 1$. Let $\{v_2, \ldots, v_n\}$ be an orthogonal basis of $F^{\perp}$. Then $\{v_1, \ldots, v_n\}$ are pairwise orthogonal. Furthermore, they are linearly independent, for if

$$a_1v_1 + \cdots + a_nv_n = 0$$

with $a_i \in k$ then we take the scalar product with v_i to get $a_iv_i^2 = 0$ whence $a_i = 0$ for all i.

Remark. We have shown in fact that if g is non-degenerate, and $v \in E$ is such that $v^2 \neq 0$ then we can complete v to an orthogonal basis of E.

Suppose that the form g is degenerate. Let E_0 be its kernel. We can write E as a direct sum

$$E = E_0 \oplus W$$

for some subspace W. The restriction of g to W is non-degenerate, otherwise there would be an element of W which is in the kernel of E, and $\neq 0$. Hence if $\{v_1, \ldots, v_r\}$ is a basis of E_0, and $\{w_1, \ldots, w_{n-r}\}$ is an orthogonal basis of W, then

$$\{v_1, \ldots, v_r, w_1, \ldots, w_{n-r}\}$$

is an orthogonal basis of E, as was to be shown.

COROLLARY. *Let $\{v_1, \ldots, v_n\}$ be an orthogonal basis of E. Assume that $v_i^2 \neq 0$ for $i \leqq r$ and $v_i^2 = 0$ for $i > r$. Then the kernel of E is equal to $(v_{r+1}, \ldots, v_n)$.*

Proof. Obvious.

If $\{v_1, \ldots, v_n\}$ is an orthogonal basis of E and if we write

$$X = x_1v_1 + \cdots + x_nv_n$$

with $x_i \in k$, then

$$X^2 = a_1x_1^2 + \cdots + a_nx_n^2$$

where $a_i = \langle v_i, v_i \rangle$. In this representation of the form, we say that it is *diagonalized*. With respect to an orthogonal basis, we see at once that the associated matrix of the form is a diagonal matrix, namely

$$\begin{pmatrix} a_1 & & & & & \\ & a_2 & & 0 & & \\ & & \ddots & & & \\ & & & a_r & & \\ & 0 & & & 0 & \\ & & & & & \ddots \\ & & & & & & 0 \end{pmatrix}.$$

§4. Hyperbolic spaces

Let E be a vector space over k, with a symmetric form. We say that E is a *hyperbolic plane* if the form is non-degenerate, if E has dimension 2, and if there exists an element $w \neq 0$ in E such that $w^2 = 0$. We say that E is a *hyperbolic space* if it is an orthogonal sum of hyperbolic planes. We also say that the form on E is hyperbolic.

Suppose that E is a hyperbolic plane, with an element $w \neq 0$ such that $w^2 = 0$. Let $u \in E$ be such that $E = (w, u)$. Then $u \cdot w \neq 0$, otherwise w would be a non-zero element in the kernel. Let $b \in k$ be such that $w \cdot bu = bw \cdot u = 1$. Then select $a \in k$ such that

$$(aw + bu)^2 = 2abw \cdot u + b^2u^2 = 0.$$

(This can be done since we deal with a linear equation in a.) Put $v = aw + bu$. Then we have found a basis for E, namely $E = (w, v)$ such that

$$w^2 = v^2 = 0 \qquad \text{and} \qquad w \cdot v = 1.$$

Relative to this basis, the matrix of our form is therefore

$$\begin{pmatrix} 0 & 1 \\ 1 & 0 \end{pmatrix}.$$

We observe that, conversely, a space E having a basis $\{w, v\}$ satisfying $w^2 = v^2 = 0$ and $w \cdot v = 1$ is non-degenerate, and thus is a hyperbolic plane. A basis $\{w, v\}$ satisfying these relations will be called a *hyperbolic pair*.

An orthogonal sum of non-degenerate spaces is non-degenerate and hence a hyperbolic space is non-degenerate. We note that a hyperbolic space always has even dimension.

LEMMA. *Let E be a vector space over k, with a non-degenerate symmetric form g. Let F be a subspace, F_0 the kernel of F, and suppose we have an orthogonal decomposition*

$$F = F_0 \perp U.$$

Let $\{w_1, \ldots, w_s\}$ be a basis of F_0. Then there exist elements $v_1, \ldots, v_s$ in E perpendicular to U, such that each pair $\{w_i, v_i\}$ is a hyperbolic pair generating a hyperbolic plane P_i, and such that we have an orthogonal decomposition

$$U \perp P_1 \perp \cdots \perp P_s.$$

Proof. Let

$$U_1 = (w_2, \ldots, w_s) \oplus U.$$

Then U_1 is contained in $F_0 \oplus U$ properly, and consequently $(F_0 \oplus U)^\perp$ is contained in $U_1^\perp$ properly. Hence there exists an element $u_1 \in U_1^\perp$ but $u_1 \notin (F_0 \oplus U)^\perp$. We have $w_1 \cdot u_1 \neq 0$, and hence (w_1, u_1) is a hyperbolic plane P_1. We have seen previously that we can find $v_1 \in P_1$ such that $\{w_1, v_1\}$ is a hyperbolic pair. Furthermore, we obtain an orthogonal sum decomposition

$$F_1 = (w_2, \ldots, w_s) \perp P_1 \perp U.$$

Then it is clear that $(w_2, \ldots, w_s)$ is the kernel of F_1, and we can complete the proof by induction.

§5. *Witt's theorem*

THEOREM 2. *Let E be a vector space over k, and let g be a non-degenerate symmetric form on E. Let F, F' be subspaces of E, and let $\sigma : F \to F'$ be an isometry. Then σ can be extended to an isometry of E onto itself.*

Proof. We shall first reduce the proof to the case when F is non-degenerate.

We can write $F = F_0 \perp U$ as in the lemma of the preceding section, and then $\sigma F = F' = \sigma F_0 \perp \sigma U$. Furthermore, $\sigma F_0 = F_0'$ is the kernel of F'. Now we can enlarge both F and F' as in the lemma to orthogonal sums

$$U \perp P_1 \perp \cdots \perp P_s \quad \text{and} \quad \sigma U \perp P_1' \perp \cdots \perp P_s'$$

corresponding to a choice of basis in F_0 and its corresponding image in F_0'. Thus we can extend σ to an isometry of these extended spaces, which are non-degenerate. This gives us the desired reduction.

We assume that F, F' are non-degenerate, and proceed stepwise.

Suppose first that $F' = F$, i.e. that σ is an isometry of F onto itself. We can extend σ to E simply by leaving every element of $F^{\perp}$ fixed.

Next, assume that $\dim F = \dim F' = 1$ and that $F \neq F'$. Say $F = (v)$ and $F' = (v')$. Then $v^2 = v'^2$. Furthermore, (v, v') has dimension 2.

If (v, v') is non-degenerate, it has an isometry extending σ, which maps v on v' and v' on v. We can apply the preceding step to conclude the proof.

If (v, v') is degenerate, its kernel has dimension 1. Let w be a basis for this kernel. There exist $a, b \in k$ such that $v' = av + bw$. Then $v'^2 = a^2v^2$ and hence $a = \pm 1$. Replacing v' by $-v'$ if necessary, we may assume $a = 1$. Replacing w by bw, we may assume $v' = v + w$. Let $z = v + v'$. We apply the lemma to the space

$$(w, z) = (w) \perp (z).$$

We can find an element $y \in E$ such that

$$y \cdot z = 0, \qquad y^2 = 0, \qquad \text{and} \qquad w \cdot y = 1.$$

The space $(z, w, y) = (z) \perp (w, y)$ is non-degenerate, being an orthogonal sum of (z) and the hyperbolic plane (w, y). It has an isometry such that

$$z \leftrightarrow z, \qquad w \leftrightarrow -w, \qquad y \leftrightarrow -y.$$

But $v = \frac{1}{2}(z - w)$ is mapped on $v' = \frac{1}{2}(z + w)$ by this isometry. We have settled the present case.

We finish the proof by induction. By the existence of an orthogonal basis (Theorem 1), every subspace F of dimension > 1 has an orthogonal decomposition into a sum of subspaces of smaller dimension. Let $F = F_1 \perp F_2$ with $\dim F_1$ and $\dim F_2 \geqq 1$. Then

$$\sigma F = \sigma F_1 \perp \sigma F_2.$$

Let $\sigma_1 = \sigma \mid F_1$ be the restriction of σ to F_1. By induction, we can extend σ_1 to an isometry

$$\bar{\sigma}_1 : E \to E.$$

Then $\bar{\sigma}_1(F_1^{\perp}) = (\sigma_1 F_1)^{\perp}$. Since σF_2 is perpendicular to $\sigma F_1 = \sigma_1 F_1$, it follows that σF_2 is contained in $\bar{\sigma}_1(F_1^{\perp})$. Let $\sigma_2 = \sigma \mid F_2$. Then the isometry

$$\sigma_2 : F_2 \to \sigma_2 F_2 = \sigma F_2$$

extends by induction to an isometry

$$\bar{\sigma}_2 : F_1^{\perp} \to \bar{\sigma}_1(F_1^{\perp}).$$

The pair $(\sigma_1, \bar{\sigma}_2)$ gives us an isometry of $F_1 \perp F_1^\perp = E$ onto itself, as desired.

COROLLARY 1. *Let E, E' be vector spaces with non-degenerate symmetric forms, and assume that they are isometric. Let F, F' be subspaces, and let $\sigma : F \to F'$ be an isometry. Then σ can be extended to an isometry of E onto E'.*

Proof. Clear.

Let E be a vector space over k, with the symmetric form g. We say that g is a *null* form, or that E is a *null* space if $\langle x, y\rangle = 0$ for all $x, y \in E$. Since we assumed that the characteristic of k is $\neq 2$, the condition $x^2 = 0$ for all $x \in E$ implies that g is a null form. Indeed,

$$4x \cdot y = (x + y)^2 - (x - y)^2.$$

Let E be a space with a symmetric form g, and let F be a null subspace. Then by the lemma of §4, we can embed F in a hyperbolic subspace H whose dimension is $2 \dim F$.

As applications of Theorem 2, we get several corollaries.

COROLLARY 2. *Let E be a vector space with a non-degenerate symmetric form. Let W be a maximal null subspace, and let W' be some null subspace. Then* $\dim W' \leqq \dim W$, *and W' is contained in some maximal null subspace, whose dimension is the same as* $\dim W$.

Proof. That W' is contained in a maximal null subspace follows by Zorn's lemma. Suppose $\dim W' \geqq \dim W$. We have an isometry of W onto a subspace of W' which we can extend to an isometry of E onto itself. Then $\sigma^{-1}(W')$ is a null subspace containing W, hence is equal to W, whence $\dim W = \dim W'$. Our assertions follow by symmetry.

Let E be a vector space with a non-degenerate symmetric form. Let W be a null subspace. By the lemma of §4, we can embed W in a hyperbolic subspace H of E such that W is the maximal null subspace of H, and H is non-degenerate. Any such H will be called a *hyperbolic enlargement* of W.

COROLLARY 3. *Let E be a vector space with a non-degenerate symmetric form. Let W and W' be maximal null subspaces. Let H, H' be hyperbolic enlargements of W, W' respectively. Then H, H' are isometric and so are $H^\perp$ and $H'^\perp$.*

Proof. We have obviously an isometry of H on H', which can be extended to an isometry of E onto itself. This isometry maps $H^\perp$ on $H'^\perp$, as desired.

COROLLARY 4. *Let g_1, g_2, h be symmetric forms on vector spaces over the field k. If $g_1 \oplus h$ is isometric to $g_2 \oplus h$, and if g_1, g_2 are non-degenerate, then g_1 is isometric to g_2.*

Proof. Let g_1 be a form on E_1 and g_2 a form on E_2. Let h be a form on F. Then we have an isometry between $F \oplus E_1$ and $F \oplus E_2$. Extend the identity $id : F \to F$ to an isometry σ of $F \oplus E_1$ to $F \oplus E_2$ by Corollary 1. Since E_1 and E_2 are the respective orthogonal complements of F in their two spaces, we must have $\sigma(E_1) = E_2$, which proves what we wanted.

If g is a symmetric form on the vector space E, we shall say that g is *definite* if $g(x, x) \neq 0$ for any $x \in E$, $x \neq 0$ (i.e. $x^2 \neq 0$ if $x \neq 0$).

COROLLARY 5. *Let g be a symmetric form on the vector space E. Then g has a decomposition as an orthogonal sum*

$$g = g_0 \oplus g_{\text{hyp}} \oplus g_{\text{def}}$$

where g_0 is a null form, g_{hyp} is hyperbolic, and g_{def} is definite. The form $g_{\text{hyp}} \oplus g_{\text{def}}$ is non-degenerate. The forms g_0, g_{hyp}, and g_{def} are uniquely determined up to isometries.

Proof. The decomposition $g = g_0 \oplus g_1$ where g_0 is a null form and g_1 is non-degenerate is unique up to an isometry, since g_0 corresponds to the kernel of g.

We may therefore assume that g is non-degenerate. If

$$g = g_h \oplus g_d$$

where g_h is hyperbolic and g_d is definite, then g_h corresponds to the hyperbolic enlargement of a maximal null subspace, and by Corollary 3, it follows that g_h is uniquely determined. Hence g_d is uniquely determined as the orthogonal complement of g_h. (By uniquely determined, we mean of course up to an isometry.)

We shall abbreviate g_{hyp} by g_h and g_{def} by g_d.

§6. *The Witt group*

Let g, φ be symmetric forms on vector spaces over k. We shall say that they are *equivalent* if g_d is isometric to φ_d. The reader will verify at once that this is an equivalence relation. Furthermore the (orthogonal) sum of two null forms is a null form, and the sum of two hyperbolic forms is hyperbolic. However, the sum of two definite forms need not be definite. We write our equivalence $g \sim \varphi$. Equivalence is preserved under orthogonal sums, and hence equivalence classes of symmetric forms constitute a monoid.

THEOREM 3. *The monoid of equivalence classes of symmetric forms (over the field k) is a group.*

Proof. We have to show that every element has an additive inverse. Let g be a symmetric form, which we may assume definite. We let $-g$ be the form such that $(-g)(x, y) = -g(x, y)$. We contend that $g \oplus -g$ is equivalent to 0. Let E be the space on which g is defined. Then $g \oplus -g$ is defined on $E \oplus E$. Let W be the subspace consisting of all pairs (x, x) with $x \in E$. Then W is a null space for $g \oplus -g$. Since $\dim(E \oplus E) = 2 \dim W$, it follows that W is a maximal null space, and that $g \oplus -g$ is hyperbolic, as was to be shown.

The group of Theorem 3 will be called the *Witt group* of k, and will be denoted by $W(k)$. It is of importance in the study of representations of elements of k by the quadratic form f arising from g [i.e. $f(x) = g(x, x)$], for instance when one wants to classify the definite forms f.

We shall now define another group, which is of importance in more functorial studies of symmetric forms, for instance in studying the quadratic forms arising from manifolds in topology.

We observe that isometry classes of non-degenerate symmetric forms (over k) constitute a monoid $M(k)$, the law of composition being the orthogonal sum. Furthermore, the cancellation law holds (Corollary 4 of Theorem 2). We let

$$\gamma : M(k) \to WG(k)$$

be the canonical map of $M(k)$ into the Grothendieck group of this monoid, which we shall call the *Witt-Grothendieck* group over k. As we know, the cancellation law implies that γ is injective.

If g is a symmetric non-degenerate form over k, we define its dimension $\dim g$ to be the dimension of the space E on which it is defined. Then it is clear that

$$\dim(g \oplus g') = \dim g + \dim g'.$$

Hence dim factors through a homomorphism

$$\dim : WG(k) \to \mathbf{Z}.$$

This homomorphism splits since we have a non-degenerate symmetric form of dimension 1.

Let $WG_0(k)$ be the kernel of our homomorphism dim. If g is a symmetric non-degenerate form we can define its determinant $\det(g)$ to be the determinant of a matrix G representing g relative to a basis, modulo squares. This is well defined as an element of k^*/k^{*2}. We define det of the 0-form to be 1. Then det is a homomorphism

$$\det : M(k) \to k^*/k^{*2},$$

and can therefore be factored through a homomorphism, again denoted by det, of the Witt-Grothendieck group, $\det : WG(k) \to k^*/k^{*2}$.

Other properties of the Witt-Grothendieck group will be given in the exercises.

§7. *Symmetric forms over ordered fields*

THEOREM 4. (Sylvester) *Let k be an ordered field and let E be a vector space over k, with a non-degenerate symmetric form g. There exists an integer $r \geqq 0$ such that, if $\{v_1, \ldots, v_n\}$ is an orthogonal basis of E, then precisely r among the n elements $v_1^2, \ldots, v_n^2$ are > 0, and $n - r$ among these elements are < 0.*

Proof. Let $a_i = v_i^2$, for $i = 1, \ldots, n$. After renumbering the basis elements, say $a_1, \ldots, a_r > 0$ and $a_i < 0$ for $i > r$. Let $\{w_1, \ldots, w_n\}$ be any orthogonal basis, and let $b_i = w_i^2$. Say $b_1, \ldots, b_s > 0$ and $b_j < 0$ for $j > s$. We shall prove that $r = s$. Indeed, it will suffice to prove that

$$v_1, \ldots, v_r, w_{s+1}, \ldots, w_n$$

are linearly independent, for then we get $r + n - s \leqq n$, whence $r \leqq s$, and $r = s$ by symmetry. Suppose that

$$x_1v_1 + \cdots + x_rv_r + y_{s+1}w_{s+1} + \cdots + y_nw_n = 0.$$

Then

$$x_1v_1 + \cdots + x_rv_r = -y_{s+1}w_{s+1} - \cdots - y_nw_n.$$

Squaring both sides yields

$$a_1x_1^2 + \cdots + a_rx_r^2 = b_{s+1}y_{s+1}^2 + \cdots + b_ny_n^2.$$

The left-hand side is $\geqq 0$, and the right-hand side is $\leqq 0$. Hence both sides are equal to 0, and it follows that $x_i = y_j = 0$, in other words that our vectors are linearly independent.

COROLLARY 1. *Assume that every positive element of k is a square. Then there exists an orthogonal basis $\{v_1, \ldots, v_n\}$ of E such that $v_i^2 = 1$ for $i \leqq r$ and $v_i^2 = -1$ for $i > r$, and r is uniquely determined.*

Proof. We divide each vector in an orthogonal basis by the square root of the absolute value of its square.

A basis having the property of the corollary is called *orthonormal.* If X is an element of E having coordinates $(x_1, \ldots, x_n)$ with respect to this basis, then

$$X^2 = x_1^2 + \cdots + x_r^2 - x_{r+1}^2 - \cdots - x_n^2.$$

We say that a symmetric form g is *positive definite* if $X^2 > 0$ for all $X \in E$, $X \neq 0$. This is the case if and only if $r = n$ in Theorem 4. We say that g is *negative definite* if $X^2 < 0$ for all $X \in E$.

COROLLARY 2. *The vector space E admits an orthogonal decomposition $E = E^+ \perp E^-$ such that g is positive definite on E^+ and negative definite on E^-. The dimension of E^+ (or E^-) is the same in all such decompositions.*

Let us now assume that the form g is positive definite and that every positive element of k is a square.

We define the *norm* of an element $v \in E$ by

$$|v| = \sqrt{v \cdot v}.$$

Then we have $|v| > 0$ if $v \neq 0$. We also have the *Schwarz inequality*

$$|v \cdot w| \leqq |v|\,|w|$$

for all $v, w \in E$. This is proved in the usual way, expanding

$$0 \leqq (av \pm bw)^2 = (av \pm bw) \cdot (av \pm bw)$$

by bilinearity, and letting $b = |v|$ and $a = |w|$. One then gets

$$\mp 2ab\; v \cdot w \leqq 2|v|^2|w|^2.$$

If $|v|$ or $|w| = 0$ our inequality is trivial. If neither is 0 we divide by $|v|\,|w|$ to get what we want.

From the Schwarz inequality, we deduce the triangle inequality

$$|v + w| \leqq |v| + |w|.$$

We leave it to the reader as a routine exercise.

When we have a positive definite form, there is a canonical way of getting an orthonormal basis, starting with an arbitrary basis $\{v_1, \ldots, v_n\}$ and proceeding inductively. Let

$$v_1' = \frac{1}{|v_1|} v_1.$$

Then v_1 has norm 1. Let

$$w_2 = v_2 - (v_2 \cdot v_1')v_1',$$

and then

$$v_2' = \frac{1}{|w_2|} w_2.$$

Inductively, we let

$$w_r = v_r - (v_r \cdot v_1')v_1' - \cdots - (v_r \cdot v_{r-1}')v_{r-1}'$$

and then

$$v_r' = \frac{1}{|w_r|} w_r.$$

Then $\{v_1', \ldots, v_n'\}$ is an orthonormal basis. The inductive process just described is known as the *Gram-Schmidt orthogonalization.*

§8. *The Clifford algebra*

Let E be a vector space over the field k, and let g be a symmetric form on E. We would like to find a universal algebra over k, in which we can embed E, and such that the square in the algebra corresponds to the value of the quadratic form in E. More precisely, by a *Clifford algebra* for g, we shall mean an algebra $C(g)$ and a linear map $\rho : E \to C(g)$ having the following property: If $\psi : E \to L$ is a linear map of E into a k-algebra L such that

$$\psi(X)^2 = f(X) \cdot 1 \qquad (1 = \text{unit element of } L)$$

for all $X \in E$, then there exists a unique algebra-homomorphism

$$C(\psi) = \psi_* : C(g) \to L$$

such that the following diagram is commutative:

$$\begin{array}{ccc} E & \xrightarrow{\rho} & C(g) \\ & \psi \searrow \quad \swarrow \psi_* & \\ & L & \end{array}$$

By abstract nonsense, a Clifford algebra for g is uniquely determined, up to a unique isomorphism. Furthermore, it is clear that if $(C(g), \rho)$ exists, then $C(g)$ is generated by the image of ρ, i.e. by $\rho(E)$, *as an algebra over* k.

We shall write $\rho = \rho_g$ if it is necessary to specify the reference to g explicitly.

We have trivially

$$\rho(X)^2 = g(X, X) \cdot 1$$

for all $X \in E$, and

$$\rho(X)\rho(Y) + \rho(Y)\rho(X) = 2g(X, Y) \cdot 1$$

as one sees by replacing X by $X + Y$ in the preceding relation.

THEOREM 5. *Let g be a symmetric bilinear form on a vector space E over k. Then the Clifford algebra $(C(g), \rho)$ exists. The map ρ is injective, and $C(g)$ has dimension 2^n over k, if $n =$ dim E.*

In order to prove Theorem 5, we shall first determine the relations which must be satisfied by an algebra L and a linear map $\psi : V \to L$ such that $\psi(X)^2 = g(X, X) \cdot 1$. We follow the arguments of Artin in *Geometric Algebra*.

Let $S_1, \ldots, S_r$ be subsets of a given set M. We define the *sum* (which will not be the union) to be the set of those elements of M which occur in an odd number of the sets S_i, $i = 1, \ldots, r$.

The following rules are easily verified:

$$(S_1 + \cdots + S_r) + S_{r+1} = S_1 + \cdots + S_{r+1},$$
$$(S_1 + \cdots + S_r) \cap T = (S_1 \cap T) + \cdots + (S_r \cap T)$$

for any subset T of M.

The empty set is denoted as usual by $\emptyset$.

Let $\{v_1, \ldots, v_n\}$ be an orthogonal basis of E over k. Let $a_i = v_i^2$. Let $\psi(v_i) = e_i$. Then by assumption,

$$e_i^2 = a_i \qquad \text{and} \qquad e_ie_j + e_je_i = 0, \qquad \text{if} \quad i \neq j.$$

Let S be a subset of $\{1, \ldots, n\}$ and let $i_1, \ldots, i_m$ be the elements of S, ordered so that $i_1 < \cdots < i_m$. Let $e_S = e_{i_1} \cdots e_{i_m}$. If S, T are subsets of $\{1, \ldots, n\}$, then it is easy to see by induction that

$$e_Se_T = \prod_{\substack{s \in S \\ t \in T}} (s, t) \prod_{i \in S \cap T} v_i^2 \, e_{S+T},$$

where the symbol (s, t) is defined to be 1 if $s \leqq t$ and -1 if $s > t$. Thus the rule for taking the product of two "monomials" in $e_1, \ldots, e_n$ is determined purely combinatorially in terms of S and T, and our given squares $v_1^2, \ldots, v_n^2$. Furthermore, the algebra generated by $\psi(E)$ is generated by $e_1, \ldots, e_n$.

We now show how the above combinatorial rule allows us to define the universal algebra.

To each subset S of $\{1, \ldots, n\}$ we associate a letter e_S, and we let $C(g)$ be the free module over k generated by these letters e_S (as S ranges over all subsets of $\{1, \ldots, n\}$). Then $C(g)$ has dimension 2^n over k. We shall define a multiplication in $C(g)$. If S, T are subsets of $\{1, \ldots, n\}$ we let

$$\alpha(S, T) = \prod_{\substack{s \in S \\ t \in T}} (s, t) \prod_{i \in S \cap T} v_i^2.$$

If $\sum_S a_Se_S$ and $\sum_T b_Te_T$ are elements of $C(g)$ with coefficients $a_S, b_T \in k$, we define their product to be

$$\Big(\sum_S a_Se_S\Big)\Big(\sum_T b_Te_T\Big) = \sum_{S,T} a_Sb_T\alpha(S, T)e_{S+T}.$$

We must show that this product is associative. For this, it will clearly suffice to prove that if S, T, R are subsets of $\{1, \ldots, n\}$, then

$$(e_Se_T)e_R = e_S(e_Te_R),$$

and this last relation will be verified by brute force.

By definition,

$$e_S e_T = \alpha(S, T) e_{S+T}.$$

We shall now prove associativity. We have

$$(e_S e_T) e_R = \prod_{\substack{s \in S \\ t \in T}} (s, t) \prod_{\substack{j \in S+T \\ r \in R}} (j, r) \prod_{i \in S \cap T} v_i^2 \prod_{\lambda \in (S+T) \cap R} v_\lambda^2 \, e_{S+T+R},$$

and we shall rewrite the right-hand side in a more symmetric form.

The right-hand side consists of products over certain signs, and then over certain squares. We first consider the signs.

If we let j range over all of S and then over all of T, then any $j \in S \cap T$ will appear twice. Thus the second product is the same as the product taken for $j \in S$ and $j \in T$; in other words, the products giving the sign can be written

$$\prod_{\substack{s \in S \\ t \in T}} (s, t) \prod_{\substack{s \in S \\ r \in R}} (s, r) \prod_{\substack{t \in T \\ r \in R}} (t, r).$$

Now for the product over the squares. We have

$$(S + T) \cap R = (S \cap R) + (T \cap R).$$

If ν belongs to all three sets S, T, R then ν is in $S \cap T$ but not in $(S \cap R) + (T \cap R)$. If ν belongs to S and T but not to R, then ν is in $S \cap T$ but not in $(S \cap R) + (T \cap R)$. If ν lies in S and R but not in T, or in T and R but not in S, then it is not in $S \cap T$, but in $(S \cap R) + (T \cap R)$. Finally, if ν is in only one of the sets S, T, R or in none of them, then ν will neither be in $S \cap T$ nor in $(S \cap R) + (T \cap R)$. Thus the last two products can be written as

$$\prod_{\nu} v_\nu^2$$

for those ν such that ν appears in *more* than one of the sets S, T, R. This product is symmetric in S, R, T.

It follows at once from what we have shown that

$$(e_S e_T) e_R = e_S (e_T e_R).$$

From this it follows immediately that the product which we defined in $C(g)$ is associative. The other ring axioms are trivially satisfied, and the elements $\{e_S\}$ forms a basis of $C(g)$, which has therefore dimension 2^n.

We have an injective linear map

$$\rho : E \to C(g)$$

such that $\rho(v_i) = e_{\{i\}}$. We write $e_i = e_{\{i\}}$. If

$$X = x_1v_1 + \cdots + x_nv_n$$

then

$$\begin{aligned}\rho(X)^2 &= (x_1e_1 + \cdots + x_ne_n)(x_1e_1 + \cdots + x_ne_n)\\ &= (x_1^2a_1 + \cdots + x_n^2a_n)e_\emptyset,\end{aligned}$$

where $e_\emptyset$ is the unit element of $C(g)$, because $e_ie_j + e_je_i = 0$ if $i \neq j$. Thus our requirements concerning squares are satisfied.

If $\psi : E \to L$ is any linear map into an algebra over k such that $\psi(X)^2 = g(X, X) \cdot 1$, then we can define a ring-homomorphism of $C(g)$ into L to make the required diagram commutative. Indeed, let $e_i' = \psi(v_i)$. Let

$$e_S' = e_{i_m}' \cdots e_{i_1}'$$

where $i_1 < \cdots < i_m$ are the elements of S with their ordering. We define

$$\psi_* : C(g) \to L$$

by

$$\sum a_Se_S \mapsto \sum a_Se_S'.$$

Since the elements $\{e_S\}$ form a basis of $C(g)$, this map is well defined, and is a linear map. The remarks at the beginning of the proof show that this map is also a ring-homomorphism, and the diagram

$$\begin{array}{ccc} E & \xrightarrow{\rho} & C(g) \\ & \psi\searrow \quad \swarrow\psi_* & \\ & L & \end{array}$$

is commutative. This proves what we wanted.

§9. *Alternating forms*

Let E be a vector space over the field k, on which we now make no restriction. We let f be an alternating form on E, i.e. a bilinear map $f : E \times E \to k$ such that $f(x, x) = x^2 = 0$ for all $x \in E$. Then

$$x \cdot y = -y \cdot x$$

for all $x, y \in E$, as one sees by substituting $(x + y)$ for x in $x^2 = 0$.

As for symmetric forms, we define a *hyperbolic plane* (for the alternating form) to be a 2-dimensional space which is non-degenerate. (This time, we get automatically an element w such that $w^2 = 0$, $w \neq 0$, so there is no need to specify this.) If P is a hyperbolic plane, and $w \in P$, $w \neq 0$, then there exists an element $y \neq 0$ in P such that $w \cdot y \neq 0$. After dividing y by some constant, we may assume that $w \cdot y = 1$. Then $y \cdot w = -1$. Hence the matrix of the form with respect to the basis $\{w, y\}$ is

$$\begin{pmatrix} 0 & 1 \\ -1 & 0 \end{pmatrix}.$$

The pair w, y is called a *hyperbolic pair* as before. Given a 2-dimensional vector space over k with a bilinear form, and a pair of elements $\{w, y\}$ satisfying the relations

$$w^2 = y^2 = 0, \qquad y \cdot w = -1, \qquad w \cdot y = 1,$$

then we see that the form is alternating, and that (w, y) is a hyperbolic plane for the form.

Given an alternating form f on E, we say that E (or f) is *hyperbolic* if E is an orthogonal sum of hyperbolic planes. We say that E (or f) is *null* if $x \cdot y = 0$ for all $x, y \in E$.

THEOREM 6. *Let f be an alternating form on the vector space E over k. Then E is an orthogonal sum of its kernel and a hyperbolic subspace. If E is non-degenerate, then E is a hyperbolic space, and its dimension is even.*

Proof. A complementary subspace to the kernel is non-degenerate, and hence we may assume that E is non-degenerate. Let $w \in E$, $w \neq 0$. There exists $y \in E$ such that $w \cdot y \neq 0$ and $y \neq 0$. Then (w, y) is non-degenerate, hence is a hyperbolic plane P. We have $E = P \oplus P^{\perp}$ and $P^{\perp}$ is non-degenerate. We complete the proof by induction.

COROLLARY 1. *All alternating non-degenerate forms of a given dimension over a field k are isometric.*

We see from Theorem 6 that there exists a basis of E such that relative to this basis, the matrix of the alternating form is

$$\begin{pmatrix} 0 & 1 & & & & & & & \\ -1 & 0 & & & & & & & \\ & & 0 & 1 & & & & & \\ & & -1 & 0 & & & & & \\ & & & & \ddots & & & & \\ & & & & & 0 & 1 & & \\ & & & & & -1 & 0 & & \\ & & & & & & & 0 & \\ & & & & & & & & \ddots \\ & & & & & & & & & 0 \end{pmatrix}.$$

For convenience of writing, we reorder the basis elements of our orthogonal sum of hyperbolic planes in such a way that the matrix of the form is

$$\begin{pmatrix} 0 & I_r & 0 \\ -I_r & 0 & 0 \\ 0 & 0 & 0 \end{pmatrix}$$

where I_r is the unit $r \times r$ matrix. The matrix

$$\begin{pmatrix} 0 & I_r \\ -I_r & 0 \end{pmatrix}$$

is called the *standard alternating* matrix.

COROLLARY 2. *Let E be a vector space over k, with a non-degenerate symmetric form denoted by $\langle\ ,\ \rangle$. Let Ω be a non-degenerate alternating form on E. Then there exists a direct sum decomposition $E = E_1 \oplus E_2$ and a symmetric automorphism A of E (with respect to $\langle\ ,\ \rangle$) having the following property. If $x, y \in E$ are written*

$$x = (x_1, x_2) \quad \textit{with} \quad x_1 \in E_1 \quad \textit{and} \quad x_2 \in E_2,$$
$$y = (y_1, y_2) \quad \textit{with} \quad y_1 \in E_1 \quad \textit{and} \quad y_2 \in E_2,$$

then

$$\Omega(x, y) = \langle Ax_1, y_2 \rangle - \langle Ax_2, y_1 \rangle.$$

Proof. Take a basis of E such that the matrix of Ω with respect to this basis is the standard alternating matrix. Let f be the symmetric non-degenerate form on E given by the dot product with respect to this basis. Then we obtain a direct sum decomposition of E into subspaces E_1, E_2 (corresponding to the first n, resp. the last n coordinates), such that

$$\Omega(x, y) = f(x_1, y_2) - f(x_2, y_1).$$

Since $\langle\ ,\ \rangle$ is assumed non-degenerate, we can find an automorphism A having the desired effect, and A is symmetric because f is symmetric.

§10. *The pfaffian*

An alternating matrix is a matrix G such that ${}^tG = -G$ and the diagonal elements are equal to 0. As we saw in Chapter XIII, §6, it is the matrix of an alternating form. We let G be an $n \times n$ matrix, and assume n is *even*. (For odd n, cf. exercises.)

We start over a field of characteristic 0. By Theorem 6, there exists a non-singular matrix C such that ${}^tCGC$ is the matrix

$$\begin{pmatrix} 0 & I_r & 0 \\ -I_r & 0 & 0 \\ 0 & 0 & 0 \end{pmatrix}$$

and hence

$$\det(C)^2 \det(G) = 1 \quad \text{or} \quad 0$$

according as the kernel of the alternating form is trivial or non-trivial. Thus in any case, we see that $\det(G)$ is a square in the field.

Now we move over to the integers $\mathbf{Z}$. Let t_{ij} $(1 \leqq i < j \leqq n)$ be $n(n-1)/2$ algebraically independent elements over $\mathbf{Q}$, let $t_{ii} = 0$ for $i = 1, \ldots, n$, and let $t_{ij} = -t_{ji}$ for $i > j$. Then the matrix $T = (t_{ij})$ is alternating, and hence $\det(T)$ is a square in the field $\mathbf{Q}(t)$ obtained from $\mathbf{Q}$ by adjoining all the variables t_{ij}. However, $\det(T)$ is a polynomial in $\mathbf{Z}[t]$, and since we have unique factorization in $\mathbf{Z}[t]$, it follows that $\det(T)$ is the square of a polynomial in $\mathbf{Z}[t]$. We can write

$$\det(T) = P(t)^2.$$

The polynomial P is uniquely determined up to a factor of ± 1. If we substitute values for the t_{ij} so that the matrix T specializes to

$$\begin{pmatrix} 0 & I_{n/2} \\ -I_{n/2} & 0 \end{pmatrix},$$

then we see that there exists a unique polynomial P with integer coefficients taking the value 1 for this specialized set of values of (t). We call P the *generic pfaffian* of size n, and write it Pf.

Let R be a commutative ring. We have a homomorphism

$$\mathbf{Z}[t] \to R[t]$$

induced by the unique homomorphism of $\mathbf{Z}$ into R. The image of the generic pfaffian of size n in $R[t]$ is a polynomial with coefficients in R, which we still denote by Pf. If G is an alternating matrix with coefficients in R, then we write $\mathrm{Pf}(G)$ for the value of $\mathrm{Pf}(t)$ when we substitute g_{ij} for t_{ij} in Pf. Since the determinant commutes with homomorphisms, we have:

THEOREM 7. *Let R be a commutative ring. Let $(g_{ij}) = G$ be an alternating matrix with $g_{ij} \in R$. Then*

$$\det(G) = (\mathrm{Pf}(G))^2.$$

Furthermore, if C is an $n \times n$ matrix in R, then

$$\mathrm{Pf}(CG{}^tC) = \det(C)\,\mathrm{Pf}(G).$$

Proof. The first statement has been proved above. The second statement will follow if we can prove it over $\mathbf{Z}$. Let u_{ij} $(i, j = 1, \ldots, n)$ be algebraically independent over $\mathbf{Q}$, and such that u_{ij}, t_{ij} are algebraically independent over $\mathbf{Q}$. Let U be the matrix (u_{ij}). Then

$$\mathrm{Pf}(UT{}^tU) = \pm \det(U)\,\mathrm{Pf}(T),$$

as follows immediately from taking the square of both sides. Substitute values for U and T such that U becomes the unit matrix and T becomes the standard alternating matrix. We conclude that we must have a $+$ sign on the right-hand side. Our assertion now follows as usual for any substitution of U to a matrix in R, and any substitution of T to an alternating matrix in R, as was to be shown.

§11. Hermitian forms

Let k_0 be an ordered field (a subfield of the reals, if you wish) and let $k = k_0(i)$, where $i = \sqrt{-1}$. Then k has an automorphism of order 2, whose fixed field is k_0.

Let E be a finite dimensional vector space over k. We shall deal with a hermitian form on E, i.e. a map

$$E \times E \to k$$

written

$$(x, y) \mapsto \langle x, y\rangle$$

which is k-linear in its first variable, k-anti-linear in its second variable, and such that

$$\langle x, y\rangle = \overline{\langle y, x\rangle}$$

for all $x, y \in E$.

We observe that $\langle x, x\rangle \in k_0$ for all $x \in E$. This is essentially the reason why the proofs of statements concerning symmetric forms hold essentially without change in the hermitian case. We shall now make the list of the properties which apply to this case.

THEOREM 8. *There exists an orthogonal basis. If the form is non-degenerate, there exists an integer r having the following property. If $\{v_1, \ldots, v_n\}$ is an orthogonal basis, then precisely r among the n elements*

$$\langle v_1, v_1\rangle, \ldots, \langle v_n, v_n\rangle$$

are > 0 and $n - r$ among these elements are < 0.

An orthogonal basis $\{v_1, \ldots, v_n\}$ such that $\langle v_i, v_i \rangle = 1$ or -1 is called an *orthonormal* basis.

COROLLARY 1. *Assume that the form is non-degenerate, and that every positive element of k_0 is a square. Then there exists an orthonormal basis.*

We say that the hermitian form is *positive definite* if $\langle x, x \rangle > 0$ for all $x \in E$. We say that it is *negative definite* if $\langle x, x \rangle < 0$ for all $x \in E$.

COROLLARY 2. *Assume that the form is non-degenerate. Then E admits an orthogonal decomposition $E = E^+ \perp E^-$ such that the form is positive definite on E^+ and negative definite on E^-. The dimension of E^+ (or E^-) is the same in all such decompositions.*

The proofs of Theorem 8 and its corollaries are identical with those of the analogous results for symmetric forms, and will be left to the reader.

We have the *polarization identity*, for any k-linear map $A : E \to E$, namely

$$\langle A(x + y), (x + y) \rangle - \langle A(x - y), (x - y) \rangle = 2[\langle Ax, y \rangle + \langle Ay, x \rangle].$$

If $\langle Ax, x \rangle = 0$ for all x, we replace x by ix and get

$$\begin{aligned} \langle Ax, y \rangle + \langle Ay, x \rangle &= 0, \\ i\langle Ax, y \rangle - i\langle Ay, x \rangle &= 0. \end{aligned}$$

From this we conclude:

If $\langle Ax, x \rangle = 0$ for all x, then $A = 0$.

This is the only statement which has no analogue in the case of symmetric forms. The presence of i in one of the above linear equations is essential to the conclusion. In practice, one uses the statement in the complex case, and one meets an analogous situation in the real case when A is symmetric. Then the statement for symmetric maps is obvious.

Assume that the hermitian form is positive definite, and that every positive element of k_0 is a square.

We have the *Schwarz inequality*, namely

$$|\langle x, y \rangle|^2 \leqq \langle x, x \rangle \langle y, y \rangle$$

whose proof comes again by expanding

$$0 \leqq \langle \alpha x + \beta y, \alpha x + \beta y \rangle$$

and setting $\alpha = \langle y, y \rangle$ and $\beta = -\langle x, y \rangle$.

We define the norm of $|x|$ to be

$$|x| = \sqrt{\langle x, x \rangle}.$$

Then we get at once the triangle inequality

$$|x + y| \leqq |x|\,|y|,$$

and for $\alpha \in k$,

$$|\alpha x| = |\alpha|\,|x|.$$

Just as in the symmetric case, given a basis, one can find an orthonormal basis by the inductive procedure of subtracting successive projections. We leave this to the reader.

§12. *The spectral theorem (hermitian case)*

Throughout this section, we let E be a finite dimensional space over $\mathbf{C}$, *of dimension* $\geqq 1$, *and we endow E with a positive definite hermitian form.*

Let $A : E \to E$ be a linear map (i.e. $\mathbf{C}$-linear map) of E into itself. For fixed $y \in E$, the map $x \mapsto \langle Ax, y\rangle$ is a linear functional, and hence there exists a unique element $y^* \in E$ such that

$$\langle Ax, y\rangle = \langle x, y^*\rangle$$

for all $x \in E$. We define the map $A^* : E \to E$ by $A^*y = y^*$. It is immediately clear that A^* is linear, and we shall call A^* the *adjoint* of A with respect to our hermitian form.

The following formulas are trivially verified, for any linear maps A, B of E into itself:

$$(A + B)^* = A^* + B^*, \qquad A^{**} = A,$$
$$(\alpha A)^* = \bar{\alpha}A^*, \qquad (AB)^* = B^*A^*.$$

A linear map A is called *self-adjoint* (or *hermitian*) if $A^* = A$.

PROPOSITION 5. *A is hermitian if and only if $\langle Ax, x\rangle$ is real for all $x \in E$.*

Proof. Let A be hermitian. Then

$$\overline{\langle Ax, x\rangle} = \overline{\langle x, Ax\rangle} = \langle Ax, x\rangle,$$

whence $\langle Ax, x\rangle$ is real. Conversely, assume $\langle Ax, x\rangle$ is real for all x. Then

$$\langle Ax, x\rangle = \overline{\langle Ax, x\rangle} = \langle x, Ax\rangle = \langle A^*x, x\rangle,$$

and consequently $\langle (A - A^*)x, x\rangle = 0$ for all x. Hence $A = A^*$ by polarization.

Let $A : E \to E$ be a linear map. An element $\xi \in E$ is called an *eigenvector* of A if there exists $\lambda \in \mathbf{C}$ such that $A\xi = \lambda\xi$. If $\xi \neq 0$, then we say that λ is an *eigenvalue* of A, belonging to ξ.

Proposition 6. *Let A be hermitian. Then all eigenvalues belonging to non-zero eigenvectors of A are real. If ξ, ξ' are eigenvectors $\neq 0$ having eigenvalues λ, λ' respectively, and if $\lambda \neq \lambda'$, then $\xi \perp \xi'$.*

Proof. Let λ be an eigenvalue, belonging to the eigenvector $\xi \neq 0$. Then $\langle A\xi, \xi\rangle = \langle \xi, A\xi\rangle$, and these two numbers are equal respectively to $\lambda\langle \xi, \xi\rangle$ and $\bar{\lambda}\langle \xi, \xi\rangle$. Since $\xi \neq 0$, it follows that $\lambda = \bar{\lambda}$, i.e. that λ is real. Secondly, assume that ξ, ξ' and λ, λ' are as described above. Then

$$\langle A\xi, \xi'\rangle = \lambda\langle \xi, \xi'\rangle = \langle \xi, A\xi'\rangle = \lambda'\langle \xi, \xi'\rangle,$$

from which it follows that $\langle \xi, \xi'\rangle = 0$.

Lemma. *Let $A : E \to E$ be a linear map, and* $\dim E \geqq 1$. *Then there exists at least one non-zero eigenvector of A.*

Proof. We consider $\mathbf{C}[A]$, i.e. the ring generated by A over $\mathbf{C}$. As a vector space over $\mathbf{C}$, it is contained in the ring of endomorphisms of A, which is finite dimensional, the dimension being the same as the ring of all $n \times n$ matrices if $n = \dim E$. Hence there exists a non-zero polynomial P with coefficients in $\mathbf{C}$ such that $P(A) = 0$. We can factor P into a product of linear factors,

$$P(X) = (X - \lambda_1) \cdots (X - \lambda_m)$$

with $\lambda_j \in \mathbf{C}$. Then $(A - \lambda_1 I) \cdots (A - \lambda_m I) = 0$. Hence not all factors $A - \lambda_j I$ can be isomorphisms, and there exists $\lambda \in \mathbf{C}$ such that $A - \lambda I$ is not an isomorphism. Hence it has an element $\xi \neq 0$ in its kernel, and we get $A\xi - \lambda\xi = 0$. This shows that ξ is a non-zero eigenvector, as desired.

Spectral Theorem (*hermitian case*). *Let E be a non-zero vector space over the complex numbers, with a positive definite hermitian form. Let $A : E \to E$ be a hermitian linear map. Then E has an orthogonal basis consisting of eigenvectors of A.*

Proof. Let ξ_1 be a non-zero eigenvector, with eigenvalue λ_1, and let E_1 be the subspace generated by ξ_1. Then A maps $E_1^\perp$ into itself, because

$$\langle AE_1^\perp, \xi_1\rangle = \langle E_1^\perp, A\xi_1\rangle = \langle E_1^\perp, \lambda_1\xi_1\rangle = \lambda_1\langle E_1^\perp, \xi_1\rangle = 0,$$

whence $AE_1^\perp$ is perpendicular to ξ_1.

Since $\xi_1 \neq 0$ we have $\langle \xi_1, \xi_1\rangle > 0$ and hence, since our hermitian form is non-degenerate (being positive definite), we have

$$E = E_1 \oplus E_1^\perp.$$

The restriction of our form to $E_1^\perp$ is positive definite (if $\dim E > 1$).

From Proposition 5, we see at once that the restriction of A to $E_1^{\perp}$ is hermitian. Hence we can complete the proof by induction.

COROLLARY 1. *Hypotheses being as in the theorem, there exists an orthonormal basis consisting of eigenvectors of A.*

Proof. Divide each vector in an orthogonal basis by its norm.

COROLLARY 2. *Let E be a non-zero vector space over the complex numbers, with a positive definite hermitian form f. Let g be another hermitian form on E. Then there exists a basis of E which is orthogonal for both f and g.*

Proof. We write $f(x, y) = \langle x, y\rangle$. Since f is non-singular, being positive definite, there exists a unique hermitian linear map A such that $g(x, y) = \langle Ax, y\rangle$ for all $x, y \in E$. We apply the theorem to A, and find a basis as in the theorem, say $\{v_1, \dots, v_n\}$. Let λ_i be the eigenvalue such that $Av_i = \lambda_i v_i$. Then

$$g(v_i, v_j) = \langle Av_i, v_j\rangle = \lambda_i\langle v_i, v_j\rangle,$$

and therefore our basis is also orthogonal for g, as was to be shown.

§13. *The spectral theorem (symmetric case)*

Let E be a vector space over the real numbers, and let g be a symmetric positive definite form on E. If $A : E \to E$ is a linear map, then we know that its transpose, relative to g, is defined by the condition

$$\langle Ax, y\rangle = \langle x, {}^tAy\rangle$$

for all $x, y \in E$. We say that A is *symmetric* if $A = {}^tA$. As before, an element $\xi \in E$ is called an eigenvector of A if there exists $\lambda \in R$ such that $A\xi = \lambda\xi$, and λ is called an eigenvalue if $\xi \neq 0$.

SPECTRAL THEOREM (*symmetric case*). *Let E be a non-zero vector space over the real numbers, with a positive definite symmetric form. Let $A : E \to E$ be a symmetric linear map. Then E has an orthogonal basis consisting of eigenvectors of A.*

Proof. We shall reduce the proof to the hermitian case. We introduce the *complexification* of E. We let

$$E_{\mathbf{C}} = E \oplus E$$

be the direct sum of E with itself. If $a + bi$ is a complex number, $a, b \in R$, and if (x, y) is an element of $E_{\mathbf{C}}$, with $x, y \in E$, then we define the operation of $\mathbf{C}$ on $E_{\mathbf{C}}$ by

$$(a + bi)(x, y) = (ax - by, bx + ay).$$

A brute force computation shows that $E_{\mathbf{C}}$ is a vector space over $\mathbf{C}$. If we identify E with the first factor, namely $(E, 0)$, then we see at once that

$$E_{\mathbf{C}} = E + iE,$$

and that the operation we defined was motivated by the fact that

$$(a + bi)(x + iy) = ax - by + i(bx + ay),$$

taking into account our identification.

If $x + iy \in E_{\mathbf{C}}$, with $x, y \in E$, then we define $A_{\mathbf{C}} : E_{\mathbf{C}} \to E_{\mathbf{C}}$ by

$$A_{\mathbf{C}}(x + iy) = Ax + iAy.$$

Then $A_{\mathbf{C}}$ becomes a $\mathbf{C}$-linear map of $E_{\mathbf{C}}$ into itself, as one sees at once from the definitions.

We now define a hermitian form on $E_{\mathbf{C}}$. If

$$v = x + iy, \qquad w = x' + iy', \qquad \text{with} \quad x, y, x', y' \in E,$$

we let

$$\langle v, w \rangle_h = \langle x, x' \rangle + \langle y, y' \rangle + i\langle y, x' \rangle - i\langle x, y' \rangle.$$

It is again immediately verified that h is hermitian, and is positive definite, because g is symmetric positive definite. Furthermore, the definitions show at once that $A_{\mathbf{C}}$ is hermitian with respect to h.

We apply the spectral theorem for hermitian maps. We can find an orthogonal basis $\{\xi_1, \ldots, \xi_n\}$ of $E_{\mathbf{C}}$ over $\mathbf{C}$ consisting of eigenvectors of $A_{\mathbf{C}}$, with real eigenvalues $\lambda_1, \ldots, \lambda_n$ respectively. Write

$$\xi_\nu = x_\nu + iy_\nu$$

with $x_\nu, y_\nu \in E$. By definition of an eigenvector, we have

$$A_{\mathbf{C}}\xi_\nu = \lambda_\nu \xi_\nu = \lambda_\nu x_\nu + i\lambda_\nu y_\nu.$$

But

$$A_{\mathbf{C}}\xi_\nu = Ax_\nu + iAy_\nu.$$

If follows that $Ax_\nu = \lambda_\nu x_\nu$. Furthermore, it is clear that $\{x_1, \ldots, x_n\}$ is an orthogonal basis for E over $\mathbf{R}$. Our theorem is proved.

Remarks. The spectral theorems are valid over a real closed field; our proofs don't need any change. Furthermore, the proofs are reasonably close to those which would be given in analysis for Hilbert spaces, and compact operators. The existence of eigenvalues and eigenvectors must however be proved differently, for instance using the Gelfand theorem which we have actually proved in Chapter XII, or using a variational principle (i.e. finding a maximum or minimum for the quadratic function depending on the operator).

Corollary 1. *Hypotheses being as in the theorem, there exists an orthonormal basis consisting of eigenvectors of A.*

Proof. Divide each vector in an orthogonal basis by its norm.

Corollary 2. *Let E be a non-zero vector space over the reals, with a positive definite symmetric form f. Let g be another symmetric form on E. Then there exists a basis of E which is orthogonal for both f and g.*

Proof. We write $f(x, y) = \langle x, y\rangle$. Since f is non-singular, being positive definite, there exists a unique symmetric linear map A such that $g(x, y) = \langle Ax, y\rangle$ for all $x, y \in E$. We apply the theorem to A, and find a basis as in the theorem. It is clearly an orthogonal basis for g (cf. the same proof in the hermitian case).

Exercises

1. Let E be a vector space over a field k and let g be a bilinear form on E. Assume that whenever $x, y \in E$ are such that $g(x, y) = 0$, then $g(y, x) = 0$. Show that g is symmetric or alternating.

2. Show explicitly how $W(k)$ is a homomorphic image of $WG(k)$.

3. Show that $WG(k)$ can be expressed as a homomorphic image of $\mathbf{Z}[k^*/k^{*2}]$. [*Hint:* Use the existence of orthogonal bases.]

4. Let E be a module over $\mathbf{Z}$. Assume that E is free, of dimension $n \geqq 1$, and let f be a bilinear alternating form on E. Show that there exists a basis $\{e_i\}$ $(i = 1, \ldots, n)$ and an integer r such that $2r \leqq n$,

$$e_1 \cdot e_2 = a_1, \quad e_3 \cdot e_4 = a_2, \quad \ldots, \quad e_{2r-1} \cdot e_{2r} = a_r$$

where $a_1, \ldots, a_r \in \mathbf{Z}$, $a_i \neq 0$, and a_i divides a_{i+1} for $i = 1, \ldots, r - 1$ and finally $e_i \cdot e_j = 0$ for all other pairs of indices $i \leqq j$. Show that the ideals $\mathbf{Z}a_i$ are uniquely determined. [*Hint:* Consider the injective homomorphism $\varphi_f : E \to E^*$ of E into the dual space over $\mathbf{Z}$, viewing $\varphi_f(E)$ as a free submodule of E^*.] Generalize to principal rings when you know the basis theorem for modules over these rings.

5. Let E be finite dimensional over R, and let g be a symmetric positive definite form on E. Let A be a symmetric endomorphism of E with respect to g. Define $A \geqq 0$ to mean $\langle Ax, x\rangle \geqq 0$ for all $x \in E$. Show that $A \geqq 0$ if and only if all eigenvalues of A belonging to non-zero eigenvectors are $\geqq 0$.

6. Prove all the properties for the pfaffian stated in *Geometric Algebra*, p. 142.

7. Witt's theorem is still true for alternating forms. Prove it (or look it up in Artin or Bourbaki).

8. Show that the pfaffian of an alternating $n \times n$ matrix is 0 when n is odd.

9. Define maps of degree > 2, from one module into another. [*Hint:* For degree 3, consider the expression

$$f(x + y + z) - f(x + y) - f(x + z) - f(y + z) + f(x) + f(y) + f(z).]$$

Generalize the statement proved for quadratic maps to these higher degree maps, i.e. the uniqueness of the various multilinear maps entering into their definitions.

10. (a) Let E be a finite dimensional space over the complex numbers, and let $h : E \times E \to \mathbf{C}$ be a hermitian form. Write

$$h(x, y) = g(x, y) + if(x, y)$$

where g, f are real valued. Show that g, f are **R**-bilinear, g is symmetric, f is alternating.

(b) Let E be finite dimensional over **C**. Let $g : E \times E \to \mathbf{C}$ be **R**-bilinear. Assume that for all $x \in E$, the map $g \mapsto g(x, y)$ is **C**-linear, and that the R-bilinear form

$$f(x, y) = g(x, y) - g(y, x)$$

is real-valued on $E \times E$. Show that there exists a hermitian form h on E and a symmetric **C**-bilinear form ψ on E such that $2ig = h + \psi$. Show that h and ψ are uniquely determined.

11. In the hermitian spectral theorem, show that E admits a direct sum decomposition over **R**, $E = F + iF$, such that E is isomorphic to the complexification of F, and A induces a linear symmetric map on F.

12. Let E be a finite dimensional space over the complex, with a positive definite hermitian form. Let S be a set of (**C**-linear) endomorphisms of E having no invariant subspace except 0 and E. (This means that if F is a subspace of E and $BF \subset F$ for all $B \in S$, then $F = 0$ or $F = E$.) Let A be a hermitian map of E into itself such that $AB = BA$ for all $B \in S$. Show that $A = \lambda I$ for some real number λ. [*Hint:* Show that there exists exactly one eigenvalue of A. If there were two eigenvalues, say $\lambda_1 \neq \lambda_2$, one could find two polynomials f and g with real coefficients such that $f(A) \neq 0$, $g(A) \neq 0$ but $f(A)g(A) = 0$. Let F be the kernel of $g(A)$ and get a contradiction.]

13. Let E be as in Exercise 12. Let T be a **C**-linear map of E into itself. Let

$$A = \tfrac{1}{2}(T + T^*).$$

Show that A is hermitian. Show that T can be written in the form $A + iB$ where A, B are hermitian, and are uniquely determined.

14. Let S be a commutative set of **C**-linear endomorphisms of E having no invariant subspace unequal to 0 or E. Assume in addition that if $B \in S$, then $B^* \in S$. Show that each element of S is of type αI for some complex number α. [*Hint:* Let $B_0 \in S$. Let

$$A = \tfrac{1}{2}(B_0 + B_0^*).$$

Show that $A = \lambda I$ for some real λ.]

15. An endomorphism B of E is said to be *normal* if B commutes with B^*. State and prove a spectral theorem for normal endomorphisms.

16. Let E be a finite dimensional vector space over the reals, and let $\langle\ ,\ \rangle$ be a symmetric positive definite form. Let Ω be a non-degenerate alternating

form on E. Show that there exists a direct sum decomposition

$$E = E_1 \oplus E_2$$

having the following property. If $x, y \in E$ are written

$$\begin{aligned} x &= (x_1, x_2) \quad \text{with} \quad x_1 \in E_1 \quad \text{and} \quad x_2 \in E_2, \\ y &= (y_1, y_2) \quad \text{with} \quad y_1 \in E_1 \quad \text{and} \quad y_2 \in E_2, \end{aligned}$$

then $\Omega(x, y) = \langle x_1, y_2 \rangle - \langle x_2, y_1 \rangle$. [*Hint:* Use Corollary 2 of Theorem 6, show that A is positive definite, and take its square root to transform the direct sum decomposition obtained in that corollary.]

17. Let E be a vector space over the reals (finite dimensional, as usual). If A is an endomorphism of E, define its norm $|A|$ to be the greatest lower bound of all numbers C such that $|Ax| \leqq C|x|$. Show that this norm satisfies the triangle inequality. Show that the series

$$\exp(A) = I + A + \frac{A^2}{2!} + \cdots$$

converges, and if A commutes with B, then $\exp(A + B) = \exp(A)\exp(B)$. If A is sufficiently close to I, show that the series

$$\log(A) = \frac{(I - A)}{1} + \frac{(I - A)^2}{2} + \cdots$$

converges, and if A commutes with B, then

$$-\log(AB) = \log A + \log B.$$

18. Let E have a fixed positive definite symmetric bilinear form. We call E a *Hilbert space* (finite dimensional). A linear automorphism A of E is said to be *Hilbertian* if it is an automorphism of the form, i.e. ${}^tAA = I$. We shall write A^* instead of ${}^tA$ for the present exercises. Let A be a symmetric endomorphism of E. We shall say that A is *positive definite* if A is a linear automorphism, and $\langle Ax, x \rangle > 0$ for all $x \in E$, $x \neq 0$. Prove: If A is symmetric (resp. alternating), then $\exp(A)$ is symmetric positive definite (resp. Hilbertian). If A is a linear automorphism of E sufficiently close to I, and is positive definite symmetric (resp. Hilbertian), then $\log A$ is symmetric (resp. alternating).

19. Using the spectral theorem, show that one can define $\log A$ when A is symmetric positive definite, not necessarily close to I. Show that any automorphism A of E can be written in a unique way as a product $A = HP$, where H is Hilbertian and P is symmetric positive definite. [*Hint:* Note that A^*A is symmetric positive definite, and let $P = (A^*A)^{1/2}$, justifying the square root by means of the spectral theorem. Let $H = AP^{-1}$, to get the existence of the product. For uniqueness, suppose $A = H_1P_1$, let $H_2 = PP_1^{-1}$. Then $I = H_2^*H_2$, and using $P^* = P$, $P_1^* = P_1$, conclude that $P^2 = P_1^2$. Take the log, divide by 2, and take the exp to conclude that $P = P_1$.]

20. (Tate) Let E, F be complete normed vector spaces over the real numbers. Let $f : E \to F$ be a map having the following property. There exists a number $C > 0$ such that for all $x, y \in E$ we have

$$|f(x + y) - f(x) - f(y)| \leqq C.$$

Show that there exists a unique linear map $g : E \to F$ such that $|g - f|$ is bounded (i.e. $|g(x) - f(x)|$ is bounded as a function of x). Generalize to the bilinear case. [*Hint:* Let

$$g(x) = \lim_{n \to \infty} \frac{f(2^n x)}{2^n} \cdot]$$

CHAPTER XV

Representation of One Endomorphism

§1. Representations

Let k be a commutative ring and E a module over k. As usual, we denote by $\mathrm{End}_k(E)$ the ring of k-endomorphisms of E, i.e. the ring of k-linear maps of E into itself.

Let R be a k-algebra (given by a ring-homomorphism $k \to R$ which allows us to consider R as a k-module). By a *representation* of R in E one means a k-algebra homomorphism $R \to \mathrm{End}_k(E)$, that is a ring-homomorphism $\rho : R \to \mathrm{End}_k(E)$ which makes the following diagram commutative:

$$\begin{array}{ccc} R & \longrightarrow & \mathrm{End}_k(E) \\ & \nwarrow \quad \nearrow & \\ & k & \end{array}$$

[As usual, we view $\mathrm{End}_k(E)$ as a k-algebra; if I denotes the identity map of E, we have the homomorphism of k into $\mathrm{End}_k(E)$ given by $a \mapsto aI$. We shall also use I to denote the unit matrix if bases have been chosen. The context will always make our meaning clear.]

We shall meet several examples of representations in the sequel, with various types of rings (both commutative and non-commutative). In this chapter, the rings will be commutative.

We observe that E may be viewed as an $\mathrm{End}_k(E)$ module. Hence E may be viewed as an R-module, defining the operation of R on E by letting

$$(x, v) \mapsto \rho(x)v$$

for $x \in R$ and $v \in E$. We usually write xv instead of $\rho(x)v$.

A subgroup F of E such that $RF \subset F$ will be said to be an *invariant* submodule of E. (It is both R-invariant and k-invariant.) We also say that it is invariant under the representation.

We say that the representation is *irreducible*, or *simple*, if $E \neq 0$, and if the only invariant submodules are 0 and E itself.

The purpose of representation theories is to determine the structure of all representations of various interesting rings, and to classify their irreducible representations. In most cases, we take k to be a field, which may or may not be algebraically closed. The difficulties in proving theorems about representations may therefore lie in the complication of the ring R, or the complication of the field k, or the complication of the module E, or all three.

A representation ρ as above is said to be *completely reducible* or *semi-simple* if E is an R-direct sum of R-submodules E_i,

$$E = E_1 \oplus \cdots \oplus E_m$$

such that each E_i is irreducible. We also say that E is completely reducible. It is not true that all representations are completely reducible, and in fact those considered in this chapter will not be in general. Certain types of completely reducible representations will be studied later.

There is a special type of representation which will occur very frequently. Let $v \in E$ and assume that $E = Rv$. We shall also write $E = (v)$. We then say that E is *principal* (over R), and that the representation is *principal*. If that is the case, the set of elements $x \in R$ such that $xv = 0$ is a left ideal $\mathfrak{a}$ of R (obvious). The map of R onto E given by

$$x \mapsto xv$$

induces an isomorphism of R-modules,

$$R/\mathfrak{a} \to E$$

(viewing R as a left module over itself, and $R/\mathfrak{a}$ as the factor module). In this map, the unit element 1 of R corresponds to the generator v of E.

As a matter of notation, if $v_1, \ldots, v_n \in E$, we let $(v_1, \ldots, v_n)$ denote the submodule of E generated by $v_1, \ldots, v_n$.

Assume that E has a decomposition into a direct sum of R-submodules

$$E = E_1 \oplus \cdots \oplus E_s.$$

Assume that each E_i is free and of dimensions $\geqq 1$ over k. Let $\mathfrak{B}_1, \ldots, \mathfrak{B}_s$ be bases for $E_1, \ldots, E_s$ respectively over k. Then $\{\mathfrak{B}_1, \ldots, \mathfrak{B}_s\}$ is a basis for E. Let $\varphi \in R$, and let φ_i be the endomorphism induced by φ on E_i. Let M_i be the matrix of φ_i with respect to the basis $\mathfrak{B}_i$. Then the matrix M of φ with respect to $\{\mathfrak{B}_1, \ldots, \mathfrak{B}_s\}$ looks like

$$\begin{pmatrix} M_1 & 0 & \cdots & 0 \\ 0 & M_2 & \cdots & 0 \\ \vdots & & \ddots & \vdots \\ 0 & \cdots & & 0 \\ 0 & \cdots & 0 & M_s \end{pmatrix}.$$

A matrix of this type is said to be decomposed into *blocks*, $M_1, \ldots M_s$. When we have such a decomposition, the study of φ or its matrix is completely reduced (so to speak) to the study of the blocks.

It does not always happen that we have such a reduction, but frequently something almost as good happens. Let E' be a submodule of E, invariant under R. Assume that there exists a basis of E' over k, say $\{v_1, \ldots, v_m\}$, and that this basis can be completed to a basis of E,

$$\{v_1, \ldots, v_m, v_{m+1}, \ldots, v_n\}.$$

This is always the case if k is a field.

Let $\varphi \in R$. Then the matrix of φ with respect to this basis has the form

$$\begin{pmatrix} M' & * \\ 0 & M'' \end{pmatrix}.$$

Indeed, since E' is mapped into itself by φ, it is clear that we get M' in the upper left, and a zero matrix below it. Furthermore, for each $j = m + 1, \ldots, n$ we can write

$$\varphi v_j = c_{j1}v_1 + \cdots + c_{jm}v_m + c_{j,m+1}v_{m+1} + \cdots + c_{mn}v_n.$$

The transpose of the matrix (c_{ji}) then becomes the matrix

$$\begin{pmatrix} * \\ M'' \end{pmatrix}$$

occurring on the right in the matrix representing φ.

Furthermore, consider an exact sequence

$$0 \to E' \to E \to E'' \to 0.$$

Let $\bar{v}_{m+1}, \ldots, \bar{v}_n$ be the images of $v_{m+1}, \ldots, v_n$ under the canonical map $E \to E''$. We can define a linear map

$$\varphi'' : E'' \to E''$$

in a natural way so that $(\overline{\varphi v}) = \varphi''(\bar{v})$ for all $v \in E$. Then it is clear that the matrix of φ'' with respect to the basis $\{\bar{v}_1, \ldots, \bar{v}_n\}$ is M''.

§2. Modules over principal rings

Throughout this section, we assume that R is a principal entire ring. All modules are over R, and homomorphisms are R-homomorphisms, unless otherwise specified.

The theorems will generalize those proved in Chapter I for abelian groups. We shall also point out how the proofs of Chapter I can be adjusted with substitutions of terminology so as to yield proofs in the present case.

Let F be a free module over R, with a basis $\{x_i\}_{i \in I}$. Then the cardinality of I is uniquely determined, and is called the dimension of F. We recall that this is proved, say by taking a prime element p in R, and observing that F/pF is a vector space over the field R/pR, whose dimension is precisely the cardinality of I. We may therefore speak of the dimension of a free module over R.

THEOREM 1. *Let F be a free module, and M a submodule. Then M is free, and its dimension is less than or equal to the dimension of F.*

Proof. For simplicity, we give the proof when F has a finite basis $\{x_i\}$, $i = 1, \dots, n$. Let M_r be the intersection of M with $(x_1, \dots, x_r)$, the module generated by $x_1, \dots, x_r$. Then $M_1 = M \cap (x_1)$ is a submodule of (x_1), and is therefore of type $(a_1 x_1)$ with some $a_1 \in R$. Hence M_1 is either 0 or free, of dimension 1. Assume inductively that M_r is free of dimension $\leqq r$. Let $\mathfrak{a}$ be the set consisting of all elements $a \in R$ such that there exists an element $x \in M$ which can be written

$$x = b_1 x_1 + \cdots + b_r x_r + a x_{r+1}$$

with $b_i \in R$. Then $\mathfrak{a}$ is obviously an ideal, and is principal, generated say by an element a_{r+1}. If $a_{r+1} = 0$, then $M_{r+1} = M_r$ and we are done with the inductive step. If $a_{r+1} \neq 0$, let $w \in M_{r+1}$ be such that the coefficient of w with respect to x_{r+1} is a_{r+1}. If $x \in M_{r+1}$ then the coefficient of x with respect to x_{r+1} is divisible by a_{r+1}, and hence there exists $c \in R$ such that $x - cw$ lies in M_r. Hence

$$M = M_r + (w).$$

On the other hand, it is clear that $M_r \cap (w)$ is 0, and hence that this sum is direct, thereby proving our theorem. We also note that the proof is valid in the infinite case, replacing induction by well-ordering.

COROLLARY. *Let E be a finitely generated module and E' a submodule. Then E' is finitely generated.*

Proof. We can represent E as a factor module of a free module F with a finite number of generators: If $v_1, \dots, v_n$ are generators of E, we take a free module F with basis $\{x_1, \dots, x_n\}$ and map x_i on v_i. The inverse image of E' in F is a submodule, which is free, and finitely generated, by the theorem. Hence E' is finitely generated. The assertion also follows using simple properties of Noetherian rings and modules.

If one wants to translate the proofs of Chapter I, then one makes the following definitions. A free 1-dimensional module over R is called *infinite cyclic*. An infinite cyclic module is isomorphic to R, viewed as module over itself. Thus every non-zero submodule of an infinite cyclic module is infinite cyclic. The proof given in Chapter I for the analogue of Theorem 1 applies without further change.

Let E be a module. We say that E is a *torsion* module if given $x \in E$, there exists $a \in R$, $a \neq 0$, such that $ax = 0$. The generalization of *finite abelian group* is *finitely generated torsion module*. An element x of E is called a *torsion element* if there exists $a \in R$, $a \neq 0$, such that $ax = 0$.

Let E be a module. We denote by E_t the submodule consisting of all torsion elements of E, and call it the *torsion submodule* of E. If $E_t = 0$, we say that E is *torsion free*.

THEOREM 2. *Let E be finitely generated. Then E/E_t is free. There exists a free submodule F of E such that E is a direct sum*

$$E = E_t \oplus F.$$

The dimension of such a submodule F is uniquely determined.

Proof. We first prove that E/E_t is torsion free. If $x \in E$, let $\bar{x}$ denote its residue class mod E_t. Let $b \in R$, $b \neq 0$ be such that $b\bar{x} = 0$. Then $bx \in E_t$, and hence there exists $c \in R$, $c \neq 0$, such that $cbx = 0$. Hence $x \in E_t$ and $\bar{x} = 0$, thereby proving that E/E_t is torsion free. It is also finitely generated. Assume now that M is a torsion free module which is finitely generated. Let $\{v_1, \ldots, v_n\}$ be a maximal set of elements of M among a given finite set of generators $\{y_1, \ldots, y_m\}$ such that $\{v_1, \ldots, v_n\}$ is linearly independent. If y is one of the generators, there exist elements $a, b_1, \ldots, b_n \in R$ not all 0, such that

$$ay + b_1v_1 + \cdots + b_nv_n = 0.$$

Then $a \neq 0$ (otherwise we contradict the linear independence of $v_1, \ldots, v_n$). Hence ay lies in $(v_1, \ldots, v_n)$. Thus for each $j = 1, \ldots, m$ we can find $a_j \in R$, $a_j \neq 0$, such that a_jy_j lies in $(v_1, \ldots, v_n)$. Let $a = a_1 \cdots a_m$ be the product. Then aM is contained in $(v_1, \ldots, v_n)$, and $a \neq 0$. The map

$$x \mapsto ax$$

is an injective homomorphism, whose image is contained in a free module. This image is isomorphic to M, and we conclude from Theorem 1 that M is free, as desired.

To get the submodule F we need a lemma.

LEMMA 1. *Let E, E' be modules, and assume that E' is free. Let $f : E \to E'$ be a surjective homomorphism. Then there exists a free submodule F of E such that the restriction of f to F induces an isomorphism of F with E', and such that $E = F \oplus \operatorname{Ker} f$.*

Proof. Let $\{x_i'\}_{i \in I}$ be a basis of E'. For each i, let x_i be an element of E such that $f(x_i) = x_i'$. Let F be the submodule of E generated by all the elements x_i, $i \in I$. Then one sees at once that the family of elements $\{x_i\}_{i \in I}$ is linearly independent, and therefore that F is free. Given $x \in E$, there exist elements $a_i \in R$ such that

$$f(x) = \sum a_i x_i'.$$

Then $x - \sum a_i x_i$ lies in the kernel of f, and therefore $E = \operatorname{Ker} f + F$. It is clear that $\operatorname{Ker} f \cap F = 0$, and hence that the sum is direct, thereby proving the lemma.

We apply the lemma to the homomorphism $E \to E/E_t$ in Theorem 2 to get our decomposition $E = E_t \oplus F$. The dimension of F is uniquely determined, because F is isomorphic to E/E_t for any decomposition of E into a direct sum as stated in the theorem.

The dimension of the free module F in Theorem 2 is called the *rank* of E.

In order to get the structure theorem for finitely generated modules over R, one can proceed exactly as for abelian groups. We shall describe the dictionary which allows us to transport the proofs essentially without change.

Let E be a module over R. Let $x \in E$. The map $a \mapsto ax$ is a homomorphism of R onto the submodule generated by x, and the kernel is an ideal, which is principal, generated by an element $m \in R$. We say that m is a *period* of x. We note that m is determined up to multiplication by a unit (if $m \neq 0$). An element $c \in R$, $c \neq 0$, is said to be an *exponent* for E (resp. for x) if $cE = 0$ (resp. $cx = 0$).

Let p be a prime element. We denote by $E(p)$ the submodule of E consisting of all elements x having an exponent which is a power p^r ($r \geqq 1$). A p-submodule of E is a submodule contained in $E(p)$.

We select once and for all a system of representatives for the prime elements of R (modulo units). For instance, if R is a polynomial ring in one variable over a field, we take as representatives the irreducible polynomials with leading coefficient 1.

Let $m \in R$, $m \neq 0$. We denote by E_m the kernel of the map $x \mapsto mx$. It consists of all elements of E having exponent m.

A module E is said to be *cyclic* if it is isomorphic to $R/(a)$ for some element $a \in R$. Without loss of generality if $a \neq 0$, one may assume that

a is a product of primes in our system of representatives, and then we could say that a is the order of the module.

Let $r_1, \dots, r_s$ be integers $\geqq 1$. A p-module E is said to be of *type*

$$(p^{r_1}, \dots, p^{r_s})$$

if it is isomorphic to the product of cyclic modules $R/(p^{r_i})$ $(i = 1, \dots, s)$. If p is fixed, then one could say that the module is of type $(r_1, \dots, r_s)$ (relative to p).

All the proofs of Chapter I, §10 now go over without change. Whenever we argue on the size of a positive integer m, we have a similar argument on the number of prime factors appearing in its prime factorization. If we deal with a prime power p^r, we can view the order as being determined by r. The reader can now check for himself that the proofs of Chapter I, §10 are applicable.

However, we shall develop the theory once again without assuming any knowledge of Chapter I, §10. Thus our treatment is self-contained.

THEOREM 3. *Let E be a finitely generated torsion module $\neq 0$. Then E is the direct sum*

$$E = \coprod_p E(p),$$

taken over all primes p such that $E(p) \neq 0$. Each $E(p)$ can be written as a direct sum

$$E(p) = R/(p^{\nu_1}) \oplus \cdots \oplus R/(p^{\nu_s})$$

with $1 \leqq \nu_1 \leqq \cdots \leqq \nu_s$. The sequence $\nu_1, \dots, \nu_s$ is uniquely determined.

Proof. Let a be an exponent for E, and suppose that $a = bc$ with $(b, c) = (1)$. Let $x, y \in R$ be such that

$$1 = xb + yc.$$

We contend that $E = E_b \oplus E_c$. Our first assertion then follows by induction, expressing a as a product of prime powers. Let $v \in E$. Then

$$v = xbv + ycv.$$

Then $xbv \in E_c$ because $cxbv = xav = 0$. Similarly, $ycv \in E_b$. Finally $E_b \cap E_c = 0$, as one sees immediately. Hence E is the direct sum of E_b and E_c.

We must now prove that $E(p)$ is a direct sum as stated. If $y_1, \dots, y_m$ are elements of a module, we shall say that they are *independent* if whenever we have a relation

$$a_1y_1 + \cdots + a_my_m = 0$$

with $a_i \in R$, then we must have $a_i y_i = 0$ for all i. (Observe that *independent* does not mean *linearly independent.*) We see at once that $y_1, \ldots, y_m$ are independent if and only if the module $(y_1, \ldots, y_m)$ has the direct sum decomposition

$$(y_1, \ldots, y_m) = (y_1) \oplus \cdots \oplus (y_m)$$

in terms of the cyclic modules (y_i), $i = 1, \ldots, m$.

We now have an analogue of Lemma 1 for modules having a prime power exponent.

LEMMA 2. *Let E be a torsion module of exponent p^r ($r \geqq 1$) for some prime element p. Let $x_1 \in E$ be an element of period p^r. Let $\overline{E} = E/(x_1)$. Let $\overline{y}_1, \ldots, \overline{y}_m$ be independent elements of $\overline{E}$. Then for each i there exists a representative $y_i \in E$ of $\overline{y}_i$, such that the period of y_i is the same as the period of $\overline{y}_i$. The elements $x_1, y_1, \ldots, y_m$ are independent.*

Proof. Let $\overline{y} \in \overline{E}$ have period p^n for some $n \geqq 1$. Let y be a representative of $\overline{y}$ in E. Then $p^n y \in (x_1)$, and hence

$$p^n y = p^s c x_1, \qquad c \in R, p \nmid c,$$

for some $s \leqq r$. If $s = r$, we see that y has the same period as $\overline{y}$. If $s < r$, then $p^s c x_1$ has period p^{r-s}, and hence y has period p^{n+r-s}. We must have

$$n + r - s \leqq r,$$

because p^r is an exponent for E. Thus we obtain $n \leqq s$, and we see that

$$y - p^{s-n} c x_1$$

is a representative for $\overline{y}$, whose period is p^n.

Let y_i be a representative for $\overline{y}_i$ having the same period. We prove that $x_1, y_1, \ldots, y_m$ are independent. Suppose that $a, a_1, \ldots, a_m \in R$ are elements such that

$$a x_1 + a_1 y_1 + \cdots + a_m y_m = 0.$$

Then

$$a_1 \overline{y}_1 + \cdots + a_m \overline{y}_m = 0.$$

By hypothesis, we must have $a_i \overline{y}_i = 0$ for each i. If p^{r_i} is the period of $\overline{y}_i$, then p^{r_i} divides a_i. We then conclude that $a_i y_i = 0$ for each i, and hence finally that $a x_1 = 0$, thereby proving the desired independence.

To get the direct sum decomposition of $E(p)$, we first note that $E(p)$ is finitely generated. We may assume without loss of generality that $E = E(p)$. Let x_1 be an element of E whose period p^{r_1} is such that r_1 is maximal. Let $\overline{E} = E/(x_1)$. We contend that dim $\overline{E}_p$ as vector space over R/pR is strictly less than dim E_p. Indeed, if $\bar{y}_1, \ldots, \bar{y}_m$ are linearly independent elements of $\bar{E}_p$ over R/pR, then Lemma 2 implies that dim $E_p \geqq m + 1$ because we can always find an element of (x_1) having period p, independent of $y_1, \ldots, y_m$. Hence dim $\bar{E}_p <$ dim E_p. We can prove the direct sum decomposition by induction. If $\bar{E} \neq 0$, there exist elements $\bar{x}_2, \ldots, \bar{x}_s$ having periods $p^{r_2}, \ldots, p^{r_s}$ respectively, such that $r_2 \geqq \cdots \geqq r_s$. By Lemma 2, there exist representatives $x_2, \ldots, x_r$ in E such that x_i has period p^{r_i} and $x_1, \ldots, x_r$ are independent. Since p^{r_1} is such that r_1 is maximal, we have $r_1 \geqq r_2$, and our decomposition is achieved.

The uniqueness will be a consequence of a more general uniqueness theorem, which we state next.

THEOREM 4. *Let E be a finitely generated torsion module, $E \neq 0$. Then E is isomorphic to a direct sum of non-zero factors*

$$R/(q_1) \oplus \cdots \oplus R/(q_r),$$

where $q_1, \ldots, q_r$ are non-zero elements of R, and $q_1 \mid q_2 \mid \cdots \mid q_r$. The sequence of ideals $(q_1), \ldots, (q_r)$ is uniquely determined by the above conditions.

Proof. Using Theorem 3, decompose E into a direct sum of p-submodules, say $E(p_1) \oplus \cdots \oplus E(p_l)$, and then decompose each $E(p_i)$ into a direct sum of cyclic submodules of periods $p_i^{r_{ij}}$. We visualize these symbolically as described by the following diagram:

$$\begin{array}{cccccc} E(p_1): & r_{11} & \leqq & r_{12} & \leqq & \cdots \\ E(p_2): & r_{21} & \leqq & r_{22} & \leqq & \cdots \\ \vdots & \vdots & & \vdots & & \vdots \\ E(p_l): & r_{l1} & \leqq & r_{l2} & \leqq & \cdots \end{array}$$

A horizontal row describes the type of the module with respect to the prime at the left. The exponents r_{ij} are arranged in increasing order for each fixed $i = 1, \ldots, l$. We let $q_1, \ldots, q_r$ correspond to the columns of the matrix of exponents, in other words

$$q_1 = p_1^{r_{11}} p_2^{r_{21}} \cdots p_l^{r_{l1}},$$

$$q_2 = p_1^{r_{12}} p_2^{r_{22}} \cdots p_l^{r_{l2}},$$

$$\cdots$$

The direct sum of the cyclic modules represented by the first column is then isomorphic to $R/(q_1)$, because, as with abelian groups, the direct sum of cyclic modules whose periods are relatively prime is also cyclic. We have a similar remark for each column, and we observe that our proof actually orders the q_j by increasing divisibility, as was to be shown.

Now for uniqueness. Let p be any prime, and suppose that $E = R/(pb)$ for some $b \in R$, $b \neq 0$. Then E_p is the submodule $bR/(pb)$, as follows at once from unique factorization in R. But the kernel of the composite map

$$R \to bR \to bR/(pb)$$

is precisely (p). Thus we have an isomorphism

$$R/(p) \approx bR/(pb).$$

Let now E be expressed as in the theorem, as a direct sum of r terms. An element

$$v = v_1 \oplus \cdots \oplus v_r, \qquad v_i \in R/(q_i)$$

is in E_p if and only if $pv_i = 0$ for all i. Hence E_p is the direct sum of the kernel of multiplication by p in each term. But E_p is a vector space over $R/(p)$, and its dimension is therefore equal to the number of terms $R/(q_i)$ such that p divides q_i.

Suppose that p is a prime dividing q_1, and hence q for each $i = 1, \ldots, r$. Let E have a direct sum decomposition into s terms satisfying the conditions of the theorem, say

$$E = R/(q'_1) \oplus \cdots \oplus R/(q'_s).$$

Then p must divide at least r of the elements q'_j, whence $r \leqq s$. By symmetry, $r = s$, and p divides q'_j for all j.

Consider the module pE. By a preceding remark, if we write $q_i = pb_i$, then

$$pE \approx R/(b_1) \oplus \cdots \oplus R/(b_r),$$

and $b_1 \mid \cdots \mid b_r$. Some of the b_i may be units, but those which are not units determine their principal ideal uniquely, by induction. Hence if $(b_1) = \cdots = (b_j) = 1$ but $(b_{j+1}) \neq (1)$, then the sequence of ideals

$$(b_{j+1}), \ldots, (b_r)$$

is uniquely determined. This proves our uniqueness statement, and concludes the proof of Theorem 4.

The ideals $(q_1), \ldots, (q_r)$ are called the *invariants* of E.

The next theorem could be regarded as a corollary of Theorem 4. We shall give an independent proof for it. It will not be used in the sequel.

THEOREM 5. *Let F be a free module over R, and let M be a finitely generated submodule $\neq 0$. Then there exists a basis $\mathcal{B}$ of F, elements $e_1, \ldots, e_r$ in this basis, and non-zero elements $a_1, \ldots, a_r \in R$ such that:*

(i) *The elements $a_1e_1, \ldots, a_re_r$ form a basis of M over R.*

(ii) *We have* $a_i \mid a_{i+1}$ *for* $i = 1, \ldots, r - 1$.

The sequence of ideals $(a_1), \ldots, (a_r)$ *is uniquely determined by the preceding conditions.*

Proof. Let λ be a functional on F, in other words, an element of $\operatorname{Hom}_R(F, R)$. We let $J_\lambda = \lambda(M)$. Then J_λ is an ideal of R. Select λ_1 such that $\lambda_1(M)$ is maximal in the set of ideals $\{J_\lambda\}$, that is to say, there is no properly larger ideal in the set $\{J_\lambda\}$.

Let $\lambda_1(M) = (a_1)$. Then $a_1 \neq 0$, because there exists a non-zero element of M, and expressing this element in terms of some basis for F over R, with some non-zero coordinate, we take the projection on this coordinate to get a functional whose value on M is not 0. Let $x_1 \in M$ be such that $\lambda_1(x_1) = a_1$. For any functional g we must have $g(x_1) \in (a_1)$ [immediate from the maximality of $\lambda_1(M)$]. Writing x_1 in terms of any basis of F, we see that its coefficients must all be divisible by a_1. (If some coefficient is not divisible by a_1, project on this coefficient to get an impossible functional.) Therefore we can write $x_1 = a_1 e_1$ with some element $e_1 \in F$.

Next we prove that F is a direct sum

$$F = Re_1 \oplus \operatorname{Ker} \lambda_1.$$

Since $\lambda_1(e_1) = 1$, it is clear that $Re_1 \cap \operatorname{Ker} \lambda_1 = 0$. Furthermore, given $x \in F$ we note that $x - \lambda_1(x)e_1$ is in the kernel of λ_1. Hence F is the sum of the indicated submodules, and therefore the direct sum.

We note that $\operatorname{Ker} \lambda_1$ is free, being a submodule of a free module (Theorem 1). We let

$$F_1 = \operatorname{Ker} \lambda_1 \quad \text{and} \quad M_1 = M \cap \operatorname{Ker} \lambda_1.$$

We see at once that

$$M = Rx_1 \oplus M_1.$$

Thus M_1 is a submodule of F_1 and its dimension is one less than the dimension of M. We can therefore complete the existence proof by induction.

As for uniqueness, we must characterize our sequence of ideals $(a_1), \ldots, (a_r)$ entirely in terms of F and M.

LEMMA 3. *Let* L_a^s *be the set of all s-multilinear alternating forms on* F. *Let* J_s *be the ideal generated by all elements* $f(y_1, \ldots, y_s)$, *with* $f \in L_a^s$ *and* $y_1, \ldots, y_s \in F$. *Then*

$$J_s = (a_1 \cdots a_s).$$

Proof. We first show that $J_s \subset (a_1 \cdots a_s)$. Indeed, an element $y \in M$ can be written in the form

$$y = c_1 a_1 e_1 + \cdots + c_r a_r e_r.$$

Hence if $y_1, \ldots, y_s \in M$, and f is multilinear alternating on E, then $f(y_1, \ldots, y_s)$ is equal to a sum of terms of type

$$c_{i_1} \cdots c_{i_s} a_{i_1} \cdots a_{i_s} f(e_{i_1}, \ldots, e_{i_s}).$$

This is non-zero only when $e_{i_1}, \ldots, e_{i_s}$ are distinct, in which case the product $a_1 \cdots a_s$ divides this term, and hence J_s is contained in the stated ideal.

Conversely, we show that there exists an s-multilinear alternating form which gives precisely this product. We deduce this from determinants. We can write F as a direct sum

$$F = (e_1, \ldots, e_r) \oplus F_r$$

with some submodule F_r. Let f_i $(i = 1, \ldots, r)$ be the linear map $F \to R$ such that $f_i(e_j) = \delta_{ij}$, and such that f_i has value 0 on F_r. For $v_1, \ldots, v_s \in F$ we define

$$f(v_1, \ldots, v_s) = \det\big(f_i(v_j)\big).$$

Then f is multilinear alternating and takes on the value

$$f(e_1, \ldots, e_s) = 1,$$

as well as the value

$$f(a_1 e_1, \ldots, a_s e_s) = a_1 \cdots a_s.$$

This proves our lemma.

The uniqueness of Theorem 5 is now obvious, since first (a_1) is unique, then $(a_1 a_2)$ is unique and the quotient (a_2) is unique, and so forth by induction. Theorem 5 is proved.

We shall call $(a_1), \ldots, (a_r)$ the *invariants* of M in F.

§3. *Decomposition over one endomorphism*

Let k be a field and E a finite dimensional vector space over k, $E \neq 0$. Let $A \in \mathrm{End}_k(E)$ be a linear map of E into itself. Let t be transcendental over k. We shall define a representation of the polynomial ring $k[t]$ in E. Namely, we have a homomorphism

$$k[t] \to k[A] \subset \mathrm{End}_k(E)$$

which is obtained by substituting A for t in polynomials. The ring $k[A]$

is the subring of $\mathrm{End}_k(E)$ generated by A, and is commutative because powers of A commute with each other. Thus if $f(t)$ is a polynomial and $v \in E$, then

$$f(t)v = f(A)v.$$

The kernel of the homomorphism $f(t) \mapsto f(A)$ is a principal ideal of $k[t]$, which is $\neq 0$ because $k[A]$ is finite dimensional over k. It is generated by a unique polynomial of degree > 0, having leading coefficient 1. This polynomial will be called the *minimal polynomial* of A over k, and will be denoted by $q_A(t)$. It is of course not necessarily irreducible.

Assume that there exists an element $v \in E$ *such that* $E = k[t]v = k[A]v$. This means that E is generated over k by the elements

$$v, Av, A^2v, \ldots .$$

We called such a module *principal*, and if $R = k[t]$ we may write $E = Rv = (v)$.

If $q_A(t) = t^d + a_{d-1}t^{d-1} + \cdots + a_0$ then the elements

$$v, Av, \ldots, A^{d-1}v$$

constitute a basis for E over k. This is proved in the same way as the analogous statement for finite field extensions. First we note that they are linearly independent, because any relation of linear dependence over k would yield a polynomial $g(t)$ of degree less than $\deg q_A$ and such that $g(A) = 0$. Second, they generate E because any polynomial $f(t)$ can be written $f(t) = g(t)q_A(t) + r(t)$ with $\deg r < \deg q_A$. Hence $f(A) = r(A)$.

With respect to this basis, it is clear that the matrix of A is of the following type:

$$\begin{pmatrix} 0 & 0 & 0 & \cdots & 0 & -a_0 \\ 1 & 0 & 0 & \cdots & 0 & -a_1 \\ 0 & 1 & 0 & \cdots & 0 & -a_2 \\ \cdots & \cdots & \cdots & \cdots & \cdots & \cdots \\ 0 & 0 & 0 & \cdots & 0 & -a_{d-2} \\ 0 & 0 & 0 & \cdots & 1 & -a_{d-1} \end{pmatrix}.$$

If $E = (v)$ is principal, then E is isomorphic to $k[t]/(q_A(t))$ under the map $f(t) \to f(A)v$. The polynomial q_A is uniquely determined by A, and does not depend on the choice of generator v for E. This is essentially obvious, because if f_1, f_2 are two polynomials with leading coefficient 1, then $k[t]/(f_1(t))$ is isomorphic to $k[t]/(f_2(t))$ if and only if $f_1 = f_2$. (Decompose each polynomial into prime powers and apply the structure theorem for modules over principal rings.)

If E is principal then we shall call the polynomial q_A above the *polynomial invariant of* E, with respect to A, or simply its *invariant*.

THEOREM 6. *Let E be a non-zero finite dimensional space over the field k, and let $A \in \text{End}_k(E)$. Then E admits a direct sum decomposition*

$$E = E_1 \oplus \cdots \oplus E_r,$$

where each E_i is a principal $k[A]$-submodule, with invariant $q_i \neq 0$ such that

$$q_1 \mid q_2 \mid \cdots \mid q_r.$$

The sequence $(q_1, \ldots, q_r)$ is uniquely determined by E and A, and q_r is the minimal polynomial of A.

Proof. The first statement is simply a rephrasing in the present language for the structure theorem for modules over principal rings. Furthermore, it is clear that $q_r(A) = 0$ since $q_i \mid q_r$ for each i. No polynomial of lower degree than q_r can annihilate E, because in particular, such a polynomial does not annihilate E_r. Thus q_r is the minimal polynomial.

We shall call $(q_1, \ldots, q_r)$ the *invariants* of the pair (E, A). Let $E = k^{(n)}$, and let A be an $n \times n$ matrix, which we view as a linear map of E into itself. The invariants $(q_1, \ldots, q_r)$ will be called the *invariants* of A (over k).

COROLLARY 1. *Let k' be an extension field of k and let A be an $n \times n$ matrix in k. The invariants of A over k are the same as its invariants over k'.*

Proof. Let $\{v_1, \ldots, v_n\}$ be a basis of $k^{(n)}$ over k. Then we may view it also as a basis of $k'^{(n)}$ over k'. (The unit vectors are in the k-space generated by $v_1, \ldots, v_n$; hence $v_1, \ldots, v_n$ generate the n-dimensional space $k'^{(n)}$ over k'.) Let $E = k^{(n)}$. Let L_A be the linear map of E determined by A. Let L'_A be the linear map of $k'^{(n)}$ determined by A. The matrix of L_A with respect to our given basis is the same as the matrix of L'_A. We can select the basis corresponding to the decomposition

$$E = E_1 \oplus \cdots \oplus E_r$$

determined by the invariants $q_1, \ldots, q_r$. It follows that the invariants don't change when we lift the basis to one of $k'^{(n)}$.

COROLLARY 2. *Let A, B be $n \times n$ matrices over a field k and let k' be an extension field of k. Assume that there is an invertible matrix C' in k' such that $B = C'AC'^{-1}$. Then there is an invertible matrix C in k such that $B = CAC^{-1}$.*

Proof. Exercise.

The structure theorem for modules over principal rings gives us two kinds of decompositions. One is according to the invariants of the preceding theorem. The other is according to prime powers.

Let $E \neq 0$ be a finite dimensional space over the field k, and let $A : E \to E$ be in $\operatorname{End}_k(E)$. Let $q = q_A$ be its minimal polynomial. Then q has a factorization,

$$q = p_1^{e_1} \cdots p_s^{e_s} \qquad (e_i \geqq 1)$$

into prime powers (distinct). Hence E is a direct sum of submodules

$$E = E(p_1) \oplus \cdots \oplus E(p_s),$$

such that each $E(p_i)$ is annihilated by $p_i^{e_i}$. Furthermore, each such submodule can be expressed as a direct sum of submodules isomorphic to $k[t]/(p^e)$ for some irreducible polynomial p and some integer $e \geqq 1$.

THEOREM 7. *Let $q_A(t) = (t - \alpha)^e$ for some $\alpha \in k$, $e \geqq 1$. Assume that E is isomorphic to $k[t]/(q)$. Then E has a basis over k such that the matrix of A relative to this basis is of type*

$$\begin{pmatrix} \alpha & 0 & \cdots & & 0 \\ 1 & \alpha & & & 0 \\ \vdots & & \ddots & & \vdots \\ 0 & & \ddots & \ddots & 0 \\ 0 & \cdots & & 1 & \alpha \end{pmatrix}.$$

Proof. Since E is isomorphic to $k[t]/q$, there exists an element $v \in E$ such that $k[t]v = E$. This element corresponds to the unit element of $k[t]$ in the isomorphism

$$k[t]/(q) \to E.$$

We contend that the elements

$$v, (t - \alpha)v, \ldots, (t - \alpha)^{e-1}v,$$

or equivalently,

$$v, (A - \alpha)v, \ldots, (A - \alpha)^{e-1}v,$$

form a basis for E over k. They are linearly independent over k because any relation of linear dependence would yield a relation of linear dependence between

$$v, Av, \ldots, A^{e-1}v,$$

and hence would yield a polynomial $g(t)$ of degree less than $\deg q$ such that $g(A) = 0$. Since $\dim E = e$, it follows that our elements form a basis for E over k. But $(A - \alpha)^e = 0$. It is then clear from the definitions that the matrix of A with respect to this basis has the shape stated in our theorem.

COROLLARY. *Let k be algebraically closed, and let E be a finite dimensional non-zero vector space over k. Let $A \in \operatorname{End}_k(E)$. Then there exists a basis of E over k such that the matrix of A with respect to this basis consists of blocks, and each block is of the type described in the theorem.*

A matrix having the form described in the preceding corollary is said to be in *Jordan canonical form.*

Remark. A matrix (or an endomorphism) N is said to be *nilpotent* if there exists an integer $d > 0$ such that $N^d = 0$. We see that in the decomposition of Theorem 7, or its corollary, the matrix M is written in the form

$$M = B + N$$

where N is nilpotent. In fact, N is a triangular matrix (i.e. it has zero coefficients on and above the diagonal), and B is a diagonal matrix, whose diagonal elements are the roots of the minimal polynomial. Such a decomposition can always be achieved whenever the field k is such that all the roots of the minimal polynomial lie in k. We observe also that the only case when the matrix N is 0 is when all the roots of the minimal polynomial have multiplicity 1. In this case, if $n = \dim E$, then the matrix M is a diagonal matrix, with n distinct elements on the diagonal.

§4. The characteristic polynomial

Let k be a commutative ring and E a free module of dimension n over k. We consider the polynomial ring $k[t]$, and a linear map $A : E \to E$. We have a homomorphism

$$k[t] \to k[A]$$

as before, mapping a polynomial $f(t)$ on $f(A)$, and E becomes a module over the ring $R = k[t]$. Let M be any $n \times n$ matrix in R (for instance the matrix of A relative to a basis of E). We define the *characteristic polynomial* $P_M(t)$ to be the determinant

$$\det(tI_n - M)$$

where I_n is the unit $n \times n$ matrix. It is an element of $k[t]$. Furthermore, if N is an invertible matrix in R, then

$$\det(tI_n - N^{-1}MN) = \det(N^{-1}(tI_n - M)N) = \det(tI_n - M).$$

Hence the characteristic polynomial of $N^{-1}MN$ is the same as that of M. We may therefore define the characteristic polynomial of A, and denote by P_A, the characteristic polynomial of any matrix M associated with A with respect to some basis. (If $E = 0$, we *define the characteristic polynomial to be* 1.)

If $\varphi : k \to k'$ is a homomorphism of commutative rings, and M is an $n \times n$ matrix in k, then it is clear that

$$P_{\varphi M}(t) = \varphi P_M(t)$$

where φP_M is obtained from P_M by applying φ to the coefficients of P_M.

THEOREM 8. (Cayley-Hamilton) *We have* $P_A(A) = 0$.

Proof. Let $\{v_1, \ldots, v_n\}$ be a basis of E over k. Then

$$tv_j = \sum_{i=1}^{n} a_{ij}v_i$$

where $(a_{ij}) = M$ is the matrix of A with respect to the basis. Let $B(t)$ be the matrix $tI_n - M$. Then $B(t)$ is a matrix with coefficients in $k[t]$. Let $\widetilde{B}(t)$ be the matrix with coefficients in $k[t]$, defined in Chapter XIII, such that

$$\widetilde{B}(t)B(t) = P_A(t)I_n.$$

Then

$$\widetilde{B}(t)B(t)\begin{pmatrix} v_1 \\ \vdots \\ v_n \end{pmatrix} = \begin{pmatrix} P_A(t)v_1 \\ \vdots \\ P_A(t)v_n \end{pmatrix} = \begin{pmatrix} 0 \\ \vdots \\ 0 \end{pmatrix}$$

because

$$B(t)\begin{pmatrix} v_1 \\ \vdots \\ v_n \end{pmatrix} = \begin{pmatrix} 0 \\ \vdots \\ 0 \end{pmatrix}.$$

Hence $P_A(t)E = 0$, and therefore $P_A(A)E = 0$. This means that $P_A(A) = 0$, as was to be shown.

Assume now that k is a field. Let E be a finite dimensional vector space over k, and let $A \in \text{End}_k(v)$. By an *eigenvector* w of A in E one means an element $w \in E$, such that there exists an element $\lambda \in k$ for which $Aw = \lambda w$. If $w \neq 0$, then λ is determined uniquely, and is called an *eigenvalue* of A. Of course, distinct eigenvectors may have the same eigenvalue.

THEOREM 9. *The eigenvalues of A are precisely the roots of the characteristic polynomial of A.*

Proof. Let λ be an eigenvalue. Then $A - \lambda I$ is not invertible in $\text{End}_k(E)$, and hence $\det(A - \lambda I) = 0$. Hence λ is a root of P_A. The arguments are reversible, so we also get the converse.

For simplicity of notation, we often write $A - \lambda$ instead of $A - \lambda I$.

THEOREM 10. *Let $w_1, \ldots, w_m$ be non-zero eigenvectors of A, having distinct eigenvalues. Then they are linearly independent.*

Proof. Suppose that we have

$$a_1w_1 + \cdots + a_mw_m = 0$$

with $a_i \in k$, and let this be a shortest relation with not all $a_i = 0$ (assuming such exists). Then $a_i \neq 0$ for all i. Let $\lambda_1, \ldots, \lambda_m$ be the eigenvalues of our vectors. Apply $A - \lambda_1$ to the above relation. We get

$$a_2(\lambda_2 - \lambda_1)w_2 + \cdots + a_m(\lambda_m - \lambda_1)w_m = 0,$$

which shortens our relation, contradiction.

COROLLARY. *If A has n distinct eigenvalues $\lambda_1, \ldots, \lambda_n$ belonging to eigenvectors $v_1, \ldots, v_n$, and* $\dim E = n$, *then $\{v_1, \ldots, v_n\}$ is a basis for E. The matrix of A with respect to this basis is the diagonal matrix:*

$$\begin{pmatrix} \lambda_1 & & & 0 \\ & \lambda_2 & & \\ & & \ddots & \\ 0 & & & \lambda_n \end{pmatrix}.$$

Warning. It is not always true that there exists a basis of E consisting of eigenvectors!

Remark. Let k be a subfield of k'. If M is a matrix in k, we can define its characteristic polynomial with respect to k, and also with respect to k'. It is clear that the characteristic polynomials thus obtained are equal. If E is a vector space over k, we shall see later how to extend it to a vector space over k'. A linear map A extends to a linear map of the extended space, and the characteristic polynomial of the linear map does not change either. Actually, if we select a basis for E over k, then $E \approx k^{(n)}$, and $k^{(n)} \subset k'^{(n)}$ in a natural way. Thus selecting a basis allows us to extend the vector space, but this seems to depend on the choice of basis. We shall give an invariant definition later.

Let $E = E_1 \oplus \cdots \oplus E_r$ be an expression of E as a direct sum of vector spaces over k. Let $A \in \operatorname{End}_k(E)$, and assume that $AE_i \subset E_i$ for all $i = 1, \ldots, r$. Then A induces a linear map on E_i. We can select a basis for E consisting of bases for $E_1, \ldots, E_r$, and then the matrix for A consists of blocks. Hence we see that

$$P_A(t) = \prod_{i=1}^{r} P_{A_i}(t).$$

Thus the characteristic polynomial is multiplicative on direct sums.

Our condition above that $AE_i \subset E_i$ can also be formulated by saying that E is expressed as a $k[A]$-direct sum of $k[A]$-submodules, or also a $k[t]$-direct sum of $k[t]$-submodules. We shall apply this to the decomposition of E given in Theorem 6.

THEOREM 11. *Let E be a finite dimensional vector space over a field k, let $A \in \operatorname{End}_k(E)$, and let $q_1, \ldots, q_r$ be the invariants of (E, A). Then $P_A(t) = q_1(t) \cdots q_r(t)$.*

Proof. We assume that $E = k^{(n)}$ and that A is represented by a matrix M. We have seen that the invariants do not change when we extend k to a larger field, and neither does the characteristic polynomial. Hence we may assume that k is algebraically closed. In view of Theorem 6 we may assume that M has a single invariant q. Write

$$q(t) = (t - \alpha_1)^{e_1} \cdots (t - \alpha_s)^{e_s}$$

with distinct $\alpha_1, \ldots, \alpha_s$. We view M as a linear map, and split our vector space further into a direct sum of submodules (over $k[t]$) having invariants

$$(t - \alpha_1)^{e_1}, \ldots, (t - \alpha_s)^{e_s}$$

respectively (this is the prime power decomposition). For each one of these submodules, we can select a basis so that the matrix of the induced linear map has the shape described in Theorem 7. From this it is immediately clear that the characteristic polynomial of the map having invariant $(t - \alpha)^e$ is precisely $(t - \alpha)^e$, and our theorem is proved.

COROLLARY. *The minimal polynomial of A and its characteristic polynomial have the same irreducible factors.*

Proof. Because q_r is the minimal polynomial, by Theorem 6.

We shall generalize our remark concerning the multiplicativity of the characteristic polynomial over direct sums.

THEOREM 12. *Let k be a commutative ring, and in the following diagram,*

$$\begin{array}{ccccccccc} 0 & \to & E' & \to & E & \to & E'' & \to & 0 \\ & & {\scriptstyle A'}\downarrow & & {\scriptstyle A}\downarrow & & {\scriptstyle A''}\downarrow & & \\ 0 & \to & E' & \to & E & \to & E'' & \to & 0 \end{array}$$

let the rows be exact sequences of free modules over k, of finite dimension, and let the vertical maps be k-linear maps making the diagram commutative. Then

$$P_A(t) = P_{A'}(t)P_{A''}(t).$$

Proof. We may assume that E' is a submodule of E. We select a basis $\{v_1, \ldots, v_m\}$ for E'. Let $\{\bar{v}_{m+1}, \ldots, \bar{v}_n\}$ be a basis for E'', and let $v_{m+1}, \ldots, v_n$ be elements of E mapping on $\bar{v}_{m+1}, \ldots, \bar{v}_n$ respectively. Then $\{v_1, \ldots, v_m, v_{m+1}, \ldots, v_n\}$ is a basis for E (same proof as Theorem 3 of Chapter III, §5) and we are in the situation discussed in §1. The matrix for A has the shape

$$\begin{pmatrix} M' & * \\ 0 & M'' \end{pmatrix}$$

where M' is the matrix for A' and M'' is the matrix for A''. Taking the characteristic polynomial with respect to this matrix obviously yields our multiplicative property.

THEOREM 13. *Let k be a commutative ring, and E a free module of dimension n over k. Let $A \in \mathrm{End}_k(E)$. Let*

$$P_A(t) = t^n + c_{n-1}t^{n-1} + \cdots + c_0.$$

Then

$$\mathrm{tr}(A) = -c_{n-1} \quad \text{and} \quad \det(A) = (-1)^n c_0.$$

Proof. For the determinant, we observe that $P_A(0) = c_0$. Substituting $t = 0$ in the definition of the characteristic polynomial by the determinant shows that $c_0 = (-1)^n \det(A)$.

For the trace, let M be the matrix representing A with respect to some basis, $M = (a_{ij})$. We consider the determinant $\det(tI_n - a_{ij})$. In its expansion according to the first column, it will contain a diagonal term

$$(t - a_{11}) \cdots (t - a_{nn}),$$

which will give a contribution to the coefficient of t^{n-1} equal to

$$-(a_{11} + \cdots + a_{nn}).$$

No other term in this expansion will give a contribution to the coefficient of t^{n-1}, because the power of t occurring in another term will be at most t^{n-2}. This proves our assertion concerning the trace.

COROLLARY. *Let the notation be as in Theorem* 12. *Then*

$$\mathrm{tr}(A) = \mathrm{tr}(A') + \mathrm{tr}(A'') \quad \text{and} \quad \det(A) = \det(A')\det(A'').$$

Proof. Clear.

We shall now interpret our results in the Euler-Grothendieck group.

Let k be a commutative ring. We consider the category whose objects are pairs (E, A), where E is a k-module, and $A \in \mathrm{End}_k(E)$. We define a morphism

$$(E', A') \to (E, A)$$

to be a k-linear map $E' \xrightarrow{f} E$ making the following diagram commutative:

$$\begin{array}{ccc} E' & \xrightarrow{f} & E \\ {\scriptstyle A'}\downarrow & & \downarrow{\scriptstyle A} \\ E' & \xrightarrow[f]{} & E \end{array}$$

Then we can define the kernel of such a morphism to be again a pair. Indeed, let E'_0 be the kernel of $f : E' \to E$. Then A' maps E'_0 into itself because $fA'E'_0 = A'fE'_0 = 0$. We let A'_0 be the restriction of A' on E'_0. The pair (E'_0, A'_0) is defined to be the kernel of our morphism.

We shall denote by f again the morphism of the pair $(E', A') \to (E, A)$. We can speak of an exact sequence

$$(E', A') \to (E, A) \to (E'', A''),$$

meaning that the induced sequence

$$E' \to E \to E''$$

is exact. We also write 0 instead of $(0, 0)$, according to our universal convention to use the symbol 0 for all things which behave like a zero element.

We observe that our pairs now behave formally like modules, and they in fact form an abelian category.

Assume that k is a field. Let $\mathfrak{A}$ consist of all pairs (E, A) where E is finite dimensional over k. *Then Theorem* 12 *asserts that the characteristic polynomial is an Euler-Poincaré map defined for each object in our category* $\mathfrak{A}$, *with values into the multiplicative monoid of polynomials with leading coefficient* 1. Since the values of the map are in a monoid, this generalizes slightly the notion of Chapter IV, when we took the values in a group. Of course, when k is a field, which is the most frequent application, we can view the values of our map to be in the multiplicative group of non-zero rational functions, so our previous situation applies.

A similar remark holds now for the trace and the determinant. *If k is a field, the trace is an Euler map into the additive group of the field, and the determinant is an Euler map into the multiplicative group of the field.* We note also that all these maps (like all Euler maps) are defined on the isomorphism classes of pairs, and are defined on the Euler-Grothendieck group.

THEOREM 14. *Let k be a commutative ring, M an $n \times n$ matrix in k, and f a polynomial in $k[t]$. Assume that $P_M(t)$ has a factorization,*

$$P_M(t) = \prod_{i=1}^{n} (t - \alpha_i)$$

into linear factors over k. Then the characteristic polynomial of $f(M)$ is given by

$$P_{f(M)}(t) = \prod_{i=1}^{n} (t - f(\alpha_i)),$$

and

$$\operatorname{tr}(f(M)) = \sum_{i=1}^{n} f(\alpha_i), \qquad \det(f(M)) = \prod_{i=1}^{n} f(\alpha_i).$$

Proof. Assume first that k is a field. Then using the canonical decomposition in terms of matrices given in Theorem 7, §3, we find that our assertion is immediately obvious. When k is a ring, we use the usual substitution argument. It is however necessary to know that if $X = (x_{ij})$ is a matrix with algebraically independent coefficients over $\mathbf{Z}$, then $P_X(t)$ has n distinct roots $y_1, \ldots, y_n$ [in an algebraic closure of $\mathbf{Q}(X)$] and that we have a homomorphism

$$Z[x_{ij}, y_1, \ldots, y_n] \to k$$

mapping X on M and $y_1, \ldots, y_n$ on $\alpha_1, \ldots, \alpha_n$. This is obvious to the reader who has read the chapter on integral ring extensions, and the reader who has not can forget about this part of the theorem.

Exercises

1. Let T be an upper triangular square matrix over a commutative ring (i.e. all the elements below and on the diagonal are 0). Show that T is nilpotent.

2. Carry out explicitly the proof that the determinant of a matrix

$$\begin{pmatrix} M_1 & & & & \\ 0 & M_2 & & & ** \\ 0 & 0 & \ddots & & * \\ \vdots & \vdots & & \ddots & \\ 0 & 0 & \cdots & 0 & M_s \end{pmatrix}$$

where each M_i is a square matrix, is equal to the product of the determinants of the matrices $M_1, \ldots, M_s$.

3. Let k be a commutative ring, and let M, M' be square $n \times n$ matrices in k. Show that the characteristic polynomials of MM' and $M'M$ are equal.

4. Show that the eigenvalues of the matrix

$$\begin{pmatrix} 0 & 1 & 0 & 0 \\ 0 & 0 & 1 & 0 \\ 0 & 0 & 0 & 1 \\ 1 & 0 & 0 & 0 \end{pmatrix}$$

in the complex numbers are $\pm 1, \pm i$.

5. Let M, M' be square matrices over a field k. Let q, q' be their respective minimal polynomials. Show that the minimal polynomial of

$$\begin{pmatrix} M & 0 \\ 0 & M' \end{pmatrix}$$

is the least common multiple of q, q'.

6. Let A be a nilpotent endomorphism of a finite dimensional vector space E over the field k. Show that $\operatorname{tr}(A) = 0$.

7. Let R be a principal entire ring. Let E be a free module over R, and let $E^* = \mathrm{Hom}_R(E, R)$ be its dual module. Then E^* is free of dimension n. Let F be a submodule of E. Show that $E^*/F^\perp$ can be viewed as a submodule of F^*, and that its invariants are the same as the invariants of F in E.

8. Let E be a finite dimensional vector space over a field k. Let $A \in \mathrm{Aut}_k(E)$. Show that the following conditions are equivalent:

(i) $A = I + N$, with N nilpotent.

(ii) There exists a basis of E such that the matrix of A with respect to this basis has all its diagonal elements equal to 1 and all elements above the diagonal equal to 0.

(iii) All roots of the characteristic polynomial of A (in the algebraic closure of k) are equal to 1.

9. Let k be a field of characteristic 0, and let M be an $n \times n$ matrix in k. Show that M is nilpotent if and only if $\mathrm{tr}(M^\nu) = 0$ for $1 \leqq \nu \leqq n$.

10. Generalize Theorem 14 to rational functions (instead of polynomials), assuming that k is a field.

11. Let E be a finite dimensional space over the field k. Let $\alpha \in k$. Let E_α be the subspace of E generated by all eigenvectors of a given endomorphism A of E, having α as an eigenvalue. Show that every non-zero element of E_α is an eigenvector of A having α as an eigenvalue.

12. Let E be finite dimensional over the field k. Let $A \in \mathrm{End}_k(E)$. Let v be an eigenvector for A. Let $B \in \mathrm{End}_k(E)$ be such that $AB = BA$. Show that Bv is also an eigenvector for A (if $Bv \neq 0$), with the same eigenvalue.

Diagonalizable endomorphisms. Let E be a finite dimensional vector space over a field k, and let $S \in \mathrm{End}_k(E)$. We say that S is *diagonalizable* if there exists a basis of E consisting of eigenvectors of S. The matrix of S with respect to this basis is then a diagonal matrix.

13. (a) If S is diagonalizable, then its minimal polynomial over k is of type $q(t) = \prod_{i=1}^{m} (t - \lambda_i)$, where $\lambda_1, \ldots, \lambda_m$ are distinct elements of k.

(b) Conversely, if the minimal polynomial of S is of the preceding type, then S is diagonalizable. [*Hint:* The space can be decomposed as a direct sum of the subspaces E_{λ_1} annihilated by $S - \lambda_1$.]

(c) If S is diagonalizable, and if F is a subspace of E such that $SF \subset F$, show that S is diagonalizable as an endomorphism of F, i.e. that F has a basis consisting of eigenvectors of S.

(d) Let S, T be endomorphisms of E, and assume that S, T commute. Assume that both S, T are diagonalizable. Show that they are simultaneously diagonalizable, i.e. there exists a basis of E consisting of eigenvectors for both S and T. [*Hint:* If λ is an eigenvalue of S, and E_λ is the subspace of E consisting of all vectors v such that $Sv = \lambda v$, then $TE_\lambda \subset E_\lambda$.]

14. Let E be a finite dimensional vector space over an algebraically closed field k. Let $A \in \mathrm{End}_k(E)$. Show that A can be written in a unique way as a sum

$$A = S + N$$

where S is diagonalizable, N is nilpotent, and $SN = NS$. Show that S, N can be expressed as polynomials in A. [*Hint:* Let $P_A(t) = \prod(t - \lambda_i)^{m_i}$ be the factorization of $P_A(t)$ with distinct λ_i. Let E_i be the kernel of $(A - \lambda_i)^{m_i}$. Then E is the direct sum of the E_i. Define S on E so that on E_i, $Sv = \lambda_i v$ for all $v \in E_i$. Let $N = A - S$. Show that S, N satisfy our requirements. To get S as a polynomial in A, let g be a polynomial such that $g(t) \equiv \lambda_i \bmod (t - \lambda_i)^{m_i}$ for all i, and $g(t) \equiv 0 \bmod t$. Then $S = g(A)$ and $N = A - g(A)$.]

15. After you have read the section on the tensor product of vector spaces, you can easily do the following exercise. Let E, F be finite dimensional vector spaces over an algebraically closed field k, and let $A : E \to E$ and $B : F \to F$ be k-endomorphisms of E, F respectively. Let

$$P_A(t) = \prod(t - \alpha_i)^{n_i} \qquad \text{and} \qquad P_B(t) = \prod(t - \beta_j)^{m_j}$$

be the factorizations of their respectively characteristic polynomials, into distinct linear factors. Then

$$P_{A \otimes B}(t) = \prod_{i,j} (t - \alpha_i\beta_j)^{n_i m_j}.$$

[*Hint:* Decompose E into the direct sum of subspaces E_i, where E_i is the subspace of E annihilated by some power of $A - \alpha_i$. Do the same for F, getting a decomposition into a direct sum of subspaces F_j. Then show that some power of $A \otimes B - \alpha_i\beta_j$ annihilates $E_i \otimes F_j$. Use the fact that $E \otimes F$ is the direct sum of the subspaces $E_i \otimes F_j$, and that $\dim_k(E_i \otimes F_j) = n_i m_j$.]

16. Let Γ be a free abelian group of dimension $n \geqq 1$. Let Γ' be a subgroup of dimension n also. Let $\{v_1, \ldots, v_n\}$ be a basis of Γ, and let $\{w_1, \ldots, w_n\}$ be a basis of Γ'. Write

$$w_i = \sum a_{ij} v_j.$$

Show that the index $(\Gamma : \Gamma')$ is equal to the absolute value of the determinant of the matrix (a_{ij}).

17. Prove the normal basis theorem for finite extensions of a finite field.

18. Let $A = (a_{ij})$ be a square $n \times n$ matrix over a commutative ring k. Let A_{ij} be the matrix obtained by deleting the i-th row and j-th column from A. Let $b_{ij} = (-1)^{i+j}\det(A_{ji})$, and let B be the matrix (b_{ij}). Show that $\det(B) = \det(A)^{n-1}$, by reducing the problem to the case when A is a matrix with variable coefficients over the integers. Use this same method to give an alternative proof of the Cayley-Hamilton theorem, that $P_A(A) = 0$.

19. Let (E, A) and (E', A') be pairs consisting of a finite-dimensional vector space over a field k, and a k-endomorphism. Show that these pairs are isomorphic if and only if their invariants are equal.

20. Let E be a finite dimensional vector space over an algebraically closed field k. Let A, B be k-endomorphisms of E which commute, i.e. $AB = BA$. Show that A and B have a common eigenvector. [*Hint:* Consider a subspace consisting of all vectors having a fixed element of k as eigenvalue.]

CHAPTER XVI

Multilinear Products

§1. Tensor product

Let k be a commutative ring. If $E_1, \dots, E_n, F$ are modules, we denote by

$$L^n(E_1, \dots, E_n; F)$$

the module of n-multilinear maps

$$f : E_1 \times \cdots \times E_n \to F.$$

We recall that a multilinear map is a map which is linear (i.e. k-linear) in each variable. We use the words "linear" and "homomorphism" interchangeably. *Unless otherwise specified, modules, homomorphisms, linear, multilinear refer to the ring k.*

One may view the multilinear maps of a fixed set of modules $E_1, \dots, E_n$ as the objects of a category. Indeed, if

$$f : E_1 \times \cdots \times E_n \to F \quad \text{and} \quad g : E_1 \times \cdots \times E_n \to G$$

are multilinear, we define a morphism $f \to g$ to be a homomorphism $h : F \to G$ which makes the following diagram commutative:

$$\begin{array}{ccc} & \overset{f}{\nearrow} & F \\ E_1 \times \cdots \times E_n & & \downarrow h \\ & \underset{g}{\searrow} & G \end{array}$$

A universal object in this category is called a *tensor product* of $E_1, \dots, E_n$ (over k).

We shall now prove that a tensor product exists, and in fact construct one in a natural way. By abstract nonsense, we know of course that a tensor product is uniquely determined, up to a unique isomorphism.

Let M be the free module generated by the set of all n-tuples $(x_1, \dots, x_n)$, $(x_i \in E_i)$, i.e. generated by the set $E_1 \times \cdots \times E_n$. Let N be the submodule generated by all the elements of the following type:

$$(x_1, \dots, x_i + x'_i, \dots, x_n) - (x_1, \dots, x_i, \dots, x_n) - (x_1, \dots, x'_i, \dots, x_n)$$

$$(x_1, \dots, ax_i, \dots, x_n) - a(x_1, \dots, x_n)$$

for all $x_i \in E_i$, $x_i' \in E_i$, $a \in k$. We have the canonical injection

$$E_1 \times \cdots \times E_n \to M$$

of our set into the free module generated by it. We compose this map with the canonical map $M \to M/N$ on the factor module, to get a map

$$\varphi : E_1 \times \cdots \times E_n \to M/N.$$

We contend that φ is multilinear and is a tensor product.

It is obvious that φ is multilinear—our definition was adjusted to this purpose. Let

$$f : E_1 \times \cdots \times E_n \to G$$

be a multilinear map. By the definition of free module generated by $E_1 \times \cdots \times E_n$ we have an induced linear map $M \to G$ which makes the following diagram commutative:

$$\begin{array}{ccc} & & M \\ & \nearrow & \downarrow \\ E_1 \times \cdots \times E_n & & \\ & \searrow_f & G \end{array}$$

Since f is multilinear, the induced map $M \to G$ takes on the value 0 on N. Hence by the universal property of factor modules, it can be factored through M/N, and we have a homomorphism $f_* : M/N \to G$ which makes the following diagram commutative:

$$\begin{array}{ccc} & & M/N \\ & \nearrow^{\varphi} & \downarrow f_* \\ E_1 \times \cdots \times E_n & & \\ & \searrow_f & G \end{array}$$

Since the image of φ generates M/N, it follows that the induced map f_* is uniquely determined. This proves what we wanted.

The module M/N will be denoted by $E_1 \otimes \cdots \otimes E_n$ or also $\bigotimes_{i=1}^{n} E_i$. We have constructed a specific tensor product in the isomorphism class of tensor products, and we shall call it THE tensor product of $E_1, \ldots, E_n$. If $x_i \in E_i$, we write

$$\varphi(x_1, \ldots, x_n) = x_1 \otimes \cdots \otimes x_n = x_1 \otimes_k \cdots \otimes_k x_n.$$

We have for all i,

$$x_1 \otimes \cdots \otimes ax_i \otimes \cdots \otimes x_n = a(x_1 \otimes \cdots \otimes x_n),$$

$$x_1 \otimes \cdots \otimes (x_i + x_i') \otimes \cdots \otimes x_n$$
$$= (x_1 \otimes \cdots \otimes x_n) + (x_1 \otimes \cdots \otimes x_i' \otimes \cdots \otimes x_n)$$

for $x_i, x_i' \in E_i$ and $a \in k$.

If we have two factors, say $E \otimes F$, then every element of $E \otimes F$ can be written as a sum of terms $x \otimes y$ with $x \in E$ and $y \in F$, because such terms generate $E \otimes F$ over k, and $a(x \otimes y) = ax \otimes y$ for $a \in k$.

Warning. The tensor product can involve a great deal of collapsing between the modules. For instance, take the tensor product over $\mathbf{Z}$ of $\mathbf{Z}/m\mathbf{Z}$ and $\mathbf{Z}/n\mathbf{Z}$ where m, n are integers > 1 and are relatively prime. Then the tensor product

$$\mathbf{Z}/n\mathbf{Z} \otimes \mathbf{Z}/m\mathbf{Z} = 0.$$

Indeed, we have $n(x \otimes y) = (nx) \otimes y = 0$ and $m(x \otimes y) = x \otimes my = 0$. Hence $x \otimes y = 0$ for all $x \in \mathbf{Z}/n\mathbf{Z}$ and $y \in \mathbf{Z}/m\mathbf{Z}$. Elements of type $x \otimes y$ generate the tensor product, which is therefore 0. We shall see later conditions under which there is no collapsing.

In many subsequent results, we shall assert the existence of certain linear maps from a tensor product. This existence is proved by using the universal mapping property of bilinear maps factoring through the tensor product. The uniqueness follows by prescribing the value of the linear maps on elements of type $x \otimes y$ (say for two factors) since such elements generate the tensor product.

We shall prove the associativity of the tensor product.

PROPOSITION 1. *Let E_1, E_2, E_3 be modules. Then there exists a unique isomorphism*

$$(E_1 \otimes E_2) \otimes E_3 \to E_1 \otimes (E_2 \otimes E_3)$$

such that

$$(x \otimes y) \otimes z \mapsto x \otimes (y \otimes z)$$

for $x \in E_1$, $y \in E_2$ and $z \in E_3$.

Proof. Since elements of type $(x \otimes y) \otimes z$ generate the tensor product, the uniqueness of the desired linear map is obvious. To prove its existence, let $x \in E_1$. The map

$$\lambda_x : E_2 \times E_3 \to (E_1 \otimes E_2) \otimes E_3$$

such that $\lambda_x(y, z) = (x \otimes y) \otimes z$ is clearly bilinear, and hence factors through a linear map of the tensor product

$$\bar{\lambda}_x : E_2 \otimes E_3 \to (E_1 \otimes E_2) \otimes E_3.$$

The map

$$E_1 \times (E_2 \otimes E_3) \to (E_1 \otimes E_2) \otimes E_3$$

such that

$$(x, \alpha) \mapsto \bar{\lambda}_x(\alpha)$$

for $x \in E_1$ and $\alpha \in E_2 \otimes E_3$ is then obviously bilinear, and factors through a linear map

$$E_1 \otimes (E_2 \otimes E_3) \to (E_1 \otimes E_2) \otimes E_3,$$

which has the desired property (clear from its construction).

PROPOSITION 2. *Let E, F be modules. Then there is a unique isomorphism*

$$E \otimes F \to F \otimes E$$

such that $x \otimes y \mapsto y \otimes x$ for $x \in E$ and $y \in F$.

Proof. The map $E \times F \to F \otimes E$ such that $(x, y) \mapsto y \otimes x$ is bilinear, and factors through the tensor product $E \otimes F$, sending $x \otimes y$ on $y \otimes x$. Since this last map has an inverse (by symmetry) we obtain the desired isomorphism.

The tensor product has various functorial properties. First, suppose that

$$f_i : E'_i \to E_i \qquad (i = 1, \ldots, n)$$

is a collection of linear maps. We get an induced map on the product,

$$\prod f_i : \prod E'_i \to \prod E_i.$$

If we compose $\prod f_i$ with the canonical map into the tensor product, then we get an induced linear map which we may denote by $T(f_1, \ldots, f_n)$ which makes the following diagram commutative:

$$\begin{array}{ccc} E'_1 \times \cdots \times E'_n & \longrightarrow & E'_1 \otimes \cdots \otimes E'_n \\ \prod f_i \downarrow & & \downarrow T(f_1, \ldots, f_n) \\ E_1 \times \cdots \times E_n & \longrightarrow & E_1 \otimes \cdots \otimes E_n \end{array}$$

It is immediately verified that T is functorial, namely that if we have a composite of linear maps $f_i \circ g_i$ $(i = 1, \ldots, n)$ then

$$T(f_1 \circ g_1, \ldots, f_n \circ g_n) = T(f_1, \ldots, f_n) \circ T(g_1, \ldots, g_n)$$

and

$$T(id, \ldots, id) = id.$$

We observe that $T(f_1, \ldots, f_n)$ is the unique linear map whose effect on an element $x'_1 \otimes \cdots \otimes x'_n$ of $E'_1 \otimes \cdots \otimes E'_n$ is

$$x'_1 \otimes \cdots \otimes x'_n \mapsto f_1(x'_1) \otimes \cdots \otimes f_n(x'_n).$$

We may view T as a map

$$\prod_{i=1}^{n} L(E'_i, E_i) \to L\left(\bigotimes_{i=1}^{n} E'_i, \bigotimes_{i=1}^{n} E_i\right),$$

and the reader will have no difficulty in verifying that this map is multilinear. We shall write out what this means explicitly for two factors, so that our map can be written

$$(f, g) \mapsto T(f, g).$$

Given homomorphisms $f : F' \to F$ and $g_1, g_2 : E' \to E$, then

$$T(f, g_1 + g_2) = T(f, g_1) + T(f, g_2),$$
$$T(f, ag_1) = aT(f, g_1).$$

In particular, select a fixed module F, and consider the functor $\tau = \tau_F$ (from modules to modules) such that

$$\tau(E) = F \otimes E.$$

Then τ gives rise to a linear map

$$\tau : L(E', E) \to L(\tau(E'), \tau(E))$$

for each pair of modules E', E, by the formula

$$\tau(f) = T(id, f).$$

Remark. By abuse of notation, it is sometimes convenient to write

$$f_1 \otimes \cdots \otimes f_n \qquad \text{instead of} \qquad T(f_1, \ldots, f_n).$$

This should not be confused with the tensor product of elements taken in the tensor product of the modules

$$L(E'_1, E_1) \otimes \cdots \otimes L(E'_n, E_n).$$

The context will always make our meaning clear.

§2. *Basic properties*

The most basic relation relating linear maps, bilinear maps, and the tensor product is the following: For three modules E, F, G,

$$\boxed{L(E, L(F, G)) \approx L^2(E, F; G) \approx L(E \otimes F, G).}$$

The isomorphisms involved are described in a natural way.

(i) $L^2(E, F; G) \to L(E, L(F, G))$.

If $f : E \times F \to G$ is bilinear, and $x \in E$, then the map

$$f_x : E \to L(F, G)$$

such that $f_x(y) = f(x, y)$ is linear. Furthermore, the map $x \mapsto f_x$ is linear, and is associated with f to get (i).

(ii) $L(E, L(F, G)) \to L^2(E, F; G)$.

Let $\varphi \in L(E, L(F, G))$. We let $f_\varphi : E \times F \to G$ be the bilinear map such that

$$f_\varphi(x, y) = \varphi(x)(y).$$

Then $\varphi \mapsto f_\varphi$ defines (ii).

It is clear that the homomorphisms of (i) and (ii) are inverse to each other and therefore give isomorphisms of the first two objects in the enclosed box.

(iii) $L^2(E, F; G) \to L(E \otimes F, G)$.

This is the map $f \mapsto f_*$ which associates to each bilinear map f the induced linear map on the tensor product. The association $f \mapsto f_*$ is injective (because f_* is uniquely determined by f), and it is surjective, because any linear map of the tensor product composed with the canonical map $E \times F \to E \otimes F$ gives rise to a bilinear map on $E \times F$.

PROPOSITION 3. *Let $E = \coprod_{i=1}^{n} E_i$ be a direct sum. Then we have an isomorphism*

$$F \otimes E \leftrightarrow \coprod_{i=1}^{n} (F \otimes E_i).$$

Proof. The isomorphism is given by abstract nonsense. We keep F fixed, and consider the functor $\tau : X \mapsto F \otimes X$. As we saw above, τ is linear. We have projections $\pi_i : E \to E$ of E on E_i. Then

$$\pi_i \circ \pi_i = \pi_i, \qquad \pi_i \circ \pi_j = 0 \quad \text{if} \quad i \neq j,$$

$$\sum_{i=1}^{n} \pi_i = id.$$

We apply the functor τ, and see that $\tau(\pi_i)$ satisfies the same relations, hence gives a direct sum decomposition of $\tau(E) = F \otimes E$. Note that $\tau(\pi_i) = id \otimes \pi_i$.

Corollary. *Let I be an indexing set, and $E = \coprod_{i \in I} E_i$. Then we have an isomorphism*

$$\left(\coprod_{i \in I} E_i\right) \otimes F \approx \coprod_{i \in I} (E_i \otimes F).$$

Proof. Let S be a finite subset of I. We have a sequence of maps

$$\left(\coprod_{i \in S} E_i\right) \times F \to \coprod_{i \in S} (E_i \otimes F) \to \coprod_{i \in I} (E_i \otimes F)$$

the first of which is bilinear, and the second is linear, induced by the inclusion of S in I. The first is the obvious map. If $S \subset S'$, then a trivial commutative diagram shows that the restriction of the map

$$\left(\coprod_{i \in S'} E_i\right) \times F \to \coprod_{i \in I} (E_i \otimes F)$$

induces our preceding map on the sum for $i \in S$. But we have an *injection*

$$\left(\coprod_{i \in S} E_i\right) \times F \to \left(\coprod_{i \in S'} E_i\right) \times F.$$

Hence by compatibility, we can define a bilinear map

$$\left(\coprod_{i \in I} E_i\right) \times F \to \coprod_{i \in I} (E_i \otimes F),$$

and consequently a linear map

$$\left(\coprod_{i \in I} E_i\right) \otimes F \to \coprod_{i \in I} (E_i \otimes F).$$

In a similar way, one defines a map in the opposite direction, and it is clear that these maps are inverse to each other, hence give an isomorphism.

Suppose now that E is free, of dimension 1 over k. Let $\{v\}$ be a basis, and consider $F \otimes E$. Every element of $F \otimes E$ can be written as a sum of terms $y \otimes av$ with $y \in F$ and $a \in k$. However, $y \otimes av = ay \otimes v$. In a sum of such terms, we can then use linearity on the left,

$$\sum_{i=1}^{n} (y_i \otimes v) = \left(\sum_{i=1}^{n} y_i\right) \otimes v, \qquad y_i \in F.$$

Hence every element is in fact of type $y \otimes v$ with some $y \in F$.

We have a bilinear map

$$F \times E \to F$$

such that $(y, av) \mapsto ay$, inducing a linear map

$$F \otimes E \mapsto F.$$

We also have a linear map $F \to F \otimes E$ given by $y \mapsto y \otimes v$. It is clear that these maps are inverse to each other, and hence that we have an isomorphism $F \otimes E \approx F$. Thus every element of $F \otimes E$ can be written *uniquely* in the form $y \otimes v$, $y \in F$.

PROPOSITION 4. *Let E be free over k, with basis $\{v_i\}_{i \in I}$. Then every element of $F \otimes E$ has a unique expression of the form*

$$\sum_{i \in I} y_i \otimes v_i, \qquad y_i \in F$$

with almost all $y_i = 0$.

Proof. This follows at once from the discussion of the one-dimensional case, and the corollary of Proposition 1.

COROLLARY. *Let E, F be free over k, with bases $\{v_i\}_{i \in I}$ and $\{w_j\}_{j \in J}$ respectively. Then $E \otimes F$ is free, with basis $\{v_i \otimes w_j\}$. We have*

$$\dim(E \otimes F) = (\dim E)(\dim F).$$

Proof. Immediate from the proposition.

We see that when E is free over k, then there is no collapsing in the tensor product. Every element of $F \otimes E$ can be viewed as a "formal" linear combination of elements in a basis of E with coefficients in F.

In particular, we see that $k \otimes E$ (or $E \otimes k$) is isomorphic to E, under the correspondence $x \mapsto x \otimes 1$.

PROPOSITION 5. *Let E, F be free of finite dimension over k. Then we have an isomorphism*

$$\text{End}_k(E) \otimes \text{End}_k(F) \to \text{End}_k(E \otimes F)$$

which is the unique linear map such that

$$f \otimes g \mapsto T(f, g)$$

for $f \in \text{End}_k(E)$ and $g \in \text{End}_k(F)$.

[We note that the tensor product on the left is here taken in the tensor product of the two modules $\text{End}_k(E)$ and $\text{End}_k(F)$.]

Proof. Let $\{v_i\}$ be a basis of E and let $\{w_j\}$ be a basis of F. Then $\{v_i \otimes w_j\}$ is a basis of $E \otimes F$. For each pair of indices (i', j') there exists a unique endomorphism $f = f_{i,i'}$ of E and $g = g_{j,j'}$ of F such that

$$f(v_i) = v_{i'} \quad \text{and} \quad f(v_\nu) = 0 \text{ if } \nu \neq i$$
$$g(w_j) = w_{j'} \quad \text{and} \quad g(w_\mu) = 0 \text{ if } \mu \neq j.$$

Furthermore, the families $\{f_{i,i'}\}$ and $\{g_{j,j'}\}$ are bases of $\text{End}_k(E)$ and $\text{End}_k(F)$ respectively. Then

$$T(f, g)(v_\nu \otimes w_\mu) = \begin{cases} v_{i'} \otimes w_{j'} & \text{if } (\nu, \mu) = (i, j) \\ 0 & \text{if } (\nu, \mu) \neq (i, j). \end{cases}$$

Thus the family $\{T(f_{i,i'}, g_{j,j'})\}$ is a basis of $\text{End}_k(E \otimes F)$. Since the family $\{f_{i,i'} \otimes g_{j,j'}\}$ is a basis of $\text{End}_k(E) \otimes \text{End}_k(F)$, the assertion of our proposition is now clear.

In Proposition 5, we see that the ambiguity of the tensor sign in $f \otimes g$ is in fact unambiguous in the important special case of free, finite dimensional modules. We shall see later an important application of Proposition 5 when we discuss the tensor algebra of a module.

PROPOSITION 6. *Let*

$$0 \to E' \xrightarrow{\varphi} E \xrightarrow{\psi} E'' \to 0$$

be an exact sequence, and F any module. Then the sequence

$$F \otimes E' \to F \otimes E \to F \otimes E'' \to 0$$

is exact.

Proof. Given $x'' \in E''$ and $y \in F$, there exists $x \in E$ such that $x'' = \psi(x)$, and hence $y \otimes x''$ is the image of $y \otimes x$ under the linear map

$$F \otimes E \to F \otimes E''.$$

Since elements of type $y \otimes x''$ generate $F \otimes E''$, we conclude that the preceding linear map is surjective. One also verifies trivially that the image of $F \otimes E' \to F \otimes E$ is contained in the kernel of

$$F \otimes E \to F \otimes E''.$$

Conversely, let I be the image of $F \otimes E' \to F \otimes E$, and let

$$f : (F \otimes E)/I \to F \otimes E''$$

be the canonical map. We shall define a linear map

$$g : F \otimes E'' \to (F \otimes E)/I$$

such that $g \circ f = id$. This obviously will imply that f is injective, and hence will prove the desired converse.

Let $y \in F$ and $x'' \in E''$. Let $x \in E$ be such that $\psi(x) = x''$. We define a map $F \times E'' \to (F \otimes E)/I$ by letting

$$(y, x'') \mapsto y \otimes x \pmod{I},$$

and contend that this map is well defined, i.e. independent of the choice of x such that $\psi(x) = x''$. If $\psi(x_1) = \psi(x_2) = x''$, then $\psi(x_1 - x_2) = 0$, and by hypothesis, $x_1 - x_2 = \varphi(x')$ for some $x' \in E'$. Then

$$y \otimes x_1 - y \otimes x_2 = y \otimes (x_1 - x_2) = y \otimes \varphi(x').$$

This shows that $y \otimes x_1 \equiv y \otimes x_2 \pmod{I}$, and proves that our map is well defined. It is obviously bilinear, and hence factors through a linear map g, on the tensor product. It is clear that the restriction of $g \circ f$ on elements of type $y \otimes x''$ is the identity. Since these elements generate $F \otimes E''$, we conclude that f is injective, as was to be shown.

It is not always true that the sequence

$$0 \to F \otimes E' \to F \otimes E \to F \otimes E'' \to 0$$

is exact. It is exact if the first sequence in Proposition 3 splits, i.e. if E is essentially the direct sum of E' and E''. This is a trivial consequence of Proposition 3, and the reader should carry out the details to get accustomed to the formalism of the tensor product.

PROPOSITION 7. *Let $\mathfrak{a}$ be an ideal of k. Let E be a module. Then the map $(k/\mathfrak{a}) \times E \to E/\mathfrak{a}E$ induced by*

$$(a, x) \mapsto ax \pmod{\mathfrak{a}E}, \qquad a \in A, x \in E$$

is bilinear and induces an isomorphism

$$(A/\mathfrak{a}) \otimes E \xrightarrow{\approx} E/\mathfrak{a}E.$$

Proof. Our map $(a, x) \mapsto ax \pmod{\mathfrak{a}E}$ clearly induces a bilinear map of $A/\mathfrak{a} \times E$ onto $E/\mathfrak{a}E$, and hence a linear map of $A/\mathfrak{a} \otimes E$ onto $E/\mathfrak{a}E$. We can construct an inverse, for we have a well-defined linear map

$$E \to A/\mathfrak{a} \otimes E$$

such that $x \mapsto \bar{1} \otimes x$ (where $\bar{1}$ is the residue class of 1 in $A/\mathfrak{a}$). It is clear that $\mathfrak{a}E$ is contained in the kernel of this last linear map, and thus that we obtain a homomorphism

$$E/\mathfrak{a}E \to A/\mathfrak{a} \otimes E,$$

which is immediately verified to be inverse to the homomorphism described in the statement of the proposition.

The association $E \mapsto E/\mathfrak{a}E \approx A/\mathfrak{a} \otimes E$ is often called a reduction map. In the next section, we shall interpret this reduction map as an extension of the base.

§3. *Extension of the base*

Let k be a commutative ring and let E be a k-module. We specify k since we are going to work with several rings in a moment. Let $k \to k'$ be a homomorphism of commutative rings, so that k' is a k-algebra, and may be viewed as a k-module also. We have a 3-multilinear map

$$k' \times k' \times E \to k' \otimes E$$

defined by the rule

$$(a, b, x) \mapsto ab \otimes x.$$

This induces therefore a k-linear map

$$k' \otimes (k' \otimes E) \to k' \otimes E$$

and hence a k-bilinear map $k' \times (k' \otimes E) \to k' \otimes E$. It is immediately verified that our last map makes $k' \otimes E$ into a k'-module, which we shall call the *extension of E over k'*, and denote by $E^{k'}$. We also say that $E^{k'}$ is obtained by *extension of the base* ring from k to k'.

Example 1. Let $\mathfrak{a}$ be an ideal of k and let $k \to k/\mathfrak{a}$ be the canonical homomorphism. Then the extension of E to $k/\mathfrak{a}$ is also called the *reduction* of E modulo $\mathfrak{a}$. This happens often over the integers, when we reduce modulo a prime p (i.e. modulo the prime ideal (p)).

Example 2. Let k be a field and k' an extension field. Then E is a vector space over k, and $E^{k'}$ is a vector space over k'. In terms of a basis, we see that our extension gives what was alluded to in the preceding chapter. This example will be expanded in the exercises.

We draw the same diagrams as in field theory:

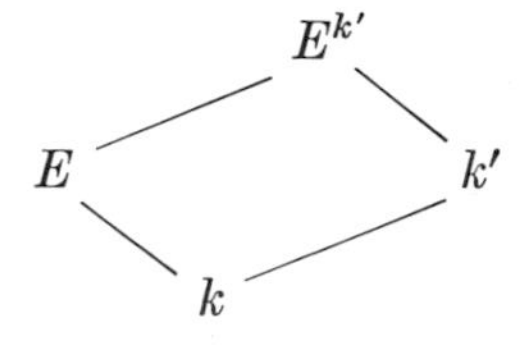

to visualize an extension of the base. From Proposition 2, we conclude:

PROPOSITION 8. *Let E be a free module over k, with basis $\{v_i\}_{i\in I}$. Let $v_i' = 1 \otimes v_i$. Then $E^{k'}$ is a free module over k', with basis $\{v_i'\}_{i\in I}$.*

We had already used a special case of this proposition when we observed that the dimension of a free module is defined, i.e. that two bases have the same cardinality. Indeed, in that case, we reduced modulo a maximal ideal of k to reduce the question to a vector space over a field.

When we start changing rings, it is desirable to indicate k in the notation for the tensor product. Thus we write

$$E^{k'} = k' \otimes E = k' \otimes_k E.$$

Then we have transitivity of the extension of the base, namely if $k \to k' \to k''$ is a succession of homomorphisms of commutative rings, then we have an isomorphism

$$k'' \otimes_k E \approx k'' \otimes_{k'} (k' \otimes_k E)$$

and this isomorphism is one of k''-modules. The proof is trivial and will be left to the reader.

If E has a multiplicative structure, we can extend the base also for this multiplication. Let $k \to A$ be a ring-homomorphism such that every element in the image of k in A commutes with every element in A (i.e. a k-algebra). Let $k \to k'$ be a homomorphism of commutative rings. We have a 4-multilinear map

$$k' \times A \times k' \times A \to k' \otimes A$$

defined by

$$(a, x, b, y) \mapsto ab \otimes xy.$$

We get an induced k-linear map

$$k' \otimes A \otimes k' \otimes A \to k' \otimes A$$

and hence an induced k-bilinear map

$$(k' \otimes A) \times (k' \otimes A) \to k' \otimes A.$$

It is trivially verified that the law of composition on $k' \otimes A$ we have just defined is associative. There is a unit element in $k' \otimes A$, namely $1 \otimes 1$. We have a ring-homomorphism of k' into $k' \otimes A$, given by $a \mapsto a \otimes 1$. In this way one sees at once that $k' \otimes A = A^{k'}$ is a k'-algebra. We note that the map $x \mapsto 1 \otimes x$ is a ring-homomorphism of A into $k' \otimes A$, and that we get a commutative diagram of ring homomorphisms,

$$\begin{array}{ccccc} & & k' \otimes A = A^{k'} & & \\ & \nearrow & & \nwarrow & \\ A & & & & k' \\ & \nwarrow & & \nearrow & \\ & & k & & \end{array}$$

§4. Tensor product of algebras

In the above considerations, the situation was unsymmetric: We could deal with a non-commutative A, but needed a commutative k'. Assume now that we deal with a symmetric situation in which everything is commutative.

PROPOSITION 9. *Coproducts exist in the category of commutative rings, and in the category of commutative algebras over a commutative ring. If $k \to A$ and $k \to B$ are two homomorphisms of commutative rings, then their coproduct over k is the homomorphism $k \to A \otimes B$ given by*

$$a \mapsto a \otimes 1 = 1 \otimes a.$$

Proof. We shall limit our proof to the case of finite coproducts, and hence by induction to the case of the coproduct of two ring homomorphisms $k \to A$ and $k \to B$. (For the infinite case, one needs a limit process, analogous to that used in the proof of the corollary to Proposition 3, and which the reader can work out as an easy exercise.)

Let A, B be commutative rings, and assume given ring-homomorphisms into a commutative ring C,

$$\varphi : A \to C \quad \text{and} \quad \psi : B \to C.$$

Then we can define a $\mathbf{Z}$-bilinear map

$$A \times B \to C$$

by $(x, y) \mapsto \varphi(x)\psi(y)$. From this we get a unique additive homomorphism

$$A \otimes B \to C$$

such that $x \otimes y \mapsto \varphi(x)\psi(y)$. We have seen above that we can define a ring structure on $A \otimes B$, such that

$$(a \otimes b)(c \otimes d) = ac \otimes bd.$$

It is then clear that our map $A \otimes B \to C$ is a ring-homomorphism. We also have two ring-homomorphisms

$$A \xrightarrow{f} A \otimes B \quad \text{and} \quad B \xrightarrow{g} A \otimes B$$

given by

$$x \mapsto x \otimes 1 \quad \text{and} \quad y \mapsto 1 \otimes y.$$

The universal property of the tensor product shows that $(A \otimes B, f, g)$ is a coproduct of our rings A and B.

If A, B, C are k-algebras, and if φ, ψ make the following diagram commutative,

$$\begin{array}{ccc} & C & \\ {}^{\varphi}\nearrow & & \nwarrow{}^{\psi} \\ A & & B \\ \nwarrow & & \nearrow \\ & k & \end{array}$$

then $A \otimes B$ is also a k-algebra (it is in fact an algebra over k, or A, or B, depending on what one wants to use), and the map $A \otimes B \to C$ obtained above gives a homomorphism of k-algebras.

A commutative ring can always be viewed as a $\mathbf{Z}$-algebra (i.e. as an algebra over the integers). Thus one sees the coproduct of commutative rings as a special case of the coproduct of k-algebras.

§5. *The tensor algebra of a module*

Let G be a commutative monoid, written additively. By a *G-graded ring*, we shall mean a ring A, which as an additive group can be expressed as a direct sum

$$A = \coprod_{r \in G} A_r,$$

and such that the ring multiplication maps $A_r \times A_s$ into A_{r+s}, for all $r, s \in G$.

In particular, we see that A_0 is a subring.

The elements of A_r are called the *homogeneous elements of degree* r.

We shall construct several examples of graded rings, according to the following pattern. Suppose given for each $r \in G$ an abelian group A_r (written additively), and for each pair $r, s \in G$ a map $A_r \times A_s \to A_{r+s}$. Assume that composition under these maps is associative and $\mathbf{Z}$-bilinear. Then the direct sum $A = \coprod_{r \in G} A_r$ is a ring: We can define multiplication in the obvious way, namely

$$\left(\sum_{r \in G} x_r\right)\left(\sum_{s \in G} y_s\right) = \sum_{t \in G}\left(\sum_{r+s=t} x_r y_s\right).$$

We shall apply these considerations when G is the monoid of natural numbers $0, 1, 2, \ldots$.

Let k be a commutative ring as before, and let E be a module (i.e. a k-module). For each integer $r \geqq 0$, we let

$$T^r(E) = \bigotimes_{i=1}^{r} E \qquad \text{and} \qquad T^0(E) = k.$$

Thus $T^r(E) = E \otimes \cdots \otimes E$ (tensor product taken r times). Then T^r is a functor, whose effect on linear maps is given as follows. If $f : E \to F$ is a linear map, then

$$T^r(f) = T(f, \dots, f)$$

in the sense of §1.

From the associativity of the tensor product, we obtain a bilinear map

$$T^r(E) \times T^s(E) \to T^{r+s}(E),$$

which is associative. Consequently, by means of this bilinear map, we can define a ring structure on the direct sum

$$T(E) = \coprod_{r=0}^{\infty} T^r(E),$$

and in fact an algebra structure (mapping k on $T^0(E) = k$). We shall call $T(E)$ the *tensor algebra* of E, over k. It is in general *not* commutative. If $x, y \in T(E)$, we shall again write $x \otimes y$ for the ring operation in $T(E)$.

Let $f : E \to F$ be a linear map. Then f induces a linear map

$$T^r(f) : T^r(E) \to T^r(F)$$

for each $r \geqq 0$, and in this way induces a map which we shall denote by $T(f)$ on $T(E)$. (There can be no ambiguity with the map of §1, which should now be written $T^1(f)$, and is in fact equal to f since $T^1(E) = E$.) It is clear that $T(f)$ is the unique linear map such that for $x_1, \dots, x_r \in E$ we have

$$T(f)(x_1 \otimes \cdots \otimes x_r) = f(x_1) \otimes \cdots \otimes f(x_r).$$

Indeed, the elements of $T^1(E) = E$ are algebra-generators of $T(E)$ over k. We see that $T(f)$ is an algebra-homomorphism. Thus T may be viewed as a functor from the category of modules to the category of graded algebras, $T(f)$ being a homomorphism of degree 0.

When E is free and finite dimensional over k, we can determine the structure of $T(E)$ completely, using Proposition 4. Let P be an algebra over k. We shall say that P is a *non-commutative polynomial algebra* if there exist elements $t_1, \dots, t_n \in P$ such that the elements

$$M_{(i)}(t) = t_{i_1} \cdots t_{i_r}$$

with $1 \leqq i_\nu \leqq n$ form a basis of P over k. We may call these elements non-commutative monomials in (t). As usual, by convention, when $r = 0$, the corresponding monomial is the unit element of P. We see that $t_1, \dots, t_n$ generate P as an algebra over k, and that P is in fact a graded algebra, where P_r consists of linear combinations of monomials $t_{i_1} \cdots t_{i_r}$

with coefficients in k. It is natural to say that $t_1, \ldots, t_n$ are *independent non-commutative variables over* k.

PROPOSITION 10. *Let E be free of dimension n over k. Then $T(E)$ is isomorphic to the non-commutative polynomial algebra on n variables over k. In other words, if $\{v_1, \ldots, v_n\}$ is a basis of E over k, then the elements*

$$M_{(i)}(v) = v_{i_1} \otimes \cdots \otimes v_{i_r}, \qquad 1 \leqq i_\nu \leqq n$$

form a basis of $T^r(E)$, and every element of $T(E)$ has a unique expression as a finite sum

$$\sum_{(i)} a_{(i)} M_{(i)}(v), \qquad a_{(i)} \in k$$

with almost all $a_{(i)}$ equal to 0.

Proof. This follows at once from Proposition 4, §2.

The tensor product of linear maps will now be interpreted in the context of the tensor algebra.

For convenience, we shall denote the module of endomorphisms $\text{End}_k(E)$ *by* $L(E)$ *for the rest of this section.*

We form the direct sum

$$(LT)(E) = \coprod_{r=0}^{\infty} L(T^r(E)),$$

which we shall also write $LT(E)$ for simplicity. (Of course, $LT(E)$ is not equal to $\text{End}_k(T(E))$, so we must view LT as a single symbol.) We shall see that LT is a functor from modules to graded algebras, by defining a suitable multiplication on $LT(E)$. Let $f \in L(T^r(E))$, $g \in L(T^s(E))$, $h \in L(T^m(E))$. We define the product $fg \in L(T^{r+s}(E))$ to be $T(f, g)$, in the notation of §1, in other words to be the unique linear map whose effect on an element $x \otimes y$ with $x \in T^r(E)$ and $y \in T^s(E)$ is

$$x \otimes y \mapsto f(x) \otimes g(y).$$

In view of the associativity of the tensor product, we obtain at once the associativity $(fg)h = f(gh)$, and we also see that our product is bilinear. Hence $LT(E)$ is a k-algebra.

We have an algebra-homomorphism

$$T(L(E)) \to LT(E)$$

given in each dimension r by the linear map

$$f_1 \otimes \cdots \otimes f_r \mapsto T(f_1, \ldots, f_r) = f_1 \cdots f_r.$$

We specify here that the tensor product on the left is taken in

$$L(E) \otimes \cdots \otimes L(E).$$

We also note that the homomorphism is in general neither surjective nor injective. When E is free finite dimensional over k, the homomorphism turns out to be both, and thus we have a clear picture of $LT(E)$ as a non-commutative polynomial algebra, generated by $L(E)$. Namely, from Proposition 5 of §2, we obtain:

PROPOSITION 11. *Let E be free, finite dimensional over k. Then we have an algebra-isomorphism*

$$T(L(E)) = T(\mathrm{End}_k(E)) \to LT(E) = \coprod_{r=0}^{\infty} \mathrm{End}_k(T^r(E))$$

given by

$$f \otimes g \mapsto T(f, g).$$

Proof. By Proposition 5 of §2, we have a linear isomorphism in each dimension, and it is clear that the map preserves multiplication.

In particular, we see that $LT(E)$ is a non-commutative polynomial algebra.

§6. *Alternating products*

We recall that an r-multilinear map $f : E^{(r)} \to F$ is said to be *alternating* if $f(x_1, \ldots, x_r) = 0$ whenever $x_i = x_j$ for some $i \neq j$.

Let $\mathfrak{a}_r$ be the submodule of $T^r(E)$ generated by all elements of type $x_1 \otimes \cdots \otimes x_r$ where $x_i = x_j$ for some $i \neq j$. We define

$$\bigwedge^r(E) = T^r(E)/\mathfrak{a}_r.$$

Then we have an r-multilinear map $E^{(r)} \to \bigwedge^r(E)$ (called canonical) obtained from the composition

$$E^{(r)} \to T^r(E) \to T^r(E)/\mathfrak{a}_r = \bigwedge^r(E).$$

It is clear that our map is alternating. *Furthermore, it is universal with respect to r-multilinear alternating maps on E.* In other words, if $f : E^{(r)} \to F$ is such a map, there exists a unique linear map $f_* : \bigwedge^r(E) \to F$ such that the following diagram is commutative:

$$\begin{array}{ccc} & & \bigwedge^r(E) \\ & \nearrow & \downarrow f_* \\ E^{(r)} & \xrightarrow{f} & F \end{array}$$

Our map f_* exists because we can first get an induced map $T^r(E) \to F$ making the following diagram commutative:

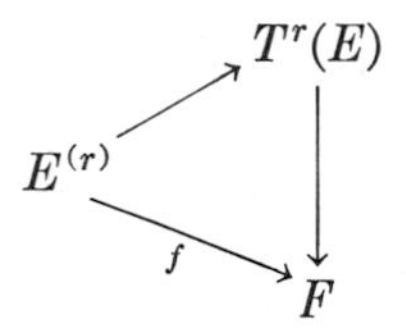

and this induced map vanishes on $\mathfrak{a}_r$, hence inducing our f_*.

In this way, $\bigwedge^r$ becomes a functor, from modules to modules.

The image of an element $(x_1, \ldots, x_r) \in E^{(r)}$ in the canonical map into $\bigwedge^r(E)$ will be denoted by $x_1 \wedge \cdots \wedge x_r$. It is also the image of $x_1 \otimes \cdots \otimes x_r$ in the factor homomorphism $T^r(E) \to \bigwedge^r(E)$.

We let $\bigwedge(E)$ be the direct sum $\coprod_{r=0}^{\infty} \bigwedge^r(E)$. We shall make $\bigwedge(E)$ into a graded k-algebra and call it the *alternating algebra* of E. We shall first discuss the general situation, with arbitrary graded rings.

Let G be an additive monoid again, and let $A = \coprod_{r \in G} A_r$ be a G-graded k-algebra. Suppose given for each A_r a submodule $\mathfrak{a}_r$, and let $\mathfrak{a} = \coprod_{r \in G} \mathfrak{a}_r$. Assume that $\mathfrak{a}$ is an ideal of A. Then $\mathfrak{a}$ is called a *homogeneous ideal*, and we can define a graded structure on $A/\mathfrak{a}$. Indeed, the bilinear map

$$A_r \times A_s \to A_{r+s}$$

sends $\mathfrak{a}_r \times A_s$ into $\mathfrak{a}_{r+s}$ and similarly, sends $A_r \times \mathfrak{a}_s$ into $\mathfrak{a}_{r+s}$. Thus using representatives in A_r, A_s respectively, we can define a bilinear map

$$A_r/\mathfrak{a}_r \times A_s/\mathfrak{a}_s \to A_{r+s}/\mathfrak{a}_{r+s},$$

and thus a bilinear map $A/\mathfrak{a} \times A/\mathfrak{a} \to A/\mathfrak{a}$, which obviously makes $A/\mathfrak{a}$ into a graded k-algebra.

We apply this to $T^r(E)$ and the modules $\mathfrak{a}_r$ defined previously. If $x_i = x_j$ $(i \neq j)$ in a product $x_1 \wedge \cdots \wedge x_r$, then for any $y_1, \ldots, y_s \in E$ we see that $x_1 \wedge \cdots \wedge x_r \wedge y_1 \wedge \cdots \wedge y_s$ lies in $\mathfrak{a}_{r+s}$, and similarly for the product on the left. Hence the direct sum $\coprod \mathfrak{a}_r$ is an ideal of $T(E)$, and we can define a k-algebra structure on $T(E)/\mathfrak{a}$. The product on homogeneous elements is given by the formula

$$((x_1 \wedge \cdots \wedge x_r), (y_1 \wedge \cdots \wedge y_s)) \mapsto x_1 \wedge \cdots \wedge x_r \wedge y_1 \wedge \cdots \wedge y_s.$$

We use the symbol $\wedge$ also to denote the product in $\bigwedge(E)$. This product is called the *alternating product*. If $x \in E$ and $y \in E$, then $x \wedge y = -y \wedge x$, as follows from the fact that $(x + y) \wedge (x + y) = 0$.

We observe that $\bigwedge$ is a functor from the category of modules to the category of graded k-algebras. To each linear map $f : E \to F$ we obtain a map

$$\bigwedge(f) : \bigwedge(E) \to \bigwedge(F)$$

which is such that for $x_1, \ldots, x_r \in E$ we have

$$\bigwedge(f)(x_1 \wedge \cdots \wedge x_r) = f(x_1) \wedge \cdots \wedge f(x_r).$$

Furthermore, $\bigwedge(f)$ is a homomorphism of graded k-algebras.

PROPOSITION 12. *Let E be free of dimension n over k. If $r > n$ then $\bigwedge^r(E) = 0$. Let $\{v_1, \ldots, v_n\}$ be a basis of E over k. If $1 \leqq r \leqq n$, then $\bigwedge^r(E)$ is free over k, and the elements*

$$v_{i_1} \wedge \cdots \wedge v_{i_r}, \qquad i_1 < \cdots < i_r$$

form a basis of $\bigwedge^r(E)$ over k. We have

$$\dim_k \bigwedge^r(E) = \binom{n}{r}.$$

Proof. We shall first prove our assertion when $r = n$. Every element of E can be written in the form $\sum a_i v_i$, and hence using the formula $x \wedge y = -y \wedge x$ we conclude that $v_1 \wedge \cdots \wedge v_n$ generates $\bigwedge^n(E)$. On the other hand, we know from the theory of determinants that given $a \in k$, there exists a unique multilinear alternating form f_a on E such that

$$f_a(v_1, \ldots, v_n) = a.$$

Consequently, there exists a unique linear map

$$\bigwedge^n(E) \to k$$

taking the value a on $v_1 \wedge \cdots \wedge v_n$. From this it follows at once that $v_1 \wedge \cdots \wedge v_n$ is a basis of $\bigwedge^n(E)$ over k.

We now prove our statement for $1 \leqq r \leqq n$. Suppose that we have a relation

$$0 = \sum a_{(i)} v_{i_1} \wedge \cdots \wedge v_{i_r}$$

with $i_1 < \cdots < i_r$ and $a_{(i)} \in k$. Select any r-tuple $(j) = (j_1, \ldots, j_r)$ such that $j_1 < \cdots < j_r$ and let $j_{r+1}, \ldots, j_n$ be those values of i which do not appear among $(j_1, \ldots, j_r)$. Take the alternating product with $v_{j_{r+1}} \wedge \cdots \wedge v_{j_n}$. Then we shall have alternating products in the sum with repeated components in all the terms except the (j)-term, and thus we obtain

$$0 = a_{(j)} v_{j_1} \wedge \cdots \wedge v_{j_r} \wedge \cdots \wedge v_{j_n}.$$

Reshuffling $v_{j_1} \wedge \cdots \wedge v_{j_n}$ into $v_1 \wedge \cdots \wedge v_n$ simply changes the right-hand side by a sign. From what we proved at the beginning of this proof, it follows that $a_{(j)} = 0$. Hence we have proved our assertion for $1 \leqq r \leqq n$.

When $r = 0$, we deal with the empty product, and 1 is a basis for $\bigwedge^0(E) = k$ over k. We leave the case $r > n$ as a trivial exercise to the reader.

The assertion concerning the dimension is trivial, considering that there is a bijection between the set of basis elements, and the subsets of the set of integers $(1, \ldots, n)$.

Remark. It is possible to give the first part of the proof, for $\bigwedge^n(E)$, without assuming known the existence of determinants. One must then show that $\mathfrak{a}_n$ admits a 1-dimensional complementary submodule in $T^n(E)$. This can be done by simple means, which we leave as an exercise. When k is a field, this exercise is even more trivial, since one can verify at once that $v_1 \otimes \cdots \otimes v_n$ does not lie in $\mathfrak{a}_n$. This alternative approach to the theorem then proves the existence of determinants.

§7. *Symmetric products*

Let $\mathfrak{S}_n$ denote the symmetric group on n letters, say operating on the integers $(1, \ldots, n)$. An r-multilinear map

$$f : E^{(r)} \to F$$

is said to be *symmetric* if $f(x_1, \ldots, x_r) = f(x_{\sigma(1)}, \ldots, x_{\sigma(r)})$ for all $\sigma \in \mathfrak{S}_r$.

In $T^r(E)$, we let $\mathfrak{b}_r$ be the submodule generated by all elements of type

$$x_1 \otimes \cdots \otimes x_r - x_{\sigma(1)} \otimes \cdots \otimes x_{\sigma(r)}$$

for all $x_i \in E$ and $\sigma \in \mathfrak{S}_r$. We define the factor module

$$S^r(E) = T^r(E)/\mathfrak{b}_r,$$

and let

$$S(E) = \coprod_{r=0}^{\infty} S^r(E)$$

be the direct sum. It is immediately obvious that the direct sum

$$\mathfrak{b} = \coprod_{r=0}^{\infty} \mathfrak{b}_r$$

is an ideal in $T(E)$, and hence that $S(E)$ is a graded k-algebra, which is called the *symmetric algebra* of E.

Furthermore, the canonical map

$$E^{(r)} \to S^r(E)$$

obtained by composing the maps

$$E^{(r)} \to T^r(E) \to T^r(E)/\mathfrak{b}_r = S^r(E)$$

is universal for r-multilinear symmetric maps. This routine has now become standard.

We observe that S is a functor, from the category of modules to the category of graded k-algebras. The image of $(x_1, \ldots, x_r)$ under the canonical map

$$E^{(r)} \to S^r(E)$$

will be denoted simply by $x_1 \cdots x_r$.

PROPOSITION 13. *Let E be free of dimension n over k. Let $\{v_1, \ldots, v_n\}$ be a basis of E over k. Viewed as elements of $S^1(E)$ in $S(E)$, these basis elements are algebraically independent over k, and $S(E)$ is therefore isomorphic to the polynomial algebra in n variables over k.*

Proof. Let $t_1, \ldots, t_n$ be algebraically independent variables over k, and form the polynomial algebra $k[t_1, \ldots, t_n]$. Let P_r be the k-module of homogeneous polynomials of degree r. We define a map of $E^{(r)} \to P_r$ as follows. If $w_1, \ldots, w_r$ are elements of E which can be written

$$w_i = \sum_{\nu=1}^{n} a_{i\nu} v_\nu, \qquad i = 1, \ldots, r,$$

then our map is given by

$$(w_1, \ldots, w_r) \mapsto (a_{11}t_1 + \cdots + a_{1n}t_n) \cdots (a_{r1}t_1 + \cdots + a_{rn}t_n).$$

It is obvious that this map is multilinear and symmetric. Hence it factors through a linear map of $S^r(E)$ into P_r:

$$\begin{array}{ccc} E^{(r)} & \to & S^r(E) \\ & \searrow \; \swarrow & \\ & P_r & \end{array}$$

From the commutativity of our diagram, it is clear that the element $v_{i_1} \cdots v_{i_r}$ in $S^r(E)$ maps on $t_{i_1} \cdots t_{i_r}$ in P_r for each r-tuple of integers $(i) = (i_1, \ldots, i_r)$. Since the monomials $M_{(i)}(t)$ of degree r are linearly independent over k, it follows that the monomials $M_{(i)}(v)$ in $S^r(E)$ are also linearly independent over k, and that our map $S^r(E) \to P_r$ is an isomorphism. One verifies at once that the multiplication in $S(E)$ corresponds to the multiplication of polynomials in $k[t]$, and thus that the map of $S(E)$ into the polynomial algebra described as above for each component $S^r(E)$ induces an algebra-isomorphism of $S(E)$ onto $k[t]$, as desired.

§8. *The Euler-Grothendieck ring*

Let k be a field. Let G be a group. By a (G, k)-module, we shall mean a pair (E, ρ), consisting of a k-space E and a homomorphism

$$\rho : G \to \mathrm{Aut}_k(E).$$

Such a homomorphism is also called a *representation* of G in E. By abuse of language, we also say that the k-space E is a G-module. The group G operates on E, and we write σx instead of $\rho(\sigma)x$. The field k will be kept fixed in what follows.

Let $\mathfrak{M}(G)$ denote the category whose objects are (G, k)-modules. A morphism in $\mathfrak{M}(G)$ is what we call a *G-homomorphism*, that is a k-linear map $f : E \to F$ such that $f(\sigma x) = \sigma f(x)$.

If E is a G-module, and $\sigma \in G$, then we have by definition a k-automorphism $\sigma : E \to E$. Since T^r is a functor, we have an induced automorphism

$$T^r(\sigma) : T^r(E) \to T^r(E)$$

for each r, and thus $T^r(E)$ is also a G-module. Taking the direct sum, we see that $T(E)$ is a G-module, and hence that T is a functor from the category of G-modules to the category of graded G-modules. Similarly for $\bigwedge^r$, S^r, and $\bigwedge$, S.

It is clear that the kernel of a G-homomorphism is a G-submodule, and that the factor module of a G-module by a G-submodule is again a G-module. We let $\mathfrak{M}_G$ be the set of G-isomorphism classes of (G, k)-modules. This set is a monoid, addition being represented on modules by direct sums. We have the Grothendieck homomorphism

$$\gamma : \mathfrak{M}_G \to K(G)$$

into the Grothendieck group, taken with respect to exact sequences. We write $K(G)$ instead of $K(\mathfrak{M}_G)$ for simplicity. If $[E]$ denotes the isomorphism class of E, we also write $\gamma(E)$ instead of $\gamma([E])$.

If E, F are G-modules, then their tensor product over k, $E \otimes F$, is also a G-module. Here again, the operation of G on $E \otimes F$ is given functorially. If $\sigma \in G$, there exists a unique k-linear map $E \otimes F \to E \otimes F$ such that for $x \in E$, $y \in F$ we have $x \otimes y \mapsto (\sigma x) \otimes (\sigma y)$. The tensor product induces a law of composition on $\mathfrak{M}_G$ because the tensor products of G-isomorphic modules are G-isomorphic. We contend that $\mathfrak{M}_G$ is also a multiplicative monoid. Our law of composition is associative because the tensor product is associative. There is a unit element, namely the class of the module k over G, defining the operation of G on k by the rule $(\sigma, a) \mapsto a$ for all $\sigma \in G$ and $a \in k$ (thus $\sigma a = a$).

The product on $\mathfrak{M}_G$ is obviously distributive with respect to our addition because the tensor product of a direct sum is the direct sum of the tensor products.

Finally, since $E \otimes F$ is G-isomorphic to $F \otimes E$, our multiplication in $\mathfrak{M}_G$ is commutative. Thus $\mathfrak{M}_G$ is a monoid under addition, a commutative monoid under the tensor product, and the multiplication is $\mathbf{Z}$-bilinear with respect to addition.

Since k is a field, we find that tensoring an exact sequence of G-modules over k with any G-module over k preserves the exactness. *From this one can define at once a product on $K(G)$, which is uniquely determined by the condition that*

$$\gamma(E)\gamma(F) = \gamma(E \otimes F)$$

for all G-modules E, F. It then follows trivially that $K(G)$ is a ring, and that γ is a homomorphism both for the additive law and the multiplicative law on $\mathfrak{M}_G$. We can therefore call $K(G)$ the *Grothendieck ring of G* (over k). Since G is fixed we shall also write K instead of $K(G)$.

If E is a G-module, we write $\lambda^i(E)$ for $\gamma(\bigwedge^i(E))$, in other words for the element in $K(G)$ which is the image under γ of the module $\bigwedge^i(E)$, or more strictly speaking, of its isomorphism class.

We shall now define a map of $\mathfrak{M}_G$ into the power series ring $K[[t]]$, namely the map λ_t such that

$$\lambda_t(E) = \sum_{i=0}^{\infty} \lambda^i(E)t^i.$$

Since $\bigwedge^0(E) = k$, it follows that $\lambda^0(E) = 1$. Thus our map is actually into the multiplicative group of power series beginning with 1. We shall write this group

$$1 + tK[[t]].$$

Thus λ_t is a map

$$\lambda_t : \mathfrak{M}_G \to 1 + tK[[t]].$$

PROPOSITION 14. *If E, F are k-modules, then we have an isomorphism*

$$\coprod_{i+j=r} \bigwedge^i(E) \otimes \bigwedge^j(F) \to \bigwedge^r(E \oplus F).$$

Proof. The proof will be left as an exercise to the reader.

COROLLARY. *The map λ_t above is a homomorphism of $\mathfrak{M}_G$ into the multiplicative group $1 + tK[[t]]$.*

In view of the universality of $K(G)$, we can extend λ_t to $K(G)$ (or more accurately, factor λ_t through $K(G)$). The induced map on $K(G)$ will again be denoted by λ_t.

We denote by $s^i(E)$ the element $\gamma(S^i(E))$, in the Grothendieck ring.

PROPOSITION 15. *For any G-module E, let*

$$s_t(E) = \sum_{i=0}^{\infty} s^i(E)t^i.$$

Then $s_t(E)\lambda_{-t}(E) = 1$.

The proof for this statement is harder, and the tools necessary to give it in a systematic way form the first chapter of any treatise dealing with the deeper aspects of the structures we have just introduced.

We conclude by an example.

Suppose that E is 1-dimensional over k. Then $\lambda^i(E) = 0$ for $i > 1$. Hence

$$\lambda_t(E) = 1 + \gamma(E)t$$

and

$$\lambda_{-t}(-\gamma(E)) = \frac{1}{1 - \gamma(E)t} = 1 + \gamma(E)t + \gamma(E)^2t^2 + \cdots.$$

When G is the trivial group, a simple proof can be given for Proposition 15, by reducing it to the 1-dimensional case.

§9. *Some functorial isomorphisms*

We begin by an abstract definition. Let $\mathfrak{A}$, $\mathfrak{B}$ be two categories. The functors of $\mathfrak{A}$ into $\mathfrak{B}$ (say covariant, and in one variable) can be viewed as the objects of a category, whose morphisms are defined as follows. If L, M are two such functors, a morphism $H : L \to M$ is a rule which to each object X of $\mathfrak{A}$ associates a morphism $H_X : L(X) \to M(X)$ in $\mathfrak{B}$, such that for any morphism $f : X \to Y$ in $\mathfrak{A}$, the following diagram is commutative:

$$\begin{array}{ccc} L(X) & \xrightarrow{H_X} & M(X) \\ {\scriptstyle L(f)}\downarrow & & \downarrow{\scriptstyle M(f)} \\ L(Y) & \xrightarrow[H_Y]{} & M(Y) \end{array}$$

We can therefore speak of isomorphisms of functors. We shall see examples of these in the theory of tensor products below. In our applications, our categories are additive, that is, the set of morphisms is an additive group, and the composition law is $\mathbf{Z}$-bilinear. In that case, a functor L is called *additive* if $L(f + g) = L(f) + L(g)$.

We let k be a commutative ring, and we shall consider additive functors from the category of k-modules into itself. For instance we may view the dual module as a functor,

$$E \mapsto E^* = L(E, k) = \mathrm{Hom}_k(E, k).$$

Similarly, we have a functor in two variables,

$$(E, F) \mapsto L(E, F) = \mathrm{Hom}_k(E, F),$$

contravariant in the first, covariant in the second, and bi-additive.

We shall give several examples of functorial isomorphisms connected with the tensor product, and for this it is most convenient to state a general theorem, giving us a criterion when a morphism of functors is in fact an isomorphism.

PROPOSITION 16. *Let L, M be two functors (both covariant or both contravariant) from the category of k-modules into itself. Assume that both functors are additive. Let $H : L \to M$ be a morphism of functors. If $H_E : L(E) \to M(E)$ is an isomorphism for every 1-dimensional free module E over k, then H_E is an isomorphism for every finite dimensional free module over k.*

Proof. We begin with a lemma.

LEMMA. *Let E and E_i $(i = 1, \ldots, m)$ be modules over a ring. Let $\varphi_i : E_i \to E$ and $\psi_i : E \to E_i$ be homomorphisms having the following properties:*

$$\psi_i \circ \varphi_i = id, \qquad \psi_i \circ \varphi_j = 0 \quad \text{if} \quad i \neq j$$

$$\sum_{i=1}^{m} \varphi_i \circ \psi_i = id.$$

Then the map

$$x \mapsto (\psi_1 x, \ldots, \psi_m x)$$

is an isomorphism of E onto the direct product $\prod_{i=1}^{m} E_i$, and the map

$$(x_1, \ldots, x_m) \mapsto \varphi_1 x_1 + \cdots + \varphi_m x_m$$

is an isomorphism of the product onto E. Conversely, if E is equal to the direct sum of submodules E_i $(i = 1, \ldots, m)$, if we let ψ_i be the inclusion of E_i in E, and φ_i the projection of E on E_i, then these maps satisfy the above-mentioned properties.

Proof. The proof is routine, and is essentially the same as that of Proposition 2 of Chapter III, §3. We shall leave it as an exercise to the reader.

We observe that the families $\{\varphi_i\}$ and $\{\psi_i\}$ satisfying the properties of the lemma behave functorially: If T is an additive contravariant functor, say, then the families $\{T(\psi_i)\}$ and $\{T(\varphi_i)\}$ also satisfy the properties of the lemma. Similarly if T is a covariant functor.

To apply the lemma, we take the modules E_i to be the 1-dimensional components occuring in a decomposition of E in terms of a basis. Let us assume for instance that L, M are both covariant. We have for each module E a commutative diagram

$$\begin{array}{ccc} L(E) & \xrightarrow{H_E} & M(E) \\ {\scriptstyle L(\varphi_i)}\uparrow & & \uparrow{\scriptstyle M(\varphi_i)} \\ L(E_i) & \xrightarrow[H_{E_i}]{} & M(E_i) \end{array}$$

and a similar diagram replacing φ_i by ψ_i, reversing the two vertical arrows. Hence we get a direct sum decomposition of $L(E)$ in terms of $L(\psi_i)$ and $L(\varphi_i)$, and similarly for $M(E)$, in terms of $M(\psi_i)$ and $M(\varphi_i)$. By hypothesis, H_{E_i} is an isomorphism. It then follows trivially that H_E is an isomorphism. For instance, to prove injectivity, we write an element $v \in L(E)$ in the form

$$v = \sum L(\varphi_i)v_i,$$

with $v_i \in L(E_i)$. If $H_E v = 0$, then

$$0 = \sum H_E L(\varphi_i)v_i = \sum M(\varphi_i)H_{E_i}v_i,$$

and since the maps $M(\varphi_i)$ $(i = 1, \ldots, m)$ give a direct sum decomposition of $M(E)$, we conclude that $H_{E_i}v_i = 0$ for all i, whence $v_i = 0$, and $v = 0$. The surjectivity is equally trivial.

When dealing with a functor of several variables, additive in each variable, one can keep all but one of the variables fixed, and then apply the proposition. We shall do this in the following corollaries.

Corollary 1. *Let E', E, F', F be free and finite dimensional over k. Then we have a functorial isomorphism*

$$L(E', E) \otimes L(F', F) \to L(E' \otimes F', E \otimes F)$$

such that

$$f \otimes g \mapsto T(f, g).$$

Proof. Keep E, F', F fixed, and view $L(E', E) \otimes L(F', F)$ as a functor in the variable E'. Similarly, view

$$L(E' \otimes F', E \otimes F)$$

as a functor in E'. The map $f \otimes g \mapsto T(f, g)$ is functorial, and thus by the lemma, it suffices to prove that it yields an isomorphism when E' has dimension 1. Assume now that this is the case; fix E' of dimension 1, and view the two expressions in the corollary as functors of the variable E. Applying the lemma again, it suffices to prove that our arrow is an isomorphism when E has dimension 1. Similarly, we may assume that F, F' have dimension 1. In that case the verification that the arrow is an isomorphism is a triviality, as desired.

Corollary 2. *Let E, F be free and finite dimensional. Then we have a functorial isomorphism*

$$\mathrm{End}_k(E) \otimes \mathrm{End}_k(F) \to \mathrm{End}_k(E \otimes F).$$

Proof. Special case of Corollary 1.

Note that Corollary 2 had already been proved before, and that we mention it here only to see how it fits with the present point of view.

Corollary 3. *Let E, F be free finite dimensional over k. There is a functorial isomorphism*

$$E^* \otimes F \to L(E, F)$$

given for $x^* \in E^*$ and $y \in F$ by the map

$$x^* \otimes y \mapsto \lambda,$$

where λ is such that for all $x \in E$, we have $\lambda(x) = \langle x, x^* \rangle y$.

The inverse isomorphism of Corollary 3 can be described as follows. Let $\{v_1, \ldots, v_n\}$ be a basis of E, and let $\{v_1^*, \ldots, v_n^*\}$ be the dual basis. If $A \in L(E, F)$, its inverse image is the element

$$\sum_{i=1}^{n} v_i^* \otimes A(v_i)$$

in the tensor product. In particular, if $E = F$, then the inverse image of the identity id_E is the element

$$\sum_{i=1}^{n} v_i^* \otimes v_i.$$

The proof of Corollary 3 is obtained by reducing it to the case when both E, F are 1-dimensional, in which case the assertion is obvious. The explicit representation of the inverse map as described above is left as an exercise.

Differential geometers are very fond of the isomorphism

$$L(E, E) \to E^* \otimes E,$$

and often use $E^* \otimes E$ when they think geometrically of $L(E, E)$, thereby emphasizing an unnecessary dualization, and an irrelevant formalism, when it is easier to deal directly with $L(E, E)$. In differential geometry, one applies various functors L to the tangent space at a point on a manifold, and elements of the spaces thus obtained are called *tensors* (of type L).

COROLLARY 4. *Let E, F be free and finite dimensional over k. There is a functorial isomorphism*

$$E^* \otimes F^* \to (E \otimes F)^*,$$

given for $x^* \in E^*$ and $y^* \in F^*$ by the map

$$x^* \otimes y^* \mapsto \lambda,$$

where λ is such that, for all $x \in E$ and $y \in F$,

$$\lambda(x \otimes y) = \langle x, x^* \rangle \langle y, y^* \rangle.$$

Proof. As before.

Finally, we leave the following results as an exercise.

PROPOSITION 17. *Let E be free and finite dimensional over k. The trace function on $L(E, E)$ is equal to the composite of the two maps*

$$L(E, E) \to E^* \otimes E \to k,$$

where the first map is the inverse of the isomorphism described in Corollary 3 of Proposition 16, and the second map is induced by the bilinear map

$$(x^*, x) \mapsto \langle x, x^* \rangle.$$

Of course, it is precisely in a situation involving the trace that the isomorphism of Corollary 3 becomes important, and that the finite dimensionality of E is used. In many applications, this finite dimensionality plays no role, and it is better to deal with $L(E, E)$ directly.

EXERCISES

1. Let k be a field and $k(\alpha)$ a finite extension. Let $f(X) = \mathrm{Irr}(\alpha, k, X)$, and suppose that f is separable. Let k' be any extension of k. Show that $k(\alpha) \otimes k'$ is a direct sum of fields. If k' is algebraically closed, show that these fields correspond to the embeddings of $k(\alpha)$ in k'.

2. Let k be a field, $f(X)$ an irreducible polynomial over k, and α a root of f. Show that $k(\alpha) \otimes k'$ is isomorphic, as a k'-algebra, to $k'[X]/(f(X))$.

3. Prove Proposition 14 by getting a natural homomorphism and arguing on the dimensions of the left- and right-hand sides of the equation.

4. Assume that the group G in the last section is trivial, and just write K instead of $K(1)$. For $x \in K$, define

$$\psi_{-t}(x) = -td \log \lambda_t(x) = -t\lambda_t'(x)/\lambda_t(x).$$

Show that ψ_{-t} is an additive and multiplicative homomorphism. Show that

$$\psi_t(E) = 1 + \gamma(E)t + \gamma(E)^2t^2 + \cdots.$$

5. Given a bilinear form on a module E over a commutative ring, show how to apply the extension of the base: If $k \to k'$ is a ring-homomorphism of commutative rings, define a natural bilinear form on $E^{k'}$ over k'.

6. Let k be a commutative ring. If E is a k-module, denote by $L_a^r(E)$ the module of r-multilinear alternating maps of E into k itself (i.e. the r-multilinear alternating forms on E). Let $L_a^0(E) = k$, and let

$$\Omega(E) = \coprod_{r=0}^{\infty} L_a^r(E).$$

Show that $\Omega(E)$ is a graded k-algebra, the multiplication being defined as follows. If $\omega \in L_a^r(E)$ and $\psi \in L_a^s(E)$, and $v_1, \ldots, v_{r+s}$ are elements of E, then

$$(\omega \wedge \psi)(v_1, \ldots, v_{r+s}) = \sum \epsilon(\sigma)\omega(v_{\sigma 1}, \ldots, v_{\sigma r})\psi(v_{\sigma(r+1)}, \ldots, v_{\sigma s}),$$

the sum being taken over all permutations σ of $(1, \ldots, r+s)$ such that $\sigma 1 < \cdots < \sigma r$ and $\sigma(r+1) < \cdots < \sigma s$.

7. Let E be a free module of dimension n over the commutative ring k. Let $f : E \to E$ be a linear map. Let $\alpha_r(f) = \operatorname{tr} \bigwedge^r(f)$, where $\bigwedge^r(f)$ is the endomorphism of $\bigwedge^r(E)$ into itself induced by f. We have

$$\alpha_0(f) = 1, \qquad \alpha_1(f) = \operatorname{tr}(f), \qquad \alpha_n(f) = \det f,$$

and $\alpha_r(f) = 0$ if $r > n$. Show that

$$\det(1 + f) = \sum_{r \geqq 0} \alpha_r(f).$$

[*Hint:* As usual, prove the statement when f is represented by a matrix with variable coefficients over the integers.] Interpret the $\alpha_r(f)$ in terms of the coefficients of the characteristic polynomial of f.

8. Let E be a finite dimensional free module over the commutative ring k. Let E^* be its dual module. For each integer $r \geqq 1$ show that $\bigwedge^r E$ and $\bigwedge^r E^*$

are dual modules to each other, under the bilinear map such that

$$(v_1 \wedge \cdots \wedge v_r, v_1' \wedge \cdots \wedge v_r') \mapsto \det(\langle v_i, v_j' \rangle)$$

where $\langle v_i, v_j' \rangle$ is the value of v_j' on v_i, as usual, for $v_i \in E$ and $v_j' \in E^*$.

9. Notation being as in the preceding exercise, let F be another k-module which is free, finite dimensional. Let $f : E \to F$ be a linear map. Relative to the bilinear map of the preceding exercise, show that the transpose of $\bigwedge^r f$ is $\bigwedge^r({}^t f)$, i.e. is equal to the r-th alternating product of the transpose of f.

10. Let P be the non-commutative polynomial algebra over a field k, in n variables. Let $x_1, \ldots, x_r$ be distinct elements of P_1 (i.e. linear expressions in the variables $t_1, \ldots, t_n$) and let $a_1, \ldots, a_r \in k$. If

$$a_1 x_1^\nu + \cdots + a_r x_r^\nu = 0$$

for all integers $\nu = 1, \ldots, r$ show that $a_i = 0$ for $i = 1, \ldots, r$. [*Hint:* Take the homomorphism on the commutative polynomial algebra and argue there.]

11. Let G be a finite set of endomorphisms of a finite dimensional vector space E over the field k. For each $\sigma \in G$, let c_σ be an element of k. Show that if

$$\sum_{\sigma \in G} c_\sigma T^r(\sigma) = 0$$

for all integers $r \geqq 1$, then $c_\sigma = 0$ for all $\sigma \in G$. [*Hint:* Use the preceding exercise, and Proposition 11.]

12. (Steinberg) Let G be a finite monoid, and $k[G]$ the monoid algebra over a field k. Let $G \to \mathrm{End}_k(E)$ be a faithful representation (i.e. injective), so that we identify G with a multiplicative subset of $\mathrm{End}_k(E)$. Show that T^r induces a representation of G on $T^r(E)$, whence a representation of $k[G]$ on $T^r(E)$ by linearity. If $\alpha \in k[G]$ and if $T^r(\alpha) = 0$ for all integers $r \geqq 1$, show that $\alpha = 0$. [*Hint:* Apply the preceding exercise.]

13. When you have read the chapter on representation of finite groups, deduce from Exercise 12 the following theorem of Burnside: Let G be a finite group, k a field of characteristic prime to the order of G, and E a finite dimensional (G, k)-space such that the representation of G is faithful. Then every irreducible representation of G appears with multiplicity $\geqq 1$ in some tensor power $T^r(E)$.

CHAPTER XVII

Semisimplicity

In many applications, a module decomposes as a direct sum of simple submodules, and then one can develop a fairly precise structure theory, both under general assumptions, and particular applications. This chapter is devoted to those results which can be proved in general. In the next chapter, we consider those additional results which can be proved in a classical and important special case.

I have more or less followed Bourbaki in the proof of the density theorem.

§1. *Matrices and linear maps over non-commutative rings*

In Chapter XIII, we considered exclusively matrices over commutative rings. For our present purposes, it is necessary to consider a more general situation.

Let K be a ring. We define a matrix (φ_{ij}) with coefficients in K just as we did for commutative rings. The product of matrices is defined by the same formula. Then we again have associativity and distributivity, whenever the size of the matrices involved in the operations makes the operations defined. In particular, the square $n \times n$ matrices over K form a ring, again denoted by $\mathrm{Mat}_n(K)$. We have a ring-homomorphism

$$K \to \mathrm{Mat}_n(K)$$

on the diagonal.

By a *division ring* we shall mean a ring with $1 \neq 0$, and such that every non-zero element has a multiplicative inverse.

If K is a division ring, then every non-zero K-module has a basis, and the cardinalities of two bases are equal. The proof is the same as in the commutative case; we never needed commutativity in the arguments. This cardinality is again called the dimension of the module over K, and a module over a division ring is called a vector space.

We can associate a matrix with linear maps, depending on the choice of a finite basis, just as in the commutative case. However, we shall consider a somewhat different situation which we want to apply to semi-simple modules.

Let R be a ring, and let

$$E = E_1 \oplus \cdots \oplus E_n, \qquad F = F_1 \oplus \cdots \oplus F_m$$

be R-modules, expressed as direct sums of R-submodules. We wish to describe the most general R-homomorphism of E into F.

Suppose first $F = F_1$ has one component. Let

$$\varphi : E_1 \oplus \cdots \oplus E_n \to F$$

be a homomorphism. Let $\varphi_j : E_j \to F$ be the restriction of φ to the factor E_j. Every element $x \in E$ has a unique expression $x = x_1 + \cdots + x_n$, with $x_j \in E_j$. We may therefore associate with x the column vector $X = {}^t(x_1, \ldots, x_n)$, whose components are in $E_1, \ldots, E_n$ respectively. We can associate with φ the row vector $(\varphi_1, \ldots, \varphi_n)$, $\varphi_j \in \mathrm{Hom}_R(E_j, F)$, and the effect of φ on the element x of E is described by matrix multiplication, of the row vector times the column vector.

More generally, consider a homomorphism

$$\varphi : E_1 \oplus \cdots \oplus E_n \to F_1 \oplus \cdots \oplus F_m.$$

Let $\pi_i : F_1 \oplus \cdots \oplus F_m \to F_i$ be the projection on the i-th factor. Then we can apply our previous remarks to $\pi_i \circ \varphi$, for each i. In this way, we see that there exist unique elements $\varphi_{ij} \in \mathrm{Hom}_R(E_j, F_i)$, such that φ has a matrix representation

$$M(\varphi) = \begin{pmatrix} \varphi_{11} & \cdots & \varphi_{1n} \\ \vdots & & \vdots \\ \varphi_{m1} & \cdots & \varphi_{mn} \end{pmatrix}$$

whose effect on an element x is given by matrix multiplication, namely

$$\begin{pmatrix} \varphi_{11} & \cdots & \varphi_{1n} \\ \vdots & & \vdots \\ \varphi_{m1} & \cdots & \varphi_{mn} \end{pmatrix} \begin{pmatrix} x_1 \\ \vdots \\ x_n \end{pmatrix}.$$

Conversely, given a matrix (φ_{ij}) with $\varphi_{ij} \in \mathrm{Hom}_R(E_j, F_i)$, we can define an element of $\mathrm{Hom}_R(E, F)$ by means of this matrix. We have an additive group-isomorphism between $\mathrm{Hom}_R(E, F)$ and this group of matrices.

In particular, let E be a fixed R-module, and let $K = \mathrm{End}_R(E)$. Then we have a ring-isomorphism

$$\mathrm{End}_R(E^{(n)}) \to \mathrm{Mat}_n(K)$$

which to each $\varphi \in \mathrm{End}_R(E^{(n)})$ *associates the matrix*

$$\begin{pmatrix} \varphi_{11} & \cdots & \varphi_{1n} \\ \vdots & & \vdots \\ \varphi_{n1} & \cdots & \varphi_{nn} \end{pmatrix}$$

determined as before, and operating on the left on column vectors of $E^{(n)}$*, with components in* E.

Remark. Let E be a one-dimensional vector space over a division ring D, and let $\{v\}$ be a basis. For each $a \in D$, there exists a unique D-linear map $f_a : E \to E$ such that $f_a(v) = av$. Then we have the rule

$$f_a f_b = f_{ba}.$$

Thus when we associate a matrix with a linear map, depending on a basis, the multiplication gets twisted. Nevertheless, the statement we just made preceding this remark is correct!! The point is that we took the φ_{ij} in $\mathrm{End}_R(E)$, and not in D, in the special case that $R = D$. Thus K is not isomorphic to D (in the non-commutative case), but anti-isomorphic. This is the only point of difference of the formal elementary theory of linear maps in the commutative or non-commutative case.

We recall that an R-module E is said to be *simple* if it is $\neq 0$ and if it has no submodule other than 0 or E.

PROPOSITION 1. *Let* E, F *be simple* R*-modules. Every non-zero homomorphism of* E *into* F *is an isomorphism. The ring* $\mathrm{End}_R(E)$ *is a division ring.*

Proof. Let $f : E \to F$ be a non-zero homomorphism. Its image and kernel are submodules, hence $\mathrm{Ker}\, f = 0$ and $\mathrm{Im}\, f = F$. Hence f is an isomorphism. If $E = F$, then f has an inverse, as desired.

(Proposition 1 has been known as *Schur's lemma.*)

The next proposition describes completely the ring of endomorphisms of a direct sum of simple modules.

PROPOSITION 2. *Let* $E = E_1^{(n_1)} \oplus \cdots \oplus E_r^{(n_r)}$ *be a direct sum of simple modules, the* E_i *being non-isomorphic, and each* E_i *being repeated* n_i *times in the sum. Then, up to a permutation,* $E_1, \ldots, E_r$ *are uniquely determined up to isomorphisms, and the multiplicities* $n_1, \ldots, n_r$ *are uniquely determined. The ring* $\mathrm{End}_R(E)$ *is isomorphic to a ring of matrices, of type*

$$\begin{pmatrix} M_1 & \cdots & 0 \\ \vdots & M_2 & \vdots \\ & \ddots & \\ 0 & \cdots & M_r \end{pmatrix}$$

where M_i *is an* $n_i \times n_i$ *matrix over* $\text{End}_R(E_i)$. *(The isomorphism is the one with respect to our direct sum decomposition.)*

Proof. The last statement follows from our previous considerations, taking into account Proposition 1.

Suppose now that we have two R-modules, with direct sum decompositions into simple submodules, and an isomorphism

$$E_1^{(n_1)} \oplus \cdots \oplus E_r^{(n_r)} \to F_1^{(m_1)} \oplus \cdots \oplus F_s^{(m_s)},$$

such that the E_i are non-isomorphic, and the F_j are non-isomorphic. From Proposition 1, we conclude that each E_i is isomorphic to some F_j, and conversely. It follows that $r = s$, and that after a permutation, $E_i \approx F_i$. Furthermore, the isomorphism must induce an isomorphism

$$E_i^{(n_i)} \to F_i^{(m_i)}$$

for each i. Since $E_i \approx F_i$, we may assume without loss of generality that in fact $E_i = F_i$. Thus we are reduced to proving: If a module is isomorphic to $E^{(n)}$ and to $E^{(m)}$, with some simple module E, then $n = m$. But $\text{End}_R(E^{(n)})$ is isomorphic to the $n \times n$ matrix ring over the division ring $\text{End}_R(E) = K$. Furthermore this isomorphism is verified at once to be an isomorphism as K-vector space. The dimension of the space of $n \times n$ matrices over K is n^2. This proves that the multiplicity n is uniquely determined, and proves our proposition.

When E admits a (finite) direct sum decomposition of simple submodules, the number of times that a simple module of a given isomorphism class occurs in a decomposition will be called the *multiplicity* of the simple module (or of the isomorphism class of the simple module).

Furthermore, if

$$E = E_1^{(n_1)} \oplus \cdots \oplus E_r^{(n_r)}$$

is expressed as a sum of simple submodules, we shall call $n_1 + \cdots + n_r$ the *length* of E. In many applications, we shall also write

$$E = n_1E_1 \oplus \cdots \oplus n_rE_r = \coprod_{i=1}^{r} n_iE_i.$$

§2. *Conditions defining semisimplicity*

Let R *be a ring. Unless otherwise specified in this section all modules and homomorphisms will be* R*-modules and* R*-homomorphisms.*

The following conditions on a module E are equivalent:

SS 1. E is the sum of a family of simple submodules.
SS 2. E is the direct sum of a family of simple submodules.

SS 3. Every submodule F of E is a direct summand, i.e. there exists a submodule F' such that $E = F \oplus F'$.

We shall now prove that these three conditions are equivalent.

LEMMA. *Let $E = \sum_{i \in I} E_i$ be a sum (non-necessarily direct) of simple submodules. Then there exists a subset $J \subset I$ such that E is the direct sum $\coprod_{j \in J} E_j$.*

Proof. Let J be a maximal subset of I such that the sum $\sum_{j \in J} E_j$ is direct. We contend that this sum is in fact equal to E. It will suffice to prove that each E_i is contained in this sum. But the intersection of our sum with E_i is a submodule of E_i, hence equal to 0 or E_i. If it is equal to 0, then J is not maximal, since we can adjoin i to it. Hence E_i is contained in the sum, and our lemma is proved.

The lemma shows that SS 1 implies SS 2. To see that SS 2 implies SS 3, take a submodule F, and let J be a maximal subset of I such that the sum $F + \coprod_{j \in J} E_j$ is direct. The same reasoning as before shows that this sum is equal to E.

To prove that SS 3 implies SS 1, we shall first prove that every non-zero module contains a simple submodule. Let E be a non-zero module, and $v \in E$, $v \neq 0$. Then by definition, Rv is a principal submodule, and the kernel of the homomorphism

$$R \to Rv$$

is a left ideal $L \neq R$. Hence L is contained in a maximal left ideal $M \neq R$ (by Zorn's lemma). Then M/L is a maximal submodule of R/L (unequal to R/L), and hence Mv is a maximal submodule of Rv, unequal to Rv, corresponding to M/L under the isomorphism

$$R/L \to Rv.$$

We can write $E = Mv \oplus M'$ with some submodule M'. Then $Rv = Mv \oplus (M' \cap Rv)$, because every element $x \in Rv$ can be written uniquely as a sum $x = \alpha v + x'$ with $\alpha \in R$ and $x' \in M'$, and $x' = x - \alpha v$ lies in Rv. Since Mv is maximal in Rv, it follows that $M' \cap Rv$ is simple, as desired.

Let E_0 be the submodule of E which is the sum of all simple submodules of E. If $E_0 \neq E$, then $E = E_0 \oplus F$ with $F \neq 0$, and there exists a simple submodule of F, contradicting the definition of E_0. This proves that SS 3 implies SS 1.

A module E satisfying our three conditions is said to be *semisimple*.

PROPOSITION 3. *Every submodule and every factor module of a semisimple module is semisimple.*

Proof. Let F be a submodule. Let F_0 be the sum of all simple submodules of F. Write $E = F_0 \oplus F_0'$. Every element x of F has a unique expression $x = x_0 + x_0'$ with $x_0 \in F_0$ and $x_0' \in F_0'$. But $x_0' = x - x_0 \in F$. Hence F is the direct sum

$$F = F_0 \oplus (F \cap F_0').$$

We must therefore have $F_0 = F$, which is semisimple. As for the factor module, write $E = F \oplus F'$. Then F' is a sum of its simple submodules, and the canonical map $E \to E/F$ induces an isomorphism of F' onto E/F. Hence E/F is semisimple.

§3. The density theorem

Let E be a semisimple R-module. Let K be the ring $\text{End}_R(E)$. Then E is also a K-module, the operation of K on E being given by

$$(\varphi, x) \mapsto \varphi(x)$$

for $\varphi \in K$ and $x \in E$. Each $\alpha \in R$ induces a K-homomorphism $f_\alpha : E \to E$ by the map $f_\alpha(x) = \alpha x$. This is what is meant by the condition

$$\varphi(\alpha x) = \alpha\varphi(x).$$

Thus we get a ring-homomorphism

$$R \to \text{End}_K(E)$$

by $\alpha \mapsto f_\alpha$. We now ask how big is the image of this ring-homomorphism. The density theorem states that it is quite big.

LEMMA. *Let E be semisimple over R. Let $K = \text{End}_R(E), f \in \text{End}_K(E)$ as above. Let $x \in E$. There exists an element $\alpha \in R$ such that $\alpha x = f(x)$.*

Proof. Since E is semisimple, we can write an R-direct sum

$$E = Rx \oplus F$$

with some submodule F. Let $\pi : E \to Rx$ be the projection. Then $\pi \in K$, and hence

$$f(x) = f(\pi x) = \pi f(x).$$

This shows that $f(x) \in Rx$, as desired.

The density theorem generalizes the lemma by dealing with a finite number of elements of E instead of just one. For the proof, we use a diagonal trick.

THEOREM 1. (Jacobson) *Let E be semisimple over R, and let $K = \mathrm{End}_R(E)$. Let $f \in \mathrm{End}_K(E)$. Let $x_1, \ldots, x_n \in E$. Then there exists an element $\alpha \in R$ such that*

$$\alpha x_i = f(x_i) \qquad \text{for} \qquad i = 1, \ldots, n.$$

Proof. For clarity of notation, we shall first carry out the proof in case E is simple. Let $f^{(n)} : E^{(n)} \to E^{(n)}$ be the product map, so that

$$f^{(n)}(y_1, \ldots, y_n) = \big(f(y_1), \ldots, f(y_n)\big).$$

Let $K' = \mathrm{End}_R(E^{(n)})$. Then K' is none other than the ring of matrices with coefficients in K. Since f commutes with elements of K in its action on E, one sees immediately that $f^{(n)}$ is in $\mathrm{End}_{K'}(E^{(n)})$. By the lemma, there exists an element $\alpha \in R$ such that

$$(\alpha x_1, \ldots, \alpha x_n) = \big(f(x_1), \ldots, f(x_n)\big),$$

which is what we wanted to prove.

When E is not simple, suppose that E is equal to a direct sum of submodules E_i (non-isomorphic), with multiplicities n_i:

$$E = E_1^{(n_1)} \oplus \cdots \oplus E_r^{(n_r)} \qquad (E_i \not\approx E_j \quad \text{if} \quad i \neq j),$$

then the matrices representing the ring of endomorphisms split according to blocks corresponding to the non-isomorphic simple components in our direct sum decomposition. Hence here again the argument goes through as before. The main point is that $f^{(n)}$ lies in $\mathrm{End}_{K'}(E^{(n)})$, and that we can apply the lemma.

The argument when E is an infinite direct sum would be similar, but the notation is disagreeable. However, in the applications we shall never need the theorem in any case other than the case when E itself is simple, and this is the reason why we first gave the proof in that case, and let the reader write out the formal details in the other cases, if he wishes.

COROLLARY 1. *Let E be a finite dimensional vector space over an algebraically closed field k, and let R be a subalgebra of $\mathrm{End}_k(E)$. If E is a simple R-module, then $R = \mathrm{End}_k(E)$.*

Proof. We contend that $\mathrm{End}_R(E) = k$. At any rate, $\mathrm{End}_R(E)$ is a division ring K, containing k as a subring and every element of k commutes with every element of K. Let $\alpha \in K$. Then $k(\alpha)$ is a field. Furthermore, K is contained in $\mathrm{End}_k(E)$ as a k-subspace, and is therefore finite dimensional over k. Hence $k(\alpha)$ is finite over k, and therefore equal to k since k is algebraically closed. This proves that $\mathrm{End}_R(E) = k$. Let now $\{v_1, \ldots, v_n\}$ be a basis of E over k. Let $A \in \mathrm{End}_k(E)$. According to the density theorem, there exists $\alpha \in R$ such that

$$\alpha v_i = A v_i \quad \text{for} \quad i = 1, \ldots, n.$$

Since the effect of A is determined by its effect on a basis, we conclude that $R = \operatorname{End}_k(E)$.

Corollary 1 is known as *Burnside's theorem.* It is used in the following situation. Let E be a finite dimensional vector space over field k. Let G be a submonoid of $GL(E)$ (multiplicative). A *G-invariant* subspace F of E is a subspace such that $\sigma F \subset F$ for all $\sigma \in G$. We say that E is *G-simple* if it has no G-invariant subspace other than 0 and E itself, and $E \neq 0$. Let $R = k[G]$ be the subalgebra of $\operatorname{End}_k(E)$ generated by G over k. Since we assumed that G is a monoid, it follows that R consists of linear combinations

$$\sum a_i \sigma_i$$

with $a_i \in k$ and $\sigma_i \in G$. Then we see that a subspace F of E is G-invariant if and only if it is R-invariant. Thus E is G-simple if and only if it is simple over R in the sense which we have been considering. We can then restate Burnside's theorem as he stated it:

COROLLARY 2. *Let E be a finite dimensional vector space over an algebraically closed field k, and let G be a (multiplicative) submonoid of $GL(E)$. If E is G-simple, then $k[G] = \operatorname{End}_k(E)$.*

When k is not algebraically closed, then we still get some result. Quite generally, let R be a ring and E a simple R-module. We have seen that $\operatorname{End}_R(E)$ is a division ring, which we denote by D, and E is a vector space over D.

Let R be a ring, and E any R-module. We shall say that E is a *faithful* module if the following condition is satisfied. Given $\alpha \in R$ such that $\alpha x = 0$ for all $x \in E$, we have $\alpha = 0$. In the applications, E is a vector space over a field k, and we have a ring-homomorphism of R into $\operatorname{End}_k(E)$. In this way, E is an R-module, and it is faithful if and only if this homomorphism is injective.

COROLLARY 3. (Wedderburn's theorem) *Let R be a ring, and E a simple, faithful module over R. Let $D = \operatorname{End}_R(E)$, and assume that E is finite dimensional over D. Then $R = \operatorname{End}_D(E)$.*

Proof. Let $\{v_1, \ldots, v_n\}$ be a basis of E over D. Given $A \in \operatorname{End}_D(E)$, by Theorem 1 there exists $\alpha \in R$ such that

$$\alpha v_i = A v_i \quad \text{for} \quad i = 1, \ldots, n.$$

Hence the map $R \to \operatorname{End}_D(E)$ is surjective. Our assumption that E is faithful over R implies that it is injective, and our corollary is proved.

§4. *Semisimple rings*

A ring R is called *semisimple* if $1 \neq 0$, and if R is semisimple as a left module over itself.

PROPOSITION 4. *If R is semisimple, then every R-module is semisimple.*

Proof. An R-module is a factor module of a free module, and a free module is a direct sum of R with itself a certain number of times. We can apply Proposition 3 to conclude the proof.

A left ideal of R is an R-module, and is thus called simple if it is simple as a module. Two ideals L, L' are called isomorphic if they are isomorphic as modules.

We shall now decompose R as a sum of its simple left ideals, and thereby get a structure theorem for R.

Let $\{L_i\}_{i \in I}$ be a family of simple left ideals, no two of which are isomorphic, and such that each simple left ideal is isomorphic to one of them. We say that this family is a family of representatives for the isomorphism classes of simple left ideals.

LEMMA. *Let L be a simple left ideal, and let E be a simple R-module. If L is not isomorphic to E, then $LE = 0$.*

Proof. We have $RLE = LE$, and LE is a submodule of E, hence equal to 0 or E. Suppose $LE = E$. Let $y \in E$ be such that

$$Ly \neq 0.$$

Since Ly is a submodule of E, it follows that $Ly = E$. The map $\alpha \mapsto \alpha y$ of L into E is a homomorphism of L into E, which is surjective, and hence non-zero. Since L is simple, this homomorphism is an isomorphism.

Let

$$R_i = \sum_{L \approx L_i} L$$

be the sum of all simple left ideals isomorphic to L_i. From the lemma, we conclude that $R_i R_j = 0$ if $i \neq j$. This will be used constantly in what follows. We note that R_i is a left ideal, and that R is the sum

$$R = \sum_{i \in I} R_i,$$

because R is a sum of simple left ideals. Hence for any $j \in I$,

$$R_j \subset R_j R = R_j R_j \subset R_j,$$

the first inclusion because R contains a unit element, and the last because R_j is a left ideal. We conclude that R_j is also a right ideal, i.e. R_j is a two-sided ideal for all $j \in I$.

We can express the unit element 1 of R as a sum

$$1 = \sum_{i \in I} e_i$$

with $e_i \in R_i$. This sum is actually finite, almost all $e_i = 0$. Say $e_i \neq 0$ for indices $i = 1, \ldots, s$, so that we write

$$1 = e_1 + \cdots + e_s.$$

For any $x \in R$, write

$$x = \sum_{i \in I} x_i, \qquad x_i \in R_i.$$

For $j = 1, \ldots, s$ we have $e_j x = e_j x_j$ and also

$$x_j = 1 \cdot x_j = e_1 x_j + \cdots + e_s x_j = e_j x_j.$$

Furthermore, $x = e_1 x + \cdots + e_s x$. This proves that there is no index i other than $i = 1, \ldots, s$ and also that the i-th component x_i of x is uniquely determined as $e_i x = e_i x_i$. Hence the sum $R = R_1 + \cdots + R_s$ is direct, and furthermore, e_i is a unit element for R_i, which is therefore a ring. Since $R_i R_j = 0$ for $i \neq j$, we find that in fact

$$R = \prod_{i=1}^{s} R_i$$

is a direct product of the rings R_i.

A ring R is said to be *simple* if it is semisimple, and if it has only one isomorphism class of left ideals. We see that we have proved a structure theorem for semisimple rings:

THEOREM 2. *Let R be semisimple. Then there is only a finite number of non-isomorphic simple left ideals, say $L_1, \ldots, L_s$. If $R_i = \sum_{L \approx L_i} L$ is the sum of all simple left ideals isomorphic to L_i, then R_i is a two-sided ideal, which is also a ring (the operations being those induced by R), and R is ring isomorphic to the direct product*

$$R = \prod_{i=1}^{s} R_i.$$

Each R_i is a simple ring. If e_i is its unit element, then $1 = e_1 + \cdots + e_s$, and $R_i = Re_i$. We have $e_i e_j = 0$ if $i \neq j$.

We shall now discuss modules.

THEOREM 3. *Let R be semisimple, and let E be an R-module $\neq 0$. Then*

$$E = \coprod_{i=1}^{s} R_i E = \coprod_{i=1}^{s} e_i E,$$

and $R_i E$ is the submodule of E consisting of the sum of all simple submodules isomorphic to L_i.

Proof. Let E_i be the sum of all simple submodules of E isomorphic to L_i. If V is a simple submodule of E, then $RV = V$, and hence $L_i V = V$ for some i. By a previous lemma, we have $L_i \approx V$. Hence E is the direct sum of $E_1, \ldots, E_s$. It is then clear that $R_i E = E_i$.

COROLLARY 1. *Let R be semisimple. Every simple module is isomorphic to one of the simple left ideals L_i.*

COROLLARY 2. *A simple ring has exactly one simple module, up to isomorphism.*

Both these corollaries are immediate consequences of Theorems 2 and 3.

§5. Simple rings

LEMMA. *Let R be a ring, and $\psi \in \mathrm{End}_R(R)$ a homomorphism of R into itself, viewed as R-module. Then there exists $\alpha \in R$ such that $\psi(x) = x\alpha$ for all $x \in R$.*

Proof. We have $\psi(x) = \psi(x \cdot 1) = x\psi(1)$. Let $\alpha = \psi(1)$.

THEOREM 4. *Let R be a simple ring. Then R is a finite direct sum of simple left ideals. There are no two-sided ideals except* 0 *and R. If L, M are simple left ideals, then there exists $\alpha \in R$ such that $L\alpha = M$. We have $LR = R$.*

Proof. Since R is by definition also semisimple, it is a direct sum of left ideals, say $\coprod_{j \in J} L_j$. We can write 1 as a finite sum $1 = \sum_{j=1}^{m} \beta_j$, with $\beta_j \in L_j$. Then

$$R = \coprod_{j=1}^{m} R\beta_j = \coprod_{j=1}^{m} L_j.$$

This proves our first assertion. As to the second, it is a consequence of the third. Let therefore L be a simple left ideal. Then LR is a left ideal, because $RLR = LR$, hence (R being semisimple) is a direct sum of simple left ideals, say

$$LR = \coprod_{j=1}^{m} L_j, \qquad L = L_1.$$

Let M be a simple left ideal. We have a direct sum decomposition

$R = L \oplus L'$. Let $\pi : R \to L$ be the projection. It is an R-endomorphism. Let $\sigma : L \to M$ be an isomorphism (it exists by Theorem 2). Then $\sigma \circ \pi : R \to R$ is an R-endomorphism. By the lemma, there exists $\alpha \in R$ such that

$$\sigma \circ \pi(x) = x\alpha \qquad \text{for all} \quad x \in R.$$

Apply this to elements $x \in L$. We find

$$\sigma(x) = x\alpha \qquad \text{for all} \quad x \in L.$$

The map $x \mapsto x\alpha$ is a R-homomorphism of L into M, is non-zero, hence is an isomorphism. From this it follows at once that $LR = R$, thereby proving our theorem.

Corollary. *Let R be a simple ring. Let E be a simple R-module, and L a simple left ideal of R. Then $LE = E$ and E is faithful.*

Proof. We have $LE = L(RE) = (LR)E = RE = E$. Suppose $\alpha E = 0$ for some $\alpha \in R$. Then $R\alpha RE = R\alpha E = 0$. But $R\alpha R$ is a two-sided ideal. Hence $R\alpha R = 0$, and $\alpha = 0$. This proves that E is faithful.

Theorem 5. (Rieffel) *Let R be a ring without two-sided ideals except 0 and R. Let L be a left ideal, $R' = \mathrm{End}_R(L)$ and $R'' = \mathrm{End}_{R'}(L)$. Then the natural map $\lambda : R \to R''$ is an isomorphism.*

Proof. The kernel of λ is a two-sided ideal, so λ is injective. Since LR is a two-sided ideal, we have $LR = R$ and $\lambda(L)\lambda(R) = \lambda(R)$. For any $x, y \in L$, and $f \in R''$, we have $f(xy) = f(x)y$, because right multiplication by y is an R-endomorphism of L. Hence $\lambda(L)$ is a left ideal of R'', so

$$R'' = R''\lambda(R) = R''\lambda(L)\lambda(R) = \lambda(L)\lambda(R) = \lambda(R),$$

as was to be shown.

Theorem 6. *Let D be a division ring, and E a finite dimensional vector space over D. Let $R = \mathrm{End}_D(E)$. Then R is simple and E is a simple R-module. Furthermore, $D = \mathrm{End}_R(E)$.*

Proof. We first show that E is a simple R-module. Let $v \in E$, $v \neq 0$. Then v can be completed to a basis of E over D, and hence, given $w \in E$, there exists $\alpha \in R$ such that $\alpha v = w$. Hence E cannot have any invariant subspaces other than 0 or itself, and is simple over R. It is clear that E is faithful over R. Let $\{v_1, \ldots, v_m\}$ be a basis of E over D. The map

$$\alpha \mapsto (\alpha v_1, \ldots, \alpha v_m)$$

of R into $E^{(m)}$ is an R-homomorphism of R into $E^{(m)}$, and is injective. Given $(w_1, \ldots, w_m) \in E^{(m)}$, there exists $\alpha \in R$ such that $\alpha v_i = w_i$ and hence R is R-isomorphic to $E^{(m)}$. This shows that R (as a module over itself) is isomorphic to a direct sum of simple modules and is therefore semisimple. Furthermore, all these simple modules are isomorphic to each other, and hence R is simple by Theorem 2.

There remains to prove that $D = \mathrm{End}_R(E)$. We note that E is a semisimple module over D since it is a vector space, and every subspace admits a complementary subspace. We can therefore apply the density theorem (the roles of R and D are now permuted!). Let $\varphi \in \mathrm{End}_R(E)$. Let $v \in E$, $v \neq 0$. By the density theorem, there exists an element $a \in D$ such that $\varphi(v) = av$. Let $w \in E$. There exists an element $f \in R$ such that $f(v) = w$. Then

$$\varphi(w) = \varphi(f(v)) = f(\varphi(v)) = f(av) = af(v) = aw.$$

Therefore $\varphi(w) = aw$ for all $w \in E$. This means that $\varphi \in D$, and concludes our proof.

Theorem 7. *Let k be a field and E a finite dimensional vector space of dimension m over k. Let $R = \mathrm{End}_k(E)$. Then R is a k-space, and*

$$\dim_k R = m^2.$$

Furthermore, m is the number of simple left ideals appearing in a direct sum decomposition of R as such a sum.

Proof. The k-space of k-endomorphisms of E is represented by the space of $m \times m$ matrices in k, so the dimension of R as a k-space is m^2. On the other hand, the proof of Theorem 6 showed that R is R-isomorphic as an R-module to the direct sum $E^{(m)}$. We know the uniqueness of the decomposition of a module into a direct sum of simple modules (Proposition 2 of §1), and this proves our assertion.

In the terminology introduced in §2, we see that the integer m in Theorem 7 is the length of R.

We can identify $R = \mathrm{End}_k(E)$ with the ring of matrices $\mathrm{Mat}_m(k)$, once a basis of E is selected. In that case, we can take the simple left ideals to be the ideals L_i $(i = 1, \ldots, m)$ where a matrix in L_i has coefficients equal to 0 except in the i-th column. An element of L_1 thus looks like

$$\begin{pmatrix} a_{11} & 0 & \cdots & 0 \\ a_{21} & 0 & \cdots & 0 \\ \vdots & \vdots & & \vdots \\ a_{m1} & 0 & \cdots & 0 \end{pmatrix}.$$

We see that R is the direct sum of the m columns.

We also observe that Theorem 6 implies the following: *If a matrix $M \in \mathrm{Mat}_m(k)$ commutes with all elements of $\mathrm{Mat}_m(k)$, then M is a scalar matrix.*

Indeed, such a matrix M can then be viewed as an R-endomorphism of E, and we know by Theorem 6 that such an endomorphism lies in k. Of course, one can also verify this directly by a brute force computation.

§6. Balanced modules

Let R be a ring and E a module. We let $R'(E) = \mathrm{End}_R(E)$ and $R''(E) = \mathrm{End}_{R'}(E)$. Let $\lambda : R \to R''$ be the natural homomorphism such that $\lambda_x(v) = xv$ for $x \in R$ and $v \in E$. If λ is an isomorphism, we shall say that E is *balanced.* We shall say that E is a *generator* (for R-modules) if every module is a homomorphic image of a (possibly infinite) direct sum of E with itself. For example, R is a generator. If E is a generator, then there is a surjective homomorphism $E^{(n)} \to R$ (we can take n finite since R is finitely generated, by one element 1).

THEOREM 8. (Morita) *Every generator E is balanced, and is finitely generated over $R'(E)$.*

Proof. (Faith) We first prove that for any module F, $R \oplus F$ is balanced. We identify R and F as the submodules $R \oplus 0$ and $0 \oplus F$ of $R \oplus F$ respectively. For $w \in F$, let $\psi_w : R \oplus F \to R \oplus F$ be the map $\psi_w(x + v) = xw$. Then any $f \in R''(R \oplus F)$ commutes with π_1, π_2, and each ψ_w. From this we see at once that $f(x + v) = f(1)(x + v)$ and hence that $R \oplus F$ is balanced. Let E be a generator, and $E^{(n)} \to R$ a surjective homomorphism. Since R is free, we can write $E^{(n)} \approx R \oplus F$ for some module F, so that $E^{(n)}$ is balanced. Let $g \in R'(E)$. Then $g^{(n)}$ commutes with every element $\varphi = (\varphi_{ij})$ in $R'(E^{(n)})$ (with components $\varphi_{ij} \in R'(E)$), and hence there is some $x \in R$ such that $g^{(n)} = \lambda_x^{(n)}$. Hence $g = \lambda_x$, thereby proving that E is balanced, since λ is obviously injective.

To prove that E is finitely generated over $R'(E)$, we have

$$R'(E)^{(n)} \approx \mathrm{Hom}_R(E^{(n)}, E) \approx \mathrm{Hom}_R(R, E) \oplus \mathrm{Hom}_R(F, E)$$

as additive groups. This relation also obviously holds as R'-modules if we define the operation of R' to be composition of mappings (on the left). Since $\mathrm{Hom}_R(R, E)$ is R'-isomorphic to E under the map $h \mapsto h(1)$, it follows that E is an R'-homomorphic image of $R'^{(n)}$, whence finitely generated over R', thereby proving the theorem.

EXERCISES

1. (a) Let R be a ring. We define the *radical* of R to be the left ideal N which is the intersection of all maximal left ideals of R. Show that $NE = 0$ for every simple R-module E. Show that N is a two-sided ideal. (b) Show that the radical of R/N is 0.

2. A ring is said to be *Artinian* if every descending sequence of left ideals $\mathfrak{a}_1 \supset \mathfrak{a}_2 \supset \cdots$ with $\mathfrak{a}_i \neq \mathfrak{a}_{i+1}$ is finite. (a) Show that a finite dimensional algebra over a field is Artinian. (b) If R is Artinian, show that every non-zero left ideal contains a simple left ideal. (c) If R is Artinian, show that every non-empty set of ideals contains a minimal ideal.

3. Let R be Artinian, and assume that its radical is 0. Show that R is semisimple. [*Hint:* Get an injection of R into a direct sum $\coprod R/M_i$ where $\{M_i\}$ is a finite set of maximal left ideals.]

4. Let R be any ring and M a finitely generated module. Let N be the radical of R. If $NM = M$ show that $M = 0$. [*Hint:* Observe that the proof of Nakayama's lemma still holds.]

5. Let R be Artinian. Show that its radical is nilpotent, i.e. that there exists an integer $r \geqq 1$ such that $N^r = 0$. [*Hint:* Consider the descending sequence of powers N^r, and apply Nakayama to suitably selected submodules of N^∞.]

6. Let R be a semisimple commutative ring. Show that R is a direct product of fields.

7. Let R be a finite dimensional commutative algebra over a field k. If R has no nilpotent element $\neq 0$, show that R is semisimple.

8. (Kolchin) Let E be a finite dimensional vector space over a field k. Let G be a subgroup of $GL(E)$ such that every element $A \in G$ is of type $I + N$ where N is nilpotent. Assume $E \neq 0$. Show that there exists an element $v \in E, v \neq 0$ such that $Av = v$ for all $A \in G$. [*Hint:* First reduce the question to the case when k is algebraically closed by showing that the problem amounts to solving linear equations. Secondly, reduce it to the case when E is a simple $k[G]$-module. Combining Burnside's theorem with the fact that $\text{tr}(A) = \text{tr}(I)$ for all $A \in G$, show that if $A_0 \in G$, $A_0 = I + N$, then $\text{tr}(NX) = 0$ for all $X \in \text{End}_k(E)$, and hence that $N = 0$, $A_0 = I$.]

9. Let E be a finite dimensional vector space over a field k. Let S be a subset of $\text{End}_k(E)$. Let R be the k-algebra generated by the elements of S. Prove that the following conditions are equivalent: R is semisimple, E is a semisimple R-module.

10. Let $A \in \text{End}_k(E)$. Then A is said to be semisimple if the set consisting of A itself satisfies the conditions of the preceding exercise. Show that an element A of $\text{End}_k(E)$ is semisimple if and only if its minimal polynomial has no factors of multiplicity > 1 over k.

11. Let E be a finite dimensional vector space over a field k, and let S be a commutative set of endomorphisms of E. Let $R = k[S]$. Assume that R is semisimple. Show that every subset of S is semisimple.

12. Prove that an R-module E is a generator if and only if it is balanced, and finitely generated projective over $R'(E)$. Show that Theorem 5 is a consequence of Theorem 8.

CHAPTER XVIII

Representations of Finite Groups

§1. *Semisimplicity of the group algebra*

Let k be a field and G a group. We form the group algebra $k[G]$. As explained in Chapter V, §1 it consists of all formal linear combinations

$$\sum_{\sigma \in G} a_\sigma \sigma$$

with coefficients $a_\sigma \in k$, almost all of which are 0. The product is taken in the natural way,

$$\left(\sum_{\sigma \in G} a_\sigma \sigma\right)\left(\sum_{\tau \in G} b_\tau \tau\right) = \sum_{\sigma, \tau} a_\sigma a_\tau \sigma\tau.$$

Let E be a k-vector space. Every algebra-homomorphism

$$k[G] \to \mathrm{End}_k(E)$$

induces a group-homomorphism

$$G \to \mathrm{Aut}_k(E),$$

and thus a representation of the ring $k[G]$ in E gives rise to a representation of the group. Given such representations, we also say that $k[G]$, or G, *operate* on E. We note that the representation makes E into a module over the ring $k[G]$.

Conversely, given a representation of the group, say $\rho : G \to \mathrm{Aut}_k(E)$, we can extend ρ to a representation of $k[G]$ as follows. Let $\alpha = \sum a_\sigma \sigma$ and $x \in E$. We define

$$\rho(\alpha)x = \sum a_\sigma \rho(\sigma)x.$$

It is immediately verified that ρ has been extended to a ring-homomorphism of $k[G]$ into $\mathrm{End}_k(E)$. We say that ρ is *faithful* on G if the map $\rho : G \to \mathrm{Aut}_k(E)$ is injective. The extension of ρ to $k[G]$ may not be faithful, however.

Given a representation of G on E, we often write simply σx instead of $\rho(\sigma)x$, whenever we deal with a fixed representation throughout a discussion. A vector space E, together with a representation ρ, will be called a *G-module*, or *G-space*, or also a (G, k)-space if we wish to specify the field k. If E, F are G-modules, we recall that a G-homomorphism $f : E \to F$ is a k-linear map such that $f(\sigma x) = \sigma f(x)$ for all $x \in E$ and $\sigma \in G$.

Given a G-homomorphism $f : E \to F$, we note that the kernel of f is a G-submodule of E, and that the k-factor space $F/f(E)$ admits an operation of G in a unique way such that the canonical map $F \to F/f(E)$ is a G-homomorphism.

If G operates on k-spaces E and F, then we can define an operation of G on $\mathrm{Hom}_k(E, F)$ in a natural way. Indeed, if $f \in \mathrm{Hom}_k(E, F)$, and $\sigma \in G$, then

$$(\sigma f)(x) = \sigma\big(f(\sigma^{-1}x)\big).$$

Then $(\sigma\tau)f = \sigma\big(\tau(f)\big)$. There will be no confusion with the iteration of σ and f. When we want to denote this iteration, we write $\sigma \circ f$ for the map such that $(\sigma \circ f)(x) = \sigma\big(f(x)\big)$, and similarly for $f \circ \sigma$. We note that f is a G-homomorphism if and only if $\sigma f = f$ for all $\sigma \in G$.

If E is a G-module, we denote by E^G the submodule consisting of all elements $x \in E$ such that $\sigma x = x$.

By a *trivial* representation $\rho : G \to \mathrm{Aut}_k(E)$, we shall mean the representation such that $\rho(G) = 1$. A representation is trivial if and only if $\sigma x = x$ for all $x \in E$. We also say in that case that G *operates trivially*. We can then write $E = E^G$.

If G is a finite group, and E is a G-module, then we can define an operation $\mathrm{Tr}_G : E \to E$ which is a k-homomorphism, namely

$$\mathrm{Tr}_G(x) = \sum_{\sigma \in G} \sigma x.$$

We observe that $\mathrm{Tr}_G(x)$ lies in E^G, i.e. is fixed under the operation of all elements of G. This is because

$$\tau\,\mathrm{Tr}_G(x) = \sum_{\sigma \in G} \tau\sigma x,$$

and multiplying by τ on the left permutes the elements of G.

In particular, if $f : E \to F$ is a k-homomorphism of G-modules, then $\mathrm{Tr}_G(f) : E \to F$ is a G-homomorphism.

PROPOSITION 1. *Let G be a finite group and let E', E, F, F' be G-modules. Let*

$$E' \xrightarrow{\varphi} E \xrightarrow{f} F \xrightarrow{\psi} F'$$

be k-homomorphisms, and assume that φ, ψ are G-homomorphisms. Then

$$\mathrm{Tr}_G(\psi \circ f \circ \varphi) = \psi \circ \mathrm{Tr}_G(f) \circ \varphi.$$

Proof. We have

$$\begin{aligned}\mathrm{Tr}_G(\psi \circ f \circ \varphi) &= \sum_{\sigma \in G} \sigma(\psi \circ f \circ \varphi) = \sum_{\sigma \in G} (\sigma\psi) \circ (\sigma f) \circ (\sigma\varphi) \\ &= \psi \circ \left(\sum_{\sigma \in G} \sigma f \right) \circ \varphi = \psi \circ \mathrm{Tr}_G(f) \circ \varphi.\end{aligned}$$

THEOREM 1. (Maschke) *Let G be a finite group of order n, and let k be a field whose characteristic does not divide n. Then the group ring $k[G]$ is semisimple.*

Proof. Let E be a G-module, and F a G-submodule. Since k is a field, there exists a k-subspace F' such that E is the k-direct sum of F and F'. We let the k-linear map $\pi : E \to F$ be the projection on F. Then $\pi(x) = x$ for all $x \in F$. Let

$$\varphi = \frac{1}{n} \mathrm{Tr}_G(\pi).$$

We have then two G-homomorphisms

$$0 \to F \underset{\varphi}{\overset{j}{\rightleftarrows}} E$$

such that j is the inclusion, and $\varphi \circ j = id$. It follows that E is the G-direct sum of F and $\mathrm{Ker}\, \varphi$, thereby proving that $k[G]$ is semisimple.

In the sequel, we assume that G is a finite group, and that all vector spaces E over k are finite dimensional. We usually denote by n the order of G, and assume throughout that the characteristic of k does not divide the order of G.

§2. *Characters*

Let $\rho : k[G] \to \mathrm{End}_k(E)$ be a representation. By the *character* χ_ρ of the representation, we shall mean the k-valued function

$$\chi_\rho : k[G] \to k$$

such that $\chi_\rho(\alpha) = \mathrm{tr}\, \rho(\alpha)$ for all $\alpha \in k[G]$. The trace here is the trace of an endomorphism, as defined in Chapter XIII, §3. If we select a basis for E over k, it is the trace of the matrix representing $\rho(\alpha)$, i.e. the sum of the diagonal elements. We have seen previously that the trace does not depend on the choice of the basis. We sometimes write χ_E instead of χ_ρ.

We also call E the *representation space* of ρ.

By the *trivial character* we shall mean the character of the representation of G on the k-space equal to k itself, such that $\sigma x = x$ for all $x \in k$. It is the function taking the value 1 on all elements of G. We denote it by χ_0 or also by 1_G if we need to specify the dependence on G.

We observe that characters are functions on G, and that the values of a character on elements of $k[G]$ are determined by its values on G (the extension from G to $k[G]$ being by k-linearity).

We say that two representations ρ, φ of G on spaces E, F are *isomorphic* if there is a G-isomorphism between E and F. We then see that if ρ, φ are isomorphic representations, then their characters are equal. (Put in another way, if E, F are G-spaces and are G-isomorphic, then $\chi_E = \chi_F$.) In everything that follows, we are interested only in isomorphism classes of representations.

If E, F are G-spaces, then their direct sum $E \oplus F$ is also a G-space, the operation of G being componentwise. If $x \oplus y \in E \oplus F$ with $x \in E$ and $y \in F$, then $\sigma(x \oplus y) = \sigma x \oplus \sigma y$.

Similarly, the tensor product $E \otimes_k F = E \otimes F$ is a G-space, the operation of G being given by $\sigma(x \otimes y) = \sigma x \otimes \sigma y$.

PROPOSITION 2. *If E, F are G-spaces, then*

$$\chi_E + \chi_F = \chi_{E \oplus F} \quad \text{and} \quad \chi_E \chi_F = \chi_{E \otimes F}.$$

Proof. The first relation holds because the matrix of an element σ in the representation $E \oplus F$ decomposes into blocks corresponding to the representation in E and the representation in F. As to the second, if $\{v_i\}$ is a basis of E and $\{w_j\}$ is a basis of F over k, then we know that $\{v_i \otimes w_j\}$ is a basis of $E \otimes F$. Let $(a_{i\nu})$ be the matrix of σ with respect to our basis of E, and $(b_{j\mu})$ its matrix with respect to our basis of F. Then

$$\sigma(v_i \otimes w_j) = \sigma v_i \otimes \sigma w_j = \sum_\nu a_{i\nu} v_\nu \otimes \sum_\mu b_{j\mu} w_\mu$$

$$= \sum_{\nu,\mu} a_{i\nu} b_{j\mu}\, v_\nu \otimes w_\mu.$$

By definition, we find

$$\chi_{E \otimes F}(\sigma) = \sum_i \sum_j a_{ii} b_{jj} = \chi_E(\sigma)\chi_F(\sigma),$$

thereby proving our proposition.

If $\rho : G \to \mathrm{Aut}_k(E)$ and $\varphi : G \to \mathrm{Aut}_k(F)$ are representations of G on E and F respectively, then we define the *sum* $\rho + \varphi$ to be the above described representation on $E \oplus F$. We see that the sum of the characters is the character of the sum of the representations. In particular, characters of G associated with representations of G on k-spaces form a monoid.

Similarly, we define the product $\rho \otimes \varphi$ to be the representation associated with the tensor product of the representation spaces for ρ and φ respectively. Thus the additive monoid of characters associated with representations has a multiplicative structure, which is distributive with respect to addition.

So far, we have defined the notion of character associated with a representation. It is now natural to form linear combinations of such characters with more general coefficients than positive integers. Thus by a *character* of G we shall mean a function on G which can be written as a linear combination of characters of representations with arbitrary integer coefficients. The characters associated with representations will be called *proper characters.* Everything we have defined of course depends on the field k, and we shall add *over* k to our expressions if we need to specify the field k.

We observe that the characters form a ring in view of Proposition 2. For most of our work we do not need the multiplicative structure, only the additive one.

By a *simple character* of G one means the character of a simple representation (i.e. the character associated with a simple $k[G]$-module).

Taking into account Theorem 1, and the results of the preceding chapter concerning the structure of simple and semisimple modules over a semisimple ring (Chapter XVII, §4) we obtain:

THEOREM 2. *There are only a finite number of simple characters of G (over k). The characters of representations of G are the linear combinations of the simple characters with integer coefficients $\geqq 0$.*

We shall use the direct product decomposition of a semisimple ring. We have

$$k[G] = \prod_{i=1}^{s} R_i$$

where each R_i is simple, and we have a corresponding decomposition of the unit element of $k[G]$:

$$1 = e_1 + \cdots + e_s,$$

where e_i is the unit element of R_i, and $e_i e_j = 0$ if $i \neq j$. Also, $R_i R_j = 0$ if $i \neq j$. *We note that $s = s(k)$ depends on k.*

If L_i denotes a typical simple module for R_i (say one of the simple left ideals), we let χ_i be the character of the representation on L_i.

We observe that $\chi_i(\alpha) = 0$ for all $\alpha \in R_j$ if $i \neq j$. This is a fundamental relation of orthogonality, which is obvious, but from which all our other relations will follow.

THEOREM 3. *Assume that k has characteristic* 0. *Then every proper character has a unique expression as a linear combination*

$$\chi = \sum_{i=1}^{s} n_i \chi_i, \qquad n_i \in \mathbf{Z},\ n_i \geqq 0,$$

where $\chi_1, \ldots, \chi_s$ are the simple characters of G over k. Two representations are isomorphic if and only if their associated characters are equal.

Proof. Let E be the representation space of χ. Then by Theorem 3 of Chapter XVII, §4,

$$E \approx \coprod_{i=1}^{s} n_i L_i.$$

The sum is finite because we assume throughout that E is finite dimensional. Since e_i acts as a unit element on L_i, we find

$$\chi_i(e_i) = \dim_k L_i.$$

We have already seen that $\chi_i(e_j) = 0$ if $i \neq j$. Hence

$$\chi(e_i) = n_i \dim_k L_i.$$

Since $\dim_k L_i$ depends only on the structure of the group algebra, we have recovered the multiplicities $n_1, \ldots, n_s$. Namely, n_i is the number of times that L_i occurs (up to an isomorphism) in the representation space of χ, and is the value of $\chi(e_i)$ divided by $\dim_k L_i$ (we are in characteristic 0). This proves our theorem.

As a matter of definition, in Theorem 3 we call n_i the *multiplicity* of χ_i in χ.

In both corollaries, we continue to assume that k has characteristic 0.

COROLLARY 1. *As functions of G into k, the simple characters*

$$\chi_1, \ldots, \chi_s$$

are linearly independent over k.

Proof. Suppose that $\sum a_i \chi_i = 0$ with $a_i \in k$. We apply this expression to e_j and get

$$0 = \left(\sum a_i \chi_i\right)(e_j) = a_j \dim_k L_j.$$

Hence $a_j = 0$ for all j.

In characteristic 0 we define the *dimension* of a proper character to be the dimension of the associated representation space.

COROLLARY 2. *The function* dim *is a homomorphism of the monoid of proper characters into* $\mathbf{Z}$.

Example. Let G be a cyclic group of order equal to a prime number p. We form the group algebra $\mathbf{Q}[G]$. Let σ be a generator of G. Let

$$e_1 = \frac{1 + \sigma + \sigma^2 + \cdots + \sigma^{p-1}}{p}, \qquad e_2 = 1 - e_1.$$

Then $\tau e_1 = e_1$ for any $\tau \in G$ and consequently $e_1^2 = e_1$. It then follows that $e_2^2 = e_2$ and $e_1 e_2 = 0$. The field $\mathbf{Q}e_1$ is isomorphic to $\mathbf{Q}$. Let $\omega = \sigma e_2$. Then $\omega^p = e_2$. Let $\mathbf{Q}_2 = \mathbf{Q}e_2$. Since $\omega \neq e_2$, and satisfies the irreducible equation

$$X^{p-1} + \cdots + 1 = 0$$

over $\mathbf{Q}_2$, it follows that $\mathbf{Q}_2(\omega)$ is isomorphic to the field obtained by adjoining a primitive p-th root of unity to the rationals. Consequently, $\mathbf{Q}[G]$ admits the direct product decomposition

$$\mathbf{Q}[G] \approx \mathbf{Q} \times \mathbf{Q}(\zeta)$$

where ζ is a primitive p-th root of unity.

As another example, let G be any finite group, and let

$$e_1 = \frac{1}{n} \sum_{\sigma \in G} \sigma.$$

Then for any $\tau \in G$ we have $\tau e_1 = e_1$, and $e_1^2 = e_1$. If we let $e_1' = 1 - e_1$ then $e_1'^2 = e_1'$, and $e_1' e_1 = e_1 e_1' = 0$. Thus for any field k (whose characteristic does not divide the order of G according to conventions in force), we see that

$$k[G] = ke_1 \times k[G]e_1'$$

is a direct product decomposition. In particular, the representation of G on the group algebra $k[G]$ itself contains a 1-dimensional representation on the component ke_1, whose character is the trivial character.

§3. *One-dimensional representations*

By abuse of language, even in characteristic $p > 0$, we say that a *character is* 1-*dimensional* if it is a homomorphism $G \to k^*$.

Assume that E is a one-dimensional vector space over k. Let

$$\rho : G \to \mathrm{Aut}_k(E)$$

be a representation. Let $\{v\}$ be a basis of E over k. Then for each $\sigma \in G$, we have

$$\sigma v = \chi(\sigma) v$$

for some element $\chi(\sigma) \in k$, and $\chi(\sigma) \neq 0$ since σ induces an automorphism of E. Then for $\tau \in G$,

$$\tau\sigma v = \chi(\sigma)\tau v = \chi(\sigma)\chi(\tau)v = \chi(\sigma\tau)v.$$

We see that $\chi : G \to k^*$ is a homomorphism, and that our 1-dimensional character is the same type of thing that occurred in Artin's theorem in Galois theory.

Conversely, let $\chi : G \to k^*$ be a homomorphism. Let E be a 1-dimensional k-space, with basis $\{v\}$, and define $\sigma(av) = a\chi(\sigma)v$ for all $a \in k$. Then we see at once that this operation of G on E gives a representation of G, whose associated character is χ.

Since G is finite, we note that

$$\chi(\sigma)^n = \chi(\sigma^n) = \chi(1) = 1.$$

Hence the values of 1-dimensional characters are n-th roots of unity. The 1-dimensional characters form a group under multiplication, and when G is a finite abelian group, we have determined its group of 1-dimensional characters in Chapter I, §11.

THEOREM 4. *Let G be a finite abelian group, and assume that k is algebraically closed. Then every simple representation of G is 1-dimensional. The simple characters of G are the homomorphisms of G into k^*.*

Proof. The group ring $k[G]$ is semisimple, commutative, and is a direct product of simple rings. Each simple ring is a ring of matrices over k (by Theorem 5, §5 of the preceding chapter), and can be commutative if and only if it is equal to k.

For every 1-dimensional character χ of G we have

$$\chi(\sigma)^{-1} = \chi(\sigma^{-1}).$$

If k is the field of complex numbers, then

$$\overline{\chi(\sigma)} = \chi(\sigma)^{-1} = \chi(\sigma^{-1}).$$

COROLLARY. *Let k be algebraically closed. Let G be a finite group. For any character χ and $\sigma \in G$, the value $\chi(\sigma)$ is equal to a sum of roots of unity with integer coefficients (i.e. coefficients in $\mathbf{Z}$ or $\mathbf{Z}/p\mathbf{Z}$ depending on the characteristic of k).*

Proof. Let H be the subgroup generated by σ. Then H is a cyclic subgroup. A representation of G having character χ can be viewed as a representation for H by restriction, having the same character. Thus our assertion follows from Theorem 4.

§4. *The space of class functions*

Let k be a field. By a *class function* of G (over k, or with values in k), we shall mean a function $f : G \to k$ such that $f(\sigma\tau\sigma^{-1}) = f(\tau)$ for all $\sigma, \tau \in G$. It is clear that characters are class functions, because for square matrices M, M' we have

$$\mathrm{tr}(MM'M^{-1}) = \mathrm{tr}(M').$$

Thus a class function may be viewed as a function on conjugacy classes.

We shall always extend the domain of definition of a class function to the group ring, by linearity. If

$$\alpha = \sum_{\sigma \in G} a_\sigma \sigma,$$

and f is a class function, we define

$$f(\alpha) = \sum_{\sigma \in G} a_\sigma f(\sigma).$$

Let $\sigma_0 \in G$. If $\sigma \in G$, we write $\sigma \sim \sigma_0$ if σ is conjugate to σ_0, that is, if there exists an element τ such that $\sigma_0 = \tau\sigma\tau^{-1}$. An element of the group ring of type

$$\gamma = \sum_{\sigma \sim \sigma_0} \sigma$$

will also be called a *conjugacy class*.

PROPOSITION 3. *Let k be any field. An element of $k[G]$ commutes with every element of G if and only if it is a linear combination of conjugacy classes with coefficients in k.*

Proof. Let $\alpha = \sum_{\sigma \in G} a_\sigma \sigma$ and assume $\alpha\tau = \tau\alpha$ for all $\tau \in G$. Then

$$\sum_{\sigma \in G} a_\sigma \tau\sigma\tau^{-1} = \sum_{\sigma \in G} a_\sigma \sigma.$$

Hence $a_{\sigma_0} = a_\sigma$ whenever σ is conjugate to σ_0, and this means that we can write

$$\alpha = \sum_{\gamma} a_\gamma \gamma$$

where the sum is taken over all conjugacy classes γ.

Remark. We note that the conjugacy classes in fact form a basis of $\mathbf{Z}[G]$ over $\mathbf{Z}$, and thus play a universal role in the theory of representations.

We observe that the conjugacy classes are linearly independent over k, and form a basis for the center of $k[G]$ over k.

Assume from now on that k is algebraically closed. Then

$$k[G] = \prod_{i=1}^{s} R_i$$

is a direct product of simple rings, and each R_i is a matrix algebra over k. In a direct product, the center is obviously the product of the centers of each factor. Let us denote by k_i the image of k in R_i, in other words,

$$k_i = ke_i,$$

where e_i is the unit element of R_i. Then the center of $k[G]$ is also equal to

$$\prod_{i=1}^{s} k_i$$

which is s-dimensional over k.

If L_i is a typical simple left ideal of R_i, then

$$R_i \approx \mathrm{End}_k(L_i).$$

We let

$$d_i = \dim_k L_i.$$

Then

$$\boxed{d_i^2 = \dim_k R_i \qquad \text{and} \qquad \sum_{i=1}^{s} d_i^2 = n.}$$

We also have the direct sum decomposition

$$R_i \approx L_i^{(d_i)}$$

as a (G,k)-space.

The above notation will remain fixed from now on.

We can summarize some of our results as follows.

PROPOSITION 4. *Let k be algebraically closed. Then the number of conjugacy classes of G is equal to the number of simple characters of G, both of these being equal to the number s above. The conjugacy classes $\gamma_1, \ldots, \gamma_s$ and the unit elements $e_1, \ldots, e_s$ form bases of the center of $k[G]$.*

The number of elements in γ_i will be denoted by h_i. The number of elements in a conjugacy class γ will be denoted by h_γ. We call it the *class number*. The center of the group algebra will be denoted by $Z_k(G)$.

We can view $k[G]$ as a G-module. Its character will be called the *regular character*, and will be denoted by χ_{reg} or r_G if we need to specify the dependence on G. The representation on $k[G]$ is called the *regular representation.* From our direct sum decomposition of $k[G]$ we get

$$\boxed{\chi_{\text{reg}} = \sum_{i=1}^{s} d_i \chi_i.}$$

We shall determine the values of the regular character.

PROPOSITION 5. *Let χ_{reg} be the regular character. Then*

$$\begin{aligned} \chi_{\text{reg}}(\sigma) &= 0 \qquad \text{if} \quad \sigma \in G,\ \sigma \neq 1 \\ \chi_{\text{reg}}(1) &= n. \end{aligned}$$

Proof. Let $1 = \sigma_1, \ldots, \sigma_n$ be the elements of G. They form a basis of $k[G]$ over k. The matrix of 1 is the unit $n \times n$ matrix. Thus our second assertion follows. If $\sigma \neq 1$, then multiplication by σ permutes $\sigma_1, \ldots, \sigma_n$, and it is immediately clear that all diagonal elements in the matrix representing σ are 0. This proves what we wanted.

We observe that we have two natural bases for the center $Z_k(G)$ of the group ring. First, the conjugacy classes of elements of G. Second, the elements $e_1, \ldots, e_s$ (i.e. the unit elements of the rings R_i). We wish to find the relation between these, in other words, we wish to find the coefficients of e_i when expressed in terms of the group elements. The next proposition does this. The values of these coefficients will be interpreted in the next section as scalar products. This will clarify their mysterious appearance.

PROPOSITION 6. *Assume again that k is algebraically closed. Let*

$$e_i = \sum_{\tau \in G} a_\tau \tau, \qquad a_\tau \in k.$$

Then

$$a_\tau = \frac{1}{n} \chi_{\text{reg}}(e_i \tau^{-1}) = \frac{d_i}{n} \chi_i(\tau^{-1}).$$

Proof. We have for all $\tau \in G$:

$$\chi_{\text{reg}}(e_i \tau^{-1}) = \chi_{\text{reg}}\left(\sum_{\sigma \in G} a_\sigma \sigma \tau^{-1}\right) = \sum_{\sigma \in G} a_\sigma \chi_{\text{reg}}(\sigma \tau^{-1}).$$

By Proposition 5, we find

$$\chi_{\text{reg}}(e_i \tau^{-1}) = n a_\tau.$$

On the other hand,

$$\chi_{\text{reg}}(e_i\tau^{-1}) = \sum_{j=1}^{s} d_j\chi_j(e_i\tau^{-1}) = d_i\chi_i(e_i\tau^{-1}) = d_i\chi_i(\tau^{-1}).$$

Hence

$$d_i\chi_i(\tau^{-1}) = na_\tau$$

for all $\tau \in G$. This proves our proposition.

COROLLARY 1. *Each e_i can be expressed in terms of group elements with coefficients which lie in the field generated over the prime field by m-th roots of unity, if m is an exponent for G.*

COROLLARY 2. *The dimensions d_i are not divisible by the characteristic of k.*

Proof. Otherwise, $e_i = 0$, which is impossible.

COROLLARY 3. *The simple characters $\chi_1, \ldots, \chi_s$ are linearly independent over k.*

Proof. The proof in Corollary 1 of Theorem 3 applies, since we now know that the characteristic does not divide d_i.

COROLLARY 4. *Assume in addition that k has characteristic 0. Then $d_i \mid n$ for each i.*

Proof. Multiplying our expression for e_i by n/d_i, and also by e_i, we find

$$\frac{n}{d_i} e_i = \sum_{\sigma \in G} \chi_i(\sigma^{-1})\sigma e_i.$$

Let ζ be a primitive m-th root of unity, and let M be the module over $\mathbf{Z}$ generated by the finite number of elements $\zeta^\nu \sigma e_i$ ($\nu = 0, \ldots, m - 1$ and $\sigma \in G$). Then from the preceding relation, we see at once that multiplication by n/d_i maps M into itself. By definition, we conclude that n/d_i is integral over $\mathbf{Z}$, and hence lies in $\mathbf{Z}$, as desired.

THEOREM 5. *Let k be algebraically closed. Let $Z_k(G)$ be the center of $k[G]$, and let $X_k(G)$ be the k-space of class functions on G. Then $Z_k(G)$ and $X_k(G)$ are the dual spaces of each other, under the pairing*

$$(f, \alpha) \mapsto f(\alpha).$$

The simple characters and the unit elements $e_1, \ldots, e_s$ form orthogonal bases to each other. We have

$$\chi_i(e_j) = \delta_{ij}\, d_i.$$

Proof. The formula has been proved in the proof of Theorem 3. The two spaces involved here both have dimension s, and $d_i \neq 0$ in k. Our proposition is then clear.

§5. Orthogonality relations

Throughout this section, we assume that k is algebraically closed.

If R is a subring of k, we denote by $X_R(G)$ the R-module generated over R by the characters of G. It is therefore the module of functions which are linear combinations of simple characters with coefficients in R. If R is the prime ring (i.e. the integers $\mathbf{Z}$ or the integers mod p if k has characteristic p), then we denote $X_R(G)$ by $X(G)$.

We shall now define a bilinear map on $X(G) \times X(G)$. If $f, g \in X(G)$, we define

$$\langle f, g \rangle = \frac{1}{n} \sum_{\sigma \in G} f(\sigma) g(\sigma^{-1}).$$

THEOREM 6. *The symbol $\langle f, g \rangle$ for $f, g \in X(G)$ takes on values in the prime ring. The simple characters form an orthonormal basis for $X(G)$, in other words*

$$\langle \chi_i, \chi_j \rangle = \delta_{ij}.$$

For each ring $R \subset k$, the symbol has a unique extension to an R-bilinear form $X_R(G) \times X_R(G) \to R$, given by the same formula as above.

Proof. By Proposition 6, we find

$$\chi_j(e_i) = \frac{d_i}{n} \sum_{\sigma \in G} \chi_i(\sigma^{-1}) \chi_j(\sigma).$$

If $i \neq j$ we get 0 on the left-hand side, so that χ_i and χ_j are orthogonal. If $i = j$ we get d_i on the left-hand side, and we know that $d_i \neq 0$ in k, by Corollary 2 of Proposition 6. Hence $\langle \chi_i, \chi_i \rangle = 1$. Since every element of $X(G)$ is a linear combination of simple characters with integer coefficients, it follows that the values of our bilinear map are in the prime ring. The extension statement is obvious, thereby proving our theorem.

Assume that k has characteristic 0. Let m be an exponent for G, and let R contain the m-th roots of unity. If R has an automorphism of order 2 such that its effect on a root of unity is $\zeta \mapsto \zeta^{-1}$, then we shall call such an automorphism a *conjugation*, and denote it by $a \mapsto \bar{a}$.

THEOREM 7. *Let k have characteristic* 0, *and let R be a subring containing the m-th roots of unity, and having a conjugation. Then the bi-*

linear form on $X(G)$ has a unique extension to a hermitian form

$$X_R(G) \times X_R(G) \to R,$$

given by the formula

$$\langle f, g \rangle = \frac{1}{n} \sum_{\sigma \in G} f(\sigma)\overline{g(\sigma)}.$$

The simple characters constitute an orthonormal basis of $X_R(G)$ with respect to this form.

Proof. The formula given in the statement of the theorem gives the same value as before for the symbol $\langle f, g \rangle$ when f, g lie in $X(G)$. Thus the extension exists, and is obviously unique.

We return to the case when k has arbitrary characteristic.

Let $Z(G)$ denote the additive group generated by the conjugacy classes $\gamma_1, \ldots, \gamma_s$ over the prime ring. It is of dimension s. We shall define a bilinear map on $Z(G) \times Z(G)$. If $\alpha = \sum a_\sigma \sigma$ has coefficients in the prime ring, we denote by α^- the element $\sum a_\sigma \sigma^{-1}$.

PROPOSITION 7. *For $\alpha, \beta \in Z(G)$, we can define a symbol $\langle \alpha, \beta \rangle$ by either one of the following expressions, which are equal:*

$$\langle \alpha, \beta \rangle = \frac{1}{n} \chi_{\text{reg}}(\alpha\beta^-) = \frac{1}{n} \sum_{\nu=1}^{s} \chi_\nu(\alpha)\chi_\nu(\beta^-).$$

The values of the symbol lie in the prime ring.

Proof. Each expression is linear in its first and second variable. Hence to prove their equality, it will suffice to prove that the two expressions are equal when we replace α by e_i and β by an element τ of G. But then, our equality is equivalent to

$$\chi_{\text{reg}}(e_i\tau^{-1}) = \sum_{\nu=1}^{s} \chi_\nu(e_i)\chi_\nu(\tau^{-1}).$$

Since $\chi_\nu(e_i) = 0$ unless $\nu = i$, we see that the right-hand side of this last relation is equal to $d_i\chi_i(\tau^{-1})$. Our two expressions are equal in view of Proposition 6. The fact that the values lie in the prime ring follows from Proposition 5: The values of the regular character on group elements are equal to 0 or n, and hence in characteristic 0, are integers divisible by n.

As with $X_R(G)$, we use the notation $Z_R(G)$ to denote the R-module generated by $\gamma_1, \ldots, \gamma_s$ over an arbitrary subring R of k.

LEMMA. *For each ring R contained in k, the pairing of Proposition 7 has a unique extension to a map*

$$Z_R(G) \times Z(G) \to R$$

which is R-linear in its first variable. If R contains the m-th roots of unity, where m is an exponent for G, and also contains $1/n$, *then* $e_i \in Z_R(G)$ *for all i. The class number* h_i *is not divisible by the characteristic of k, and we have*

$$e_i = \sum_{\nu=1}^{s} \langle e_i, \gamma_\nu \rangle \frac{1}{h_\nu} \gamma_\nu.$$

Proof. We note that h_i is not divisible by the characteristic because it is the index of a subgroup of G (the isotropy group of an element in γ_i when G operates by conjugation), and hence h_i divides n. The extension of our pairing as stated is obvious, since $\gamma_1, \ldots, \gamma_s$ form a basis of $Z(G)$ over the prime ring. The expression of e_i in terms of this basis is only a reinterpretation of Proposition 6 in terms of the present pairing.

Let E be a free module over a subring R of k, and assume that we have a bilinear symmetric (or hermitian) form on E. Let $\{v_1, \ldots, v_s\}$ be an orthogonal basis for this module. If

$$v = a_1 v_1 + \cdots + a_s v_s$$

with $a_i \in R$, then we call $a_1, \ldots, a_s$ the *Fourier coefficients* of v with respect to our basis. In terms of the form, these coefficients are given by

$$a_i = \frac{\langle v, v_i \rangle}{\langle v_i, v_i \rangle}$$

provided $\langle v_i, v_i \rangle \neq 0$.

We shall see in the next theorem that the expression for e_i in terms of $\gamma_1, \ldots, \gamma_s$ is a Fourier expansion.

THEOREM 8. *The conjugacy classes* $\gamma_1, \ldots, \gamma_s$ *constitute an orthogonal basis for* $Z(G)$. *We have* $\langle \gamma_i, \gamma_i \rangle = h_i$. *For each ring R contained in k, the bilinear map of Proposition 7 has a unique extension to a R-bilinear map*

$$Z_R(G) \times Z_R(G) \to R.$$

Proof. We use the lemma. By linearity, the formula in the lemma remains valid when we replace R by k, and when we replace e_i by any element of $Z_k(G)$, in particular when we replace e_i by γ_i. But $\{\gamma_1, \ldots, \gamma_s\}$ is a basis of $Z_k(G)$, over k. Hence we find that $\langle \gamma_i, \gamma_i \rangle = h_i$ and $\langle \gamma_i, \gamma_j \rangle = 0$ if $i \neq j$, as was to be shown.

COROLLARY. *If G is commutative, then*

$$\frac{1}{n} \sum_{\nu=1}^{n} \chi_\nu(\sigma)\chi_\nu(\tau^{-1}) = \begin{cases} 0 & \text{if } \sigma \text{ is not equal to } \tau \\ 1 & \text{if } \sigma \text{ is equal to } \tau. \end{cases}$$

Proof. When G is commutative, each conjugacy class has exactly one element, and the number of simple characters is equal to the order of the group.

We consider the case of characteristic 0 for our $Z(G)$ just as we did for $X(G)$. Let k have characteristic 0, and R be a subring of k containing the m-th roots of unity, and having a conjugation. Let $\alpha = \sum_{\sigma \in G} a_\sigma \sigma$ with $a_\sigma \in R$. We define

$$\bar{\alpha} = \sum_{\sigma \in G} \bar{a}_\sigma \sigma^{-1}.$$

THEOREM 9. *Let k have characteristic 0, and let R be a subring of k, containing the m-th roots of unity, and having a conjugation. Then the pairing of Proposition 7 has a unique extension to a hermitian form*

$$Z_R(G) \times Z_R(G) \to R$$

given by the formulas

$$\langle \alpha, \beta \rangle = \frac{1}{n} \chi_{\text{reg}}(\alpha\bar{\beta}) = \frac{1}{n} \sum_{\nu=1}^{s} \chi_\nu(\alpha)\overline{\chi_\nu(\beta)}.$$

The conjugacy classes $\gamma_1, \ldots, \gamma_s$ form an orthogonal basis for $Z_R(G)$. If R contains $1/n$, then $e_1, \ldots, e_s$ lie in R and also form an orthogonal basis for $Z_R(G)$. We have $\langle e_i, e_i \rangle = d_i^2/n$.

Proof. The formula given in the statement of the theorem gives the same value as the symbol $\langle \alpha, \beta \rangle$ of Proposition 7 when α, β lie in $Z(G)$. Thus the extension exists, and is obviously unique. Using the second formula in Proposition 7, defining the scalar product, and recalling that $\chi_\nu(e_i) = 0$ if $\nu \neq i$, we see that

$$\langle e_i, e_i \rangle = \frac{1}{n} \chi_i(e_i)\overline{\chi_i(e_i)},$$

whence our assertion follows.

We observe that the Fourier coefficients of e_i relative to the basis $\gamma_1, \ldots, \gamma_s$ are the same with respect to the bilinear form of Theorem 8, or the hermitian form of Theorem 9. This comes from the fact that $\gamma_1, \ldots, \gamma_s$ lie in $Z(G)$, and form a basis of $Z(G)$ over the prime ring.

§6. *Induced characters*

The notation is the same as in the preceding section. However, we don't need all the results proved there, all we need is the bilinear pairing on $X(G)$, and its extension to

$$X_R(G) \times X_R(G) \to R.$$

The symbol $\langle\ ,\ \rangle$ may be interpreted either as the bilinear extension, or the hermitian extension according to Theorem 7.

Let S be a subgroup of G. We have an R-linear map called the restriction

$$\mathrm{Res}_S^G : X_R(G) \to X_R(S)$$

which to each class function on G associates its restriction to S. It is a ring-homomorphism. We sometimes let f_S denote the restriction of f to S.

We shall define a map in the opposite direction,

$$\mathrm{Tr}_G^S : X_R(S) \to X_R(G),$$

which we call the *induction map*, or *transfer* map. If $g \in X_R(S)$, we extend g to g_0 on G by letting $g_0(\sigma) = 0$ if $\sigma \notin S$. Then we define

$$\mathrm{Tr}_G^S(g)(\sigma) = \frac{1}{(S:1)} \sum_{\tau \in G} g_0(\tau\sigma\tau^{-1}).$$

Then $\mathrm{Tr}_G^S(g)$ is a class function on G. If we do not need to refer to S or G in the notation, we often write g^* instead of $\mathrm{Tr}_G^S(g)$, and call it the *induced function*. It is clear that Tr_G^S is R-linear.

Since we deal with two groups S and G, we shall denote the scalar product by $\langle\ ,\ \rangle_S$ and $\langle\ ,\ \rangle_G$ when it is taken with these respective groups. The next theorem shows among other things that the restriction and transfer are adjoint to each other with respect to our form.

THEOREM 10. *Let S be a subgroup of G. Then the following rules hold:*

(i) (Frobenius reciprocity) *For $f \in X_R(G)$, and $g \in X_R(S)$ we have*

$$\langle \mathrm{Tr}_G^S(g), f\rangle_G = \langle g, \mathrm{Res}_S^G(f)\rangle_S.$$

(ii) $\mathrm{Tr}_G^S(g)f = \mathrm{Tr}_G^S(gf_S)$.

(iii) *If $T \subset S \subset G$ are subgroups of G, then*

$$\mathrm{Tr}_G^S \circ \mathrm{Tr}_S^T = \mathrm{Tr}_G^T.$$

(iv) *If $\sigma \in G$ and g^σ is defined by $g^\sigma(\tau^\sigma) = g(\tau)$, where $\tau^\sigma = \sigma^{-1}\tau\sigma$, then*

$$\mathrm{Tr}_G^S(g) = \mathrm{Tr}_G^{S^\sigma}(g^\sigma).$$

(v) *If ψ is a proper character of S then $\mathrm{Tr}_G^S(\psi)$ is a proper character of G.*

Proof. Let us first prove (ii). In the star notation, we must show that $g^*f = (gf_S)^*$. We have

$$(g^*f)(\tau) = \frac{1}{(S:1)} \sum_{\sigma \in G} g_0(\sigma\tau\sigma^{-1})f(\tau) = \frac{1}{(S:1)} \sum_{\sigma \in G} g_0(\sigma\tau\sigma^{-1})f(\sigma\tau\sigma^{-1}).$$

The last expression just obtained is equal to $(gf_S)^*$, thereby proving (ii). Let us sum over τ in G. The only non-zero contributions in our double sum will come from those elements of S which can be expressed in the form $\sigma\tau\sigma^{-1}$ with $\sigma, \tau \in G$. The number of pairs (σ, τ) such that $\sigma\tau\sigma^{-1}$ is equal to a fixed element of G is equal to n (because for every $\lambda \in G$, $(\sigma\lambda, \lambda^{-1}\tau\lambda)$ is another such pair, and the total number of pairs is n^2). Hence our expression is equal to

$$(G:1)\frac{1}{(S:1)}\sum_{\lambda\in S} g(\lambda)f(\lambda).$$

Our first rule then follows from the definitions of the scalar products in G and S respectively.

Now let $g = \psi$ be a proper character of S, and let $f = \chi$ be a simple character of G. From (i) we find that the Fourier coefficients of g^* are integers $\geqq 0$ because $\mathrm{Res}_S^G(\chi)$ is a proper character of S. Therefore the scalar product

$$\langle \psi, \mathrm{Res}_S^G(\chi)\rangle_S$$

is $\geqq 0$. Hence ψ^* is a proper character of G, thereby proving (v).

In order to prove the transitivity property, it is convenient to use the following notation.

Let $\{c\}$ denote the set of *right* cosets of S in G. For each right coset c, we select a fixed coset representative denoted by $\bar{c}$. Thus if $\bar{c}_1, \ldots, \bar{c}_r$ are these representatives, then

$$G = \bigcup_c c = \bigcup_c S\bar{c} = \bigcup_{i=1}^{r} S\bar{c}_i.$$

LEMMA. *Let g be a class function on S. Then*

$$\mathrm{Tr}_G^S(g)(\xi) = \sum_{i=1}^{r} g_0(\bar{c}_i\xi\bar{c}_i^{-1}).$$

Proof. We can split the sum over all $\sigma \in G$ in the definition of the induced function into a double sum

$$\sum_{\sigma\in G} = \sum_{\sigma\in S}\sum_{i=1}^{r}$$

and observe that each term $g_0(\sigma\bar{c}\xi\bar{c}^{-1}\sigma^{-1})$ is equal to $g_0(\bar{c}\xi\bar{c}^{-1})$ if $\sigma \in S$, because g is a class function. Hence the sum over $\sigma \in S$ is enough to cancel the factor $1/(S:1)$ in front, to give the expression in the lemma.

If $T \subset S \subset G$ are subgroups of G, and if

$$G = \bigcup S\bar{c}_i \qquad \text{and} \qquad S = \bigcup T\bar{d}_j$$

are decompositions into right cosets, then $\{\bar{d}_j \bar{c}_i\}$ form a system of representatives for the right cosets of T in G. From this the transitivity property (iii) is obvious.

We shall leave (iv) as an exercise (trivial, using the lemma).

§7. *Induced representations*

Let G be a finite group and S a subgroup. Let F be an S-module. We consider the category $\mathcal{C}$ whose objects are S-homomorphism $\varphi : F \to E$ of F into a G-module E. (We note that a G-module E can be regarded as an S-module by restriction.) If $\varphi' : F \to E'$ is another object in $\mathcal{C}$, we define a morphism $\varphi' \to \varphi$ in $\mathcal{C}$ to be a G-homomorphism $\eta : E' \to E$ making the following diagram commutative:

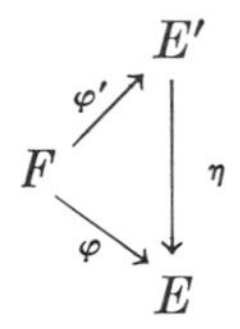

A universal object in $\mathcal{C}$ is determined up to a unique G-isomorphism. It will be denoted by

$$\varphi_G^S : F \to \mathrm{Tr}_G^S(F).$$

The symbols Tr are intended to suggest a trace. We shall see below the justification for this notation.

We shall prove below that a universal object always exists. If $\varphi : F \to E$ is a universal object, we call E *an induced module*. It is uniquely determined, up to a unique G-isomorphism making a diagram commutative. For convenience, we shall select one induced module such that φ is an inclusion. We shall then call this particular module $\mathrm{Tr}_G^S(F)$ *the* G-module *induced* by F.

Let $f : F' \to F$ be an S-homomorphism. If

$$\varphi_G^S : F' \to \mathrm{Tr}_G^S(F')$$

is a G-module induced by F', then there exists a unique G-homomorphism $\mathrm{Tr}_G^S(F') \to \mathrm{Tr}_G^S(F)$ making the following diagram commutative:

$$\begin{array}{ccc} F' & \xrightarrow{\varphi_G^S} & \mathrm{Tr}_G^S(F') \\ f\downarrow & \searrow & \downarrow \mathrm{Tr}_G^S(f) \\ F & \xrightarrow[\varphi_G^S]{} & \mathrm{Tr}_G^S(F) \end{array}$$

It is simply the G-homomorphism corresponding to the universal property for the S-homomorphism $\varphi_G^S \circ f$, represented by a dashed line in our diagram. *Thus* Tr_G^S *is a functor, from the category of S-modules to the category of G-modules.*

From the universality and uniqueness of the induced module, we get some formal properties:

Tr_G^S *commutes with direct sums: If we have an S-direct sum* $F \oplus F'$, *then*

$$\mathrm{Tr}_G^S(F \oplus F') \approx \mathrm{Tr}_G^S(F) \oplus \mathrm{Tr}_G^S(F'),$$

the direct sum on the right being a G-direct sum.

If $f, g : F' \to F$ *are S-homomorphisms, then*

$$\mathrm{Tr}_G^S(f + g) = \mathrm{Tr}_G^S(f) + \mathrm{Tr}_G^S(g).$$

If $T \subset S \subset G$ *are subgroups of G, and F is a T-module, then*

$$\mathrm{Tr}_G^S \circ \mathrm{Tr}_S^T(F) \approx \mathrm{Tr}_G^T(F).$$

In all three cases, the equality between the left member and the right member of our equations follows at once by using the uniqueness of the universal object. We shall leave the verifications to the reader.

To prove the existence of the induced module, we let $M_G^S(F)$ be the additive group of functions $f : G \to F$ satisfying

$$\sigma f(\xi) = f(\sigma \xi)$$

for $\sigma \in S$ and $\xi \in G$. We define an operation of G on $M_G^S(F)$ by letting

$$(\sigma f)(\xi) = f(\xi\sigma)$$

for σ, $\xi \in G$. It is then clear that $M_G^S(F)$ is a G-module.

PROPOSITION 8. *Let* $\varphi : F \to M_G^S(F)$ *be such that* $\varphi(x) = \varphi_x$ *is the map*

$$\varphi_x(\tau) = \begin{cases} 0 & \text{if} \quad \tau \notin S \\ \tau x & \text{if} \quad \tau \in S. \end{cases}$$

Then φ *is an S-homomorphism,* $\varphi : F \to M_G^S(F)$ *is universal, and* φ *is injective. The image of* φ *consists of those elements* $f \in M_G^S(F)$ *such that* $f(\tau) = 0$ *if* $\tau \notin S$.

Proof. Let $\sigma \in S$ and $x \in F$. Let $\tau \in G$. Then

$$(\sigma\varphi_x)(\tau) = \varphi_x(\tau\sigma).$$

If $\tau \in S$, then this last expression is equal to $\varphi_{\sigma x}(\tau)$. If $\tau \notin S$, then $\tau\sigma \notin S$, and hence both $\varphi_{\sigma x}(\tau)$ and $\varphi_x(\tau\sigma)$ are equal to 0. Thus φ is an S-homomorphism, and it is immediately clear that φ is injective. Furthermore, if $f \in M_G^S(F)$ is such that $f(\tau) = 0$ if $\tau \notin S$, then from the definitions, we conclude that $f = \varphi_x$ where $x = f(1)$.

There remains to prove that φ is universal. To do this, we shall analyze more closely the structure of $M_G^S(F)$.

PROPOSITION 9. *Let $G = \bigcup_{i=1}^{r} S\bar{c}_i$ be a decomposition of G into right cosets. Let F_1 be the additive group of functions in $M_G^S(F)$ having value 0 at elements $\xi \in G$, $\xi \notin S$. Then*

$$M_G^S(F) = \coprod_{i=1}^{r} \bar{c}_i^{-1} F_1,$$

the direct sum being taken as an abelian group.

Proof. For each $f \in M_G^S(F)$, let f_i be the function such that

$$f_i(\xi) = \begin{cases} 0 & \text{if} \quad \xi \notin S\bar{c}_i \\ f(\xi) & \text{if} \quad \xi \in S\bar{c}_i. \end{cases}$$

For all $\sigma \in S$ we have $f_i(\sigma\bar{c}_i) = (\bar{c}_i f_i)(\sigma)$. It is immediately clear that $\bar{c}_i f_i$ lies in F_1, and

$$f = \sum_{i=1}^{r} \bar{c}_i^{-1}(\bar{c}_i f_i).$$

Thus $M_G^S(F)$ is the sum of the subgroups $\bar{c}_i^{-1} F_1$. It is clear that this sum is direct, as desired.

We note that $\{\bar{c}_1^{-1}, \ldots, \bar{c}_r^{-1}\}$ form a system of representatives for the *left* cosets of S in G. The operation of G on $M_G^S(F)$ is defined by the preceding direct sum decomposition. We see that G permutes the factors transitively. The factor F_1 is S-isomorphic to the original module F, as stated in Proposition 8.

THEOREM 11. *Let $\{\lambda_1, \ldots, \lambda_r\}$ be a system of left coset representatives of S in G. There exists a G-module E containing F as an S-submodule, such that*

$$E = \coprod_{i=1}^{r} \lambda_i F$$

is a direct sum (as abelian group). Let $\varphi : F \to E$ be the inclusion mapping. Then φ is universal in our category $\mathcal{C}$, i.e. E is an induced module.

Proof. By the usual set-theoretic procedure of replacing F_1 by F in $M_G^S(F)$, we obtain a G-module E containing F as a S-submodule, and having the desired direct sum decomposition. Let $\varphi' : F \to E'$ be an S-

homomorphism into a G-module E'. We define

$$h : E \to E'$$

by the rule

$$h(\lambda_1 x_1 + \cdots + \lambda_r x_r) = \lambda_1 \varphi'(x_1) + \cdots + \lambda_r \varphi'(x_r)$$

for $x_i \in F$. This is well defined since our sum for E is direct. We must show that h is a G-homomorphism. Let $\sigma \in G$. Then

$$\sigma\lambda_i = \lambda_{\sigma(i)}\tau_{\sigma,i}$$

where $\sigma(i)$ is some index depending on σ and i, and $\tau_{\sigma,i}$ is an element of S, also depending on σ, i. Then

$$h(\sigma\lambda_i x_i) = h(\lambda_{\sigma(i)}\tau_{\sigma,i}x_i) = \lambda_{\sigma(i)}\varphi'(\tau_{\sigma,i}x_i).$$

Since φ' is an S-homomorphism, we see that this expression is equal to

$$\lambda_{\sigma(i)}\tau_{\sigma,i}\varphi'(x_i) = \sigma h(\lambda_i x_i).$$

By linearity, we conclude that h is a G-homomorphism, as desired.

Suppose that instead of considering arbitrary modules, we had taken a fixed ground field k, and considered only k-spaces on which we have a representation of G. Then it is clear that all our constructions and definitions can be applied in this context. Therefore, if we have a representation of S on a k-space F, we obtain an induced representation of G on $\mathrm{Tr}_G^S(F)$.

PROPOSITION 10. *Let ψ be the character of the representation of S on the k-space F. Let E be the space of an induced representation. Then the character χ of E is equal to the induced character ψ^*, i.e. is given by the formula*

$$\chi(\xi) = \sum_c \psi_0(\bar{c}\xi\bar{c}^{-1}),$$

where the sum is taken over the right cosets c of S in G, $\bar{c}$ is a fixed coset representative for c, and ψ_0 is the extension of ψ to G obtained by setting $\psi_0(\sigma) = 0$ if $\sigma \notin S$.

Proof. Let $\{w_1, \ldots, w_m\}$ be a basis for F over k. We know that

$$E = \coprod \bar{c}^{-1}F.$$

Let σ be an element of G. The elements $\{\overline{c\sigma}^{-1}w_j\}_{c,j}$ form a basis for E over k.

We observe that $\bar{c}\sigma\overline{c\sigma}^{-1}$ is an element of S because

$$S\bar{c}\sigma = Sc\sigma = S\overline{c\sigma}.$$

We have

$$\sigma(\overline{c\sigma}^{-1}w_j) = \bar{c}^{-1}(\bar{c}\sigma\overline{c\sigma}^{-1})w_j.$$

Let

$$(\bar{c}\sigma\overline{c\sigma}^{-1})_{\mu j}$$

be the components of the matrix representing the effect of $\bar{c}\sigma\overline{c\sigma}^{-1}$ on F with respect to the basis $\{w_1, \ldots, w_m\}$. Then the action of σ on E is given by

$$\begin{aligned}\sigma(\overline{c\sigma}^{-1}w_j) &= \bar{c}^{-1}\sum_{\mu}(\bar{c}\sigma\overline{c\sigma}^{-1})_{\mu j}w_\mu \\ &= \sum_{\mu}(\bar{c}\sigma\overline{c\sigma}^{-1})_{\mu j}(\bar{c}^{-1}w_\mu).\end{aligned}$$

By definition,

$$\chi(\sigma) = \sum_{c\sigma=c}\sum_{j}(\bar{c}\sigma\overline{c\sigma}^{-1})_{jj}.$$

But $c\sigma = c$ if and only if $\bar{c}\sigma\bar{c}^{-1} \in S$. Furthermore,

$$\psi(\bar{c}\sigma\bar{c}^{-1}) = \sum_{j}(\bar{c}\sigma\bar{c}^{-1})_{jj}.$$

Hence

$$\chi(\sigma) = \sum_{c}\psi_0(\bar{c}\sigma\bar{c}^{-1}),$$

as was to be shown.

The next three sections, which are essentially independent of each other, give examples of induced representations. In each case, we show that certain representations are either induced from certain well-known types, or are linear combinations with integral coefficients of certain well-known types. The most striking feature is that we obtain all characters as linear combinations of induced characters arising from 1-dimensional characters. Thus the theory of characters is reduced to the study of 1-dimensional, or abelian characters.

§8. Positive decomposition of the regular character

Let G be a finite group and let k be the complex numbers. We let 1_G be the trivial character, and r_G denote the regular character.

Proposition 11. *Let H be a subgroup of G, and let ψ be a character of H. Let ψ^* be the induced character. Then the multiplicity of 1_H in ψ is the same as the multiplicity of 1_G in ψ^*.*

Proof. We have by Theorem 10 (i),

$$\langle\psi, 1_H\rangle_H = \langle\psi^*, 1_G\rangle_G.$$

These scalar products are precisely the multiplicities in question.

Proposition 12. *The regular representation is the representation induced by the trivial character on the trivial subgroup of G.*

Proof. This follows at once from the definition of the induced character

$$\psi^*(\tau) = \sum_{\sigma \in G} \psi_0(\sigma\tau\sigma^{-1}),$$

taking $\psi = 1$ on the trivial subgroup.

Corollary. *The multiplicity of* 1_G *in the regular character* r_G *is equal to* 1.

We shall now investigate the character

$$u_G = r_G - 1_G.$$

Theorem 12. (Brauer) *The character* nu_G *is a linear combination with positive integer coefficients of characters induced by* 1*-dimensional characters of cyclic subgroups of* G.

The proof consists of two propositions, which give an explicit description of the induced characters. I am indebted to Serre for the exposition, derived from Brauer's.

If A is a cyclic group of order a, we define the function θ_A on A by the conditions:

$$\theta_A(\sigma) = \begin{cases} a & \text{if } \sigma \text{ is a generator of } A \\ 0 & \text{otherwise.} \end{cases}$$

We let $\lambda_A = \varphi(a)r_A - \theta_A$ (where φ is the Euler function), and $\lambda_A = 0$ if $a = 1$.

The desired result is contained in the following two propositions.

Proposition 13. *Let* G *be a finite group of order* n. *Then*

$$nu_G = \sum \lambda_A^*,$$

the sum being taken over all cyclic subgroups of G.

Proof. Given two class functions χ, ψ on G, we have the usual scalar product:

$$\langle \psi, \chi \rangle_G = \frac{1}{n} \sum_{\sigma \in G} \psi(\sigma)\overline{\chi(\sigma)}.$$

Let ψ be any class function on G. Then:

$$\begin{aligned} \langle \psi, nu_G \rangle &= \langle \psi, nr_G \rangle - \langle \psi, n1_G \rangle \\ &= n\psi(1) - \sum_{\sigma \in G} \psi(\sigma). \end{aligned}$$

On the other hand, using the fact that the induced character is the transpose of the restriction, we obtain

$$\begin{aligned}\sum_A \langle \psi, \lambda_A^* \rangle &= \sum_A \langle \psi \mid A, \lambda_A \rangle \\ &= \sum_A \langle \psi \mid A, \varphi(a) r_A - \theta_A \rangle \\ &= \sum_A \varphi(a)\psi(1) - \sum_A \frac{1}{a} \sum_{\sigma \text{ gen } A} a\psi(\sigma) \\ &= n\psi(1) - \sum_{\sigma \in G} \psi(\sigma).\end{aligned}$$

Since the functions on the right and left of the equality sign in the statement of our proposition have the same scalar product with an arbitrary function, they are equal. This proves our proposition.

PROPOSITION 14. *If $A \neq \{1\}$, the function λ_A is a linear combination of irreducible non-trivial characters of A with positive integral coefficients.*

Proof. If A is cyclic of prime order, then by Proposition 13, we know that $\lambda_A = nu_A$, and our assertion follows from the standard structure of the regular representation.

In order to prove the assertion in general, it suffices to prove that the Fourier coefficients of λ_A with respect to a character of degree 1 are integers $\geqq 0$. Let ψ be a character of degree 1. We take the scalar product with respect to A, and obtain:

$$\begin{aligned}\langle \psi, \lambda_A \rangle &= \varphi(a)\psi(1) - \sum_{\sigma \text{ gen}} \psi(\sigma) \\ &= \varphi(a) - \sum_{\sigma \text{ gen}} \psi(\sigma) \\ &= \sum_{\sigma \text{ gen}} (1 - \psi(\sigma)).\end{aligned}$$

The sum $\sum \psi(\sigma)$ taken over generators of A is an algebraic integer, and is in fact a rational number (for any number of elementary reasons), hence a rational integer. Furthermore, if ψ is non-trivial, all real parts of

$$1 - \psi(\sigma)$$

are > 0 if $\sigma \neq id$ and are 0 if $\sigma = id$. From the last two inequalities, we conclude that the sums must be equal to a positive integer. If ψ is the trivial character, then the sum is clearly 0. Our proposition is proved.

§9. *Supersolvable groups*

Let G be a finite group. We shall say that G is *supersolvable* if there exists a sequence of subgroups

$$\{1\} \subset G_1 \subset G_2 \subset \cdots \subset G_m$$

such that each G_i is normal in G, and G_{i+1}/G_i is cyclic of prime order.

From the theory of p-groups, we know that every p-group is supersolvable, and so is the direct product of a p-group with an abelian group.

PROPOSITION 15. *Every subgroup and every factor group of a supersolvable group is supersolvable.*

Proof. Obvious, using the standard homomorphism theorems.

PROPOSITION 16. *Let G be a non-abelian supersolvable group. Then there exists a normal abelian subgroup which contains the center properly.*

Proof. Let C be the center of G, and let $\overline{G} = G/C$. Let $\overline{H}$ be a normal subgroup of prime order in $\overline{G}$ and let H be its inverse image in G under the canonical map $G \to G/C$. If $\bar{\sigma}$ is a generator of $\overline{H}$, then an inverse image σ of $\bar{\sigma}$, together with C, generate H. Hence H is abelian, normal, and contains the center properly.

THEOREM 13. (Blichfeldt) *Let G be a supersolvable group, let k be an algebraically closed field. Let E be a simple (G, k)-space. If $\dim_k E > 1$, then there exists a proper subgroup H of G and a simple H-space F such that E is induced by F.*

Proof. Since a simple representation of an abelian group is 1-dimensional, our hypothesis implies that G is not abelian.

We shall first give the proof of our theorem under the additional hypothesis that E is faithful. (This means that $\sigma x = x$ for all $x \in E$ implies $\sigma = 1$.) It will be easy to remove this restriction at the end.

LEMMA *Let G be a finite group, and k an algebraically closed field. Let E be a simple, faithful G-space over k. Assume that there exists a normal abelian subgroup H of G containing the center of G properly. Then there exists a subgroup H_1 of G containing H, and a simple H_1-space F such that E is the induced module of F from H_1 to G.*

Proof. We view E as an H-space. It is a direct sum of simple H-spaces, and since H is abelian, such simple H-space is 1-dimensional.

Let $v \in E$ generate a 1-dimensional H-space. Let ψ be its character. If $w \in E$ also generates a 1-dimensional H-space, with the same character ψ, then for all $a, b \in k$ and $\tau \in H$ we have

$$\tau(av + bw) = \psi(\tau)(av + bw).$$

If we denote by F_ψ the subspace of E generated by all 1-dimensional H-subspaces having the character ψ, then we have an H-direct sum decomposition

$$E = \coprod_{\psi} F_\psi.$$

We contend that $E \neq F_\psi$. Otherwise, let $v \in E$, $v \neq 0$, and $\sigma \in G$. Then $\sigma^{-1}v$ is a 1-dimensional H-space by assumption, and has character ψ. Hence for $\tau \in H$,

$$\tau(\sigma^{-1}v) = \psi(\tau)\sigma^{-1}v$$
$$(\sigma\tau\sigma^{-1})v = \sigma\psi(\tau)\sigma^{-1}v = \psi(\tau)v.$$

This shows that $\sigma\tau\sigma^{-1}$ and τ have the same effect on the element v of E. Since H is not contained in the center of G, there exist $\tau \in H$ and $\sigma \in G$ such that $\sigma\tau\sigma^{-1} \neq \tau$, and we have contradicted the assumption that E is faithful.

We shall prove that G permutes the spaces F_ψ transitively.

Let $v \in F_\psi$. For any $\tau \in H$ and $\sigma \in G$, we have

$$\tau(\sigma v) = \sigma(\sigma^{-1}\tau\sigma)v = \sigma\psi(\sigma^{-1}\tau\sigma)v = \psi_\sigma(\tau)\sigma v,$$

where ψ_σ is the function on H given by $\psi_\sigma(\tau) = \psi(\sigma^{-1}\tau\sigma)$. This shows that σ maps F_ψ into F_{ψ_σ}. However, by symmetry, we see that σ^{-1} maps F_{ψ_σ} into F_ψ, and the two maps σ, σ^{-1} give inverse mappings between F_{ψ_σ} and F_ψ. Thus G permutes the spaces $\{F_\psi\}$.

Let $E' = GF_{\psi_0} = \sum \sigma F_{\psi_0}$ for some fixed ψ_σ. Then E' is a G-subspace of E, and since E was assumed to be simple, it follows that $E' = E$. This proves that the spaces $\{F_\psi\}$ are permuted transitively.

Let $F = F_{\psi_1}$ for some fixed ψ_1. Then F is an H-subspace of E. Let H_1 be the subgroup of all elements $\tau \in G$ such that $\tau F = F$. Then $H_1 \neq G$ since $E \neq F_\psi$. *We contend that F is a simple H_1-subspace, and that E is the induced space of F from H_1 to G.*

To see this, let $G = \bigcup H_1\bar{c}$ be a decomposition of G in terms of right cosets of H_1. Then the elements $\{\bar{c}^{-1}\}$ form a system of left coset representatives of H_1. Since

$$E = \sum_{\sigma \in G} \sigma F$$

it follows that

$$E = \sum_{c} \bar{c}^{-1}F.$$

We contend that this last sum is direct, and that F is a simple H_1-space.

Since G permutes the spaces $\{F_\psi\}$, we see by definition that H_1 is the isotropy group of F for the operation of G on this set of spaces, and hence that the elements of the orbit are precisely $\{\bar{c}^{-1}F\}$, as c ranges over all the cosets. Thus the spaces $\{\bar{c}^{-1}F\}$ are distinct, and we have a direct sum decomposition

$$E = \coprod_c \bar{c}^{-1}F.$$

If W is a proper H_1-subspace of F, then $\coprod \bar{c}^{-1}W$ is a proper G-subspace of E, contradicting the hypothesis that E is simple. This proves our assertions.

We can now apply Theorem 11 to conclude that E is the induced module from F, thereby proving Theorem 13, in case E is assumed to be faithful.

Suppose now that E is not faithful. Let G_0 be the normal subgroup of G which is the kernel of the representation $G \to \mathrm{Aut}_k(E)$. Let $\overline{G} = G/G_0$. Then E gives a faithful representation of $\overline{G}$. As E is not 1-dimensional, then $\overline{G}$ is not abelian and there exists a proper normal subgroup $\overline{H}$ of $\overline{G}$ and a simple $\overline{H}$-space F such that

$$E = \mathrm{Tr}_{\overline{G}}^{\overline{H}}(F).$$

Let H be the inverse image of $\overline{H}$ in the natural map $G \to \overline{G}$. Then $H \supset G_0$, and F is a simple H-space. In the operation of $\overline{G}$ as a permutation group of the k-subspaces $\{\sigma F\}_{\sigma \in G}$, we know that $\overline{H}$ is the isotropy group of one component. Hence H is the isotropy group in G of this same operation, and hence applying Theorem 11 again, we conclude that E is induced by F in G, i.e.

$$E = \mathrm{Tr}_G^H(F),$$

thereby proving Theorem 13.

COROLLARY. *Let G be a product of a p-group and a cyclic group, and let k be an algebraically closed field. If E is a simple (G, k)-space, and is not 1-dimensional, then E is induced by a 1-dimensional representation of some subgroup.*

Proof. We apply the theorem step by step using the transitivity of induced representations until we get a 1-dimensional representation of a subgroup.

§10. *Brauer's theorem*

We let $k = \mathbf{C}$ be the field of complex numbers. We let R be a subring of k. We shall deal with $X_R(G)$, i.e. the ring consisting of all linear combinations with coefficients in R of the simple characters of G over k. (It is a ring by Proposition 2, §2.)

Let $H = \{H_\alpha\}$ be a fixed family of subgroups of G, indexed by indices $\{\alpha\}$. We let $V_R(G)$ be the additive subgroup of $X_R(G)$ generated by all the functions which are induced by functions in $X_R(H_\alpha)$ for some H_α in our family. In other words,

$$V_R(G) = \sum_\alpha \mathrm{Tr}_G^{H_\alpha}(X_R(H_\alpha)).$$

We could also say that $V_R(G)$ is the subgroup generated over R by all the characters induced from all the H_α.

LEMMA 1. *$V_R(G)$ is an ideal in $X_R(G)$.*

Proof. This is immediate from Theorem 10, (ii) of §6.

For many applications, the family of subgroups will consists of "elementary" subgroups: Let p be a prime number. By a *p-elementary group* we shall mean the product of a p-group and a cyclic group (whose order may be assumed prime to p, since we can absorb the p-part of a cyclic factor into the p-group). An element $\sigma \in G$ is said to be *p-regular* if its period is prime to p, and *p-singular* if its period is a power of p. Given $x \in G$, we can write in a unique way

$$x = \sigma\tau$$

where σ is p-singular, τ is p-regular, and σ, τ commute. Indeed, if $p^r m$ is the period of x, with m prime to p, then $1 = \nu p^r + \mu m$ whence $x = (x^m)^\mu (x^{p^r})^\nu$ and we get our factorization. It is clearly unique, since the factors have to lie in the cyclic subgroup generated by x. We call the two factors the *p-singular* and *p-regular factors* of x respectively.

The above decomposition also shows:

PROPOSITION 17. *Every subgroup and every factor group of a p-elementary group is p-elementary. If S is a subgroup of the p-elementary group $P \times C$, where P is a p-group, and C is cyclic, of order prime to p, then $S = (S \cap P) \times (S \cap C)$.*

Proof. Clear.

Our purpose is to show, among other things, that if our family $\{H_\alpha\}$ is such that every p-elementary subgroup of G is contained in some H_α, then $V_R(G) = X_R(G)$ for every ring R. It would of course suffice to do it for $R = \mathbf{Z}$, but for our purposes, it is necessary to prove the result first using a bigger ring. The main result is contained in Theorems 15 and 16, due to Brauer. We shall give an exposition of Brauer-Tate (*Annals of Math.*, July 1955).

We let R be the ring $\mathbf{Z}[\zeta]$ where ζ is a primitive n-th root of unity. There exists a basis of R as a $\mathbf{Z}$-module, namely $1, \zeta, \ldots, \zeta^{N-1}$ for some integer N. This is a trivial fact, and we can take N to be the degree of the irreducible polynomials of ζ over $\mathbf{Q}$. This irreducible polynomial has leading coefficient 1, and has integer coefficients, so the fact that

$$1, \zeta, \ldots, \zeta^{N-1}$$

form a basis of $\mathbf{Z}[\zeta]$ follows from the Euclidean algorithm. We don't need to know anything more about this degree N.

We shall prove our assertion first for the above ring R. The rest then follows by using the following lemma.

LEMMA 2. *If $d \in \mathbf{Z}$ and the constant function $d.1_G$ belongs to V_R then $d.1_G$ belongs to $V_{\mathbf{Z}}$.*

Proof. We contend that $1, \zeta, \ldots, \zeta^{N-1}$ are linearly independent over $X_{\mathbf{Z}}(G)$. Indeed, a relation of linear dependence would yield

$$\sum_{\nu=1}^{s} \sum_{j=0}^{N-1} c_{\nu j} \chi_\nu \zeta^j = 0$$

with integers $c_{\nu j}$ not all 0. But the simple characters are linearly independent over k. The above relation is a relation between these simple characters with coefficients in R, and we get a contradiction. We conclude therefore that

$$V_R = V_{\mathbf{Z}} \oplus V_{\mathbf{Z}}\zeta \oplus \cdots \oplus V_{\mathbf{Z}}\zeta^{N-1}$$

is a direct sum (of abelian groups), and our lemma follows.

If we can succeed in proving that the constant function 1_G lies in $V_R(G)$, then by the lemma, we conclude that it lies in $V_{\mathbf{Z}}(G)$, and since $V_{\mathbf{Z}}(G)$ is an ideal, that $X_{\mathbf{Z}}(G) = V_{\mathbf{Z}}(G)$.

To prove our theorem, we need a sequence of lemmas.

Two elements x, x' of G are said to be *p-conjugate* if their p-regular factors are conjugate in the ordinary sense. It is clear that p-conjugacy is an equivalence relation, and an equivalence class will be called a *p-conjugacy class*, or simply a *p-class*.

LEMMA 3. *Let $f \in X_R(G)$, and assume that $f(\sigma) \in \mathbf{Z}$ for all $\sigma \in G$. Then f is constant* mod p *on every p-class.*

Proof. Let $x = \sigma\tau$, where σ is p-singular, and τ is p-regular, and σ, τ commute. It will suffice to prove that

$$f(x) \equiv f(\tau) \pmod{p}.$$

Let H be the cyclic subgroup generated by x. Then the restriction of f to H can be written

$$f_H = \sum a_j \psi_j$$

with $a_j \in R$, and ψ_j being the simple characters of H, hence homomorphisms of H into k^*. For some power p^r we have $x^{p^r} = \tau^{p^r}$, whence $\psi_j(x)^{p^r} = \psi_j(\tau)^{p^r}$, and hence

$$f(x)^{p^r} \equiv f(\tau)^{p^r} \pmod{pR}.$$

We now use the following lemma.

Lemma 4. *Let $R = \mathbf{Z}[\zeta]$ be as before. If $a \in \mathbf{Z}$ and $a \in pR$ then $a \in p\mathbf{Z}$.*

Proof. This is immediate from the fact that R has a basis over $\mathbf{Z}$ such that 1 is a basis element.

Applying Lemma 4, we conclude that $f(x) \equiv f(\tau) \pmod{p}$, because $b^{p^r} \equiv b \pmod{p}$ for every integer b.

Lemma 5. *Let τ be p-regular in G, and let T be the cyclic subgroup generated by τ. Let C be the subgroup of G consisting of all elements commuting with τ. Let P be a p-Sylow subgroup of C. Then there exists an element $\psi \in X_R(T \times P)$ such that the induced function $f = \psi^*$ has the following properties:*

(1) $f(\sigma) \in \mathbf{Z}$ *for all* $\sigma \in G$.

(2) $f(\sigma) = 0$ *if σ does not belong to the p-class of τ.*

(3) $f(\tau) = (C : P) \not\equiv 0 \pmod{p}$.

Proof. We note that the subgroup of G generated by T and P is a direct product $T \times P$. Let $\psi_1, \ldots, \psi_r$ be the simple characters of the cyclic group T, and assume that these are extended to $T \times P$ by composition with the projection:

$$T \times P \to T \to k^*.$$

We denote the extensions again by $\psi_1, \ldots, \psi_r$. Then we let

$$\psi = \sum_{\nu=1}^{r} \overline{\psi_\nu(\tau)}\psi_\nu.$$

The orthogonality relations for the simple characters of T show that

$$\psi(\tau y) = \psi(\tau) = (T : 1) \quad \text{for} \quad y \in P$$
$$\psi(\sigma) = 0 \quad \text{if} \quad \sigma \in TP, \quad \text{and} \quad \sigma \notin \tau P.$$

We contend that ψ^* satisfies our requirements.

First, it is clear that ψ lies in $X_R(TP)$.

We have for $\sigma \in G$:

$$\psi^*(\sigma) = \frac{1}{(TP : 1)} \sum_{x \in G} \psi_0(x\sigma x^{-1}) = \frac{1}{(P : 1)} \mu(\sigma)$$

where $\mu(\sigma)$ is the number of elements $x \in G$ such that $x\sigma x^{-1}$ lies in τP. The number $\mu(\sigma)$ is divisible by $(P : 1)$ because if an element x of G moves σ into τP by conjugation, so does every element of Px. Hence the values of ψ^* lie in $\mathbf{Z}$.

Furthermore, $\mu(\sigma) \neq 0$ only if σ is p-conjugate to τ, whence our condition (2) follows.

Finally, we can have $x\tau x^{-1} = \tau y$ with $y \in P$ only if $y = 1$ (because the period of τ is prime to p). Hence $\mu(\tau) = (C : 1)$, and our condition (3) follows.

LEMMA 6. *Assume that the family of subgroups $\{H_\alpha\}$ covers G (i.e. every element of G lies in some H_α). If f is a class function on G taking its values in $\mathbf{Z}$, and such that all the values are divisible by $n = (G : 1)$, then f belongs to $V_R(G)$.*

Proof. Let γ be a conjugacy class, and let p be prime to n. Every element of G is p-regular, and all p-subgroups of G are trivial. Furthermore, p-conjugacy is the same as conjugacy. Applying Lemma 5, we find that there exists in $V_R(G)$ a function taking the value 0 on elements $\sigma \notin \gamma$, and taking an integral value dividing n on elements of γ. Multiplying this function by some integer, we find that there exists a function in $V_R(G)$ taking the value n for all elements of γ, and the value 0 otherwise. The lemma then follows immediately.

THEOREM 14. (Artin) *Every character of G is a linear combination with rational coefficients of induced characters from cyclic subgroups.*

Proof. In Lemma 6, let $\{H_\alpha\}$ be the family of cyclic subgroups of G. The constant function $n.1_G$ belongs to $V_R(G)$. By Lemma 2, this function belongs to $V_{\mathbf{Z}}(G)$, and hence $nX_{\mathbf{Z}}(G) \subset V_{\mathbf{Z}}(G)$. Hence

$$X_{\mathbf{Z}}(G) \subset \frac{1}{n} V_{\mathbf{Z}}(G),$$

thereby proving the theorem.

LEMMA 7. *Let p be a prime number, and assume that every p-elementary subgroup of G is contained in some H_α. Then there exists a function $f \in V_R(G)$ whose values are in $\mathbf{Z}$, and $\equiv 1 \pmod{p^r}$.*

Proof. We apply Lemma 5 again. For each p-class γ, we can find a function f_γ in $V_R(G)$, whose values are 0 on elements outside γ, and $\not\equiv 0 \bmod p$ for elements of γ. Let $f = \sum_\gamma f_\gamma$, the sum being taken over all p-classes. Then $f(\sigma) \not\equiv 0 \pmod{p}$ for all $\sigma \in G$. Taking $f^{(p-1)p^r}$ gives what we want.

LEMMA 8. *Let p be a prime number and assume that every p-elementary subgroup of G is contained in some H_α. Let $n = n_0 p^r$ where n_0 is prime to p. Then the constant function $n_0.1_G$ belongs to $V_{\mathbf{Z}}(G)$.*

Proof. By Lemma 2, it suffices to prove that $n_0.1_G$ belongs to $V_R(G)$. Let f be as in Lemma 7. Then

$$n_0.1_G = n_0(1_G - f) + n_0 f.$$

Since $n_0(1_G - f)$ has values divisible by $n_0 p^r = n$, it lies in $V_R(G)$ by Lemma 6. On the other hand, $n_0 f \in V_R(G)$ because $f \in V_R(G)$. This proves our lemma.

THEOREM 15. (Brauer) *Assume that for every prime number p, every p-elementary subgroup of G is contained in some H_α. Then $X(G) = V_{\mathbf{Z}}(G)$. Every character of G is a linear combination, with integer coefficients, of characters induced from subgroups H_α.*

Proof. Immediate from Lemma 8, since we can find functions $n_0.1_G$ in $V_{\mathbf{Z}}(G)$ with n_0 relatively prime to any given prime number.

COROLLARY. *A class function f on G belongs to $X(G)$ if and only if its restriction to H_α belongs to $X(H_\alpha)$ for each α.*

Proof. Assume that the restriction of f to H_α is a character on H_α for each α. By the theorem, we can write

$$1_G = \sum_\alpha c_\alpha \mathrm{Tr}_G^{H_\alpha}(\psi_\alpha)$$

where $c_\alpha \in \mathbf{Z}$, and $\psi_\alpha \in X(H_\alpha)$. Hence

$$f = \sum_\alpha c_\alpha \mathrm{Tr}_G^{H_\alpha}(\psi_\alpha f_{H_\alpha}),$$

using Theorem 10 (ii) of §6. If $f_{H_\alpha} \in X(H_\alpha)$, we conclude that f belongs to $X(G)$. The converse is of course trivial.

THEOREM 16. (Brauer) *Every character of G is a linear combination with integer coefficients of characters induced by 1-dimensional characters of subgroups.*

Proof. By Theorem 15, and the transitivity of induction, it suffices to prove that every character of a p-elementary group has the property stated in the theorem. But we have proved this in the preceding section, Corollary of Theorem 13.

§11. *Field of definition of a representation*

Let k be a field and G a group. Let E be a k-space and assume we have a representation of G on E. Let k' be an extension field of k. Then G operates on $k' \otimes_k E$ by the rule

$$\sigma(a \otimes x) = a \otimes \sigma x$$

for $a \in k'$ and $x \in E$. This is obtained from the bilinear map on the product $k' \times E$ given by

$$(a, x) \mapsto a \otimes \sigma x.$$

We view $E' = k' \otimes_k E$ as the extension of E by k', and we obtain a representation of G on E'.

PROPOSITION 18. *Let the notation be as above. Then the characters of the representations of G on E and on E' are equal.*

Proof. Let $\{v_1, \ldots, v_m\}$ be a basis of E over k. Then

$$\{1 \otimes v_1, \ldots, 1 \otimes v_m\}$$

is a basis of E' over k'. Thus the matrices representing an element σ of G with respect to the two bases are equal, and consequently the traces are equal.

Conversely, let k' be a field and k a subfield. A representation of G on a k'-space E' is said to be *definable over* k if there exists a k-space E and a representation of G on E such that E' is G-isomorphic to $k' \otimes_k E$.

PROPOSITION 19. *Let E, F be simple representation spaces for the finite group G over k. Let k' be an extension of k. Assume that E, F are not G-isomorphic. Then no k'-simple component of $E^{k'}$ appears in the direct sum decomposition of $F^{k'}$ into k'-simple subspaces.*

Proof. Consider the direct product decomposition

$$k[G] = \prod_{\mu=1}^{s(k)} R_\mu(k)$$

over k, into a direct product of simple rings. Without loss of generality, we may assume that E, F are simple left ideals of $k[G]$, and they will belong to distinct factors of this product by assumption. We now take the tensor product with k', getting nothing else but $k'[G]$. Then we obtain a direct product decomposition over k'. Since $R_\nu(k)R_\mu(k) = 0$ if $\nu \neq \mu$, this will actually be given by a direct product decomposition of each factor $R_\mu(k)$:

$$k'[G] = \prod_{\mu=1}^{s(k)} \prod_{i=1}^{m(\mu)} R_{\mu i}(k').$$

Say $E = L_\nu$ and $F = L_\mu$ with $\nu \neq \mu$. Then $R_\mu E = 0$. Hence $R_{\mu i}E^{k'} = 0$ for each $i = 1, \ldots, m(\mu)$. This implies that no simple component of $E^{k'}$ can be G-isomorphic to any one of the simple left ideals of $R_{\mu i}$, and proves what we wanted.

COROLLARY. *The simple characters $\chi_1, \ldots, \chi_{s(k)}$ of G over k are linearly independent over any extension k' of k.*

Proof. This follows at once from the proposition, together with the linear independence of the k'-simple characters over k'.

Propositions 18 and 19 are essentially general statements of an abstract nature. The next theorem uses Brauer's theorem in its proof.

THEOREM 17. (Brauer) *Let G be a finite group of exponent m. Every representation of G over the complex numbers (or an algebraically closed field of characteristic 0) is definable over the field $\mathbf{Q}(\zeta_m)$ where ζ_m is a primitive m-th root of unity.*

Proof. Let χ be the character of a representation of G over $\mathbf{C}$, i.e. a proper character. By Theorem 16, we can write

$$\chi = \sum_j c_j \mathrm{Tr}_G^{S_j}(\psi_j), \qquad c_j \in \mathbf{Z},$$

the sum being taken over a finite number of subgroups S_j, and ψ_j being a 1-dimensional character of S_j. It is clear that each ψ_j is definable over $\mathbf{Q}(\zeta_m)$. Thus the induced character ψ_j^* is definable over $\mathbf{Q}(\zeta_m)$. Each ψ_j^* can be written

$$\psi_j^* = \sum_\mu d_{j\mu}\chi_\mu, \qquad d_{j\mu} \in \mathbf{Z}$$

where $\{\chi_\mu\}$ are the simple characters of G over $\mathbf{Q}(\zeta_m)$. Hence

$$\chi = \sum_\mu \left(\sum_j c_j d_{j\mu} \right) \chi_\mu.$$

The expression of χ as a linear combination of the simple characters over k is unique, and hence the coefficient

$$\sum_j c_j d_{j\mu}$$

is $\geqq 0$. This proves what we wanted.

EXERCISES

The first exercises develop orthogonality relations for the coefficients of matrix representations. These are slightly more general than the character relations. The proofs are independent of those given in the text and hence give an alternative way of obtaining these results, *independent of the preceding chapter*. Only Schur's lemma and complete reducibility are used.

1. Let G be a finite group and k any field. Let E, F be simple (G,k)-spaces. Let λ be a k-linear functional on E, let $x \in E$ and $y \in F$. If E, F are not isomorphic, show that

$$\sum_{\sigma \in G} \lambda(\sigma x)\sigma^{-1}y = 0.$$

[*Hint:* For fixed y, the map $x \mapsto \sum \lambda(\sigma x)\sigma^{-1}y$ is a G-homomorphism of E into F.] In particular, for any functional μ on F,

$$\sum_{\sigma \in G} \lambda(\sigma x)\mu(\sigma^{-1}y) = 0.$$

2. Show that the result of Exercise 1 can be applied to each coefficient of a matrix representation of G. Assuming k algebraically closed of characteristic not dividing the order of G, deduce the orthogonality relation $\langle \chi, \psi \rangle = 0$ for two distinct irreducible characters χ, ψ of G over k, where the scalar product of two functions f, g on G is defined by

$$\langle f, g \rangle = \frac{1}{n} \sum_{\sigma \in G} f(\sigma)g(\sigma^{-1}).$$

As usual, n always denotes the order of G.

3. Let k be an algebraically closed field, and let E be a simple (G,k)-space. Then any G-endomorphism of E is equal to a scalar multiple of the identity. [*Hint:* The division algebra $\mathrm{End}_{G,k}(E)$ is finite dimensional over k, hence equal to k.]

4. Let k be algebraically closed, and assume that its characteristic does not divide the order of G. Let E be a vector space of dimension d over k.

(a) Let λ be a functional on E and let $x \in E$. Let $\varphi_{\lambda,x} \in \mathrm{End}_k(E)$ be such that, for all $y \in E$,

$$\varphi_{\lambda,x}(y) = \lambda(y)x.$$

Show that $\mathrm{tr}(\varphi_{\lambda,x}) = \lambda(x)$. [*Hint:* If $x \neq 0$, extend x to a suitable basis of E and compute the trace relative to this basis.]

(b) Let $\rho : G \to \mathrm{Aut}_k(E)$ be a simple representation of G. Let $x, y \in E$. Then the characteristic of k does not divide d, and

$$\sum_{\sigma \in G} \lambda(\sigma x)\sigma^{-1}y = \frac{n}{d}\lambda(y)x.$$

[*Hint:* For fixed y, the map

$$x \mapsto \sum_{\sigma \in G} \lambda(\sigma x)\sigma^{-1}y$$

is a G-homomorphism of E into itself, hence equal to cI for some $c \in k$. In fact, it is equal to

$$\sum_{\sigma \in G} \sigma^{-1} \circ \varphi_{\lambda,y} \circ \sigma.$$

We have written σ instead of $\rho(\sigma)$ for simplicity. The trace of this expression is equal to $n \cdot \operatorname{tr}(\varphi_{\lambda,y})$, and also to dc. Taking λ, y such that $\lambda(y) = 1$ shows that the characteristic does not divide d, and one can then solve for c as desired.]

(c) If λ, μ are functionals on E, then

$$\sum_{\sigma \in G} \lambda(\sigma x)\mu(\sigma^{-1}y) = \frac{n}{d} \lambda(y)\mu(x).$$

5. (a) Let χ be the character of the representation in Exercise 4. Show that $\langle \chi, \chi \rangle = 1$. [*Hint:* Viewing ρ as a matrix representation, we have

$$\chi = \rho_{11} + \cdots + \rho_{dd}.]$$

In particular, if $\chi_1, \ldots, \chi_s$ are the simple characters, and if we *define*

$$e_i = \frac{d_i}{n} \sum_{\sigma \in G} \chi_i(\sigma)\sigma^{-1},$$

then $\chi_j(e_i) = \delta_{ij}d_i$.

(b) Granting that $\chi_{\text{reg}}(\sigma) = 0$ if $\sigma \neq 1$ and $\chi_{\text{reg}}(1) = n$, show that $\chi_{\text{reg}} = \sum d_i\chi_i$, where d_i is the dimension of χ_i. [*Hint:* Write $\chi_{\text{reg}} = \sum m_j\chi_j$, and take the scalar product with χ_i, computing it from the orthogonality relations, and also from its definition.] The values of the regular representation are of course obvious.

(c) Show that each e_i can be expressed as a sum of conjugacy classes with coefficients in k, and hence lies in the center of $k[G]$.

(d) Let E_i be any representation space of χ_i, and let ρ_i be the representation of G (or $k[G]$) on E_i. For $\alpha \in k[G]$ we let $\rho_i(\alpha) : E_i \to E_i$ be the map such that $\rho_i(\alpha)x = \alpha x$ for all $x \in E_i$. Show that $\rho_i(e_i) = id$, and that $\rho_i(e_j) = 0$ if $i \neq j$. [*Hint:* The map $x \mapsto e_i x$ is a G-homomorphism of E_j into itself by (c), and is therefore a scalar multiple of the identity by Exercise 3. This scalar is trivially computed to be 1 or 0 by taking the trace and using the orthogonality relations between simple characters.]

(e) Show that $\sum_{i=1}^{s} e_i = 1$.

(f) Let α be in the center of $k[G]$. Then for any i, $\rho_i(\alpha)$ is a scalar multiple of the identity on E_i, say

$$\rho_i(\alpha) = c_i\rho_i(e_i) = c_i \cdot id_{E_i}, \qquad c_i \in k.$$

Conclude that $\alpha = c_1e_1 + \cdots + c_se_s$, and hence that the center of $k[G]$ has dimension precisely s. In particular, there are exactly s conjugacy classes $\gamma_1, \ldots, \gamma_s$, which form a basis of the center $Z_k(G)$ of $k[G]$ over k. [*Hint:* The linear combination $c_1e_1 + \cdots + c_se_s$ has the same effect as α on each E_i. Since $k[G]$ itself is isomorphic to a direct sum $\coprod d_iE_i$, it follows that α is equal to this linear combination.]

6. Let f be a class function. Show that

$$f = \sum_{i=1}^{s} \langle f, \chi_i \rangle \chi_i.$$

Deduce Plancherel's formula, for two class functions f, g, namely:

$$\langle f, g\rangle = \sum_{i=1}^{s} \langle f, \chi_i\rangle\langle \chi_i, g\rangle.$$

Taking for f, g characteristic functions of classes, find as a special case the orthogonality relation of Theorem 9, namely $\langle \gamma_i, \gamma_j\rangle = h_i\delta_{ij}$.

7. Let $\rho^{(i)}$ denote a matrix representation on E_i, and let $\rho_{\nu\mu}^{(i)}$ be the coefficient functions of this matrix, $i = 1, \ldots, s$ and $\nu, \mu = 1, \ldots, d_i$. Show that these functions $\{\rho_{\nu\mu}^{(i)}\}$ form an orthogonal basis for the space of functions on G, and hence that for any function f (not necessarily a class function), we have

$$f = \sum_{i=1}^{s} \sum_{\nu,\mu} \frac{1}{d_i} \langle f, \rho_{\nu\mu}^{(i)}\rangle \rho_{\nu\mu}^{(i)}.$$

8. The following formalism is the analogue of Artin's formalism of L-series in number theory. Cf. Artin's "Zur Theorie der L-Reihen mit allgemeinen Gruppencharakteren", Collected papers, and also S. Lang, "*L-series of a covering*", *Proc. Nat. Acad. Sc. USA* (1956).

We consider a category with objects $\{U\}$. As usual, we say that a finite group G operates on U if we are given a homomorphism $\rho : G \to \mathrm{Aut}(U)$. We then say that U is a G-object, and also that ρ is a representation of G in U. We say that G operates trivially if $\rho(G) = id$. For simplicity, we omit the ρ from the notation. By a G-morphism $f : U \to V$ between G-objects, one means a morphism such that $f \circ \sigma = \sigma \circ f$ for all $\sigma \in G$.

We shall assume that for each G-object U there exists an object U/G on which G operates trivially, and a G-morphism $\pi_{U,G} : U \to U/G$ having the following universal property: If $f : U \to U'$ is a G-morphism, then there exists a unique morphism $f/G : U/G \to U'/G$ making the following diagram commutative:

$$\begin{array}{ccc} U & \xrightarrow{f} & U' \\ \downarrow & & \downarrow \\ U/G & \xrightarrow[f/G]{} & U'/G \end{array}$$

In particular, if H is a normal subgroup of G, show that G/H operates in a natural way on U/H.

Let k be an algebraically closed field of characteristic 0. We assume given a functor E from our category to the category of finite dimensional k-spaces. If U is an object in our category, and $f : U \to U'$ is a morphism, then we get a homomorphism

$$E(f) = f_* : E(U) \to E(U').$$

(The reader may keep in mind the special case when we deal with the category of reasonable topological spaces, and E is the homology functor in a given dimension.)

If G operates on U, then we get an operation of G on $E(U)$ by functoriality.

Let U be a G-object, and $F : U \to U$ a G-morphism. If $P_F(t) = \prod (t - \alpha_i)$ is the characteristic polynomial of the linear map $F_* : E(U) \to E(U)$, we define

$$Z_F(t) = \prod (1 - \alpha_i t),$$

and call this the zeta function of F. If F is the identity, then $Z_F(t) = (1 - t)^{B(U)}$ where we define $B(U)$ to be $\dim_k E(U)$.

Let χ be a simple character of G. Let d_χ be the dimension of the simple representation of G belonging to χ, and $n = \text{ord}(G)$. We define a linear map on $E(U)$ by letting

$$e_\chi = \frac{d_\chi}{n} \sum_{\sigma \in G} \chi(\sigma^{-1})\sigma_*.$$

Show that $e_\chi^2 = e_\chi$, and that for any positive integer μ we have $(e_\chi \circ F_*)^\mu = e_\chi \circ F_*^\mu$.

If $P_\chi(t) = \prod (t - \beta_j(\chi))$ is the characteristic polynomial of $e_\chi \circ F_*$, define

$$L_F(t, \chi, U/G) = \prod (1 - \beta_j(\chi)t).$$

Show that the logarithmic derivative of this function is equal to

$$-\frac{1}{N} \sum_{\mu=1}^{\infty} \text{tr}(e_\chi \circ F_*^\mu) t^{\mu-1}.$$

Define $L_F(t, \chi, U/G)$ for any character χ by linearity. If we write $V = U/G$ by abuse of notation, then we also write $L_F(t, \chi, U/V)$. Then for any χ, χ' we have by definition,

$$L_F(t, \chi + \chi', U/V) = L_F(t, \chi, U/V) L_F(t, \chi', U/V).$$

We make one additional assumption on the situation:
Assume that the characteristic polynomial of

$$\frac{1}{n} \sum_{\sigma \in G} \sigma_* \circ F_*$$

is equal to the characteristic polynomial of F/G on $E(U/G)$. Prove the following statement:

(a) If $G = \{1\}$ then

$$L_F(t, 1, U/U) = Z_F(t).$$

(b) Let $f = F/G$. Then

$$L_F(t, 1, U/V) = Z_f(t).$$

(c) Let H be a subgroup of G and let ψ be a character of H. Let $W = U/H$, and let ψ^* be the induced character from H to G. Then

$$L_F(t, \psi, U/W) = L_F(t, \psi^*, U/V).$$

(d) Let H be normal in G. Then G/H operates on $U/H = W$. Let ψ be a character of G/H, and let χ be the character of G obtained by composing ψ with the canonical map $G \to G/H$. Let $\varphi = F/H$ be the morphism induced on $U/H = W$. Then

$$L_\varphi(t, \psi, W/V) = L_F(t, \chi, U/V).$$

(e) If $V = U/G$ and $B(V) = \dim_k E(V)$, show that $(1 - t)^{B(V)}$ divides $(1 - t)^{B(U)}$. Use the regular character to determine a factorization of $(1 - t)^{B(U)}$.

and call this the [illegible] function of G. If X is the identity [illegible]

where we denote [illegible]

Let X be a simple chara[illegible] of G. Let [illegible] the dimension of [illegible] representation of G [illegible]

[illegible] by letting

[illegible]

APPENDIX

The Transcendence of e and π

The proof which we shall give here follows the classical method of Gelfond and Schneider, properly formulated. It is based on a theorem concerning values of functions satisfying differential equations, and it had been recognized for some time that such values are subject to severe restrictions, in various contexts. Here, we deal with the most general algebraic differential equation. The literature on the subject is still very small, and the reader will find most of it in the following monographs:

A. O. Gelfond, *Transcendental and algebraic numbers*, translated by Leo F. Boron, Dover, New York, 1960.

T. Schneider, *Enführung in die transcendenten Zahlen*, Springer, Berlin, 1957.

C. L. Siegel, *Transcendental numbers*, Annals of Math. Princeton, 1949.

Applications and generalizations of the theorem stated in this Appendix will be found in my two papers *Transcendental points on group varieties*, and *Algebraic values of meromorphic functions*, Topology, 1963 and 1965.

We shall assume that the reader is acquainted with elementary facts concerning functions of a complex variable. Let f be an entire function (i.e. a function which is holomorphic on the complex plane). For our purposes, we say f is of order $\leqq \rho$ if there exists a number $C > 1$ such that for all large R we have

$$|f(z)| \leqq C^{R^{\rho}}$$

whenever $|z| \leqq R$. A meromorphic function is said to be of order $\leqq \rho$ if it is a quotient of entire functions of order $\leqq \rho$.

THEOREM. *Let K be a finite extension of the rational numbers. Let $f_1, \ldots, f_N$ be meromorphic functions of order $\leqq \rho$. Assume that the field $K(f_1 \ldots, f_N)$ has transcendence degree $\geqq 2$ over K, and that the derivative $D = d/dz$ maps the ring $K[f_1, \ldots, f_N]$ into itself. Let $w_1, \ldots, w_m$ be distinct complex numbers not lying among the poles of the f_i, such that*

$$f_i(w_\nu) \in K$$

for all $i = 1, \ldots, N$ and $\nu = 1, \ldots, m$. Then $m \leqq 10\rho[K : \mathbf{Q}]$.

COROLLARY 1. (Hermite-Lindemann) *If α is algebraic (over* $\mathbf{Q}$*) and* $\neq 0$*, then e^α is transcendental. Hence π is transcendental.*

Proof. Suppose that α and e^α are algebraic. Let $K = \mathbf{Q}(\alpha, e^\alpha)$. The two functions z and e^z are algebraically independent over K (trivial), and the ring $K[z, e^z]$ is obviously mapped into itself by the derivative. Our functions take on algebraic values in K at $\alpha, 2\alpha, \ldots, m\alpha$ for any m, contradiction. Since $e^{2\pi i} = 1$, it follows that $2\pi i$ is transcendental.

COROLLARY 2. (Gelfond-Schneider) *If α is algebraic $\neq 0, 1$ and if β is algebraic irrational, then $\alpha^\beta = e^{\beta \log \alpha}$ is transcendental.*

Proof. We proceed as in Corollary 1, considering the functions $e^{\beta t}$ and e^t which are algebraically independent because β is assumed irrational. We look at the numbers $\log \alpha, 2 \log \alpha, \ldots, m \log \alpha$ to get a contradiction as in Corollary 1.

Before giving the main arguments proving the theorem, we state some lemmas. The first two, due to Siegel, have to do with integral solutions of linear homogeneous equations.

LEMMA 1. *Let*

$$a_{11}x_1 + \cdots + a_{1n}x_n = 0$$
$$\cdots$$
$$a_{r1}x_1 + \cdots + a_{rn}x_n = 0$$

be a system of linear equations with integer coefficients a_{ij}, and $n > r$. Let A be a number such that $|a_{ij}| \leqq A$ for all i, j. Then there exists an integral, non-trivial solution with

$$|x_j| \leqq 2(2nA)^{r/(n-r)}.$$

Proof. We view our system of linear equations as a linear equation $L(X) = 0$, where L is a linear map, $L : \mathbf{Z}^{(n)} \to \mathbf{Z}^{(r)}$, determined by the matrix of coefficients. If B is a positive number, we denote by $\mathbf{Z}^{(n)}(B)$ the set of vectors X in $\mathbf{Z}^{(n)}$ such that $|X| \leqq B$ (where $|X|$ is the maximum of the absolute values of the coefficients of X). Then L maps $\mathbf{Z}^{(n)}(B)$ into $\mathbf{Z}^{(r)}(nBA)$. The number of elements in $\mathbf{Z}^{(n)}(B)$ is $\geqq B^n$ and $\leqq (2B + 1)^n$. We seek a value of B such that there will be two distinct elements X, Y in $\mathbf{Z}^{(n)}(B)$ having the same image, $L(X) = L(Y)$. For this, it will suffice that $B^n > (2nBA)^r$, and thus it will suffice that $B = (2nA)^{r/(n-r)}$. We take $X - Y$ as the solution of our problem.

Let K be a finite extension of $\mathbf{Q}$, and let I_K be the integral closure of $\mathbf{Z}$ in K. From Exercise 6 of Chapter IX, we know that I_K is a free module over $\mathbf{Z}$, of dimension $[K : \mathbf{Q}]$. We view K as contained in the complex numbers. If $\alpha \in K$, a conjugate of α will be taken to be an element $\sigma\alpha$, where σ is an embedding of K in $\mathbf{C}$. By the *size* of a set of elements of K

we shall mean the maximum of the absolute values of all conjugates of these elements.

By the size of a vector $X = (x_1, \ldots, x_n)$ we shall mean the size of the set of its coordinates.

Let $\omega_1, \ldots, \omega_M$ be a basis of I_K over $\mathbf{Z}$. Let $\alpha \in I_K$, and write

$$\alpha = a_1\omega_1 + \cdots + a_M\omega_M.$$

Let $\omega'_1, \ldots, \omega'_M$ be the dual basis of $\omega_1, \ldots, \omega_M$ with respect to the trace. Then we can express the (Fourier) coefficients a_j of α as a trace,

$$a_j = \mathrm{Tr}(\alpha\omega'_j).$$

The trace is a sum over the conjugates. Hence the size of these coefficients is bounded by the size of α, times a fixed constant, depending on the size of the elements ω'_j.

LEMMA 2. *Let K be a finite extension of $\mathbf{Q}$. Let*

$$\begin{aligned} \alpha_{11}x_1 + \cdots + \alpha_{1n}x_n &= 0 \\ &\cdots \\ \alpha_{r1}x_1 + \cdots + \alpha_{rn}x_n &= 0 \end{aligned}$$

be a system of linear equations with coefficients in I_K, and $n > r$. Let A be a number such that $\mathrm{size}(\alpha_{ij}) \leqq A$, *for all i, j. Then there exists a nontrivial solution X in I_K such that*

$$\mathrm{size}(X) \leqq C_1(C_2 nA)^{r/(n-r)},$$

where C_1, C_2 are constants depending only on K.

Proof. Let $\omega_1, \ldots, \omega_M$ be a basis of I_K over $\mathbf{Z}$. Each x_j can be written

$$x_j = \xi_{j1}\omega_1 + \cdots + \xi_{jM}\omega_M$$

with unknowns $\xi_{j\lambda}$. Each α_{ij} can be written

$$\alpha_{ij} = a_{ij1}\omega_1 + \cdots + a_{ijM}\omega_M$$

with integers $a_{ij\lambda} \in \mathbf{Z}$. If we multiply out the $\alpha_{ij}x_j$, we find that our linear equations with coefficients in I_K are equivalent to a system of rM linear equations in the nM unknowns $\xi_{j\lambda}$, with coefficients in $\mathbf{Z}$, whose size is bounded by CA, where C is a number depending only on M and the size of the elements ω_λ, together with the products $\omega_\lambda\omega_\mu$, in other words where C depends only on K. Applying Lemma 1, we obtain a solution in terms of the $\xi_{j\lambda}$, and hence a solution X in I_K, whose size satisfies the desired bound.

The next lemma has to do with estimates of derivatives. By the size of a polynomial with coefficients in K, we shall mean the size of its set of coefficients. A *denominator* for a set of elements of K will be any positive rational integer whose product with every element of the set is an algebraic integer. We define in a similar way a denominator for a polynomial with coefficients in K. We abbreviate "denominator" by den.

Let

$$P(T_1, \ldots, T_N) = \sum \alpha_{(\nu)} M_{(\nu)}(T)$$

be a polynomial with complex coefficients, and let

$$Q(T_1, \ldots, T_N) = \sum \beta_{(\nu)} M_{(\nu)}(T)$$

be a polynomial with real coefficients $\geqq 0$. We say that Q *dominates* P if $|\alpha_{(\nu)}| \leqq \beta_{(\nu)}$ for all (ν). It is then immediately verified that the relation of dominance is preserved under addition, multiplication, and taking partial derivatives with respect to the variables $T_1, \ldots, T_N$.

LEMMA 3. *Let K be of finite degree over $\mathbf{Q}$. Let $f_1, \ldots, f_N$ be functions, holomorphic on a neighborhood of a point $w \in \mathbf{C}$, and assume that $D = d/dz$ maps the ring $K[f_1, \ldots, f_N]$ into itself. Assume that $f_i(w) \in K$ for all i. Then there exists a number C_1 having the following property. Let $P(T_1, \ldots, T_N)$ be a polynomial with coefficients in K, of degree $\leqq r$. If we set $f = P(f_1, \ldots, f_N)$, then we have, for all positive integers k,*

$$\operatorname{size}(D^k f(w)) \leqq \operatorname{size}(P) r^k k! C_1^{k+r}.$$

Furthermore, there is a denominator for $D^k f(w)$ bounded by $\operatorname{den}(P) C_1^{k+r}$.

Proof. There exist polynomials $P_i(T_1, \ldots, T_N)$ with coefficients in K such that

$$Df_i = P_i(f_1, \ldots, f_N).$$

Let h be the maximum of their degrees. There exists a unique derivation $\overline{D}$ on $K[T_1, \ldots, T_N]$ such that $\overline{D} T_i = P_i(T_1, \ldots, T_N)$. For any polynomial P we have

$$\overline{D}(P(T_1, \ldots, T_N)) = \sum_{i=1}^{N} (D_i P)(T_1, \ldots, T_N) \cdot P_i(T_1, \ldots, T_N),$$

where $D_1, \ldots, D_N$ are the partial derivatives. The polynomial P is dominated by

$$\operatorname{size}(P)(1 + T_1 + \cdots + T_N)^r,$$

and each P_i is dominated by $\operatorname{size}(P_i)(1 + T_1 + \cdots + T_N)^h$. Thus $\overline{D}P$ is dominated by

$$\operatorname{size}(P) C_2 r (1 + T_1 + \cdots + T_N)^{r+h}.$$

Proceeding inductively, one sees that $\overline{D}^k P$ is dominated by

$$\text{size}(P)C_3^k r^k k!(1 + T_1 \ldots + T_N)^{r+kh}.$$

Substituting values $f_i(w)$ for T_i, we obtain the desired bound on $D^k f(w)$. The second assertion concerning denominators is proved also by a trivial induction.

We now come to the main part of the proof of our theorem. Let f, g be two functions among $f_1, \ldots, f_N$ which are algebraically independent over K. Let r be a positive integer divisible by $2m$. We shall let r tend to infinity at the end of the proof.

Let

$$F = \sum_{i,j=1}^{r} b_{ij} f^i g^j$$

have coefficients b_{ij} in K. Let $n = r^2/2m$. We can select the b_{ij} not all equal to 0, and such that

$$D^k F(w_\nu) = 0$$

for $0 \leqq k < n$ and $\nu = 1, \ldots, m$. Indeed, we have to solve a system of mn linear equations in $r^2 = 2mn$ unknowns. Note that

$$mn/(2mn - mn) = 1.$$

We multiply these equations by a denominator for the coefficients. Using the estimate of Lemma 3, and Lemma 2, we can in fact take the b_{ij} to be algebraic integers, whose size is bounded by

$$O(r^n n! C_1^{n+r}) \leqq O(n^{2n})$$

for $n \to \infty$.

Since f, g are algebraically independent over K, our function F is not identically zero. We let s be the smallest integer such that all derivatives of F up to order $s - 1$ vanish at all points $w_1, \ldots, w_m$, but such that $D^s F$ does not vanish at one of the w, say w_1. Then $s \geqq n$. We let

$$\gamma = D^s F(w_1) \neq 0.$$

Then γ is an element of K, and by Lemma 3, it has a denominator which is bounded by $O(C_1^s)$ for $s \to \infty$. Let c be this denominator. The norm of $c\gamma$ from K to $\mathbf{Q}$ is then a non-zero rational integer. Each conjugate of $c\gamma$ is bounded by $O(s^{5s})$. Consequently, we get

$$(1) \qquad 1 \leqq |N_{\mathbf{Q}}^K(c\gamma)| \leqq O(s^{5s})^{[K:\mathbf{Q}]-1}|\gamma|,$$

where $|\gamma|$ is the fixed absolute value of γ, which will now be estimated very well by global arguments.

Let θ be an entire function of order $\leqq \rho$, such that θf and θg are entire, and $\theta(w_1) \neq 0$. Then $\theta^{2r}F$ is entire. We consider the entire function

$$H(z) = \frac{\theta(z)^{2r}F(z)}{\prod_{\nu=1}^{m} (z - w_\nu)^s}.$$

Then $H(w_1)$ differs from $D^sF(w_1)$ by obvious factors, bounded by $C_4^s s!$. By the maximum modulus principle, its absolute value is bounded by the maximum of H on a large circle of radius R. If we take R large, then $z - w_\nu$ has approximately the same absolute value as R, and consequently, on the circle of radius R, $H(z)$ is bounded in absolute value by an expression of type

$$\frac{s^{3s}C_5^{2rR^\rho}}{R^{ms}}.$$

We select $R = s^{1/2\rho}$. We then get the estimate

$$|\gamma| \leqq \frac{s^{4s}C_6^s}{s^{ms/2\rho}}.$$

We now let r tend to infinity. Then both n and s tend to infinity. Combining this last inequality with inequality (1), we obtain the desired bound on m. This concludes the proof.

Of course, we made no effort to be especially careful in the powers of s occurring in the estimates, and the number 10 can obviously be decreased by exercising a little more care in the estimates.

The theorem we proved should only be the simplest in an extensive theory dealing with problems of transcendence degree. In some sense, the theorem is best possible without additional hypotheses. For instance, if $P(t)$ is a polynomial with integer coefficients, then $e^{P(t)}$ will take the value 1 at all roots of P, these being algebraic. Furthermore, the functions

$$t, e^t, e^{t^2}, \ldots, e^{t^n}$$

are algebraically independent, but take on values in $\mathbf{Q}(e)$ for all integral values of t.

However, one expects rather strong results of algebraic independence to hold. Lindemann proved that *if* $\alpha_1, \ldots, \alpha_n$ *are algebraic numbers, linearly independent over* $\mathbf{Q}$, *then*

$$e^{\alpha_1}, \ldots, e^{\alpha_n}$$

are algebraically independent.

More generally, Schanuel has made the following conjecture: If $\alpha_1, \ldots, \alpha_n$ are complex numbers, linearly independent over $\mathbf{Q}$, then the transcendence degree of

$$\alpha_1, \ldots, \alpha_n, e^{\alpha_1}, \ldots, e^{\alpha_n}$$

should be $\geqq n$. (It is conceivable that one has to make some mild restriction on the numbers $\alpha_1, \ldots, \alpha_n$, which would however be of no consequence for the applications, since all classical numbers would be allowable.)

From this one would deduce at once the algebraic independence of e and π (looking at 1, $2\pi i$, e, $e^{2\pi i}$), and all other independence statements concerning the ordinary exponential function and logarithm which one feels to be true, for instance, the statement that π cannot lie in the field obtained by starting with the algebraic numbers, adjoining values of the exponential function, taking algebraic closure, and iterating these two operations. Such statements have to do with values of the exponential function lying in certain fields of transcendence degree $<n$, and one hopes that by a suitable deepening of Theorem 1, one will reach the desired results.

Index

INDEX

ABCDE6987